New Trends in Materials Chemistry

NATO ASI Series

Advanced Science Institutes Series

A Series presenting the results of activities sponsored by the NATO Science Committee, which aims at the dissemination of advanced scientific and technological knowledge, with a view to strengthening links between scientific communities.

The Series is published by an international board of publishers in conjunction with the NATO Scientific Affairs Division

A	Life Sciences	Plenum Publishing Corporation
B	Physics	London and New York
C	Mathematical and Physical Sciences	Kluwer Academic Publishers
D	Behavioural and Social Sciences	Dordrecht, Boston and London
E	Applied Sciences	
F	Computer and Systems Sciences	Springer-Verlag
G	Ecological Sciences	Berlin, Heidelberg, New York, London,
H	Cell Biology	Paris and Tokyo
I	Global Environmental Change	

PARTNERSHIP SUB-SERIES

1.	Disarmament Technologies	Kluwer Academic Publishers
2.	Environment	Springer-Verlag / Kluwer Academic Publishers
3.	High Technology	Kluwer Academic Publishers
4.	Science and Technology Policy	Kluwer Academic Publishers
5.	Computer Networking	Kluwer Academic Publishers

The Partnership Sub-Series incorporates activities undertaken in collaboration with NATO's Cooperation Partners, the countries of the CIS and Central and Eastern Europe, in Priority Areas of concern to those countries.

NATO-PCO-DATA BASE

The electronic index to the NATO ASI Series provides full bibliographical references (with keywords and/or abstracts) to more than 50000 contributions from international scientists published in all sections of the NATO ASI Series.
Access to the NATO-PCO-DATA BASE is possible in two ways:

– via online FILE 128 (NATO-PCO-DATA BASE) hosted by ESRIN,
Via Galileo Galilei, I-00044 Frascati, Italy.

– via CD-ROM "NATO-PCO-DATA BASE" with user-friendly retrieval software in English, French and German (© WTV GmbH and DATAWARE Technologies Inc. 1989).

The CD-ROM can be ordered through any member of the Board of Publishers or through NATO-PCO, Overijse, Belgium.

Series C: Mathematical and Physical Sciences – Vol. 498

New Trends in Materials Chemistry

edited by

Richard Catlow

The Royal Institution of Great Britain,
London, U.K.

and

Anthony Cheetham

Materials Research Laboratory,
University of California,
Santa Barbara, U.S.A.

Springer-Science+Business Media, B.V.

Proceedings of the NATO Advanced Study Institute on
New Trends in Materials Chemistry
Il Ciocco, Lucca, Italy
September 1995

A C.I.P. Catalogue record for this book is available from the Library of Congress

ISBN 978-94-010-6347-0   ISBN 978-94-011-5570-0 (eBook)
DOI 10.1007/978-94-011-5570-0

Printed on acid-free paper

CONTENTS

vi

PREFACE

Materials Chemistry is rapidly emerging as a key component of contemporary science. The strongly interdisciplinary nature of the field requires input from all branches of chemistry, from crystallography, from solid state physics and from computational and theoretical techniques. This book aims to give a coherent survey of the field by considering all the major aspects of the current study of the chemistry of materials. Early chapters emphasise basic principles and techniques. Strong emphasis is given to new techniques and technologies, for example, the opportunities opened up by new synchrotron sources in crystallography, and new computational techniques in simulation studies of complex materials. Characterisation techniques including crystallographic, microscopic and spectroscopic techniques are then described. Key contemporary themes such as atomic transport, reactivity and catalysis are reviewed. Later chapters focus on specific classes of material, including solid state ionics, ceramics (including giant magneto–resistance and high temperature superconducting solids), microporous and molecular materials. We hope that the book provides a snapshot of the scientific and technological challenges in this fast developing field.

The editors would like to thank the NATO Scientific Affairs Division for funding the School on which this volume is based; financial contribution from Johnson Matthey Technology Centre is also gratefully acknowledged. We are most grateful to Mrs Jean Conisbee for all her efforts in preparing the manuscript.

Richard Catlow
Anthony K. Cheetham

WHAT IS A MATERIAL?

P. DAY

Davy Faraday Laboratory
The Royal Institution of Great Britain, 21 Albemarle Street,
London W1X 4BS, UK

Abstract

Materials are what everything is made of, but that is not a helpful definition for a meeting on materials chemistry. The significant feature of materials science is that it deals with matter in bulk, and hence with properties which are those of the aggregate, not just the sum of the constituents. That has made if harder for chemists than for physicists to comprehend structure-property relations because the former instinctively view solids as built from individual atoms, or molecules. But involvement of chemists in materials science has brought much greater variety and complexity in the range of substances being prepared and studied, both with respect to structures and properties.

As an adjective, 'material' is the opposite of 'spiritual', but as a noun the same word has come to mean something much narrower than a tangible substance. All the solids around us, natural and man-made, are materials but

1

C.R.A. Catlow and A. Cheetham (eds.), New Trends in Materials Chemistry, 1–17.

when we speak of the science of materials, it is to be construed as the effort to understand the relationships between structure and properties so that structures can be selected and constructed to have the properties desired. The properties in question are nearly always physical ones: chemical reactions of, on or within solids are usually thought of as the subject matter of chemistry proper, though perhaps an exception could be made in the case of ionic migration.

It is because the emphasis of materials science has traditionally been on physical aspects of solid state behaviour that chemists have not been involved with it very much till recently. Also the concepts needed to rationalise solid state physical properties are taken from the vocabulary of physics, which is alien to most chemists, being based as it is on the translational symmetry of the lattice and its associated wave functions. Chemists prefer to visualise a crystal as built up from constituent ions or molecules rather than as a periodic potential. Another way of putting it is that they are more at home in real space than reciprocal or momentum space. Still, over the last ten years or so chemistry has come to make an increasingly important contribution to the science of materials, so that the phrase *materials chemistry* came into being.

Chemistry has made its impact on the world of materials in two ways. First, it has brought new insights and methods to bear on the synthesis of solids. Traditional methods of making ceramics, glasses, semiconductors and so on were optimised largely by empirical means, some going back hundreds (if not thousands) of years: 'heat it and beat it' is a phrase encapsulating much of the older ceramics industry, for example. Close study of the chemical reaction mechanisms involved in forming solid phases from precursors has made it possible to use much lower temperatures to make ceramics (the so-called 'chimie douce', or gentle chemistry route) which, with sol-gel processing, has revolutionised production of these materials. [1] Similarly, to deposit thin films (a ubiquitous necessity not just in the semiconductor industry but in anti-abrasion surface hardening and optical coating) selectively, decomposing carefully designed organometallic molecules has proved a notable advance over the 'engineering' approach of flinging atoms at a cold surface in an ultra high vacuum.

The second way in which chemistry has widened the perspectives of materials science will, in the long run, certainly prove more influential even than the control of synthesis. It lies at the core of the whole discipline of chemistry, and consists quite simply in enlarging the range of the possible by synthesising new architectures of atoms and molecules not previously found in the natural or manmade worlds. Such architectures can be much more complex than hitherto, providing unit cells bordering on the mesoscopic, and provide access to properties not previously observed. Just one aspect of this novelty is the notion of *self-assembly*. Borrowed from biology, this phrase denotes that the three-dimensional structure of an aggregate is implicit in the shape and charge distribution of the individual molecules that make it up. When more fully realised, it will enable 'one pot' synthesis of complex lattices. In the following sections of this general introduction, we indicate some of the properties being sought, with special emphasis on molecular-based materials.

1. Degrees of Order

Classifying solids in terms of the degree of structural order that they exhibit, we can evoke the notion of a *correlation function*, defined by the average value of the product of two vectors. One vector represents the position or orientation of the contents of a reference unit cell and the other the corresponding arrangement at a distance more or less remote from the first. [2] If the positions and orientations of the contents of the remote region map completely on to that of the reference cell, the average value of the vector product is unity, while if there is no correlation between them the correlation function is zero. We can define both short and long range correlation functions and classify solids in terms of the length scale over which correlation exists. What we call *amorphous* solids show correlation between atom positions over short range but not long. Glasses (which may be formed by polyhedral SiO_4 and PO_4 sharing vertices, or even by random atomic distributions in metals) are isotropic on length scales much larger than interatomic. However, other classes of solid may be disordered in more complicated ways. *Liquid crystals*, for example, show good long range structural correlation in one or two dimensions simultaneously with only short range correlation in the third. Solid polymers, too, having covalent backbones, show anisotropic atom-atom correlations which lead to unusual dynamical behaviour.

Infinite correlation lengths in three dimensions characterise the crystalline state where, in practice, the infinitely repeating units can be not only single atoms but also molecules. In a so-called continuous lattice solid, like SiO_2 or Al_2O_3, no molecular units can be distinguished, but a molecular crystal like CO_2 or naphthalene dissolves or vaporises into molecules rather than atoms. A specially interesting class of crystals, which has become more and more important in recent years, are ones that combine the characteristics of continuous and molecular in the same lattice, for example continuous layers interleaved by molecular entities, as in the layer perovskite salts $(C_nH_{2n+1}NH_3)_2MX_4$. [3] A final category of solid is the '*composite*', strictly speaking a two-phase material in which exceptional mechanical properties are conferred by small particles of one kind of material within the matrix of another on a mesoscopic scale, that is, in the range 10-100 nm. Thus one talks of metal-matrix or polymer-matrix composites, for instance incorporating carbon fibres.

2. Bonding (Simple and Composite)

This is not the place to develop models of chemical bonding; suffice it to say that the classical paradigms of ionic, covalent, metallic and van der Waals bonding are augmented in real solids by many other more interesting possibilities. Table 1 lists a number of cases in which two quite distinct bonding modes coexist in the same crystal lattice. For instance in a crystal of C_{60} intramolecular C–C bonds are certainly covalent but the interaction between C_{60} is of van der Waals type, although at low temperature a more specific interaction develops between C–C double bonds on one C_{60} and the C_5 rings on its neighbours (Figure 1(a)). These differences are relevant to the dynamics of the C_{60} crystal: free rotation of the molecules at high temperature being replaced by finite jumps as the temperature is reduced, inducing a series of structural phase transitions. [4]

TABLE 1. Some examples of solids showing single and composite modes of chemical bonding

	Ionic	Covalent	Metallic	Van der Waals
Ionic	NaCl	K_3C_{60}	$K_2Pt(CN)_4Br_{0.30}2H_2O$	TaS_2
Covalent		Si		C_{60}
Metallic			Cu	$(BEDT-TTF)_2X$
Van der Waals				Xe

When C_{60} forms the superconducting compound K_3C_{60}, ionic (Coulomb) interactions are added to the covalent and van der Waals ones (Figure 1(b)). [5] More unusual are cases such as $K_2Pt(CN)_4Br_{0.30}2H_2O$ (commonly called KCP) where metallic bonding in one dimension is combined with ionic bonding in directions orthogonal to the chain of metal atoms. [6] Even more peculiar are the superconducting molecular charge

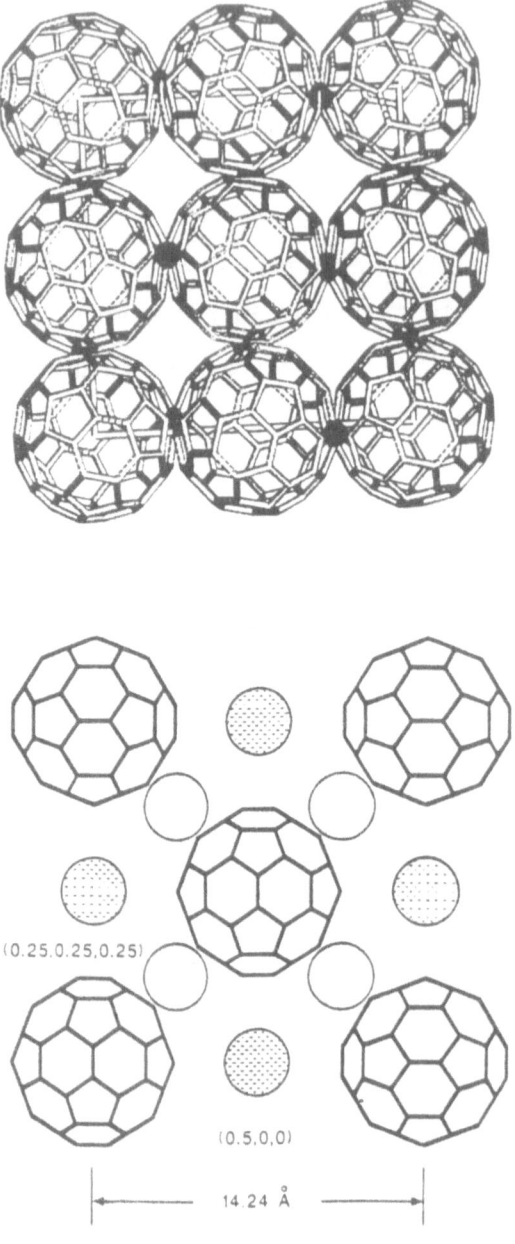

(0.25,0.25,0.25)

(0.5,0,0)

14.24 Å

Figure 1. The crystal structure of (a) C_{60}; (b) K_3C_{60}

transfer salts like $(BEDT-TTF)_2 X$ where BEDT-TTF is bisethylenedithiotetrathiafulvalene, a quasi-planar organo-sulphur molecule, and X is an inorganic anion such as I_3^-, $AuBr_2^-$, $Cu(NCS)_2^-$ etc. The metallic character arises from overlap of frontier molecular orbitals on neighbouring molecules which, however, are charged and so interact primarily by Coulomb forces with the polyatomic anions. [7] The crystal structure of one such compound is shown in Figure 2.

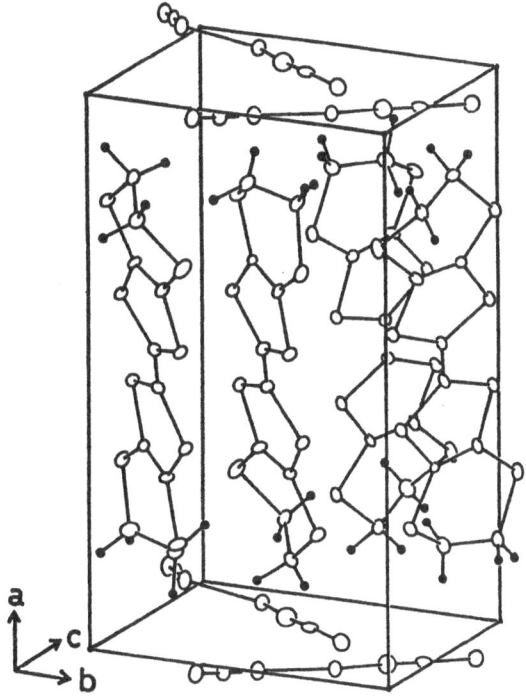

Figure 2. The crystal structure of $(BEDT-TTF)_2Cu(NCS)_2$ [8]

3. Properties (Alone and in Combination)

The physical properties of solids are rarely the sum of properties attributable to the individual components. One has only to speak of a molecular superconductor to realise that any notion of the superconductivity of one molecule in the lattice is a nonsense – the property is a *collective* one, belonging to the aggregate. In a rough way one could make the analogy with extrinsic and intrinsic thermodynamic properties: temperature, for instance, is not something you can ascribe separately to each unit in a solid ensemble.

The properties of interest to chemists are exactly those studied in simple prototypical solids for many years by the physics community, though with the proviso that (because chemists are able to make much more complex ensembles) new properties or combinations may arise that challenge the simple paradigms. Properties divide themselves into those which arise from the cohesive forces binding the crystal and those that can be classified as *'electronic'*. Among the former are thermal properties such as specific heat, and mechanical properties like compressibility. It might appear that hardness would also come into the same category, but in practice it is rarely an intrinsic property, being determined more by isolated and extended defects and dislocations. Chemists have not thought much about mechanical properties, certainly not with a view to designing such properties into the structure. However, matching the observed compressibility (and its pressure dependence) is a sensitive method of calibrating the interionic and intermolecular potentials used in the structure simulations like those described by Catlow in the present volume. [9]

Electronic properties denote ones determined by occupancy of specific electronic orbitals, in contrast to cohesive energies which arise from the sum of all occupied levels. One should distinguish between those electronic properties that have their origin in the ground state and those which involve excitation to an excited state. Most important among the former are magnetism and electrical conductivity, both of which rely on partial occupancy of the highest filled (what organic chemists call 'frontier') orbitals. In electrical insulators, which form most of the subject matter of solid state chemistry, any partly filled frontier orbitals (which are usually d- or f-type) are localised on particular atom centres.

This is not the place to introduce either of these very large topics, but simply refer the reader to the many standard sources [10, 11]. Excited state properties of great importance in materials science are electronic absorption and luminescence, the former in dyestuffs, filters and laser Q-switches, and the latter in displays and laser materials. All the latter properties arise from resonance between photon energy and the energy difference between the ground state and discrete excited states: another category of property arises from excited states which are not in resonance and, furthermore, may embrace the totality of excited states up to (and perhaps including) the ionisation continuum. The latter include the refractive index and also non-linear optical properties of great importance in telecommunications [12].

Just as we found that many of the most interesting solids had more than one mode of chemical bonding in their lattices, so it is the case that many interesting properties arise not alone but by combining physical phenomena. Thus, for example, thermally induced phase transformations combine with differences in optical behaviour in two phases to give the property of thermochromism. Application of stress, too, influences other physical properties, leading to the technologically important properties of ferroelectricity, ferroelasticity and magnetostriction. Likewise photon absorption gives rise to photochromism or photoconductivity, and even changes in magnetism, though photomagnetic effects are less well known. Finally, electrical conductivity can be influenced by magnetic fields, most spectacularly in the so-called 'giant magnetoresistance' effects in $Ln_xSr_{1-x}MnO_3$ [13].

In the remainder of this introductory survey I want to exemplify some of the points made above by considering two special topics. Since most of the subject matter of this volume deals with the continuous lattice solid state, I will take the opportunity to describe some of the special features associated

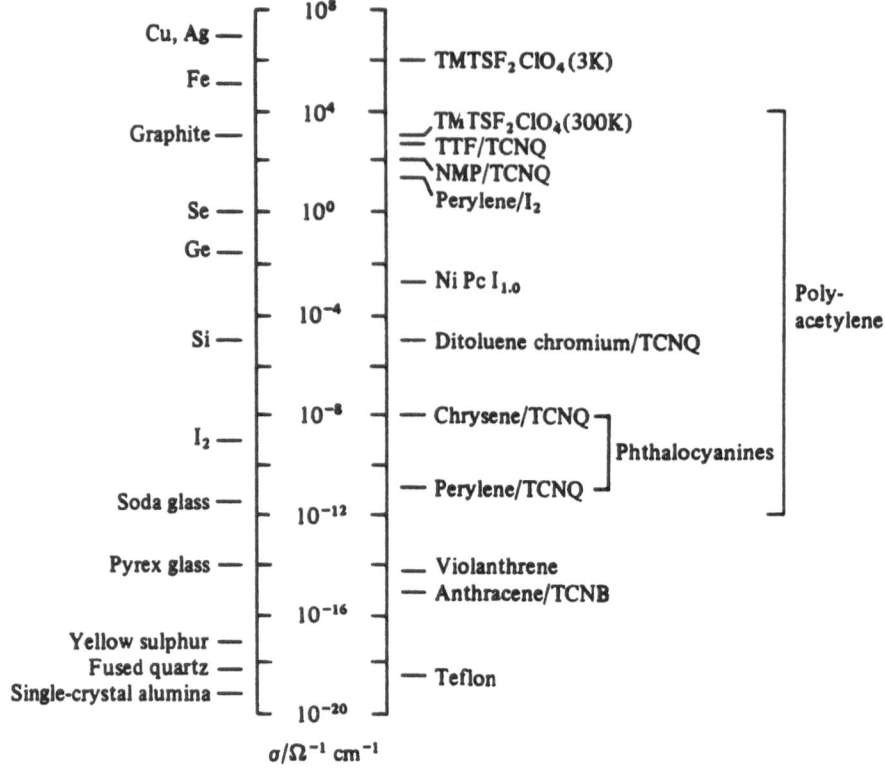

Figure 3. Electrical conductivity of continuous and molecular solids

with the less well developed field of molecular materials. Second, as an illustration of the diverse materials that can exhibit a single property, and of some of the characterisation problems that they pose, some examples are given from the field of superconductivity.

4. The Molecular Solid State: a New Horizon

The paradigms of physical behaviour that have shaped our view of the solid state are naturally based on prototypes with quite simple crystal structures, allied to chemical formulae that indicate the presence of only one to three atoms in the chemical unit cell. Such is the astonishing variety of electronic structure accessible through the permutations of 90 stable elements in the Periodic Table, however, that all the extremes of bonding type are already available to us even in this apparently limited subset of elements and compounds. In respect of one physical property (electronic conductivity) the situation is encapsulated in Figure 3. Conductivity is one of the properties that spans the largest number of orders of magnitude of any that could be imagined in the universe: 28 from the most conducting metal to the most insulating solid substance. As one can see from the left hand side of Figure 3, the majority of the examples are in fact elements. On the right hand side, though, is another group of materials whose conductivities span just as wide a range as the simple ones, but whose formulae are distinctly more elaborate. They are all molecular solids. Among them are metals just as conducting as copper, and also superconducting (which copper is not); semiconductors that can be doped in a similar way to silicon, and insulators as resistive as the purest ionic crystals. So parallel to the more familiar solid state of continuous lattice materials is the world of molecular solids.

Given that so many fascinating and useful properties arise in such profusion from continuous lattice solids, you may ask, quite reasonably, why one should bother with these much more complex ones. Several convincing answers can be given. First, they are prepared in quite different ways from conventional metals and ceramics, at or close to room temperature, and usually from solution. Thus they give the solid state chemist access to the wider world inhabited by the coordination, organometallic and organic chemists. The implication is that contemporary synthetic virtuosity can be harnessed to the solid state.

A second obvious feature distinguishing molecular arrays is that of orientational order: either whole molecules or their sidechains, can change their relative orientation as a function of temperature, pressure, stress, applied fields etc, bringing about phase transitions that cause macroscopic changes in physical properties. One has only to think of liquid crystal displays to see the implications. Allied to orientation is the issue of anisotropy: low-dimensional conductivity [14], and deposition of oriented thin films (for example by Langmuir Blodgett dipping) are examples.

5. Superconductivity: a Case for Chemistry

From its first discovery in 1910 up to the 1980s, superconductivity was not a property of much interest to chemists. The materials that exhibited this bizarre property lay firmly in the realm of the metallurgist and the materials scientist, while the theory was couched in terms of wave vectors and phonons that made it accessible only to physicists. Even the theoretical prediction that one needed to maximise the electronic density of states at the Fermi surface and increase the electron-phonon coupling was scarcely a recipe for action by a synthetic chemist. Since then, of course, superconductivity has entered chemistry in a big way: an object lesson in materials chemistry.

It is only necessary to look at the evolution of the most visible parameter of superconductivity, the critical temperature T_c, with time (Figure 4) to observe how important it has been to the progress of the subject to examine new categories of material. Within each set of materials in Figure 4, chronologically the first steps were the most significant so that, over time, further steps within the same material class gave rise to more modest incremental progress. Thus, for example, exhausting the entire material class consisting of the elements yields a maximum T_C of 9.2 K, while extending the field to binary phases took T_C up to 23 K (though it took 30 years to do so). In my view, the probability is that the line representing the mixed valency copper oxides is close to saturating, while with the discovery of the superconducting fulleride salts such as the one in Figure 1(b), the trend for molecular materials remains firmly upwards.

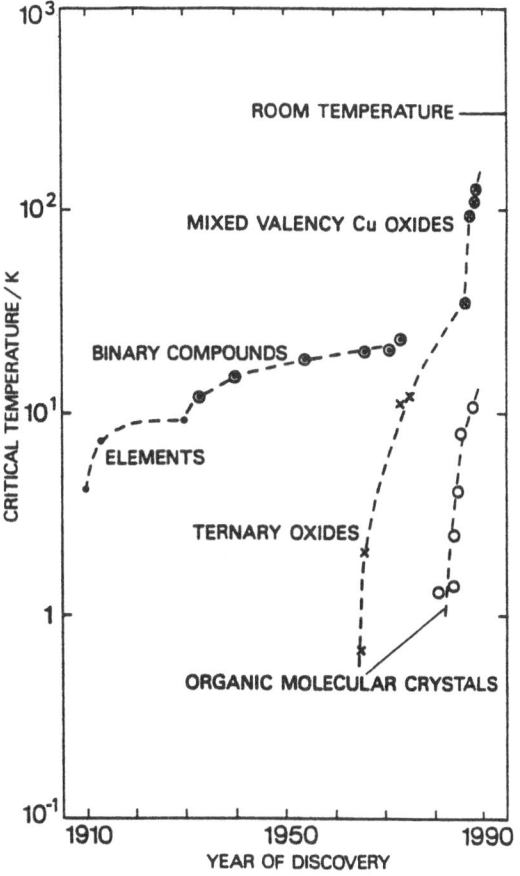

Figure 4. Superconducting critical temperatures for various classes of material

After the epoch making discovery by Bednorz and Müller [18] (who are actually physicists) of superconductivity in a phase mixture containing La,

14

Ba, Cu and O, whose composition they did not know, many solid state physicists converged on the problem with the result that a lot of physical properties were measured on distinctly ill characterised samples. Only with knowledge of preparative solid state chemistry methods could single phase samples of $La_{2-x}Ba_xCuO_{4-\delta}$ and $YBa_2Cu_3O_{7-\delta}$ be prepared, so that the structures and properties [17] could be studied in a definitive way. An instance of the subtlety of these structures is shown in Figure 5, which also illustrates how important it is to determine them in real space, by electron microscopy, and not only in reciprocal space by diffraction.

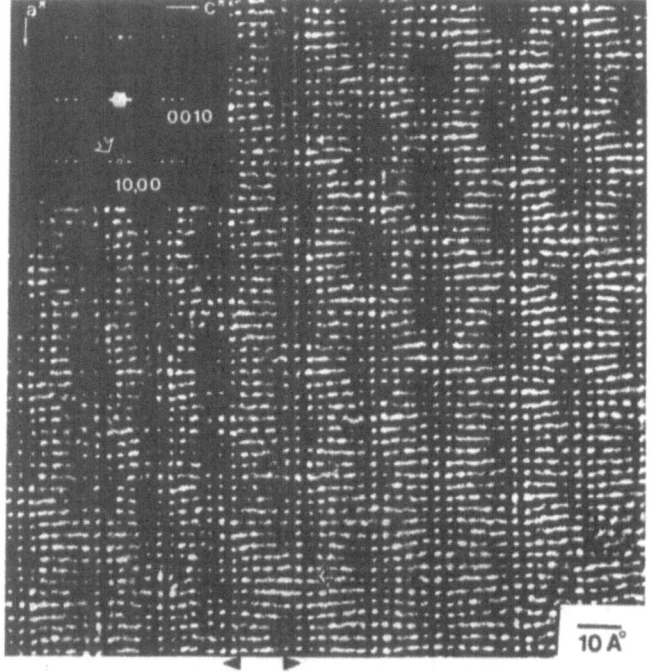

Figure 5. Lattice image of $Bi_2Sr_{2+x}Ca_{1-x}Cu_2O_8$ observed by high resolution transmission electron microscopy [18]

In Figure 5, the resolution is about 1.5 Å, so the black dots are images of individual columns of atoms, projected parallel to the direction of the electron beam. The darker atoms are the ones with the largest concentration of electrons, i.e. Bi, which clearly form double layers. The lighter O atoms scarcely contribute. Two distinct features stand out, showing how tricky it can be to prepare and characterise such phases. First, as one looks along the Bi layers it becomes clear that the Bi–Bi distances are not equal but undergo a sinusoidal modulation. This is an example of an incommensurate structure, so called because the periodicity of the modulation is not a rational multiple of the Bi–Bi separation. Second, as one scans the large area of the crystal visible in Figure 5 it is also apparent that the separations between the Bi layers are not equal. Actually there are two different separations: 15Å and 19Å, the first corresponding to two CuO coordination polyhedra between BiO and the latter to three. Such an arrangement is called an *intergrowth* because what we are looking at is a mixture of two phases on an atomic scale. Inspection shows that the sequence, 15, 19... etc. is not alternating, but random. Consequently, to define a pure phase so that one could carry out definitive measurements of its physical properties would mean the most careful control of preparative conditions.

6. Conclusion

The foregoing example shows how closely chemistry has become woven into the science of materials. The other superconductors in Figures 1 and 2 bear out the same message. Similar examples could be given from other classes of physical property, and other chapters in this book enlarge on many of the themes outlined in this introductory survey. The conclusion is that 'materials chemistry' is now firmly of age.

References

1. Rouxel, J., Tournoux, M. and Brec, R. (1994) *Soft Chemistry Routes to New Materials*, Trans Tech Publications, Switzerland.

2. Fisher, M.E. (1965) *Lectures in Theoretical Physics, Vol. VII C*, University of Colorado Press, Boulder, p. 61.

3. Day, P. (1985) Organic-inorganic layer compounds: physical properties and chemical reactions, *Phil. Trans. Roy. Soc.* **A314**, 145-158.

4. David, W.I.F., Ibberson, R.M., Matthewman, J.C., Prassides, K., Dennis, T.J.S., Hare, J.P., Kroto, H.W., Taylor, R. and Walton, D.R.M. (1991) Crystal Structure and Bonding of Ordered C_{60}, *Nature* **353**, 147-149.

5. Stephens, P.W., Mihaly, L., Lee, P.L., Whetten, R.L., Huang, S-M., Kauer, R., Diederich, F. and Holczer, K. (1991) Structure of single-phase superconducting K_3C_{60}, *Nature* **351**, 632-634.

6. Williams, J.M. and Schultz, A.J. (1979) One-dimensional partially oxidised tetracyanoplatinate metals, in W.E. Hatfield (ed.), *Molecular Metals*, Plenum Press, New York, pp. 337-368.

7. Day, P. (1994) Charge transfer salts, in D. Bloor, R.J. Brook, M.C. Flemings and S. Mahajan (eds.), *Encyclopedia of Advanced Materials*, Pergamon Press, Oxford, pp. 417-421.

8. Urayama, H. Yamochi, H., Saito, G., Sato, S., Kawaryoto, A., Tanaka, J., Mori, T., Maruyama, Y. and Inokuchi, H. (1988) Crystal structure of organic superconductor $(BEDT-TTF)_2Cu(NCS)_2$, *Chem. Lett.*, 463-466.

9. Catlow, C.R.A. (1996), chapter in this book.

10. Kittel, C. (1976) *Introduction to Solid State Physics*, John Wiley, New York.

11. Cheetham, A.K. and Day, P. (eds.) (1987) *Solid State Chemistry: Techniques*, Clarendon Press, Oxford.

12. Prasad, P.N. (1990) Photonics and non-linear optics, in R.M. Metzger, P. Day and G. Papavassiliou (eds.), *Lower Dimensional Systems and Molecular Electronics*, Plenum Press, New York, pp. 563-572.

13. Maignan, A., Caignaert, V., Simon, Ch., Hervieu, M. and Ravean, B. (1995) Giant magnetoresistance properties of manganese perovskites, *J. Mat. Chem.* **5**, 1089-1091.

14. Day, P. (1983) Low-dimensional solids, *Chemistry in Britain* **19**, 306-314.

15. Bednorz, J.G. and Müller, K.A. (1986) Possible high T_C superconductivity in the Ba-La-Cu-O system, *Z. Phys.* **B64**, 189-193.

16. e.g. David, W.I.F., Harrison, W.T.A., Gunn, J.M.F., Moze, O., Soper, A.K., Day, P., Jorgensen, J.D., Hinks, D.G., Beno, M.A., Soderholm, L., Capone II, D.W., Schuller, I.K., Segre, C.U., Zhang, K. and Grace, J.D. (1987) Structure and crystal chemistry of the high-T_C superconductor $YBa_2Cu_3O_{7-x}$, *Nature* **327**, 310-312.

17. e.g. Rosseinsky, M.J., Prassides, K. and Day, P. (1991) $La_{2-x}Sr_xCuO_{4-\delta}$: structural, magnetic and transport measurements on antiferromagnets, insulators and superconductors, *J. Mat. Chem.* **1**, 597-610.

18. Gai, P.L. and Day, P. (1988) Microstructural modulations in superconducting $Bi(Sr, Ca)_{1.33}CuO_x$, *Physica C* **152**, 154-156.

DIFFRACTION EXPERIMENTS IN MATERIALS SCIENCE.

H. GIES
Ruhr University Bochum, Institute of Mineralogy,
Universitat Strasse 150, D4630-BOCHUM, Germany

1. Introduction

Diffraction of X-rays from crystalline matter was discovered by von Laue and coworkers (1) more than 80 years ago. The basic theory of diffraction was then soon developed by von Laue(1) and the Braggs(2). The impact of the diffraction experiment for the investigation of the structure of crystalline solids had been recognised by the Braggs(3). They advanced the theory of diffraction by systematically using the experiment for the solution of new crystal structures. The history of the rise of the diffraction experiment to a routine analytical technique is well documented in numerous text books (*e.g.*(4)) and monographs(5). (Some valuable old editions have been reprinted and are still available(6)). Modern, state of the art diffraction analysis is dominated by the advancement of the development of the technical equipment such as radiation sources, monochromators, detectors, and computers, which open new frontiers for applications in materials science.

This chapter will provide a survey on attractive experiments for the materials scientist familiar with routine diffraction experiments. It will deal with the more general aspects of diffraction using single crystal experiments based on the kinematical theory. The advantages of modern equipment and the limits of diffraction experiment will be discussed in this chapter. In following chapters other diffraction techniques and experiments such as electron diffraction, neutron and X-ray powder diffraction, and the structure solution from powder diffraction data will be discussed in detail.

2. Basic Diffraction Theory

Diffraction experiments probe bulk properties of, in general, crystalline solids. Using high energy radiation from neutron-, electron-, or X-ray sources in the range of *ca.* 0.5 - 2Å the long range periodicity of the solid gives typical interference pattern related to the geometrical and chemical constitution of the material. Since X-ray radiation sources are available in almost every analytical division in university, as well as in industry, and since the analysis is non-destructive, diffraction experiments are most widely used for

C.R.A. Catlow and A. Cheetham (eds.), New Trends in Materials Chemistry, 19–33.
© *1997 Kluwer Academic Publishers.*

the investigation of crystalline matter. The basic equations describing the interaction of the diffracted beam and the crystalline material are, of course, the Bragg-equation:

$$n\lambda = 2d \sin\theta$$

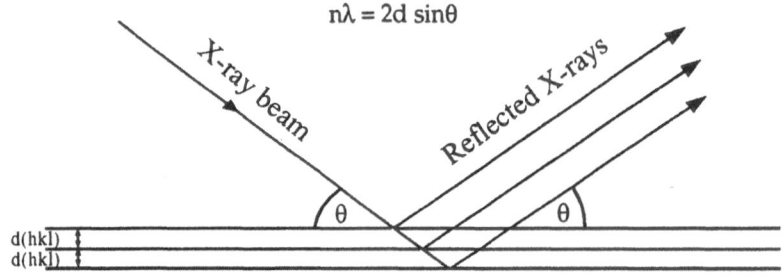

Figure 1: Braggs's law shown as constructive interference of X-rays with a set of parallel hkl-planes of a crystalline solid.

and the Laue equation:

$$a(\cos\psi - \cos\phi) = h\lambda$$

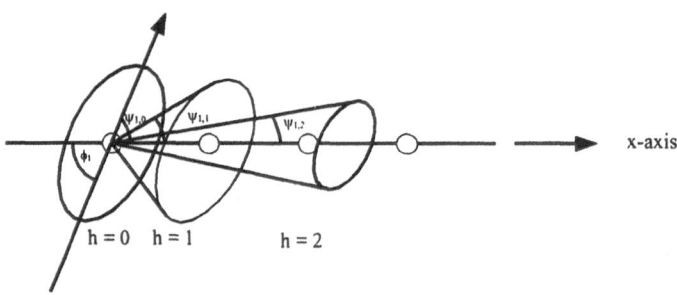

Figure 2: The Laue-equation shown as interaction of X-ray with a 1-dimensional crystal.

The nature of the radiation used for the experiment, however, greatly influences its result. X-rays are used in the 1 Å range and interact with the electrons of the atoms of the material. The result is a picture of the electron charge density distribution usually given as centres of the atomic charge density. Dependent on the absorption coefficients, the penetration of the X-ray in the material is considerable so that the atomic structure of the volume of the crystal dominates the diffraction pattern. Because of its abundant availability and the routine interpretation of the peak intensities based on the kinematic theory, X-ray diffraction is the standard technique for structure solution and refinement of crystalline solids.

For electron diffraction(7) the wavelength used is much shorter (in the range of 0.01 Å). Since the interaction with the atomic electron shell is very strong the diffraction experiment is very efficient. For single crystal studies the size of the specimen might be reduced to 100 Å or less. However, the quantitative interpretation of the diffraction patter is still hampered by multiple diffraction effects, the increased influence of dynamic diffraction theory in the diffraction experiment from small, perfect single crystals, and the experimental difficulties in recording quantitatively the peak intensities. So far, electron diffraction is mainly used as a powerful probe for lattice periodicities.

Neutron diffraction experiments(8) are restricted to large scale research facilities close to neutron sources. The wavelength used for diffraction experiments is in the range of that of X-rays. Since the neutron beam is scattered by the atomic nucleus (centre of mass) the picture obtained from the solid is complementary to the X-ray experiment (centre of electron density). There is no decay in intensity with respect to the scattering angle and, therefore, the light elements of the periodic table are detected with much higher accuracy. In addition the scattering length of the individual nuclei in the neutron diffraction experiment further enhances the scattered intensity for the light elements, in particular for hydrogen. The strength of the interaction between the neutron beam and the sample is very low requiring about 1000 times more sample than for the X-ray experiment. The neutron beam is also sensitive to the nuclear spin and probes the magnetic structure of the solid. In addition there are a number of neutron spectroscopic experiments which are complementary to e.g. IR-experiments and NMR-experiments (for details see chapters 5 and 6 in this book). Because of its restricted availability neutron diffraction experiments are not routine However, for the magnetic structure of a crystalline material and for the analysis of light elements, neutron diffraction is an invaluable tool.

The condition for constructive interference of the diffracted beam is clearly dependent on the periodicity of the structure of the material. The d-spacing in the Bragg-equation connects the diffraction experiment to the 3-dimensional periodicity of the crystal lattice and its symmetry. The diffraction data, therefore, represent a time and space average of the crystal structure of the entire sample used in the experiment. Disturbances of the periodicity in the crystal structure along one, two or three directions as in the stacking disorder of sheet-like arrays for one dimensional disorder lead to diffuse reflections along the corresponding axis in reciprocal space whereas reflections in the sheet remain sharp. Dynamic disorder results in averaged electron densities which are dependent on the temperature of the data collection.

The Bragg-equation allows for a geometrical analysis of the material. Precise d-spacings and lattice parameters can be calculated using data from high diffraction angles and the symmetry of the material can be analysed. The evaluation of the intensities of the diffraction peaks yield information on all atoms of the material taking part in the diffraction process. Because of the periodicity of the atomic arrangement it is sufficient to consider the content of the unit cell. The structure factor F(hkl) is the relevant quantity which describes the contribution of each atom of the unit cell to every reflection hkl:

$$F(hkl) = \sum_{j-1}^{j=n} g_j e^{2\pi i(hx_j+ky_j+lz_j)}$$

The graphical representation of the structure factor as shown below demonstrates that every atom j of the asymmetric unit contributes to the intensity of every reflection hkl of the diffraction data set.

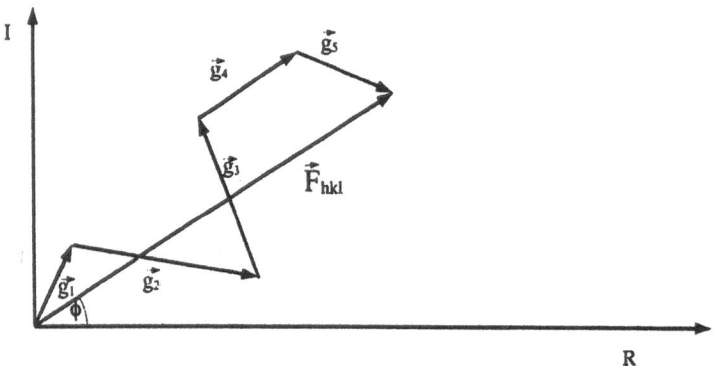

Figure 3: Graphic representation of the structure factor F_{hkl} as sum of individual atomic form factors g for a general reflection hkl.

Here j symbolises the j-th atom in the unit cell and x, y, z the fractional coordinates of the respective atom. The atomic form factor g is already corrected for the so-called thermal vibration T which might be better understood as an atomic displacement parameter.

$$g_j = f_{j,\theta} * T_{j,\theta}$$

The angular dependence of the contribution of every atom to the intensity of each reflection leads to discrimination between different atoms being most effective at low diffraction angles. The fall off of the scattering power of the atoms with higher angles is even more pronounced for those atoms with large displacement parameters or affected by dynamic disorder. For a reliable analysis of the electron density distribution in a crystal structure it is, therefore, most important to measure the low angle reflections very carefully.

The phase angle ϕ of the structure factor can take any value between 0 and 360 degree in noncentrosymmetric structures. However, in centrosymmetric materials ϕ is limited to values of 0 and 180 degree. This has important implications for the calculation of the electron density using the Fourier transformation. The electron density ρ in the crystal may be calculated from the structure factor as follows:

$$\rho(x,y,z) = \frac{1}{V_c} \sum_{-\infty(h,k,l)}^{+\infty(h,k,l)} F(hkl)e^{-2\pi i(hx+ky+lz)}$$

In the case of centrosymmetric structures, an approximate model is already sufficient to calculate precise electron densities since, as noted, the phase angle only takes the values of 0 or 180 degrees. The calculated electron density then represents the true result of the diffraction experiment which is extremely useful for the detailed interpretation of the crystal structure.

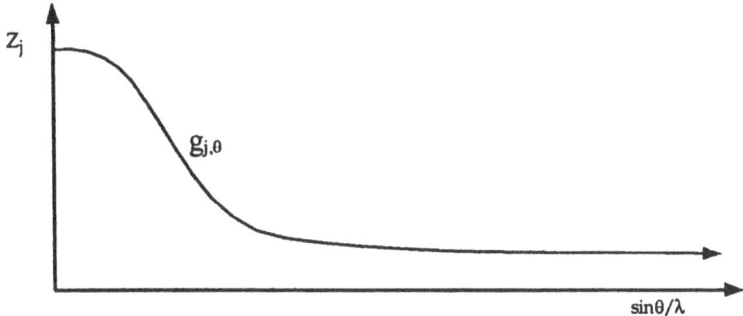

Figure 4: Graphical display of the dependence of the atomic form factor g on the scattering angle θ. The scattering power of the atom j depends on the number of electrons Z and is strongest for low scattering angles.

3.   Modern crystallographic equipment

Advances in technology have dramatically changed both routine and more specialised diffraction analysis. New radiation sources, monochromators, diffractometers, detectors, and analysis software have been a major factor in the success of modern diffraction crystallography.

High energy X-ray light sources such as synchrotrons and rotating anode generators have greatly increased the efficiency of the diffraction experiment(9). In addition to the high energy of the X-ray beam the synchrotron offers a tuneable wavelength which is useful in particular for anomalous dispersion experiments in the structural analysis of macromolecular systems. The number of synchrotron facilities dedicated to structure analysis and spectroscopy is increasing and access is no longer restricted to the specialist community. However, if the quality of the material to be investigated is the limiting factor (mosaic spread in single crystals, beam sensitivity, crystallinity, FWHM for powder samples) and

The design of monochromators has also been optimised for higher beam intensity(9). Whereas double crystal monochromators are used for synchrotron radiation based diffractometers which yield monochromatic and highly parallel X-rays, curved X-ray mirrors with grated d-spacing multilayer coatings have been introduced to laboratory X-ray generators. Together with a parallel X-ray beam an intensity gain of about two orders of magnitude is achieved.

Within the past 5 years the standard laboratory X-ray diffractometer has also been revolutionised. The standard 4-circle diffractometer which has, for more than 25 years, served the crystallographic community as the workhorse for intensity–data collection from X-ray diffraction experiments of single crystals is about to be replaced by more modern technology. Image plate detectors and CCD detectors will replace the standard diffractometer because of their superior efficiency in the data collection process(9). Whereas the CCD detectors have been optimised for smaller unit cells and high precision structure analysis, the image plate detector is mainly used in macromolecular crystallography of biological materials. Both systems are about 5-10 times faster than conventional diffractometers and will transform the routine single crystal experiment into a standard analytical technique.

The ever increasing performance of desk top computers and the implementation of robust "Direct Method" programs on PCs(10) has converted the solution of crystal structures into a job of a few seconds at least for small crystal structures. The self explanatory and self diagnostic structure solution software no longer requires an expert for handling routine data sets. The subsequent structure refinement with full matrix least squares fitting is also performed on PC's including the graphical presentation of the crystal structure so that crystal structure analysis has become a desk–top procedure available also for the non expert user(11). In particular, since the time for data collection has become so short and since most of the analysis software is available as public domain software, every scientist working on structure related properties of materials should himself take care of the crystal structure analysis.

4. Advantages and limits of the diffraction experiment

The diffraction experiment clearly develops its full strength when investigating   perfect crystalline solids (albeit with mosaic spread). Dynamic and/or static disorder of atoms or groups of atoms, non stoichiometry, incommensurate ordering of composites, and other deviations from the ideal crystalline solid lead to complications in solving the crystal structure and refining the structural model with standard software. In order to recognise such effects it is essential to analyse diffraction data recorded at least in 2 dimensions, *e.g.* on conventional X-ray film or 2-D detector devices. However, in many cases a simplified picture of the average structure can be derived by treating the intensity data set as from an ideal structure. Therefore, depending on the complexity of the problem, one should carefully set the experimental conditions for the diffraction experiment. There is a set of rules which provides a very helpful guide(12). If the crystals are transparent, their birefrinction and extinction behaviour in the optical microscope gives clear information on the quality of the crystal. Many crystals grow with a distinct morphology which is a macroscopic indicator of their point group symmetry. Extinction behaviour and the crystal morphology might be used to identify twinning or intergrowth of crystals. The X-ray powder pattern allows us to judge the crystallinity of the material and also gives information on possible defects. Traditionally, single crystal Weissenberg or precession experiments have been conducted in order to check the quality and periodicity of the crystal. These time consuming experiments will certainly become obsolete with the advent of modern 2-dimensional detector technology. This final check might then become part of the set–up routine for the intensity–data collection.

4.1   RESOLUTION IN THE DIFFRACTION EXPERIMENT

The natural limit of resolution in the diffraction experiment is set by the choice of the wavelength. Using Bragg's equation the minimum d-spacing is obtained for $\theta = 90^\circ$.

$$d_{min} = \lambda/2$$

Since the scattering power declines with increasing diffraction angle θ one should chose the wavelength for the experiment well below the resolution limit. In order to achieve atomic resolution the typical wavelength on a laboratory diffractometer is

$$\lambda = 0.7107\text{Å},$$

corresponding to the characteristic wavelength of Mo Kα. However, with increasing complexity of the crystal structure and increasing volume of the unit cell content the number of reflections within the diffraction sphere increases and their spatial separation decreases.   On conventional diffractometers lattice parameters of 50Å and more are difficult to resolve with Mo Kα radiation and require longer wavelength or modern detector systems which can be adapted for better spatial resolution. On the other hand, synchrotron X-ray radiation is ideally suited for more complicated structures since the wavelength for the experiment is easy to modify and the high intensity of the primary X-ray beam generates measurable reflection intensities at much higher angles.

Parameters for the data collection should be chosen in the light of the type of data analysis intended. For structure solution with direct methods, it is very important to record as many reflections as possible with intensities (I) better than 2σ(I). A conventional structure refinement with anisotropic displacement parameters requires about 15 reflections per variable with usually 9 variables for every atom in the asymmetric unit on a general site. For high precision electron density calculations, the low angle peaks should be measured carefully including symmetry equivalent ones for better statistics; whereas precise geometrical structure refinement should be performed with as many high angle peaks as possible.

4.2 STRUCTURE SOLUTION

Most structures are solved today with direct method programs which are available as public domain software or from the manufactures of diffraction equipment. However, there are also natural limits to the technique summarised in 'Sheldrick's rule'(13): *In order to solve a structure ab initio ~ 70% of the data up to 1.2Å resolution should be observed with I > 2σ(I)*. This requires good quality crystals and carefully measured intensities. In addition there should be a sufficient number of high normalised intensities (E-values) since the probability of the sign propagation from starting phases depends on large E-values. However, the percentage of high E-values decreases with the increasing number of atoms in the asymmetric unit.

In macromolecular crystallography, therefore, a number of more sophisticated techniques is employed in order to solve the phase problem. Anomalous scattering and multi-wavelength experiments, based on the same theory, are used in order to find starting phases or partial structures which are subsequently used for the elucidation of the whole crystal structure(14).

For the solution of macromolecular structures a maximum entropy based technique was introduced several years ago(15). It uses a 'likelihood function' for the evaluation of propagated signs. The extensive computing time required for this process still restricts the usage, and the power and success of the method is difficult to judge. However, this might change in near future with the ever increasing power of computers.

4.3 STRUCTURE REFINEMENT

The least squares refinement of the crystal structure is a routine process which is highly automated by modern computer programs. In general, the data treatment is straight forward and the structure analysis is reliable. However, if the calculated result is not in agreement with expected values, an explanation for the observation should be given.

As already mentioned, at least in the case of centrosymmetric structures, the electron density distribution is usually an unambiguous result of the structure analysis and should be interpreted as such. The structural model is the numerical fit to the experimental data and will be calculated according to the rules of least squares analysis. For example, static or dynamic disorder will produce averaged values, distances that are too low and angles that are too high, while giving, at the same time, unrealistically high displacement parameters. The electron density map will clearly reveal this as smeared out electron clouds indicating time or space averaged arrangements. Similarly, the choice of a wrong space group symmetry leads to unexpected geometrical results in the structure refinement.

4.4   HIGH RESOLUTION EXPERIMENTS

The accuracy of the diffraction experiment with X-rays depends on reliability of the measured intensities and on the order of the reflection indices h, k, and l which have been recorded. As noted, precise intensities from low order reflections are essential for electron density maps and should be measured carefully including all symmetry equivalent ones. On the other hand, high order reflections with d-values below 1Å are most important for the precision of the atomic positions; however, they suffer, in general, from low intensity values. Since the high resolution data analysis requires both, information from low and high order reflections, only high quality crystals can be used for such investigations which should be checked in preliminary experiments(16).

5.   Diffraction experiments relevant in material science

As part of the structure determination experiment and the structure refinement, a number quantities of the material are determined which contain important information on its properties;   they are, however, usually not discussed in detail. The materials scientist should be aware of this and exploit the diffraction experiment to his advantage. Today and even more in the future, crystal structure analysis is and will be a routine analysis available to every solid state scientist, in particular when modern laboratory diffractometers collect complete single crystal diffraction data sets within a few hours. The analysis of the data set will then be up to the scientist who himself made the material or is interested in structural properties.

5.1   ANALYSIS OF SYMMETRY PROPERTIES

The determination of the crystal system, the point group symmetry and the space group symmetry is an important result of the diffraction analysis. According to Neumann's principle, the symmetry of the vectorial properties behave at least as the point group symmetry of the material(17). In addition, when a centre of symmetry is present in the point group (*i.e.* in all Laue classes) properties described with tensors of uneven rank vanish. The symmetry information on a material, therefore, is extremely important for *e.g.* ferroic properties and piezoelectricity, and allows us to select materials as potential candidates from crystal structure data bases for particular investigations.

The symmetry determined in the diffraction experiment, however, is averaged in time and space. It is most important for the description of the physical properties of material to use complementary experiments detecting the local order in the solid, *e.g.* IR- or NMR-experiments, which may lead to apparently contradictory results in the specific case. The nature of local symmetry in a material is also most important for computational studies, *e.g.* quantum chemical or semi-empirical simulations which rely on the true local symmetry and not the macroscopic average given by the diffraction experiment.

5.2 ANALYSIS OF THERMAL BEHAVIOUR

Diffraction experiments at different temperatures give an insight into the temperature dependence of the physical properties of a material. A simple experiment is the determination of the principal thermal expansion coefficient, α_{ij}, which describes the linear part of the anisotropic variation of the lattice parameters[17,18]. The most precise experiment is done with single crystals where a large number of d-value can be used to calculate directly the expansion coefficients. Knowing the thermal expansion the strain tensor ε_{ij} can be derived[19]:

$$\varepsilon_{ij} = \alpha_{ij} \, \Delta T,$$

The thermal expansivity of materials plays an important role in all areas where composites are used over a temperature range. The thermal expansion coefficients are also useful quantities to compare with the results of theoretical simulations of temperature dependent properties provided by free energy minimisationtechniques discussed in Chapter 7.

The crystal structure analysis also provides information on the static and dynamic disorder in crystals[20]. An indicator of such behaviour are unusual displacement parameters, u_{ij}. Quite often angles and distances of the corresponding atoms are too large and too short respectively, which is indicative of an averaging process. Experiments performed at different temperatures should markedly vary the displacement parameters in the case of dynamic disorder, whereas static disorder would result in changes typical for the temperature difference. From a series of temperature dependent experiments, the activation energy for the dynamic process might be calculated and a reaction path deduced.

5.3   ANALYSIS OF PHASE TRANSITION

Phase transitions are, of course, a drastically non-linear feature of the temperature or composition dependence of a crystal[21]. In the course of the transition, physical properties of the material change in a discontinuous way. The analysis of the crystalline state, before and after the transition, is the basis for the understanding and modelling of the atomic processes describing the mechanism of the phase transition, *e.g.* employing Landau theory. Although the picture of the crystal structure is static in the diffraction experiment and provides only a snapshot in the reaction path, the time-resolved complementary techniques such as spectroscopic experiments rely for their interpretation on the results of the structure refinement.

The change of properties of a material as function of temperature or composition is widely used in sensor technology. In the case of ferroic properties such as ferroelectricity, ferroelasticity, piezoelectricity which are described as tensors of uneven rank, the symmetry analysis is most important since they exist only in noncentrosymmetric space groups. Therefore, the geometrical interpretation of the diffraction data suffices as a first qualitative result, instead of a time consuming complete crystal structure refinement.

5.4   ANALYSIS OF DEFECTS

The crystallographic description of a crystal is based, in general, on an ideal periodic array of atoms. The real crystal contains defects in 0-, 1-, 2-, or even 3-dimensions(22). Every class of defects leads to typical phenomena in the diffraction pattern which can be used for the quantitative analysis of the nature of the defect. Whereas defects in 0-dimensions show up in intensity modulations of sharp Bragg-peaks, defects of higher dimensions lead to diffuse intensities of equal dimension which overlap with dynamic features in the diffraction data. Satellite reflections of commensurate or incommensurate superstructures are another feature in diffraction experiments, providing information on periodic disorder in the crystalline solid.

5.5   ANALYSIS OF DYNAMIC OR STATIC DISORDER

As already discussed previously, a detailed analysis of a high quality diffraction data set gives insight into local static or dynamic disorder in the solid state. In either case, unrealistic high displacement parameters show up in the structure refinement which are temperature dependent if the disorder is dynamic.

Long range disorder, however, yields satellite reflections for incommensurate superstructures or diffuse intensities in one or two dimensions depending on the nature of the stacking disorder of structural entities(23). In any case the quantitative analysis of the diffraction data becomes much more complicated since the structural analysis of the commensurate Bragg peaks gives only the average or superposition structure and the details of the disorder are hidden in the intensities and diffraction angles of the satellite reflections and the diffuse intensities respectively.

The analysis of disordered materials has greatly benefited from the discovery of quasi crystals(24). Although these crystallise in periodic arrays in 3-dimensional space they have high local symmetry and give a regular diffraction pattern. The aperiodicity extends in one, two or three dimensions depending on the composition of the material. With the concept of supersymmetry, however, these phases might be described as regular in higher dimensional space groups and treated as projection from four, five, or six dimensional space onto three dimensional space depending on the nature of aperiodicity. There are computer programs available now which allow extended disorder to be included into the structure refinement for a most complete picture of the true three dimensional structure.

5.6 ELECTRON CHARGE DENSITY DISTRIBUTION

High precision diffraction data sets allow for the analysis of the accurate electron distribution in the crystal. The experiment has been used to give an insight into the nature of the chemical bond and as a complement for theoretical calculations(25). However, the experiment is difficult and requires careful evaluation of the result of the refinement. In order to achieve highest resolution in the diffraction experiment, the size and quality of the crystals must be exceptional.

The intensity-data collection is conducted with high accuracy from low order reflections to high order reflections. Whereas the low order reflections contain most information on the contribution of the electron cloud and provide the experimental basis for the calculation of the charge distribution, the high order reflections are essential for the determination of the atomic sites with high precision. The structure refinement is then performed with multipole functions for the core electrons and subtracted from the experimental electron density distribution. As result the deformation electron density map is obtained with details of the spatial distribution of the valence electrons. Depending on the nature of the radiation, we may distinguish between X-X- and X-N-experiments. The advantage of the neutron experiment is the excellent accuracy of the atomic sites which gives, in combination with the X-ray data set, a most complete picture of the valence electron distribution in the solid. Because of the restricted availability of neutron sources and the larger sample volume required for the diffraction experiment, X-X-analyses have become more popular.

Considerable effort was made in the seventies and eighties to link experimental results with theoretical studies; the results, however, were unsatisfactory and sometimes even contradictory. However, with the ever growing power of computers and theoretical techniques and the developments in diffraction instrumentation, material scientists interested in electronic properties of crystalline solids should pursue further joint computational / experimental studies of electron densities in crystals.

5.7 ANALYSIS OF ORDERING PROCESSES

Element segregation in crystalline compounds is typically observed in the intensities of the diffraction maxima. Since every reflection contains contributions of the diffracted beam of all atoms of the asymmetric unit the processes of ordering is revealed in the crystal structure refinement by adjusting the occupancy factors of the atomic sites(26).

Perfect ordering usually demands high symmetry in the crystal lattice whereas deviations from perfect ordering require symmetry changes to lower symmetry space groups. Depending on the difference in scattering power, partial ordering is difficult to analyse and care must be taken in the determination of the space group symmetry. Since the data reduction is performed including the symmetry information, the calculation performed thereafter uses averaged intensities for the refinement process leading inevitably to wrong results. Conventional film data of single crystals are often very helpful in symmetry analysis because they detect scattered intensity in two dimensions including weak superstructure reflections. With modern area detectors, this also becomes available for automated diffractometers speeding up the analysis by at least an order of magnitude. However, for centrosymmetric structures, as already noted, correct phases can be determined with a partial structural model leading to fault free electron density maps which should be used in the low symmetry space group for the evaluation of the ordering process and the determination of space group symmetry.

In cases where the scattering contrast of the elements involved in the ordering process is weak, complementary experiments with other diffraction techniques or spectroscopic methods are very helpful. Electron diffraction, Neutron diffraction and MAS NMR have been used extensively for the study of the ordering processes, *e.g.* of acidic centres, cation sites and sorbate molecule in zeolites, the cation distribution and oxygen occupancies in high temperature superconductors, the nature of the coordination sphere in alloys, in particular of icosahedral materials.

6 Conclusion

Diffraction experiments have become an indispensable tool available to solid state and materials scientists for routine analysis as well as for the elucidation of complex structural problems. In our review of the state of single crystal X-ray diffraction experiments we have aimed to show the non-routine user the wide range of information that can be obtained from a crystal structure refinement. In addition, the material scientists should increasingly use the technique himself since public domain software is available and hardware is widely distributed in research institutions. For the specialist X-ray crystallographer there is still the need to develop further the field of structure analysis much of which has been so successfully developed into routine procedures.

References

(1)   Friedrich, W., Knipping, P., von Laue, M.T.F.: Interferenzerscheinungen bei Röntgenstrahlen. Sitzungsber. Math. Phys. Kl. K. Bayr. Akad. Wiss. München, 303(1912).

(2)   a) Bragg, W.L.: The specular reflection of X-rays, *Nature* 90, 410(1912).
        b) Bragg, W.L.: The diffraction of short electromagnetic waves by a crystal, *Proc. Cambridge Philos. Soc.*, 17, 43(1993).
        c) Bragg, W.L.: The structure of some crystals as indicated by their diffraction of X-rays, *Proc. R. Soc. London, Ser. A.* 89, 248(1993).

(3)   Bragg, W.H., Bragg, W.L.: "X-rays and crystal stucture." G. Bell, London, 1915.

(4)   Giacovazzo, C., *et al.*: "Fundamentals of crystallography". Editor: C. Giacovazzo, Oxford University Press, Oxford, 1992.

(5)   "International tables for crystallography", Vol. A - C, Kluwer Academic Publishers, Dorrecht, 1983.

(6)   a) Bloss, F.D.: "Crystallography and crystal chemistry", Mineralogical Society of America, Washington, 1994.
        b) Dunitz, J.D.: "X-ray analysis and the structure of organic molecules", Helvetia Chimica Acta, Basel / VCH, Weinheim 1995.

(7)   Dorset, D.L.: "Structural electron crystallography", Plenum Press, New York, 1995.

(8)   Krivoglaz, M.A.: "X-ray and Neutron diffraction in nonideal crystals", Springer, Berlin, 1996.

(9)   Monaco, H.L.: "Experimental methods in X-ray crystallography" in Giacovazzo, C., *et al.*, Fundamentals of crystallography, C. Giacovazo, ed., Oxford University Press, Oxford, 1992, pp 229-318.

(10)   Sheldrick, G.: SHELXS 86
        For a survey of Direct Method programs see also the web site of the International Union of Crystallography, http://www.unige.ch/crystal/w3vlc/crystal.index.html

(11)   Sheldrick G.: SHELX93
        For a survey of structure refinement programs and crystal structure graphics programs see also The web site of the International Union of Crystallography, http://www.unige.ch/crystal/w3vlc/crystal.index.html

(12)   Ladd, M.F.C., Palmer, R.A.: "Structure determination by X-ray crystallography", Plenum Press, New York, 1994, pp 117-182.

(13)   Sheldrick, G., *Acta Cryst.* A46, 467(1990).

(14)   Hendrickson, W.A., *Science* 254, 51(1991).

(15)   a) Bricogne, G.: in "Crystallographic cmputing 5", eds. D. Moras, A.D. Podjarny, J.C. Thierry, Oxford University Press, 1991, pp 257-297.
        b) Gilmore, C.J., Bricogne, G.: in "Crystallographic cmputing 5", edts. D. Moras, A.D. Podjarny, J.C. Thierry, Oxford University Press, 1991, pp 298-307.

(16)   Vainshtein, B.K.: "Modern Crystallography I: Fundamentals of crystals", Springer, Berlin, 1994.

(17)   Nye, J.F.: "Physical properties of crystals", Oxford University Press, 1985, pp 3-32.

(18)   Salje, E.K.H.: "Phase transitions in ferroelastic and coelastic crystals", Cambridge University Press, 1990.

(19)   Nye, J.F.: "The physical properties of crystals", Oxford University Press, 1988, pp 82-109.

(20)   Dunitz, J.D.: "X-ray analysis and the crystal structure of organic molecules", Helvetica Chimica Acta, Basel / VCH, Weinheim, 1995, pp 225-264.

(21)   Tolédano, J.-C., Tolédano, P.: "The Landau theory of phase transitions", World Scientific, Singapore, 1987.

(22)   a) Bohm, J.: "Realstruktur von Kristallen", E. Schweizerbartsche   Verlagsbuchhandlung, 1995.
        b) Catti, M.: in Giacovazzo, C., *et al*.: "Fundamentals of crystallography". Editor: C. Giacovazzo, Oxford University Press, Oxford, 1992, pp 599-643.

(23)   Drits, V.A., Tchoubar, C.: "X-ray diffraction by disordered lamellar structures", Springer, Berlin, 1990.

(24)   Janot, C.: "Quasicrystals. A Primer", Clarendon Press, Oxford, 1992.

(25)   Craven, B.M.: in "Crystallographic computing 4", eds. N.W. Isaacs, M.R. Taylor, Oxford University Press, 1988, pp 211-220.

(26)   a) Kelly, A., Groves, G.W.: "Crystallography and crystal defects", Longman, London, 1970.
        b) Vaishtein, B.K., Fridkin, V.M., Indenbom, V.L.: "Modern crystallography 2: Structure of crystals", Springer, 1994, pp 330-399.

STRUCTURE DETERMINATION FROM POWDER DIFFRACTION DATA

ANTHONY K. CHEETHAM
Materials Research Laboratory, University of California
Santa Barbara, CA 93106, USA

1. Introduction

1.1 MOTIVATIONS

The explosion of interest in powder diffraction methods during the last 25 years has been driven by a number of factors. The major one was most certainly the development of the Rietveld method (1) in the late 1960s, since, at a stroke, this extended the scope of powder techniques from simple, high symmetry materials to compounds of substantial complexity in any space group. Within 5 years, for example, the method was being used to refine the structures of orthorhombic and monoclinic materials with as many as 22 atoms in the asymmetric unit (2), and by 1977, Cheetham and Taylor were able to review the application of the Rietveld method to over 150 compounds (3). The majority of these early applications involved the use of neutrons, but the field received a further boost in the late 1970s and early 1980s with the extension of the Rietveld method to X-ray data (4), time-of-flight neutron data (5), and then synchrotron X-ray data (6). These instrumental advances were accompanied by software developments, such as the availability of the DBWS (7) and GSAS (8) packages, making possible the analysis of data from complex mixtures or the simultaneous analysis of more than one dataset. Another major area of interest has been the development of methods for solving unknown structures from powder diffraction data, a subject that is the primary focus of this chapter.

A quite different driver for the developments in powder diffraction, however, has been the growing need for tools that are able to probe the structures of materials that are only available in powder form, or can only be studied as powders (*e.g.* under difficult *in situ* conditions). Such materials include many zeolite catalysts, as well as certain high T_c cuprates and fullerenes. Table 1 lists some of the many areas in which powder diffraction methods have had a major impact; clearly, modern materials science, and many other areas, have been major beneficiaries of the developments in this area during the last 25 years, and this trend will surely continue into the 21st century.

C.R.A. Catlow and A. Cheetham (eds.), New Trends in Materials Chemistry, 35–51.

Table 1: Impact of powder diffraction methods in materials science and other areas

Hydrogen storage	Superconductivity
Metal hydrides	High Tc cuprates
Magnets	Batteries / Fuel cells
Magnetoresistance, GMR	β-alumina solid electrolytes
Heterogeneous catalysts	Ferroelectrics
Zeolites, clays	$PbTiO_3$ etc.
Ceramics	Electro-optics
Zirconias	Non-linear optics e.g $KTiPO_4$
Novel materials	Biominerals
C60 fullerenes	Apatites
Coordination compounds	Organic materials
Homogeneous catalysts	Pharmaceuticals

The aim of this introductory chapter is to trace the key developments in powder diffraction methods from their discovery in the early years of this century to the present day. We shall emphasise the evolution of tools for solving unknown structures, remembering, of course, that the refinement step is also an important component of this process.

2. Early history of powder diffraction

The possibility of using powder diffraction methods to study materials was recognised shortly after the discovery of X-ray diffraction by Laue and von Knipping in 1910. In particular, the construction of a simple powder diffractometer was described by Hull in 1917 (9), and the instrument was used to obtain patterns from a number of simple materials such as diamond, graphite and iron. Even at this early stage, the use of metal foils to remove $K\beta$ radiation from the X-ray beam was well understood. Within a few years, many others, including the Braggs and Pauling, had exploited the powder method to study a wide range of materials, including metals, minerals, and simple organic solids. It could reasonably be argued that the first *ab initio* structure determinations were performed during this period, since the crystal structures of many simple materials, *e.g.* rocksalt, were obtained from powder diffraction data, alone.

The first systematic attempts to determine unknown structures of non-cubic materials were probably those of Zachariasen, reported in the late 1940s. For example, the hexagonal structure of UCl_3, in space group $P6_3/m$, was determined by first placing the heavy atom and then estimating the position of the chlorine by careful inspection of the intensities of different classes of reflections (10). In the same issue of *Acta Crystallographica*, a second paper by Zachariasen describes the structures of 8 uranium halides and oxohalides from X-ray powder data (11), and in the following year a similar approach was used by Mooney to solve the tetragonal structure of UCl_4 (12).

These early approaches might be regarded as trial-and-error methods, though they reveal great insight into the relationships between trends in the integrated intensities of different classes of reflections and the locations of the scattering centres. They certainly laid the foundations for the systematic approaches that evolved during subsequent decades.

3. Early *ab initio* approaches

There are at least two papers in the 1960s that describe systematic attempts to use the structure solving tools of modern crystallography, direct methods and Patterson techniques, to solve structure from powder data. In a remarkable paper by Zachariasen and Ellinger in 1963 (13), the monoclinic structure of β-plutonium, in space group I2/m, was solved by using a manual direct methods phasing procedure. There are seven Pu atoms in the asymmetric unit, underlining the complexity of this task. A particularly interesting aspect of this work was the clever use of the anisotropic thermal expansion of β-plutonium to unscramble the individual Bragg intensities of overlapping reflections from patterns collected at different temperatures. As will become clear later, the treatment of overlapping reflections remains one of the major issues in structure determination from powder data.

Another eye-catching paper from the 1960s is that by Debets (14) in which the orthorhombic structure of UO_2Cl_2 in space group Pnma was determined by Patterson methods. As in the work of Zachariasen and Ellinger, their approach is not radically different from that which has been used widely in the late 1980s and 1990s. An interesting difference between these early studies and the more recent work, however, is that the structure refinement step did not take advantage of least-squares methods, which, of course, are used routinely today. Nevertheless, the essential correctness of the UO_2Cl_2 structure has since been confirmed by Taylor and Wilson (15).

4. Pre-Rietveld refinement methods

The development of least-squares crystallographic structure refinement methods in the 1960s, which was facilitated by the growing availability of digital computers, was applied not only to single crystal data but also to powder data. A number of laboratories, such as U.K.A.E.A, Harwell, made widespread use of single crystal codes for refining structures from powder data, and some of the codes were adapted to handle groups of inequivalent overlapping reflections that could not be resolved experimentally. Table 2 shows an example of such a refinement, carried out at Harwell shortly before the availability of a Rietveld programme that would run on the computer there.

Table 2: Integrated intensity structure refinement for $Fe_{0.923}O$ at 800 C, based upon powder neutron diffraction data. There were 12 observations and four variable parameters: the scale factor, the occupancy number for the tetrahedral interstitial site, and independent isotropic temperature factors for the iron and oxygen atoms (16).

hkl	$I(abs)$	$I(calc)$
111	6092	6315
200	89766	89604
220	79186	79285
311	3378	3484
222	26340	26139
400	12697	13094
331	1540	1377
420	31104	31301
422	23695	23438
333/511	1061	898
440	5831	6042
600/442	10394	10308

R(I) = 0.78%

The paucity of data and the poor observation-to-parameter ratio make it hard to believe that this was essentially the state-of-the-art in the late 1960s, but it is important to stress that such studies played an important role at the time in the quantitative structural characterisation of high symmetry inorganic materials. A particular class of materials that benefited from this approach was non-stoichiometric compounds, which are typically high symmetry phases that are found at high temperatures. Nevertheless, the limitation of the integrated intensity method, at the time, was that it could not be applied to the complex patterns obtained from low symmetry materials (Figure 1). The advent of the Rietveld refinement method, however, was soon to solve this problem.

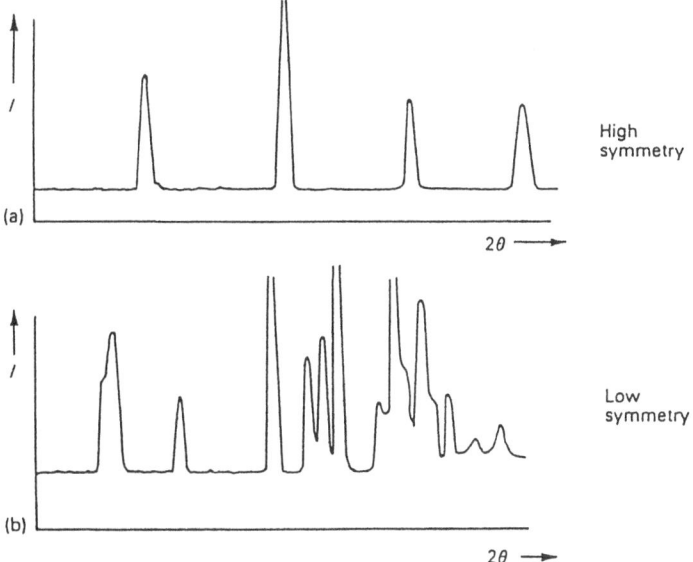

Figure 1: Schematic powder diffraction patterns of samples with (a) high symmetry and (b) low symmetry. In the latter case, the intensities of individual Bragg reflections may be difficult or impossible to determine due to peak overlap.

5. Rietveld refinement

In response to the need to develop enhanced procedures for obtaining structural information from powder samples, Rietveld (1) proposed, in the late 1960s, a method for analysing the more complex patterns obtained from low symmetry materials by means of a curve-fitting procedure in which the least-squares refinement minimises the difference between the observed and calculated profiles, rather than individual reflections. In the first instance, this procedure was carried out with neutrons rather than X-rays because of the simpler peak shape of the Bragg reflections. With neutrons, it can normally be assumed that the reflections are Gaussian in shape, and the calculated intensity at each point (say, $0.05°$ 2θ steps) on the profile is obtained by summing the contributions from the Gaussian peaks that overlap at that point. In addition to the conventional parameters in the least-squares (*i.e.* scale factor, atomic coordinates and temperature factors), additional parameters are required: the lattice parameters (which determine the positions of the reflections), a zero-point error for the counter setting, and three parameters that describe the variation of the Gaussian half-width (full width at half maximum intensity) with scattering angle. The technique has been applied to a wide range of solid-state problems and has been reviewed by several authors during the last 20 years (3,17,18). A typical example is shown in Figure 2 (19).

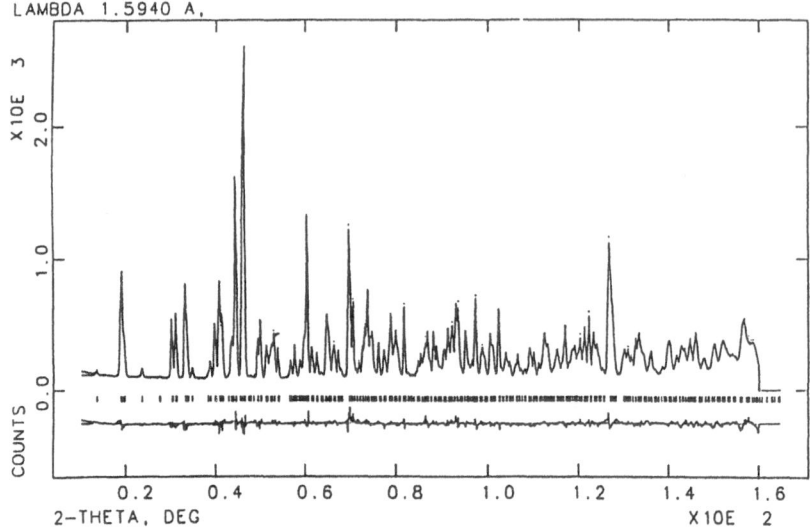

Figure 2: The observed (dots), calculated (smooth curve) and difference (lower curve) profiles resulting from the Rietveld analysis of powder neutron diffraction data collected on Sr_3CuPtO_6. Reflection markers are also shown.

The application of the Rietveld method to neutron data in the early 70s was soon followed by its extension to laboratory, X-ray diffractometer data (4). The problem of the more complex peak shape was resolved by employing alternative peak-shape functions, such as the Lorentzian and the pseudo-Viogt. Other problems that can plague X-ray studies include preferred orientation and poor powder averaging (graininess), both of which arise from the fact that X-rays probe a smaller sample volume than neutrons; these were addressed by paying closer attention to the data collection strategy.

The accuracy and precision of a structure refinement from X-ray data can normally be optimised by collecting high resolution data at a synchrotron source (6). Figure 3 shows an example of such a refinement (20). The resolution of the powder diffractometers at the second and third generation sources is so good that sample imperfections now play a major role in determining the shape of the Bragg peaks. This presents both challenges and opportunities. For the crystallographer, the subtle variations in the peaks shapes from one class of reflection to another (which may stem from, say, anisotropic particle size or strain effects) may be an irritation if the sole aim is to obtain a high-quality refinement of the crystal structure. However, the materials scientist may be delighted to retrieve a wealth of additional information pertaining to the microstructure of the sample.

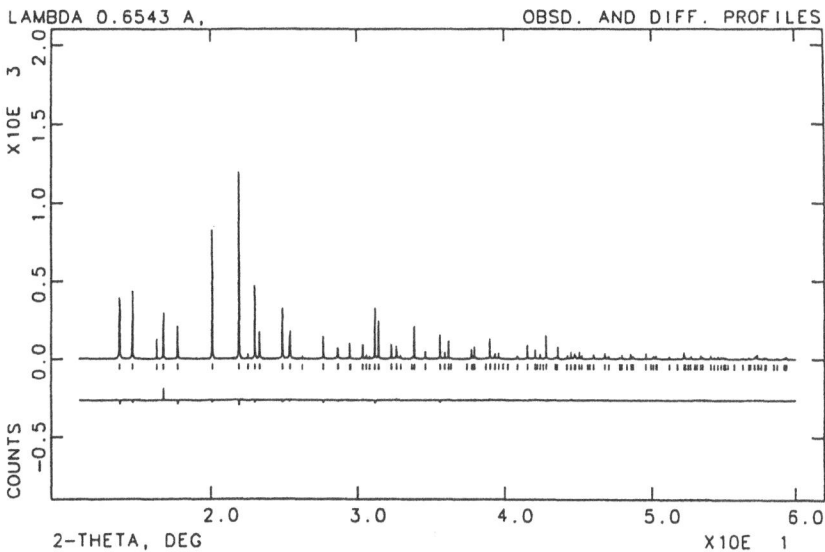

Figure 3: The observed, calculated and difference patterns resulting from the Rietveld analysis of the synchrotron X-ray powder pattern of $LiNbO_3$ (21). The lower curve is the difference profile; reflection markers are shown. The data were collected on beamline X7A at the NSLS, Brookhaven National Laboratory.

Refinement by the Rietveld method is now commonplace with both laboratory and synchrotron X-ray data, although it is not, in general, as precise as the neutron method (Table 3). There are three reasons for this. First, as mentioned above, it is more difficult to collect X-ray data that are essentially free from systematic errors. Consequently, it is not unusual to find that the precision, as measured in terms of the e.s.d.s, gives a misleading impression of the real accuracy of the structure. Second, the fall-off of intensity with scattering angle due to the X-ray form factor reduces the quality of the information that can be retrieved from the high angle region of an X-ray pattern. Third, the wide variation in X-ray scattering factors between elements from different parts of the periodic table leads to considerable differences in the sensitivities with which atoms can be located; in particular, heavy atoms will be better defined than light atoms. In Table 3, for example, the y coordinate of Cr(2) is determined with greater precision than the y coordinate of O(2) by powder X-ray diffraction. This problem does not arise to the same extent with neutrons because their scattering amplitudes (or scattering lengths, as they are known) fall within a relatively narrow range of values (23).

Table 3: Structural parameters for α-CrPO$_4$ refined using synchrotron X-ray (Marked X) and neutron (N) data in Imma (No. 74) with e.s.d.s in parenthesis (21). Values from the reported single crystal study (marked S) are given for comparison (22).

Atom	x	y	z	B_{iso}[a]
Cr(1)	1/2	1/2	0	0.3(2)N
				0.283(6)S
Cr(2)	1/4	0.3660(3)X	1/4	
		0.3650(4)N		0.0(1)N
		0.36611(3)S		0.316(4)S
P(1)	1/2	1/4	0.0819(12)X	
			0.0790(8)N	0.0(1)N
			0.0825(2)S	0.30(1)S
P(2)	1/4	0.5738(4)X	1/4	
		0.5739(2)N		0.47(8)N
		0.57358(5)S		0.345(7)S
O(1)	0.3790(10)X	1/4	0.2269(17)X	
	0.3766(3)N		0.2280(5)N	0.53(8)N
	0.3773(2)S		0.2268(3)S	0.42(2)S
O(2)	0.3603(6)X	0.4914(5)X	0.2145(11)X	
	0.3610(2)N	0.4907(1)N	0.2142(3)N	0.62(6)N
	0.3611(1)S	0.4902(1)S	0.2146(2)S	0.42(1)S
O(3)	0.2263(6)X	0.6352(5)X	0.576(10)X	
	0.2240(1)N	0.6368(2)N	0.0546(3)N	0.68(5)N
	0.2238(1)S	0.6363(1)S	0.0552(2)S	0.56(1)S
O(4)	1/2	0.3509(8)S	-0.0457(15)X	
		0.3486(2)N	-0.0422(4)N	0.31(/7)N
		0.3496(2)S	-0.0432(3)S	0.50(2)S

a: for the powder X-ray refinement, overall $B_{iso} = 0.24(7)$Å^2

The Rietveld method was also extended to the analysis of time-of-flight neutron powder data collected at pulsed sources. The procedure is essentially the same as that used in constant-wavelength experiments, except that the peak shape function is considerably more complex, due in part to the shape of the neutron pulse, and wavelength-dependent corrections (*e.g* absorption and extinction) must be taken into account (5). One advantage of time-of-flight powder methods is that the whole diffraction pattern is collected simultaneously since the counter, or bank of counters, remains stationary, making it an attractive way of following structural changes that evolve as a function of time, temperature or pressure. In addition, since the incident and scattered beams can pass through small apertures in, say, a high-pressure apparatus, the design of special environments is clearly easier for such measurements (24). A further advantage is that it is relatively easy to obtain high resolution data by using a long incident flight path and placing the detectors in the back scattering position.

The Rietveld method is a powerful tool, but it is limited by the same drawback that affects powder methods in general: the loss of information that arises from the compression of the three-dimensional diffraction pattern into a single dimension. It is also important to underline the fact that the Rietveld method, though an excellent technique for *refining* structures, requires a good starting model if it is to converge successfully and does not, by itself, constitute a method for *structure determination*. We shall now return to the question of solving unknown structures and examine the state-of-the-art in this area.

6. Solving unknown structures from powder data

There has been a great deal of interest concerning the determination of unknown structures from powder diffraction data during the last decade and there have been several reviews of the subject (25,26,27). The process may conveniently be broken down into a series of steps, though there may be considerable overlap between the different stages:

(i)   Determination of the unit cell.

(ii)   Decomposition of the powder pattern into integrated intensities, I_{hkl}.

(iii) Assignment of space group from systematic absences

(iv)   Solution of the phase problem

(v)   Refinement of the structure, typically by the Rietveld method.

Most of these stages are discussed in detail in more specialised chapters in this book, so only a few general comments will be made at this stage.

Step (i), the indexing of the powder pattern to yield a unit cell, is normally carried out by auto-indexing methods, for which a number of powerful computer programs are now available. These include ITO (28), TREOR (29) and DICVOL (30). Access to more than one of these programs is desirable since they work in different ways and successful indexing may not always be achieved with one particular program; the key point is that successful indexing is facilitated by collecting high quality data. In practice, it seems that very few structure determinations now stumble at the point of determining the unit cell.

We should also note that the identity of the space group may become apparent following the auto-indexing stage, though uncertainties frequently remain and must be resolved at a later stage (see below). In cases where the unit cell determination is proving difficult, it might be worthwhile to obtain selected area electron diffraction patterns from microcrystals, noting that an electron diffraction pattern is the equivalent of a zero-level precession photograph with X-rays. The feasibility of this approach will depend upon the stability of the sample in the beam, but the extra effort that it entails may be rewarded, especially if there is a subtle superstructure to which powder X-ray methods may not be sensitive.

The second step of the structure determination, the decomposition of the pattern into individual integrated intensities, is often the most challenging one because it is here that severe ambiguities may arise due to overlapping of peaks. Such overlapping may be accidental or may be an unavoidable consequence of the symmetry (*e.g.* the exact overlap of non-equivalent reflections in certain high symmetry Laue groups). A number of powerful single-step strategies that have been developed to address the pattern-decomposition problem are now embodied in computer programs such as ALLHKL (31), WPPF (32), GSAS (8, incorporating the Le Bail method (33)), LSQPROF (34) and EXTRA (35). The earliest development in this area, due to Pawley (31), was based upon a Rietveld fitting procedure in which the integrated intensities were refined in addition to the lattice parameters, peak shape parameters, *etc.* Le Bail's method (33) is closely related, but is somewhat more robust in its treatment of overlapping data.

From this stage onwards, the analysis can mirror that of a single crystal study. In stage (iii), the possible space groups can be assigned from the systematic absences, although in cases of uncertainty it may be useful to carry out the pattern decomposition in a number of alternative space groups (or to obtain a series of electron diffraction patterns, as discussed above). Uncertainties often remain (as they do with single crystal methods) and may have to be resolved during the structure solution and/or refinement steps. The phase problem is then solved in stage (iv) by conventional crystallographic methods, *ie.* Patterson or Direct Methods, the choice being dictated by the chemical nature of the material. Early work in this area utilised programmes that had been developed for the analysis of single crystal data, but some direct methods codes that are optimised for powder data are now available, including SIRPOW (36) and SIMPEL (37). Patterson determinations are also benefiting from vector-search algorithms.

As with the solution of structures from single crystal data, light atom problems will normally respond better to Direct Methods, whilst structures containing a subset of heavy atoms will be more amenable to Patterson techniques. The principal difficulty is that, even if there are no ambiguities due to peak overlap, the data set will be considerably smaller than that obtained in a single crystal study and the phasing procedure will be less straightforward. It is a tribute to the robustness of modern structure-solving techniques that it is still possible to determine structures under these unfavourable circumstances. Once a suitable starting model has been obtained, stage (v), the refinement of the structure, can proceed by using the Rietveld method. Quite commonly, the starting model will be incomplete and additional atoms will be found during the refinement procedure by using difference Fourier methods.

More recently, the probability of solving a structure from powder data has improved because there has been a move towards the development of pattern-decomposition methods that are more sophisticated. Typically, these new methods are not single-step procedures, but involve an iteration between the pattern decomposition step and the subsequent Patterson or direct methods calculations. For example, the observed intensities that are obtained from a successful pattern decomposition should yield a Patterson map that fulfils certain requirements, *e.g.* it should be positive at all points. Some of the codes that link the pattern decomposition and the structure-solving stage are DOREES (38) and FIPS (39), and those based upon Maximum Entropy (41,41) and Bayesian fitting procedures (42). A particularly powerful approach is the use of Entropy Maximisation and Likelihood Ranking (43), a method that has been used in other areas of crystallography and has now been adapted for powder data in the code MICE (44).

7. Trial-and-error and simulation methods

In addition to the systematic approaches described above, there has been a long-standing tradition of solving unknown structures from powder data by trial-and-error methods. A typical example can be seen in the work of Titcomb *et al.* (45), who solved the superstructure of the metal hydride phase, CeH_{2+x}, by exploring all of the possible arrangements of the interstitial hydrogens in the fluorite-related parent structure. The fluorite-related structure of Bi_3ReO_8 was solved in a similar manner (46). The starting models obtained by trial-and-error were then refined by the Rietveld method. In this approach, it is clearly advantageous (and often essential) to use information that may be available from other studies on the material of interest. In Bi_3ReO_8, for example, it was know from infra-red measurements that the oxygen coordination around the rhenium atom was tetrahedral. Similarly, model-building, together with information from electron microscopy and ^{29}Si magic angle spinning NMR, was used to elucidate the structure of the molecular sieve zeolite, ZSM-23 (47).

The manual trial-and-error strategy is not very attractive, since it can be very time-consuming and the chances of success are not particularly high. However, modern simulation methods, together with the power of modern computers, can be used to remove much of the labour and uncertainty from this approach by automating the way in which previous knowledge of a system, or related systems, is used. In the zeolite area, for example, Deem and Newsam (48) have developed a simulated annealing method that can be used to predict unknown zeolite structures from a knowledge of the unit cell, the space group, and the number of tetrahedral Si/Al (T) sites in the cell (in cases where the space group or number of T sites is uncertain, the calculation of sufficiently fast that alternative possibilities can be tested). The simulation procedure employs cost functions that depend upon the T-T distances and T-T-T angles in a large body of known zeolitic structures.

Table 4: Some examples of *ab initio* structure determinations from synchrotron X-ray powder data.

Compound	Space Group	Number of atoms in Asymmetric Unit	Reference
α-CrPO$_4$	Imma	8	[54]
I$_2$O$_4$	P2$_1$/c	6	[55]
Al$_2$Y$_4$O$_9$	P2$_1$/c	15	[55]
MnPO$_4$.H$_2$O	C2/c	6	[56]
PbC$_2$O$_4$	P1	7	[57]
Clathrasil, Sigma-2	I4$_1$/amd	17	[58]
LaMo$_5$O$_8$	P2$_1$/a	14	[59]
BeH$_2$	Ibam	4	[60,61]
UPd$_2$Sn	Pnma	4	[62]
C$_5$H$_{11}$NO$_2$	Pna2$_1$	19	[63]
NaCD$_3$	I222	10	[64,65]
C$_{10}$N$_6$SH$_{16}$	P2$_1$/n	33	[66]
BaBiO$_{2.5}$	P2$_1$/c	5	[67]
(VO)$_3$(PO$_4$)$_2$.9H$_2$O	P2$_1$/n	13 non-H	[68]
CuPt$_3$O$_6$	Pn2$_1$m	10	[69]
Ga$_2$(HPO$_3$)$_3$.4H$_2$O	P2$_1$	29	[70]
LaTiAlO		60	[71]

Molecular crystals, too, lend themselves naturally to simulation methods, since their molecular structures (or fragments thereof) are often known with some confidence. We can use prior knowledge of the molecular structure (or an energy-minimised molecular structure obtained by quantum mechanical calculations) and move the molecular fragment by translations and rotations within the unit cell using Monte Carlo methods. A knowledge of the space group is again required, of course. The crystal structure can then be predicted by using energy functions based upon appropriate inter- and intra-molecular potentials, or by comparison between the calculated and observed X-ray powder patterns. Approximate models can then be refined by the Rietveld method. Examples of structures solved in this manner include piracetam,$C_6H_{10}N_2O_2$ (49) and 1-methylfluorene, $C_{14}H_{12}$ (50). The approach is very straightforward for rigid molecules, but becomes considerably more difficult as the number of degrees of freedom increases.

8. Some examples of structure determination from powder data

A great deal of the development work in the field of structure determination from powder data has relied on the use of conventional, laboratory X-ray sources, and there were a number of important early successes in the area (51,52). However, synchrotron X-ray data has profound advantages over conventional X-ray data for structure determination. In particular, the combination of the high brightness and excellent vertical collimation can be harnessed to construct diffractometers with unparalleled angular resolution, as in the case of Cox's instrument at the National Synchrotron Light Source (NSLS), Brookhaven National Laboratory (53), where the resolution at the focusing position is <0.02° in 2θ. This is partly achieved by constructing the instrument in the vertical plane, since the vertical divergence, θ_v, is only ~0.01° at 2.5 GeV. With high resolution data, the solution of structures from powder data is greatly facilitated because ambiguities due to peak overlap are minimised and the information content of the dataset is optimised. Many successful structure solutions and refinements have now been performed, a selection of which is given in Table 4. On the other hand, very few structures have been solved from powder neutron diffraction data. This is partly a consequence of the lower resolution of most neutron diffractometers, but it is mainly due to the near equivalence of the neutron scattering lengths for most elements, as a result of which the phase problem cannot be solved on the basis of locating a small sub-set of atoms.

In the first example of a structure solved from synchrotron X-ray powder data, carried out in 1986 by Attfield et al. (54), the orthorhombic structure of α-CrPO$_4$, with 8 atoms in the asymmetric unit, was solved by Patterson methods using a vector search procedure; 68 well-resolved peaks were utilised. A relatively poor R_{pr}-factor (19.3%) was obtained for the final Rietveld refinement with the synchrotron data, no doubt due to problems with preferred orientation and hkl-dependent line broadening effects, but a subsequent medium resolution neutron study (on D1a at ILL Grenoble) gave an excellent fit (R_{pr} = 8.3%), confirming the correctness of the X-ray model. A comparison of the coordinates obtained from the X-ray and neutron refinements, and a subsequent single crystal study, was given in Table 3. In particular, we note that the neutron refinement gives improved precision for almost all atoms, in spite of the modest resolution of the neutron data.

During the last decade, there has been widespread use of synchrotron powder methods to solve unknown structures (Table 4), the most striking development being the extension of the method to systems of very considerable complexity, with as many as 60 atoms in the asymmetric unit cell (71). Such complex structures normally require a combination of both synchrotron X-ray and neutron data for their solution and refinement, since they lie at the limit of what can currently be done with a single dataset. The ease of access to good laboratory diffractometers, however, has encouraged an even greater effort with laboratory data. Structures of great complexity have been solved (e.g. β-Ba_3AlF_9, with 29 atoms in the asymmetric unit (72)), and there has been extensive use of the methodologies in the areas of molecular organic crystals, coordination compounds and organometallic materials (most of the effort in the powder diffraction area has traditionally been in the realm of non-molecular inorganic materials). Eye-catching examples include recent work on bipyridyl complexes of nickel and copper (73) and the carbonyl cluster compound, $[HgRu(CO)_4]_4$ (74).

9. Conclusions

The rôle of powder diffraction in the structural characterisation of materials has expanded dramatically during the last 30 years. A number of developments have played important roles: (i) the advent of the Rietveld refinement method; (ii) improvements in laboratory X-ray instrumentation; (iii) the availability of high resolution powder diffractometers at pulsed neutron sources and synchrotron sources; (iv) advances in computational methods for structure solution, and (v) improvements in computer hardware, e.g. personal computers that are capable of running Rietveld codes. The power of powder techniques is such that they have had an impact in most of the major developments in the field of new materials during recent years; solid electrolytes, high temperature superconductors, fullerenes, zeolites and giant magnetoresistance (GMR) materials are obvious examples. As a consequence, powder diffraction has been transformed from the ugly duckling of crystallography into one of the most exciting and fast-moving areas.

Notwithstanding the remarkable progress, much work remains to be done. The solution of unknown structures from powder data is by no means routine, and the methods need to be further automated before they can be used by non-specialists, even those with crystallographic experience. Furthermore, there is considerable scope for advances in refinement procedures, in spite of the power of the Rietveld method. For example, some of the complex structures that are now being solved are at the limit of what can be refined by current procedures. As a consequence, the accuracy of many of these more complex structures falls well short of what we would hope for and aspire towards. This can easily be seen by looking at the bond lengths that are obtained from refinements of complex organic or zeolitic materials; it is not unusual to find interatomic distances that are clearly outside the range that would be considered to be chemically acceptable.

The solution to this problem will no doubt come from several areas. First, it will become possible to collect better data, and more of it, especially with access to short X-ray wavelengths at the 3rd generation synchrotron sources (systematic errors in X-ray data can be dramatically reduced at shorter wavelengths). Second, the simultaneous analysis of X-ray and neutron data is already having an impact, but we shall no doubt see the use of data from other techniques such as solid state NMR and EXAFS. In addition, advances in computation are taking us towards a scenario where energy minimisation will become a part of the refinement procedure. For example, it is already clear that we can sometimes calculate the structure of an all-silica zeolite with better accuracy than we can determine it experimentally by powder X-ray diffraction (75). Finally, we shall see the use of more subtlety in the refinement process, such as the more extensive use of maximum entropy methods (76,77).

This overview would not be complete without reference to the developments in single crystal methods that may have an impact on powder crystallography. The construction of 3rd generation synchrotron sources has, once again, focused attention on the possibility of collecting X-ray data from micron-size crystals. Progress in this area has not been as rapid as many had expected, but a recent example from ESRF in Grenoble (78) may offer a glimpse of future possibilities in this area. Nor should we forget the power of the electron microscope for interrogating small crystals. There have been several examples (*e.g.* 79,80) of structure refinements by using higher order Laue zones (HOLZ) from convergent beam electron diffraction patterns, and this area is likely to attract further attention. Nevertheless, the current capabilities and the exciting opportunities for the future can leave us in no doubt that powder diffraction will continue to play a dominant role in this area for the foreseeable future.

References

1.   H.M. Rietveld, J. Appl. Crystallogr., 1969, **2**, 65.

2.   R.B. von Dreele and A.K. Cheetham, Proc. Roy. Soc., 1974, **A338**, 311.

3.   A.K. Cheetham and J.C. Taylor, J. Solid State Chem., 1977, **21**, 253.

4.   G. Malmros and J.O. Thomas, *J. Appl. Crystallogr.*, 1977, **10**, 7; R.A. Young, P.E. Mackie, and R.B. Von Dreele, *J. Appl. Crystallogr.*, 1977, **10**, 262.

5.   R.B. Von Dreele, J.D. Jorgensen , and C.G. Windsor, *J. Appl. Crystallogr.*, 1982, **15**, 581.

6.   D.E. Cox, J.B. Hastings, W. Thomlinson, and C.T. Prewitt, *Nucl. Instrum. Method*, 1983, **208**, 573.

7.   R.B. von Dreele and A. Larson,

8.   D.B. Wiles and R.A. Young,

9.   Hull,

10.   W.H. Zachariasen, *Acta Cryst.*, 1948, **1**, 265.

11.   W.H. Zachariasen, *Acta Cryst.*, 1948, **1**, ???.

12.   R. Mooney, *Acta Cryst.*, 1948, **2**, 189.

50

13. W.H. Zachariasen and Ellinger, *Acta Cryst.*, 1963, 16, 369.

14. ?. Debets, *Acta Cryst.*, 1968, B24, 400.

15. J.C. Taylor and P.W. Wilson, *Acta Cryst.*, 1973, B29, 1073.

16. A.K. Cheetham, B.E.F. Fender and R.I. Taylor, *J. Phys. C (Solid State Phys.)*, 1970, 4, 2160.

17. A.W. Hewat, *Chem. Scripta.*, 1986, 26A, 119.

18. *"The Rietveld Method"*, (R.A. Young, Editor), Oxford University Press, 1993.

19. A.P. Wilkinson, A.K. Cheetham, W. Kunnman, and Å. Kvick, *Eur. J. Solid State Inorg. Chem.*, 1991, 28, 453.

20. A.P. Wilkinson and A.K. Cheetham, unpublished results.

21. J.P. Attfield, A.K. Cheetham, D.E. Cox, and A.W. Sleight, *J. Appl. Crystallogr.*, 1988, 21, 452.

22. R. Glaum, R. Gruehn, and M. Moller, *Z. Anorg. Allg. Chem.*, 1986, 543, 111.

23. G.E. Bacon, *"Neutron Diffraction"*, 3rd ed., Oxford, 1975.

24. J.D. Jorgensen, in *"Chemical Crystallography with Pulsed Neutrons and Synchrotron X-Rays"*, NATO Ad. Study Ins. Ser., Series C, vol. 221, (Eds. M.A. Carrondo and G.A. Jeffrey) D. Reidel, Dordrecht, 1988, 159.

25. A.K. Cheetham, *Mater. Sci. Forum,*, 1986, 9, 103.

26. A.K. Cheetham in "The Rietveld Method", (R.A. Young, Editor), Oxford University Press, 1993.

27. K.D.M. Harris and M. Tremayne, *Chem. Mater.*, 1996, 8, 2554.

28. J.W. Visser, *J. Appl. Crystallogr.*, 1969, 2, 89.

29. P-E. Werner, L. Eriksson, and M.J. Westdahl, *J. Appl. Crystallogr.*, 1985, 18, 367.

30. A. Boultif and D. Louer, *J. Appl. Crystallogr.*, 1991, 24, 987.

31. G.S. Pawley, J. Appl. Crystallogr., 1981, 14, 357.

32. H. Toraya, J. Appl. Crystallogr., 1986, 19, 440.

33. A. Le Bail, H. Duroy, and J.L. Fourquet, *Mater. Res. Bull.*, 1988, 23, 447.

34. J. Jansen, R. Peschar, and H. Schenk, *J. Appl. Crystallogr.*, 1992, 25, 231.

35. A. Altomare, M.C. Burla, G. Cascarano, A. Guagliardi, A.G.G. Moliterni, and G. Polidori, *J. Appl. Crystallogr.*, 1995, 28, 842.

36. G. Cascarano, L. Favia, and C. Giacovazzo, *J. Appl. Crystallogr.*, 1992, 25, 310.

37. J. Jansen, R. Peschar, and H. Schenk, *Z. Krist*, 1993, 206, 33.

38. J. Jansen, R. Peschar, and H. Schenk, *J. Appl. Crystallogr.*, 1992, 25, 237.

39. M. A. Estermann, L. B. McCusker, and C. Baerlocher, *J. Appl. Crystallogr.*, 1992, 25, 539.

40. W. I. F. David, *J. Appl. Crystallogr.*, 1987, 20, 316.

41. W. I. F. David, *Nature*, 1990, 346, 731.

42. D. S. Sivia and W. I. F. David, *Acta Cryst.*, 1994, A50, 703.

43. G. Bricogne and C. J. Gilmore, *Acta Cryst.*, 1990, A46, 285.

44. C. J. Gilmore, G. Bricogne and C. Bannister, *Acta Cryst.*, 1990, A46, 297.

45. C. G. Titcomb, A. K. Cheetham and B. E. F. Fender, *J. Phys. C, Solid State Phys.*, 1974, 7, 2409.

46. A. K. Cheetham and A. R. Rae-Smith, *Acta Cryst.*, 1985, B41, 225.

47. P. A. Wright, J. M. Thomas, G. R. Millward, S. Ramdas, and S. A. I. Barri, *J. Chem. Soc. Chem. Comm.*, 1985, 117.

48. M. W. Deem and J. M. Newsam, *Nature*, 1989, 342, 360.

49. D. Loüer, M. Loüer, V. A. Dzyabchenko, V. Agafonov, and R. Ceolin, *Acta Cryst.*, 1995, B51, 182.

50. M. Tremayne, B. M. Kariuki, and K. D. M. Harris, *J. Mater. Chem.*, 1996, 6, 1601.

51. J.-E. Berg and P.-E. Werner, *Z. Krist.*, 1977, 145, 310.

52. P. Rudolf and A. Clearfield, *Inorg. Chem.*, 1984, 23, 4679.

53. D.E. Cox, J.B. Hastings, L. P. Cardoso, and L. W. Finger, *Mater. Sci. Forum,*, 1986, 9, 1.

54. J.P. Attfield, A.W. Sleight, A.K. Cheetham, *Nature (London)*, 1986, **322**, 620.

55. M.S. Lehmann, A.N. Christensen, H. Fjellvag, R. Feidenhans'l, and M. Nielsen, *J. Appl. Crysallogr.*, 1987, **20**, 123.

56. P. Lightfoot, A.K. Cheetham, and A.W. Sleight, *Inorg. Chem.*, 1987, **26**, 3544.

57. A.N. Christensen, D.E. Cox, and M.S. Lehmann, *Acta Chem.Scand.*, 1989, **43**, 19.

58. L.B. McCusker, *J.Appl.Crystallogr.*, 1988, **21**, 305.

59. S.J. Hibble, A.K. Cheetham, A.R.L. Bogle, H.R. Wakerley, and D.E. Cox, *J. Am.Chem. Soc.*, 1988, **110**, 3295.

60. G.S. Smith, Q.C. Johnson, D.E. Cox, R.L. Snyder, D.K. Smith, and A. Zalkin, *Adv. X-ray Anal.*, 1987, **30**, 383.

61. G.S. Smith, Q.C. Johnson, D.K. Smith, D.E. Cox, R.L. Snyder, and R-S. Zhou, *Solid State Commun.*, 1988, **67**, 491.

62. M. Marezio, D.E. Cox, C. Rossel, and M.B. Maple, *Solid State Commun.*, 1988, **67**, 831.

63. M. Kurahshi, K. Honda, and M. Goto, *Photon factory activity report* 1989, **7**, 170.

64. E. Weiss, S. Corbelin, J.K. Cockcroft, and A.N. Fitch, Angew. Chem. Int. Ed. Engl. 1990, **29**, 650; *Angew. Chem.*, 1990, **102**, 728.

65. E. Weiss, S. Corbelin, J.K. Cockcroft, and A.N. Fitch, *Chem. Ber.* 1990, **123**, 1629.

66. R. Cernik, A.K. Cheetham, C.K. Prout, D.J. Watkin, A.P. Wilkinson, and B.T.M. Willis, *J. Appl. Crystallogr.* 1991, **24**, 222.

67. P. Lightfoot, J.A. Hriljac, S. Pei, Y. Zheng, A.W. Mitchell, D.R. Richards, B. Dabrowski, J.D. Jorgenson, and D.G. Hinks, *J. Solid State Chem.* 1991, **92**, 473.

68. R.G. Teller, P. Blum, E. Kostiner and J.A. Hriljac, manuscript in preparation.

69. J.A. Hriljac, J.B. Parise, G.H. Kwei, R. Shannon, and K. Schwartz, *J. Phys. Chem. Solids*, 1991, **52**, 1273.

70. R.E. Morris, W.T.A. Harrison, J.M. Nicol, A.P. Wilkinson and A.K. Cheetham, *Nature (London)*, 1992, **359**, 519.

71. R.E. Morris, J.J. Owen, J.K. Stalick and A.K. Cheetham, *J Solid State Chem.*, 111, 52-57 (1994).

72. A. Le Bail, *J. Solid State Chem.*, 1993, **103**, 287.

73. N. Masciocchi, P. Cairati, L. Carlucci, G. Mezza, *J. Chem. Soc.-Dalton Trans.*, 1996, **13**, 2739.

74. N. Masciocchi, P. Cairati, F. Ragaini, and A. Sironi, *Organometallics*, 1993, **12**, 4499.

75. A. K. Cheetham, L. M. Bull, and N. J. Henson, Studies in *Surface Science & Catalysis*, 1997, **105**, 2267.

76. M. Sakata, S. Mori, S. Kumazawa, M. Takata, and M. Toraya, *J.Appl.Crystallogr.* 1990, **23**, 526.

77. M. Sakata, U. Tatsuya, M. Takata, and C. J. Howard, *J.Appl.Crystallogr.* 1993, **26**, 159.

78. P. A. Wright *et al.* (to be published).

79. R. Vincent, D. M. Bird and J. W. Steeds, *Philos. Mag.*, 1984, **A50**, 745, 765.

80. K. Tsuda and M. Tanaka, *Acta Cryst.*, 1995, **A51**, 7.

SOME APPLICATIONS OF ELECTRON MICROSCOPY AND RELATED TECHNIQUES IN MATERIALS CHEMISTRY

M.Á. ALARIO-FRANCO
Facultad de Ciencias Quimicas, Universidad Complutense, 28040 MADRID, Spain

Abstract

One of the more obviously important aspects of Materials Chemistry is to establish the structure of solids down to the microstructure level. Indeed, the very nature of solids means that the microstructure can play a controlling rôle in the important properties that made a solid useful, *i.e. a material*. For these reasons, electron diffraction and electron microscopy and a plethora of different techniques have become essential tools in the study of the Solid State aspects of Materials Chemistry.

In the present article, after a brief description of the main techniques that can be used within a modern transmission electron microscope: TEM, ED, X-EDS and EELS, examples will be given of the use of those techniques to tackle a few typical cases including defects, superstructure ordering, microdomains and modulated structures.

1. Introduction

Once Louis de Broglie established the dual nature of waves and particles[1], the road was open to invent the Transmission Electron Microscope (TEM) and shortly after, in 1931, Max Knoll and Ernest Ruska built the first one, for which Ruska received the Nobel Prize for Physics 55 years later, sharing it with the discoverers and developers of the Scanning Tunnelling Microscope (STM). Ruska's machine became a commercial instrument with a resolution of the order of 20Å, even before World War II; in fact, according to van Dyck[2], the first pictures of biological objects were obtained by Barton in 1932. However, the real impact of electron microscopy in Materials Science only starts in the late fifties and early sixties when machines working at a stable accelerating potential of 100kV became commercially available.

The working principle of the electron microscope is indeed similar to its optical (visible light) counterpart but, due to the specific characteristic of

C.R.A. Catlow and A. Cheetham (eds.), New Trends in Materials Chemistry, 53–78.
© *1997 Kluwer Academic Publishers.*

electrons relative to photons, there are important differences in the design and use of these analogous techniques.

To start with, electrons are charged particles, so that they can only be handled in high vacuum; yet this inconvenience is more than compensated for by the fact that, for that very reason, electrons can be deflected — *i.e. focused* —and also their interaction with matter, both nuclei and electrons, is very strong. This implies that only very small amounts of sample are required to work in an electron microscope: of the order of $10^{-15}mm^3$ as compared to ~1 mm^3 in single crystal X-ray diffraction and over a cubic centimetre in single crystal neutron diffraction.

Electrons can be accelerated so that their energy (and obviously, their wavelength) can be varied and controlled. The relation between the wavelength λ and the accelerating voltage Φ is given by:

$$\lambda = h / \{2me\Phi(1 + e\Phi / 2mc^2)\}^{1/2}, \qquad (1)$$

where h is Planck's constant, e and m the electron charge and mass, and c the velocity of light in vacuum. Table 1 shows some values of λ for different values of Φ.

TABLE 1. Electron wavelength l as a function of the accelerating potential

Φ (volt)	λ (Å)
1	12.26
10^3	0.38
10^5	0.0370
4×10^5	0.0602
10^6	0.00807
6×10^6	0.0019

This means that, as the resolving power of a microscope is proportional to the wavelength (see below) electron microscopes are orders of magnitude more powerful than optical ones in this respect since visible light has a wavelength of the order of several thousand Angstroms. Moreover, interatomic distances are usually in the Angstrom range, so that with electron microscopy one can aim at, and achieve resolution of atomic features, and even atoms. In principle, then, the higher the accelerating voltage, the higher the resolving power and the more information that can be obtained. Nevertheless, complicated practical aspects both in their use, such as voltage stability and radiation damage, and in the construction of very high voltage machines, mean that there are very few operating around or above a million volts (notably in Tsukuba and Berkeley). These same factors have limited the commercially available instruments to 400 kV with, at present, a tendency to restrict the voltage to 300 kV and to improve some of the crucial technical characteristics of this type of instrument, such as the illumination system, using field emission guns, and the quality, in particular the spherical aberration, of the objective lens. Yet these instruments are capable of a resolution of the order of 1.6-2Å, which are really well suited to study many of the most important aspects of the *microstructure* of materials.

1.1 HOW DOES THE TRANSMISSION ELECTRON MICROSCOPE (TEM) WORK?

In the usual design of an electron microscope, Figure 1, an electron beam produced in an electron gun is condensed through two lenses (condenser lenses) that make a parallel and coherent electron beam, that is then targeted on the sample.

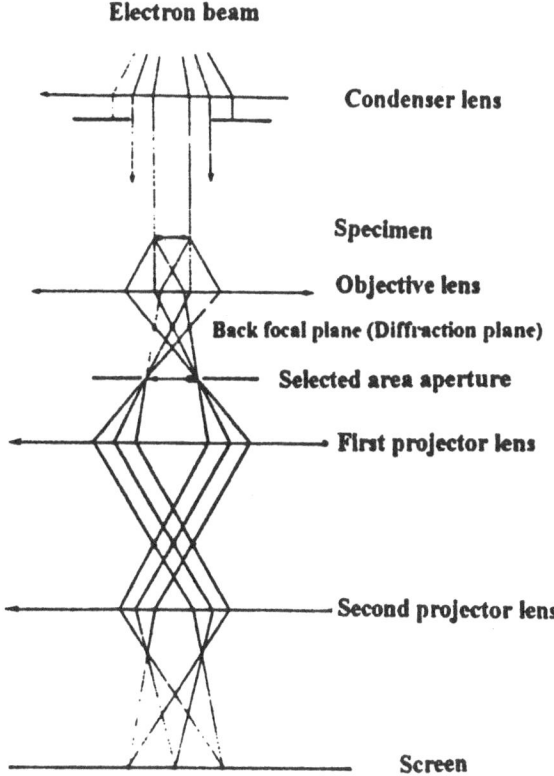

Figure 1: Electron path in an electron microscope operating in the diffraction mode.

In the case of a sufficiently thin sample (~100-300Å, that is periodic) with a periodicity comparable in extent to the electron wavelength (λ) and, in the most favourable case, single -crystalline, although most of the beam will be transmitted without deviation or energetic attenuation, part of it will be scattered, that is, *diffracted*. This means that rays will be produced in certain regions of the space, as a result of the principle of interference between the rays leaving the sample. Diffraction is, of course, controlled by Bragg's law:

$$2d_{hkl}\sin \Theta = n\lambda , \qquad (2)$$

where d_{hkl} is the spacing of the lattice planes, of Miller indices hkl, scattering in a direction which makes and angle Θ (Bragg's angle) with the transmitted beam direction, and $n = 0, \pm1, \pm2, ...$ corresponds to the diffraction order.

Now if a lens capable of deviating electrons, nowadays usually an electromagnetic lens, is put ahead of the sample, the scattered radiation is brought to a focus in the back focal plane of the lens. It can be seen on Figure 1 that all rays diffracted in the same direction, *i.e.* all parallel rays converge in the same point in the back focal plane. What one observes there is, in fact, the diffraction pattern of the sample, a typical sample of which appears on Figure 2.

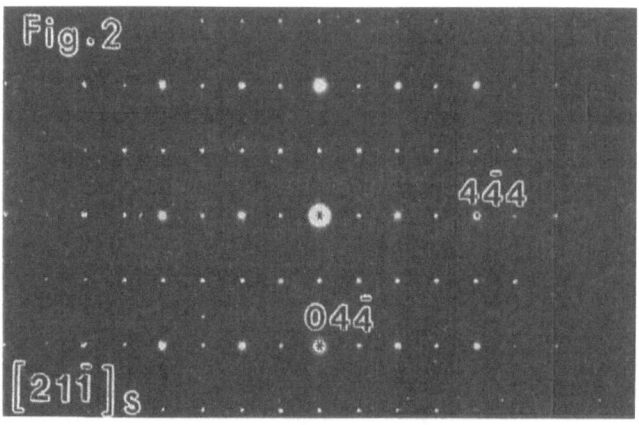

Figure 2: Typical electron diffraction pattern (crystal of Mn_2ErS_4. Zone axis [211]; courtesy Dr Otero–Diaz).

The diffraction process is actually a rather complex one. Even in the case of kinematic diffraction, that is, for the unusual case of very thin and weakly scattering materials, such as small crystals of solid hydrogen as cited by Cowley[3]. In these cases, the total scattered amplitude in a given direction is given by the amplitudes scattered by all the atoms in the crystal. Now, in the diffraction process both the amplitude and the phase of the impinging wave are modified. The total scattered amplitude is given by:

$$\psi(u) = \sum_j f_j \exp(2\pi i u r) \qquad (3)$$

where f_j is the scattering factor for an atom located at position r_j and u is a vector indicating the change in direction of the original beam along S_0, after being diffracted along S_1. From Bragg's law, this 'scattering vector' u has the magnitude $(2\sin\Theta)/\lambda$, and length $1/d_{hkl}$ and is perpendicular to the family of planes (hkl) giving that scattering amplitude. The ensemble of all points which are the end points of all u vectors constitutes a lattice of points, called *reciprocal lattice*, in which points diffracted intensity, *i.e.* the permitted Bragg reflections, appear. Consequently a diffraction pattern is a planar section of the reciprocal lattice of the crystal. Yet in this, the most simple case, we have left out a number of important issues related to both the interaction of the electrons and sample, and the modifications that the necessarily imperfect lens produces in the electron wave. Dynamical effects and aberrations are the main categories of these problems, and the bibliography cited should be consulted for a detailed account of these matter.

It is also clear that if two crystals overlap along the electron path, two overlapping diffraction patterns will be observed; at the limit, if an infinite number of crystals, of the same material, were placed within the electron path, instead of a single crystal type pattern, an 'infinite' number would be observed in the back focal plane: in these cases, a series of rings will appear, each one due to a characteristic Bragg reflection; a case like this is shown in Figure 3.

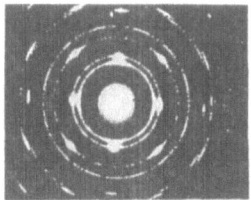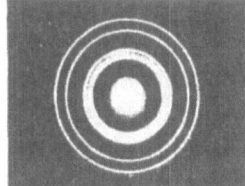

Figure 3: Electron diffraction pattern corresponding to a single crystal (left), a polycrystalline sample with small crystal size (right) and an intermediate situation (centre).

Also, a spotty ring pattern does indicate an intermediate situation with some relatively large crystals distributed within a polycrystalline powder. In this manner, one can study polycrystalline materials by means of electron diffraction, by measuring the Bragg spacings from which we can obtain the unit cell. In the case of an orthogonal crystal, there exists a very simple, well known relation between the so-called *d-spacings* and the unit cell parameters a, b and c:

$$1/d^2_{hkl} = (h/a)^2 + (l/c)^2 , \qquad (4)$$

and corresponding expressions for non-orthogonal crystals[4].

The geometry of the intensities in the diffraction pattern allows, in fact, a number of other important pieces of information to be obtained. Both Convergent Beam Electron Diffraction (CBED)[5] and Microdiffraction[6] are of much help in the determination of crystallographic properties such as crystal systems and Bravais lattices and, in favourable circumstances, of point and space groups.

Although from the diffracted intensities it is, in principle, possible to deduce the crystal structure of the corresponding crystal, as shown among others by Vainstein[7], the most common way nowadays of using the electron microscope in Solid State Chemistry and Materials Science is through the electron microscope images.

The next step in the process is the formation of an image in the so-called image plane situated further ahead of the back focal plane. It can be seen in Figure 1 that all beams that originate from one point of the sample recombine in the same point of the image plane, reforming an image of the object.

However, the process of image formation can be performed in different ways by the use of an *aperture*, that is a perforated disk which allows one to select the beams that contribute to the image.

In the simplest, the so-called *bright field* case, only the transmitted beam contributes to the image. Alternatively, in the so-called *dark field* case, one of the diffracted beams can be used. Obviously there will be some contrast in the image, since not all the electrons contribute, and this is the denominated *diffraction contrast*, In fact, as will be seen later, combining bright and dark field images in suitable conditions under a two beam situation, one can obtain interesting information about defects in the samples such as dislocations, stacking faults, crystallographic shear planes.

If the image is formed with large numbers of beams, it is obvious that more information will be present in the image provided that the resolution of the microscope is adequate; high resolution images can be obtained in this way. In the limit, all the beams can be used to make the image; yet some difficulties arise in terms of the usual problems of spherical and chromatic aberrations that introduce modifications in the observed image.

The resolution of an electron microscope[8] at the optimum defocus condition, the so-called Scherzer focus, is given by:

$$R = 0.65C_s^{1/4}\lambda^{3/4} \qquad (5)$$

where C_S is the spherical aberration constant, that is, a characteristic of each objective lens and λ the wavelength. This formula indicates that it is not sufficient to increase the acceleration voltage, *i.e.* decrease the wavelength, automatically to increase the resolution, if the construction of the objective lens does not keep the lowest possible spherical aberration. At present, for the most common 200 kV electron microscope without special attachments, a resolution of ~2.6Å is easily achieved. At 400 kV, on the other hand, 1.6-1.8Å can routinely be obtained.

1.2   IMAGE SIMULATION IN THE HIGH RESOLUTION ELECTRON MICROSCOPE

Although in favourable but exceptional cases, one can visualise a crystal structure in an electron micrograph, the commonest situation is the opposite, and electron microscopes are very sensitive to experimental parameters such as thickness, defocus and crystal orientation, which all affect the contrast of the observed image. For these reasons, an inevitable complement of the electron microscope experimental images is the simulation of the interaction between the electron wave and a model for the expected structure of the solid, taking into account the experimental conditions within the microscope. The problem consists in establishing the electron wave function first at the exit surface of the crystal and then, after it has gone through the objective lens and into its back focal plane and finally, at the image plane.

The usual way to tackle the first problem is to calculate the electron amplitudes and phases along these three steps, of which the first one is the most tedious. Once the electron wave at the exit surface of the crystal is known, a double Fourier transformation, taking into account defocus and aberration, provides the simulated images. The wave at the exit surface can be calculated by a number of ways, of which the so-called multi-slice method introduced by Cowley and Moodie[9,10] is the more commonly used. In this procedure, the specimen is divided into slices perpendicular to the beam and for each slice the effects of propagation (Fresnel Diffraction) and transmission (effect of the specimen potential) are treated at first separately and then combined in a convolution operation so that after the i-slice along the z direction of the beam, the wave function is:

$$\psi_i = [\psi_{i-1}(xy)*p_{i-1}(xy)].q_i(xy) , \qquad (6)$$

where $q_i(xy)$ is the transmission function for the i-slice, $p_{i-1}(xy)$ represents the diffraction effects between two consecutive slices and * stands for the convolution operation. In practice, it is easier to work in reciprocal space, since for crystalline materials the transmission function only exists at the reciprocal lattice points. The transmission function in reciprocal space is the Fourier transform of the corresponding function in real space. Analogously, the reciprocal space multi-slice equation is the Fourier transform of equation (6), that is:

$$\Phi_i(uv) = [\Phi_{i-1}(uv).P_{i-1}(uv)*Q_i(uv) , \qquad (7)$$

where u and v are the reciprocal space coordinates corresponding to the real-space coordinates x and y respectively.

If one does not take into account the inelastically scattered electrons, the transmission function is a phase grating of the form:

$$q_i(xy) = \exp[-y\sigma\phi_i(xy)\Delta z_i] , \qquad (8)$$

where Δz_i is the slice thickness of the i-th slice, $\phi_i(xy)$ is the projected potential of this slice per unit length along the beam direction and σ the interaction constant for electrons, a parameter that depends on the electron wavelength. The propagation of the function in reciprocal space is given by:

$$P(hk) = \exp[-2\pi i \zeta(hk)\Delta z] \qquad (9)$$

With these assumptions, the multi-slice iteration can be written as:

$$\Phi_n(hk) = \sum_{h'k'} \Phi_{n-1}(h'k')P_{n-1}(h'k')Q_n(h-h',k-k') \qquad (10)$$

for all beams included within the aperture.

In practice, one has to simulate many images as a function of different experimental parameters and, in particular, for different specimen thickness and defocus value; an example of the type of results obtained is given in Section 2.4. An image matching process with the experimental image is then made. For details of this procedure, the references given at the end should be consulted.

1.3 RELATED SPECTROSCOPY

The interaction of matter with electron waves not only gives rise to diffraction, and through this to imaging as a result of the different processes that take place when a high energy electron impinges on a piece of matter; the study of solids by means of an electron microscope allows the simultaneous use of many related techniques that are usually collectively ranged under the heading of 'Analytical Microscopy' or, often as 'Microanalysis'.

These techniques are spectroscopic since they are based on the measurement of energy exchanges between the beam and the solid, and not simply on changes in the direction of the transmitted electrons.

Among the large number of spectroscopy based techniques[11] that can be used in connection with the Transmission Electron Microscope, two are most common, both in terms of easiness of use and interest of the information they allow us to gather. These are 'energy dispersive X-ray spectroscopy' (X-EDS or EDS) and 'electron energy loss spectroscopy' (EELS). The first one measures the X-ray spectra that are inevitably produced as a result of the electronic radiation of the sample, so that a spectrometer, commonly based on a silicon or germanium detector, analyses these X-rays and, in principle, gives an unequivocal signature of the elements contained on it through the uniqueness of the atomic energy level characteristic of each element[12]. The detection range spans up to 20 keV and in the case of thin crystals, an energy resolution of ~100-150 eV and a lateral resolution of the order of ~100Å can be achieved; this means that for crystals of the order of 500Å in diameter, one can get, in the most favourable circumstances, a precision of the order of 1% in the atomic composition. Amongst the most important limitations of the X-EDS technique is the difficulty of determining light elements (lighter than say, sodium) although in more recent windowless type detectors, elements down to boron can be detected and eventually quantified, Yet for the light elements, EELS is often a more reasonable choice.

The EELS technique[13,14] measures the energy lost by the electrons that pass through the sample. Obviously this energy is lost in interaction processes with the electrons of the atoms in the solid giving information on which atoms are in the sample. To this end, an electron spectrometer is located after the sample, usually at the bottom of the microscope column, which separates the incident electrons as a function of their kinetic energy. This can in fact be done in two ways. Either a 'series acquisition' is used, in which one scans the energy spectrum in a sequential way (which may take some time) or a 'parallel acquisition' in which, with an appropriate wide angle detector, the whole spectrum is obtained simultaneously. The latter technique is known as Parallel Electron Energy Loss Spectroscopy (PEELS).

In both cases, the EELS spectra show the dispersed intensity as a function of the loss in kinetic energy experienced by the electron beam. Three regions can be differentiated in these spectra:

a) *Zero loss*, corresponding to the elastic peak, that is the transmitted beam plus the elastically scattered electrons and the presence of those electrons that have experienced losses below the detection limit, *e.g.* in phonon production.

b) *Low loss region*, up to ~50eV, in which the intensity of the signals are obviously far lower than in the zero loss region (~5-10% of it). They are produced by the energy losses due to the interaction of the beam with the collective electrons in the solid (valence and/or conduction bands, *e.g.* plasmon type excitations).

c) *High loss region*, or 'core loss', which corresponds to losses of the order of >50eV and although the signal intensities are much lower than those of the other two regions, this is the most fruitful part of the spectrum in analytical terms. In fact, from the absorption edges, especially the K-edges, one can deduce the presence of light elements, since the absorption cross section increases as the atomic number decreases.

2.   **Examples of the use of TEM, ED, X-EDS, and EELS in some problems of materials science**

1)   Characterisation of polycrystalline powders by TEM: β–FeOOH

2)   Extended defects in the system TiO_2–V_2O_3

3)   A chemical reaction produced by electron irradiation in the electron microscope: $Mn(Mn,Er)_2S_4$

4)   A modulated structure in $K_{2-x}La_xCuO_4$

5)   Carbon/copper ordering in HTSC materials

2.1   CHARACTERISATION OF POLYCRYSTALLINE POWDERS BY TEM: β-FeOOH

One of the earliest and commonest applications of TEM and electron diffraction was the characterisation of powders such as pigments and catalysts. Particle size, shape and porosity, as well as the evolution of these properties in the course of the thermal treatment or decomposition have been commonly studied, especially since a very high resolution instrument is not needed to produce interesting results. As an example, we will consider the thermal decomposition of β-FeOOH, an iron oxihydroxide[15] with the mineral name of *Akaganéite.*

β-FeOOH crystallises in a Hollandite-type structure in which pairs of edge sharing octahedra form strings parallel to the c axis that, by sharing corners, leave open quite big tunnels. In Hollandite itself, one of the multiple derivatives of manganese dioxide, the tunnels are occupied by potassium ions. In β-FeOOH, water molecules play a similar rôle and it is formulated as β-FeOOH.nH_2O, with n around one[16]. Figure 4a shows an electron micrograph of the cigar-like crystals characteristic of β iron oxihydroxide. When these crystals are evacuated at 150°C, most water is lost without affecting the crystallinity of the sample. Figure 4b shows a micrograph in which fringes corresponding to the (110) planes of the structure, d~7.4Å, are well resolved. When the material is treated at increasingly higher temperatures, the general morphology of the crystal is maintained but important changes take place. After heating the sample at 210°C in vacuum, a number of pores in the crystals develop as shown in Figure 4c. This, together with IR resonance spectroscopy and X-ray and electron diffraction evidence, suggests that the decomposition of akagneite has started. Yet up to ~310°C in vacuum, the main characteristics of the structure are still present and Figure 4d shows that the (110) fringes are well preserved in the non porous areas. However, at ~350°C in vacuum, dehydration has fully taken place:

$$\beta\text{-FeOOH}.nH_2O \rightarrow \gamma\text{-Fe}_2O_3 + (n + 1/2)H_2O \tag{11}$$

and the phase observed corresponds to the spinel-like γ-Fe$_2$O$_3$. Indeed, Figure 4e shows that each crystal presents a big pore around which the (100) fringes of γ-Fe$_2$O$_3$, d~8.3Å, are resolved.

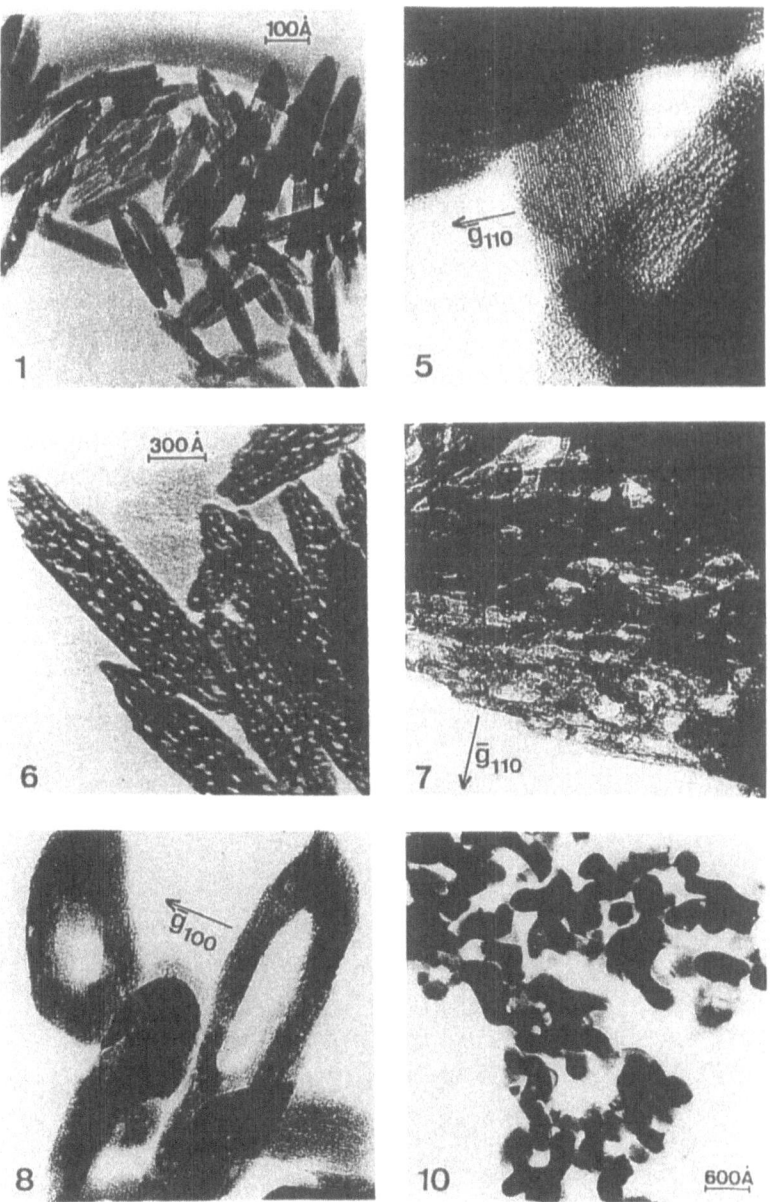

Figure 4: Different stages in the decomposition of Akaganétite as observed by TEM (courtesy of Professor González–Calbet); see text for details.

64

At still higher temperatures (≥510°C) in vacuum or in air, even the morphology changes due to sintering, as can be seen in Figure 4f, where welded particles with rounded shapes are predominant. Simultaneously, the most stable iron oxide at this temperature, α-Fe$_2$O$_3$ is formed. Complementary studies by gas adsorption and other techniques allowed us, quite comprehensively, to characterise the evolution of β-FeOOH.nH$_2$O, a process in which dehydration, dehydroxylation and structural transformation successively takes place. It is clear that, in this interesting process, TEM played the prominent rôle in the understanding of the evolution of the texture and microstructure in the course of the dehydration/decomposition of β-iron oxihydroxide.

2.2   EXTENDED DEFECTS IN THE SYSTEM TiO$_2$--V$_2$O$_3$

By reacting TiO$_2$ with V$_2$O$_3$ in sealed evacuated quartz ampoules[17] at T ~1100K, extended defects[18] are produced. These are mainly of two types: Antiphase Boundaries (APB) on the (110) planes and Crystallographic Shear Planes (CSP) parallel to (132). The presence of planar defects in a crystal produces a deformation of the diffraction spots. This *streaking* appears in a direction perpendicular to the fault plane. When more than one type of fault is present, it is possible to establish the orientation of the planar faults from their relative orientation with respect to the diffraction pattern.

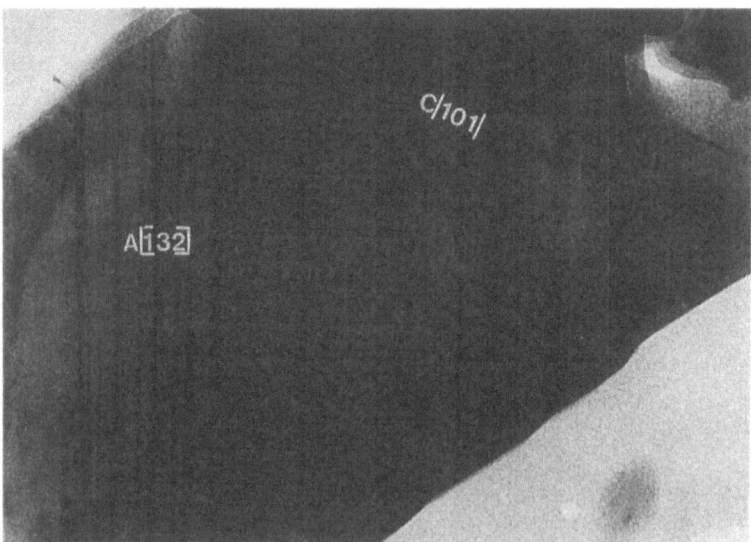

Figure 5: Coexisting A) {132}$_R$ and C) {101}$_R$ faults in a sample of TiO$_2$ doped with ~3% V$_2$O$_3$. Both types of faults are visible under two–beam conditions for g$\langle 130 \rangle$.

Figure 5 shows an electron microscope image of a crystal containing on average ~3% V$_2$O$_3$, where both types of fault are visible. Planar faults inclined to the electron beam originate a series of light and dark fringes. Under two beam conditions[19], *i.e.* when only the transmitted beam and one of the diffracted beams are operating, the fault contrast depends on the phase shift of the diffracted beam as given by the phase angle:

$$\alpha = 2\pi g.r , \qquad (12)$$

where $g(hkl)$ is the reciprocal vector corresponding to the diffracted beam and $r(uvw)$ is the *fault vector*, that is, the vector which transforms the part of the crystal at one side of the fault to the one at the other. Obviously,

$$\alpha = 2\pi(hu+kv+lw) = N\pi. \qquad (13)$$

Faults are visible when N is an odd number. They are, however, invisible when N is even. Figure 6 shows different pictures, under two beam conditions, of the same crystal of vanadium doped rutile where the (101) APB and (132) CSP faults are, respectively, both visible, (101) visible and (132) invisible, (101) invisible and (132) visible, and both invisible.

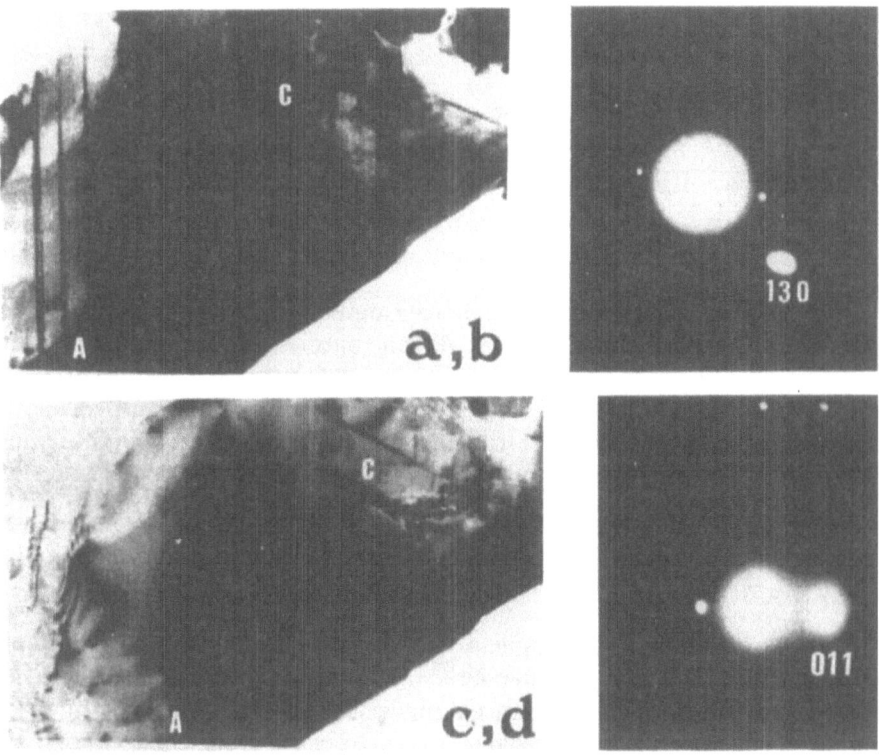

Figure 6: a, b: Both types of faults are visible under two beam conditions for the g_{130} reflection.
c, d: (101) visible and (132) invisible for g_{011}.

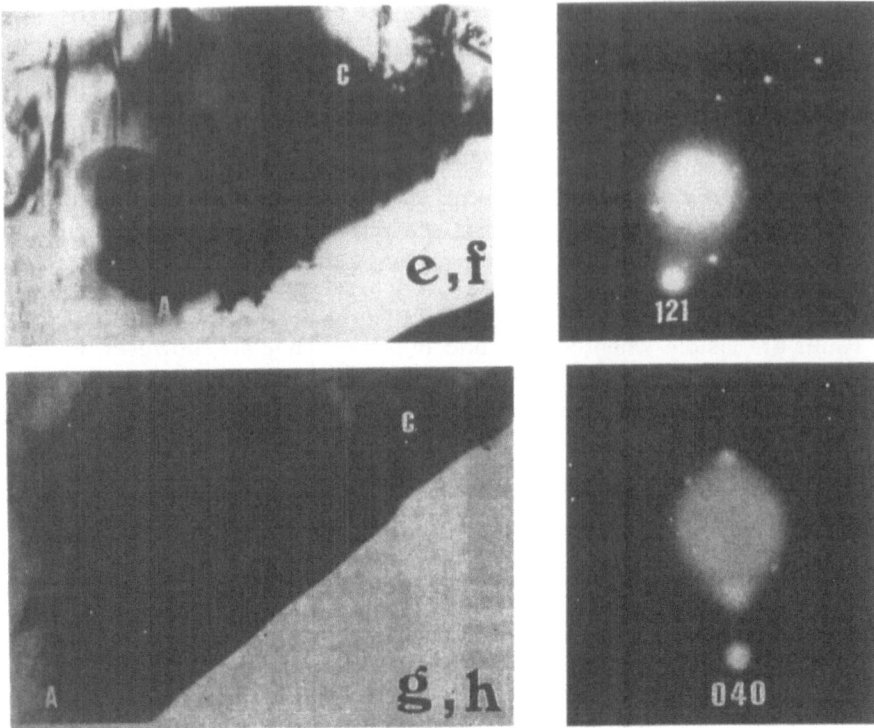

Figure 6: e, f: (101) extinct and (132) visible for g_{121}.

g, h: Both types of faults are invisible for g_{040}.

From these and many other observations[20], and on the basis of previous work by Eikum and Smallman[21] about ABP in pure rutile, and of Bursill and Hyde[22] on ABP and CSP in reduced TiO_{2-x}, we were able to assign to these faults the vectors1/2[101] and 1/2[001] respectively. If the fault plane is parallel to the beam, just one fringe is seen and, if the resolution of the microscope is adequate, the structure of the defect can, in favourable cases, be directly established.

2.3 A CHEMICAL REACTION PRODUCED BY ELECTRON IRRADIATION IN THE ELECTRON MICROSCOPE: $Mn(Mn,Er)_2S_4$

Electron irradiation of the sample in the electron microscope is a permanent risk in the observation of materials[23]. The resulting radiation damage is particularly problematic in the case of molecular materials, but can also be a problem with non-molecular systems. Indeed, electron beam damage is easy to produce but difficult to control. Yet there are some instances in which one can use it to follow reactions in the electron microscope. One such case recently studied by Landa and Otero[24,] refers to the decomposition of a mixed manganese erbium sulphite spinel into two other phases under the electron beam, a reaction produced by increasing the illumination by momentarily removing the condenser aperture.

Figure 7 shows a micrograph of the starting spinel that, according to the diffraction pattern shown in the inset, is orientated along the [110] zone axis; the image contrast suggests that it is a well ordered crystal and fringes corresponding to the {111}, {200} and {220} families of lattice planes are clearly seen.

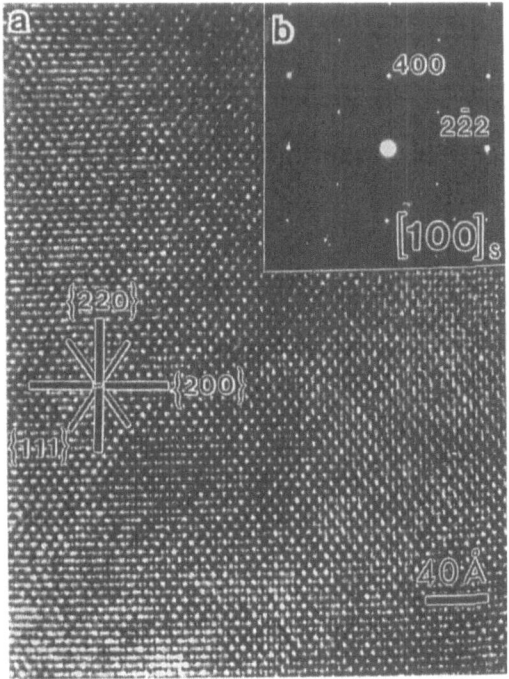

Figure 7: a) High resolution electron micrograph from a crystal of the Mn_2ErS_4 spinel; zone axis [110]; b) corresponding diffraction pattern (courtesy Dr Landa–Cánovas and Dr Otero–Díaz).

After a short period of observation (~15 sec.) with the aperture removed, streaking is apparent in the electron diffraction pattern. Figure 8a from another crystal along [211] shows such streaking and, additionally, three types of spots can be differentiated:

i)   the strongest spots corresponding to a rock-salt like structure (a-Mn(Er)S) along [211] with lattice parameter a≈5.3Å;

ii) weaker spots corresponding to the [001] zone axis of the orthorhombic $Mn(Mn,Er)_2S_4$;

iii) a third group of still weaker spots forming an ensemble of reflections parallel to [113] of the NaCl type structure as in i) and belonging to the monoclinic Mn(Mn,Er)$_4$S$_7$.

Figure 8c gives the indexing of such an elaborate pattern in terms of the three structures.

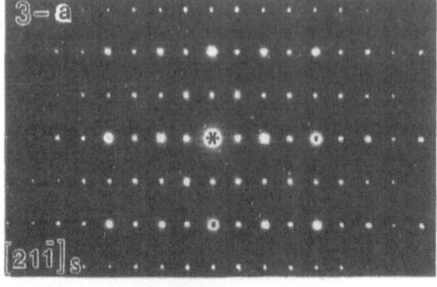

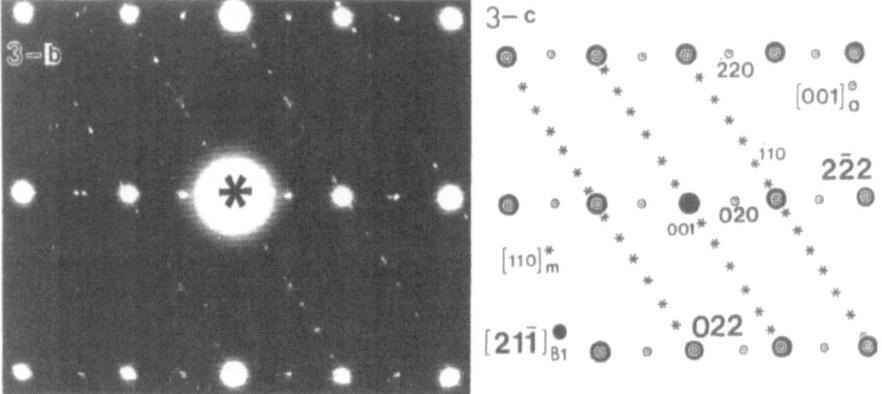

Figure 8 Consequences of electron irradiation in the Mn$_2$ErS$_4$ spinel (courtesy Dr Landa–Cánovas and Dr Otero–Díaz).

Although there is some streaking indicative of disorder in this structure, a seven-fold superstructure along [113] of the B1 type cell is clearly apparent; this periodicity corresponds to d≈11.2Å, in good correspondence with d(001) (≈11.0Å, of the monoclinic phase.

Figure 9 shows a high resolution low magnification micrograph of the corresponding crystal where two slabs of different phases are intergrown within the B1 type Mn(Er)S matrix. Lattice fringes of lamella A correspond to d(001)≈11.0Å of the monoclinic phase while those measured in lamella B can be assigned to the (020) planes of the orthorhombic Mn(Mn,Er)$_2$S$_4$.

Electron irradiation of this sulphide spinel results in a ready decomposition to a NaCl-type phase but in the process (and obviously to compensate for the stoichiometric change) slabs of orthorhombic Mn(Mn,Er)$_2$S$_4$ with the CaTi$_2$O$_4$ type structure, and monoclinic

Mn(Mn,Er)$_4$S$_7$ with the Y$_5$S$_7$ structure are present intergrown within the predominant B1 matrix.

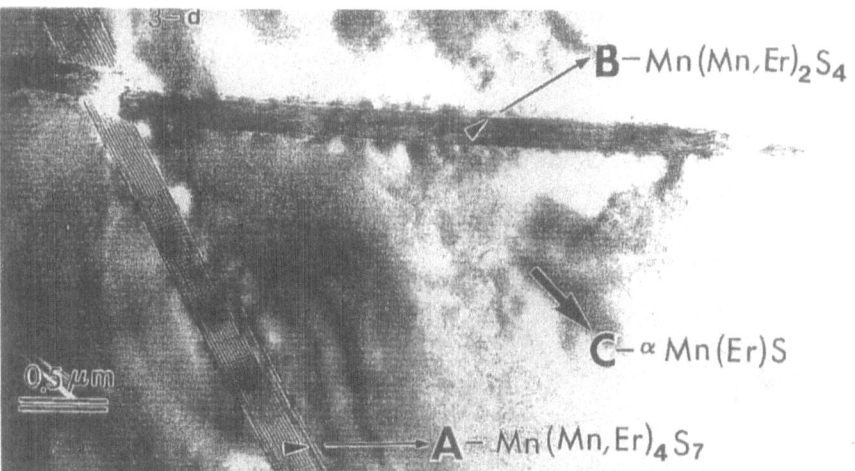

Figure 9: High resolution electron micrograph of a crystal of the Mn$_2$ErS$_4$ spinel after electron irradiation. An intergrowth of three different, but related, phases is apparent (courtesy Dr Landa–Cánovas and Dr Otero–Díaz).

2.4   A MODULATED STRUCTURE IN K$_{2-x}$La$_x$CuO$_4$

After the discovery of superconducting properties by Bednorz and Muller[25] in alkaline earth doped La$_2$CuO$_4$, there were many studies of other ionic substitutions. In the case of potassium substitution, that is in the system La$_{2-x}$K$_x$CuO$_4$, although the critical temperatures observed[26,27] were not very high (~20K) an interesting microstructure has been observed by HREM[28].

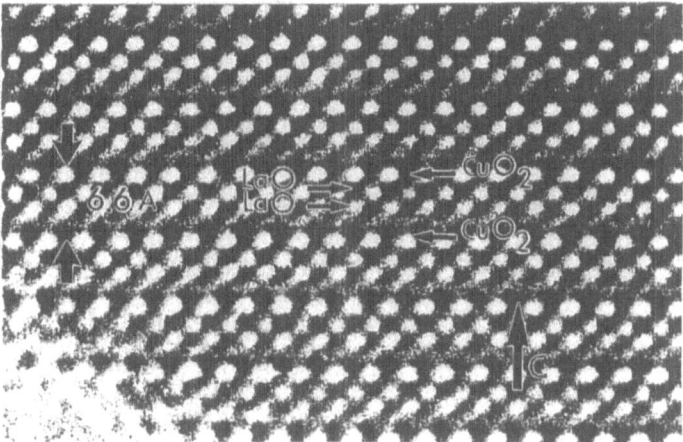

Figure 10: HREM micrograph of pristine La$_2$CuO$_4$.

Figure 10 shows a HREM micrograph of pristine La_2CuO_4: a sequence of CuO_2 and (2 x) La–O planes is clearly imaged along the b axis.

However, electron diffraction patterns of $La_{2-x}K_xCuO_4$ samples with compositions around x = 0.27 (Figure 11) do show three types of diffraction maxima:

i) strong spots which can be indexed on the basis of an La_2CuO_4 type cell in space group $I4/mmm$ and parameters a = 3.77 and c = 13.3Å;

ii) extra reflections of weaker intensity appearing on both sides of the main reflections along c^*.

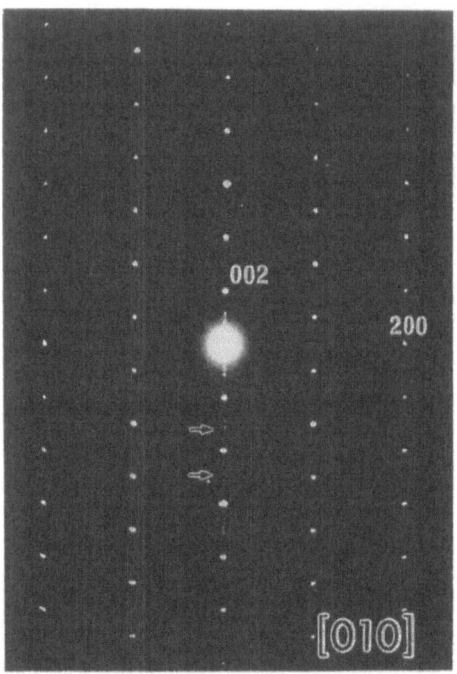

Figure 11: Electron diffraction pattern of potassium doped La_2CuO_4. An incommensurate modulation with period $1/15.7A^{-1}$ is present.

These are the first order satellites of the main reflections and their presence suggests the existence of a *modulated structure*. This is a crystal structure in which the occupation factors and/or the positions of a certain type of atom progressively change along one direction, the modulation direction[29]. In this particular case, the period of the modulation d^*_{sat} changes from crystal to crystal, even in the same sample and, moreover, its value is not an integral multiple of the c axis of the basic structure. In other words, it is an *incommensurate modulated structure*. In fact, closer observation of the satellites by means of tilting experiments[30] indicated that they are not points but rods extending in a direction perpendicular to the c^* axis and with a length of the order of 4.3-2.5Å-1 in reciprocal space, suggesting that they correspond to ordering states of ~25-40Å in real space. The reciprocal lattice corresponding to this system appears in Figure 12.

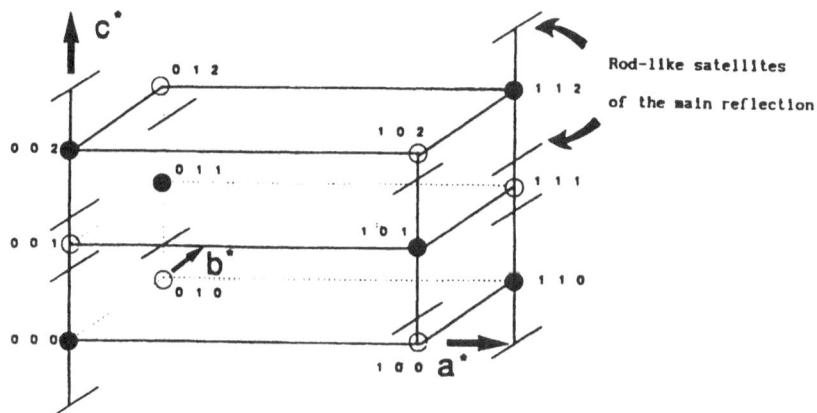

Figure 12: Reciprocal lattice unit cell characteristic of $La_{2-x}K_xCuO_4$ (empty circles, extinct reflections; full circles, allowed reflections; rods, satellite reflections of the main ones).

A high resolution micrograph along the(010) zone axis (Figure 13) clearly reveals the atomic (CuO_2) planes and, in between them, the two potassium doped (LaO) planes; it is clear that these show a markedly different contrast, one being much brighter and resembling the adjacent (CuO_2) plane. Considering the important difference in electron scattering for La and K, we attribute this contrast difference to the alternating substitution of La by K in every other (LaO) type plane. In an attempt to confirm such an ordering case, an image simulation was performed[30] for the crystal appearing in Figure 13. To this aim, one has to suppose different atomic arrangements that one can *a priori* presume may be responsible for those contrast differences. In principle, a very large number of different arrangements can be imaged, yet if one restricts the case to differences between lanthanum and potassium occupation factors, the number of possibilities is greatly reduced. In this case, the most obvious sequences of La and K within the unit cell were considered to be:

$$-CuO_2-LaO-LaO-CuO_2-LaO-KO-CuO_2- \qquad (14)$$

$$-CuO_2-LaO-LaO-CuO_2-KO-KO-CuO_2- \qquad (15)$$

$$-CuO_2-LaO-KO-CuO_2-KO-LaO-CuO_2- \qquad (16)$$

$$-CuO_2-LaO-KO-CuO_2-LaO-KO-CuO_2- \qquad (17)$$

In every one of these simulated models, Figure 14, it was obvious that a (KO) plane has a very different contrast from that of a pure (LaO) plane, but is similar to a (CuO_2) one.

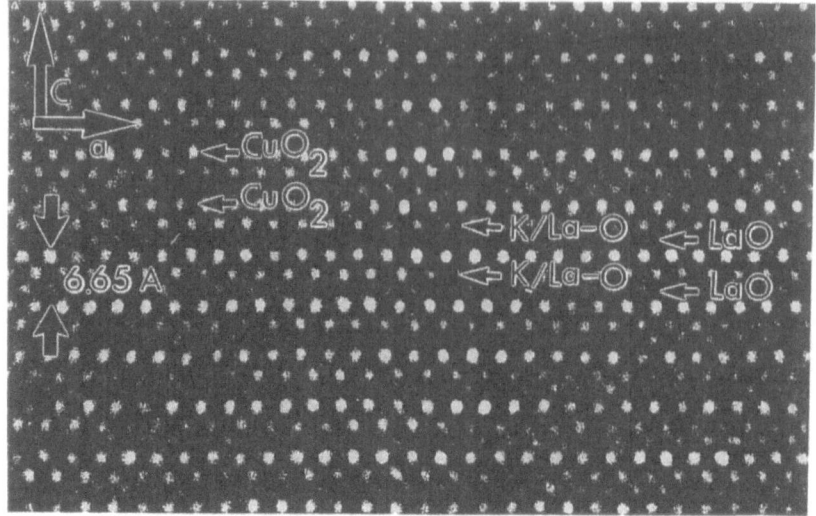

Figure 13: HREM micrograph of a crystal of $La_{2-x}K_xCuO_4$

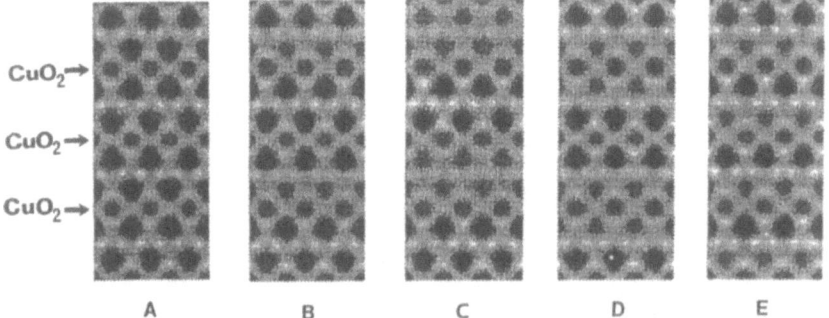

Figure 14: Image simulation for different models of La/K ordering in $La_{2-x}K_xCuO_4$

A comparison between through focus simulated images for different crystal values and the experimental pictures, suggested that the last indicated sequence (Equation 17), in which a (KO) plane substitutes every other (LaO) plane, gives the best fit. Consequently it was concluded that, far from being random, the substitution of lanthanum by potassium in La_2CuO_4 takes place in a quite elaborate ordering situation in which the sequence of atomic composition planes along the c axis can be represented as:

$$-CuO_2-LaO-(K/La)O-CuO_2-LaO-(K/La)O-CuO_2- \qquad (18)$$

This type of ordering raises a number of questions. In the first instance, the fact that K^+ and La^{3+} *are* ordered at the microstructure level can most likely be attributed to the differences in charge and radii of these two cations. But the most remarkable feature of this ordering is certainly the fact that it has different coherence length along the three space directions. As seen by electron diffraction and as represented by the reciprocal lattice (Figure 12) this ordering is *long range* along both (001) and the other direction perpendicular to the rod length, but in the first case it is *incommensurate*, and it is relatively *short range* (~20-40Å) along the rod direction. The reason for this is not entirely clear, though it is probably related to the kinetics of sample synthesis.

The substitution of lanthanum by potassium in La_2CuO_4, far from being a simple process, produces marked changes in the physico-chemical properties, including the induction of superconductivity and, most remarkably in structural terms, it is the origin of a very elaborate ordering process between the two big cations. Whether there is a causal relation between both phenomena or they simply coincide due to the substitution process remains an open question. Similarly, it remains to be seen whether the other well known substitutions in the same structure, in particular those of barium and strontium that gave rise to the High Temperature Superconducting Materials 'estampida' also order the structure of the lanthanum cuprate.

3. Carbon/Copper ordering in HTSC Materials

The continuous search for HTSC materials has recently produced a family of very interesting mercury-based superconducting cuprates[31] of general formula:

$$HgBa_2Ca_{n-1}Cu_nO_{2n+2+\delta}, \qquad (20)$$

for which critical temperatures of the order of 135K at room pressure[32] and of 160K under pressure[33,34] have been obtained. These constitute the present record in terms of critical temperature for any type of material. However, in view of the toxicity of mercury, attempts are being made to replace it by different cations. In this way, and by high pressure synthesis, another family of mercury free materials with the general formula:

$$(Cu_x/C)Ba_2Ca_{n-1}Cu_nO_{2n+2+\delta} \qquad (21)$$

were produced[35,36] with critical temperatures of the order of 117K, the highest attained so far with non-toxic materials. However, there are some doubts concerning the presence of carbon in theses species. Initially a copper/copper vacancy ordering was suggested on the basis of electron diffraction evidence[33] where a doubling of the usual *single charge reservoir layer (CRL)*[37] structure along the a axis was observed. Figure 15 shows an electron diffraction pattern of the n = 3 member of the above family along the (010) zone axis.

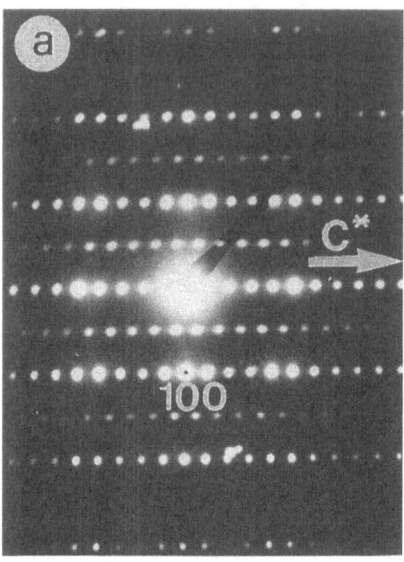

Figure 15: Electron diffraction pattern of a crystal of the n = 3 member of the $(Cu_x/C_{1-x})Ba_2Ca_{n-1}Cu_nO_{2n+2+\delta}$ family along the b axis.

The extra rows marked by arrows indicate that the a axis is doubled. However, high resolution electron microscopy by Matsui *et al.*[38] using a high voltage electron microscope (1200 kV), suggested copper/carbon ordering. The high resolution image on Figure 16 with a resolution better than 1.6Å allows the structure to be observed along the b axis so that, besides the general layer stacking of the structure along c, with a layer sequence of the type:

$$-MO-BaO-CuO-CaO-CuO-CaO-CuO-BaO-MO- \qquad (22)$$

where MO stands for the charge reservoir layer, the doubling of the a axis referred to above can be observed. Indeed, the main interest of the picture resides in the fact that along the a axis, and in this *CRL* layer, a series of dark spots with different intensities alternate, with a distance along any two of them of ~7.6Å; this alteration was interpreted by Matsui *et al.* as responsible for the x axis doubling, first observed by electron diffraction and attributed to a Cu/C ordering along a. The fact that copper, with atomic number z = 29, has a much higher electron density than carbon, with z = 6, justifies the differences in darkness observed along a. Nevertheless, as it is

clear that the interpretation of that evidence as due to a Cu/C ordering along
a is somewhat tenuous, we attempted to confirm it by means of EELS.

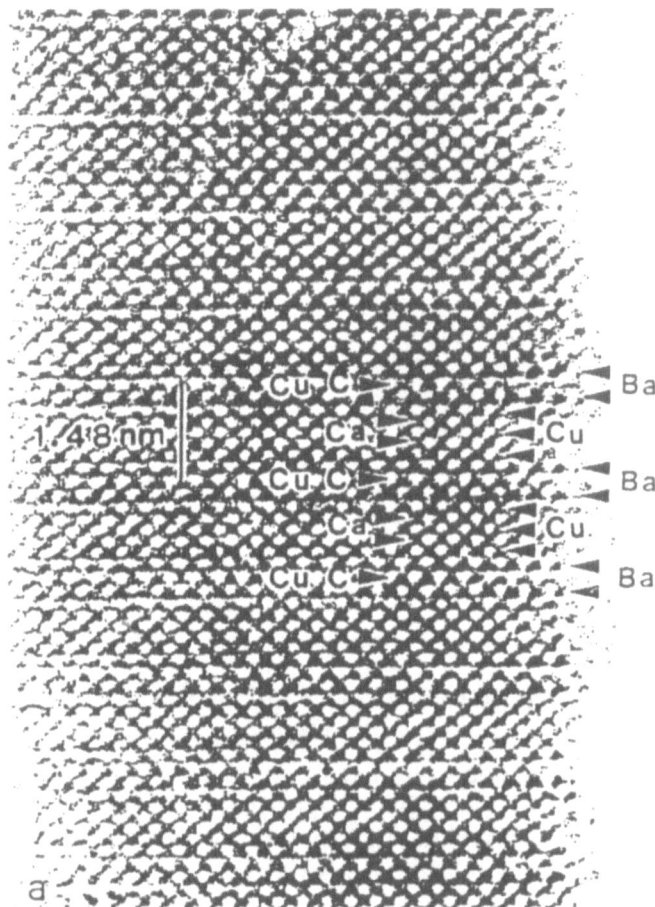

Figure 16: High resolution electron micrograph of a crystal of the cupro/carbonate family showing
the copper (darker) / carbon (lighter) ordering (Courtesy Professor Matsui).

Figure 17 shows a typical EELS spectrum for one of the samples[39]. It can be
seen that, besides the very clear Ba and Ca peaks, there is a small but
noticeable carbon peak. Even more, when the spectrum was obtained along a
crystal direction for over 60Å (Figure 17) the signals corresponding to the
different ions are seen to be rather constant, indicating that the composition is
homogeneous. When this result was quantified to the amount of copper

present, it was established that the most probable composition of the crystal was;

$$(Cu_x/C_{1-x})Ba_2Ca_2Cu_3O_y,\qquad(23)$$

with x ~0.5.

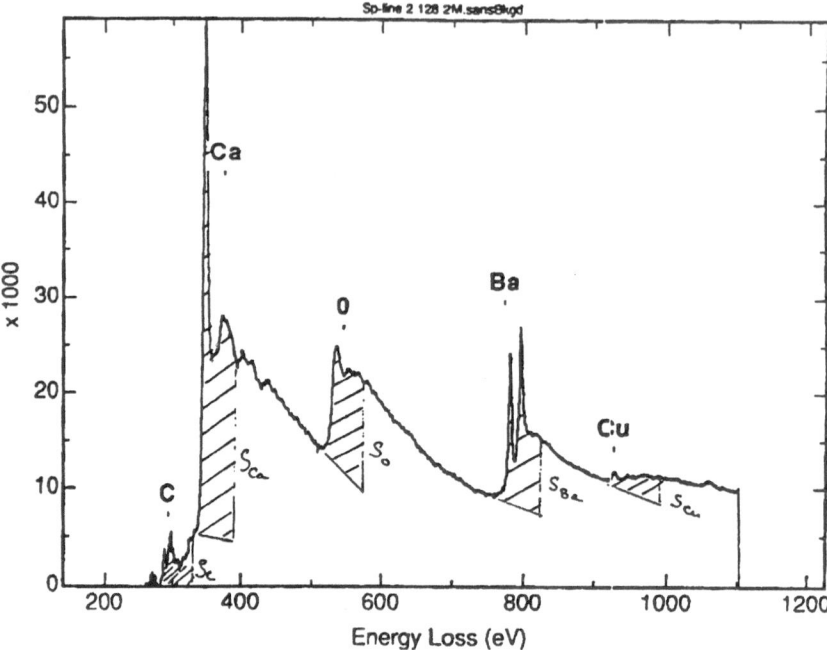

Figure 17: EELS spectrum showing the presence of carbon in the crystals (Courtesy Dr Colliex)

The combination of HRTEM and EELS has thus allowed the composition of the recently found non-toxic family of HTSC materials to be established with a certain amount of confidence. Nevertheless, a full crystal structure determination is needed before the presence of carbon in the structure can be ascertained. This is, however, not particularly easy to perform since these, being high pressure materials, the requirement of either good single crystals or sufficient amounts of powder are difficult to meet in practice.

4. Acknowledgements

The author thanks his colleagues and collaborators for their contribution to the present work. In particular he would like to mention Drs J.M. González-Calbet, L.C. Otero-Diaz, A. Landa-Cánovas, M.A. Señaris-Rodriguez from the Universidad Complutense and Dr C. Chaillout and the authors of reference 39 from the Laboratoire de Cristalographie and CRTBT (CNRS, Grenoble, France). Dr C. Barba and the technical staff of the Servicio de

Microscopia Electrónica are also thanked, as is CICYT for financial support through Grants Mat/ 86, 89 and 92.

REFERENCES

1. Broglie, L. de, (1925) *Annales de Physique* , 10(3), 22.

2. van Dyck, D., (1986) in *Microscopia Elettronica in transmissione nella Scienza dei Materialli*, Edizione Enda, Rome.

3. Cowley, J.M. (1992) in P. Buseck, J.M. Cowley and L.Eyring (eds.) *High Resolution Transmission Electron Microscopy and Associated Techniques*, Oxford Science Publishers, Oxford, 58.

4. McKie, D. and McKie, Ch. (1974) *Crystalline Solids*, J. Wiley, New York.

5. Tunaka, M. and Takuechi, M. (1985) *Convergent Beam Electron Diffraction*, Jeol, Tokyo.

6. Moniroli, J.P. and Steeds, J.W. (1992) *Ultramicroscopy*, 45, 219.

7. Vainstein, B.K. (1964) *Structure Analysis by Electron Diffraction*, Pergamon, Oxford.

8. Cowley, J.M. (1992) in P. Buseck, J.M. Cowley and L.Eyring (eds.) *High Resolution Transmission Electron Microscopy and Associated Techniques*, Oxford Science Publishers, Oxford. p. 306.

9. Cowley, J.M. (1981) *Diffractio Physics*, 2nd Edition, North Holland, Amsterdam.

10. Self, P.G. and O'Keefe, M. (1992) in Cowley, J.M. (1992) in P. Buseck, J.M. Cowley and L.Eyring (eds.) *High Resolution Transmission Electron Microscopy and Associated Techniques*, Oxford Science Publishers, Oxford. Chapter 8.

11. Spence, J.H. (1992) in Cowley, J.M. (1992) in P. Buseck, J.M. Cowley and L.Eyring (eds.) *High Resolution Transmission Electron Microscopy and Associated Techniques*, Oxford Science Publishers, Oxford. Chapter 7.

12. Hren, J.J., Goldstein, J.I. and Joy, D.C. (eds.) *Introduction to Analytcal Spectroscopy*, Plenum, New York.

13. Egerton, R.F. (1986) *Electron Energy Loss Spectroscopy in the Electron Microscope*, Plenum, New York.

14. Colliex, Ch. (1984) in Barrer, R and Cosslet, V.E. (eds.) *Advances in Optical and Electron Microscopy*, Vol. 9, p.65, Academic Press, New York.

15. González–Calbet, J.M. (1991) *Textura porosa de la Akaganeita Sintética*, Tesis Doctoral, Universidad Complutense, Madrid.

16. González–Calbet, J.M., Alario–Franco, M.A. and Gayoso–Andrade, M. (1981) *J. Inorg. Nucl. Chem.*, 43, 257.

17. Otero Diaz, L.C. (1989) *Estudio de Defectos Extensos y su ordenación en el sistema Titanio-Vanadio-Oxigeno*, Tesis Doctoral, Universidad Complutense, Madrid.

18. Alario–Franco, M.A. (1987) *Cryst. Lattice Defects*, 14, 357.

19. Hasimoto, H., Howie, A. and Whelan, M. (1962) *Proc. Roy. Soc. London*, A269, 80.

20. Otero–Diaz, L.C. and Alario–Franco, M.A. (1983) *Anales Quim.*, 79B, 510.

21. Eikum, A. and Smallman, R.E. (1965), *Phil. Mag.*, 11 627.

22. Bursill, L.A. and Hyde, B.G. (1970), *Proc Roy. Soc. London*, A320, 147.

23. Murr, L.E. (1970) *Electron Optical Applications in Materials Science*, McGraw Hill, London, p.424.

24. Landa, A.R. and Ptero, L.C. (1993) *Solid State Ionics*, 63–65, 378.

25. Bednorz, J.G. and Muller, A.K. (1986) *Z. Physik B, Cond. Matt.* 64, 189.

26. Stoll, S.L. (1992), Ph.D. Thesis, University of California, Berkeley.

27. Senaris–Rodriguez, M.A., Alario–Franco, M.A., Stoll, S.L. and Stacy, A. (1993) *Solid State Ionics*, 63–65, 945.

28. Senaris–Rodriguez and M.A., Alario–Franco, M.A. (1993) HREM study of the La/K ordering in superconducting $La_{2-x}K_xCuO_4$.

78

29. Buseck, P.R. and Cowley, J.M. (1983) *Am. Mineral.*, 68, 18.

30. Senaris–Rodriguez, M.A. (1992) *Estudio Microestructural de superconductores de alta temperatura*, Tesis Doctoral, Universidad Complutense, Madrid.

31. Putilin, S.N., Antipov, E.V., Chmaissem, O. and Marezio, M. (1993) *Nature*, 362, 226.

32. Cantona, M., Schilling, A., Nissen, H–U and Ott, H.R. (1993) *Physica C*, 215, 11–18.

33. Chu, C.W., Gao, L., Chen, F., Huan, F.J., Meng, R.L., Xue, Y.Y. (1993) *Nature*, 365, 323.

34. Nunez–Regueiro, M., Tholence, J–L., Antipov, E.V., Cappioni, J.J. and maerzio, M. (1993) *Science*, 262, 97.

35. Alario–Franco, M.A., Chaillout, C., Cappioni, J.J., Tholence, J.–L. and Souletie, B. (1994) *Physica C*, 222, 52.

36. Wu, S.J., Adachi, S., Jin, C.–Q., Yamauchi, H. and Tanaka, S. (1994) *Pysica C*, 223, 243.

37. Alario–Franco, M.A. (1994) *Adv. Materials*, 7(2), 229.

38. Kawashima, T., Matsui, Y. and Takayama–Muromachi, E., (1994) *Physica C*, 224, 69.

39. Alario–Franco, M.A., Bordet, P., Capponi, J.J., Chaillout, C., Chenavas, J., Fournier, T., Marezio, M., Souletie, B., Sulpice, A., Tholence, J.–L., Colliex, C., Argoud, R., Baldonedo, J.L., Gorius, M.F. and Perroux, M. (1994) *Physica C*, 231, 103.

MICROCRYSTALLINE MATERIALS CHARACTERISED BY INFRARED SPECTROSCOPY

L. MARCHESE, G. MARTRA AND S. COLUCCIA
Dipartimento di Chimica Inorganica, Chimica Fisica e
Chimica dei Materiali dell'Università di Torino
via P. Giuria, 7 - 10125 Torino - ITALY

Abstract

Infrared spectroscopy has contributed significantly to the study of microcrystalline materials and, in particular, valuable information on their surface properties are obtained by this technique. In this context, we review IR case-studies concerned with the nature and structure of surface centres on pure MgO and on finely divided nickel supported on magnesium oxide (Ni/MgO). Attention is focused on the vibrational spectrum of adsorbed carbon monoxide (employed as probe molecule), which is shown to be a powerful tool to investigate surface physico-chemical features of the above mentioned dispersed materials (morphology, nature and structure of surface sites, acid-base character, reactivity). Some examples showing the advantages of using Fourier Transform Infrared spectroscopy (FTIR) in the field of surface science of microcrystalline materials will be also shown.

1. Introduction

It was in the fifties that Terenin *et al.* [1] with their studies on hydroxyls on oxides and Eischen *et al.* [2] with their spectra of CO adsorbed on metals supported on silica showed the opportunities that infrared spectroscopy offered in surface studies and definitely established it as the master spectroscopic technique for surface characterisation of highly dispersed systems. The main experimental factor favouring this technique was the availability commercially of instruments which could be used for studying powders without any modification. Photon scattering, which increases with the frequency of incident light, is extremely severe in the UV-NIR range ($50.000-4000$ cm^{-1}) and requires reflectance attachment to be adopted, whereas it is much lower in the medium IR ($4000-400$ cm^{-1}) and the familiar transmission technique can still be easily used.

C.R.A. Catlow and A. Cheetham (eds.), New Trends in Materials Chemistry, 79–109.
© 1997 *Kluwer Academic Publishers.*

IR bands are associated with vibrations in the system and, as frequencies, intensities and widths are strictly related to the nature of chemical bonds, molecular structures and intermolecular interactions, they are sensitive to any change in these parameters. Moreover, though all vibrations in chemical species should be properly treated as a motion of all atoms in the system, it has long been known empirically that given groups of atoms are constantly associated with bands in typical frequency regions. The possibility of recognising group frequencies greatly simplifies the interpretation of complex infrared spectra [3]. Rich varieties of surface species have been identified. The first review on the subject was published by Eischens and Pliskin in 1958 [4] and the first comprehensive book by Little in 1966 [5]. Many books and reviews summarising the increasing amount of data have been periodically published [6-11] and the current wide use of the IR spectroscopy in the field of the heterogeneous catalysis is testified by the very recent reviews quoted in references 12-21.

In terms of gas-solid systems, including both surface and bulk vibrations, the types of information given by infrared spectra may be roughly listed as follows:

a)   nature and structure of adsorbed species;

b)   interactions in the adsorbed layer;

c)   nature and structure of adsorbing sites;

d)   adsorbate-adsorbent bonds;

e)   vibrational surface states of the solid and their modification upon adsorption;

f)   bulk vibrations of the solid and the effect on them, if any, of adsorption.

These areas have not been investigated to the same extent because they are not equally accessible and the vast majority of data have been collected, by infrared transmission spectroscopy, on topics a)-c). The main practical reason is that the infrared bands associated with the species formed on the surface when gases or vapours are allowed to interact with the powder are observed in a frequency range which is mostly free from absorptions due to the solid. Intense bands due to bulk vibrational modes of microcrystals may obscure the spectrum up to 1200-1000 cm^{-1} and only weak overtones and combination bands are found in the background in the region extending from that low frequency edge to the high frequency limit at 4000 cm^{-1} (Figure 1).

Nowadays the possibility of studies on topics d)-f) is growing rapidly because experiments in the Far-IR region (< 400 cm^{-1}) are becoming more suitable. In fact, much progress has been made in the field of the instrumentation associated with the adoption of the Fourier transform (FT) of conventional sources and with the availability of unconventional very powerful sources such as synchrotron radiation. This review will, however, concentrate on laboratory based experiments.

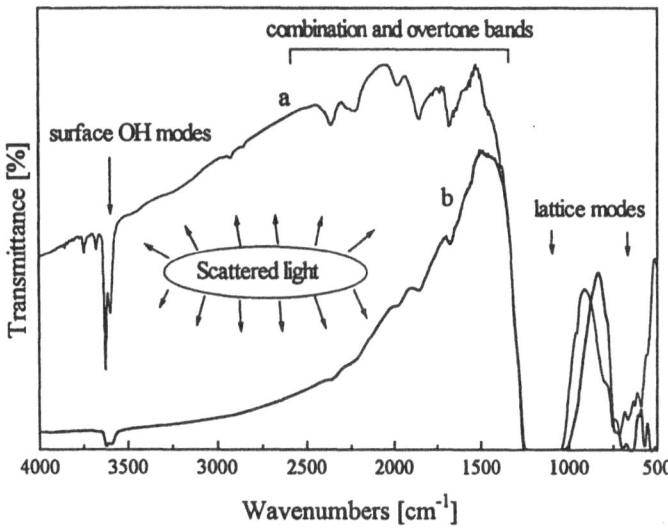

Figure 1: IR spectra of crystalline microporous Si-substituted aluminophosphates. Curve (a): spectrum of a SAPO-18 which has crystallites of 0.5 to 1 μm; curve (b): spectrum of a SAPO-34 which has crystallites of 5 to 10 μm.

2. Experimental Setup and Recent Applications of IR Spectroscopy

Many techniques have been either adopted or specifically developed to provide vibrational spectra of dispersed samples. Examples are emission spectroscopy, internal and external reflection IR spectroscopy, Raman spectroscopy, electron scattering techniques. They are needed in many special cases [6-8]; for example, photoacoustic infrared spectroscopy is useful with extremely opaque sample like coke [6, 22-23].

However, conventional IR transmission techniques, requiring self-supporting pellets obtained by pressing the powder, are still the most widely used. It is mainly the development of FTIR spectrometers which has enhanced the application of infrared spectroscopy to dispersed systems. Indeed, some of the general advantages of interferometers [23], a) *short scanning time* (< 0.1 s), b) *high signal/noise ratio*, c) *high spectral resolution* and d) *precision in the overall range of transmittance*, as compared with the dispersive instruments, are particularly relevant for powders. For instance, opaque materials (*e.g.* those which greatly scatter the IR light because of the presence of large grains) can be now investigated in transmission mode.

Thermal treatments "in situ" and experiments on the adsorption-desorption of gases and vapours can be easily done in properly designed cells. Figure 2 illustrates a recently designed cell [24] for low-temperature IR measurements.

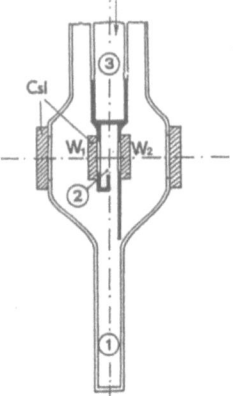

Figure 2: IR cell for measurements at the liquid nitrogen temperature; the material under investigation can be pre-treated at temperatures up to 1173K before recording the spectra. The sample is heated in position 1 by means of an external furnace and, after cooling to room temperature, the sample is transferred to a copper cage (position 2) by properly inclining and rotating the cell. Liquid nitrogen is then fed into the Kovar-pyrex tubing 3. The two CsI windows in close contact with the copper cage (W1 and W2 in Figure 2) form a chamber where the temperature of the pellet gradually reaches the temperature of *ca.* 100 K (reprinted from ref. 24. Copyright 1993 The Royal Society of Chemistry).

This represents one of the simplest procedures allowing both very low temperatures (77-100K) and very high temperatures (1173 K) to be reached in the same equipment.

Sophisticated models for specific needs such as very low temperature spectra, reflection techniques, are described in references 6 and 25 and variable-temperature cells which operate down to the He boiling point are described in reference 26.

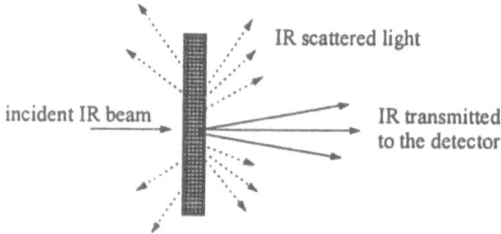

Figure 3: IR beam through a pellet of a microcrystalline material.

The effects occurring when IR radiation interacts with a pellet of a microcrystalline sample are represented schematically in Figure 3; a fraction of IR light is scattered (photon scattering) and, consequently, does not reach the detector, thereby reducing the signal. The extent of this process depends not only on the frequency of the incident radiation, as shown by the shape of the spectra in Figure 1, but also on the dimension of the particles as well as on their packing: the larger the empty spaces between the crystallites in the pellets, the greater is the scattered light [27].

Figure 1 shows two FTIR spectra of microcrystalline materials which have particles with very different sizes. The two solids are of the family of the Si-substituted aluminophosphates (SAPOs) and both have similar microporous zeolite-type structures and a similar chemical composition [28]. Curve (a) is a spectrum of a SAPO-18 which has particles with size in the 0.5 to 1 μm range whereas curve (b) is that of a SAPO-34 with particles of 5 to 10 μm: the scattering increases with the particle size and, consequently, the signal loss in the high frequency region of the IR spectrum may become very severe. This fact is relevant because OH surface groups absorb in the 4000 - 3000 cm^{-1} range and, because the knowledge of the nature of these species largely determines the surface properties of any catalytic material; the possibility of collecting data in this region is of particular interest.

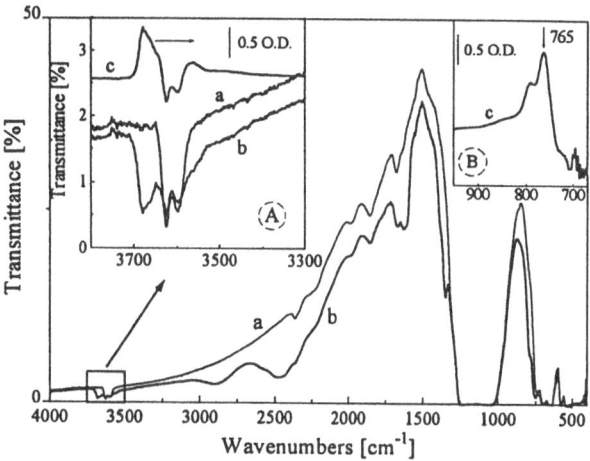

Figure 4: IR spectra of H$_2$O adsorbed on a microcrystalline SAPO-34 with a very high scattering profile (the same as figure 1,b). Curve (a) is the dehydrated sample whereas curve (b) is the sample with adsorbed water. Inset A: magnified view of the OH stretching region; curve (c) is a difference spectrum (in absorbance scale) obtained subtracting (b) from (a). Inset B: magnified view of the 950-650 cm^{-1} region where only the difference spectrum is reported.

2.1   HIGHLY SCATTERING MATERIALS

OH groups of materials with particles up to 10 μm size can be easily detected by means of IR spectroscopy in transmission mode. In the inset A of Figure 4 the OH stretching region of the SAPO-34 is magnified (curve a); from this spectrum it is evident that even when the transmittance is lower than 2% the OH bands are very clearly detected because both of the high signal/noise ratio and of the very high precision in the overall range of transmittance of FTIR spectrometers. Furthermore, the bands produced upon the interaction between adsorbed molecules and OH groups in highly scattering materials are easily and reliably detected. For instance, the adsorption of H_2O on SAPO-34 (Figure 4, curve b) produces H_2O/H^+ adducts (water H-bonded to protons) and H_3O^+ (and/or $H_5O_2^+$), as indicated by related vibrational bands which were identified by subtracting the spectra of the bare sample (curve c in the insets A and B). It is worth stressing that some of the bands formed upon interaction of H_2O with protons in SAPO-34 were clearly revealed [29] also in the 1000-700 cm^{-1} region, obscured by very intense lattice modes (see the inset B of Figure 4).

2.2   FRAMEWORK AND CATIONIC VIBRATIONS OF ZEOLITES

The nature of the vibrational modes of zeolite framework has been discussed in many papers and reviewed by de Man and van Santen [30] and van Santen and Vogel [31]. The lattice vibrations absorbing at wavenumbers lower than 1000 cm^{-1}, though very intense, can be studied accurately by means of FTIR spectroscopy if very thin pellets are used. It was demonstrated that some of these vibrations are very much perturbed by the presence of adsorbates [32-33]. Moreover, the possibility of studying the cationic vibrations exploring the far infrared region ($v < 400$ cm^{-1}) has enhanced our knowledge of the adsorbate-cation interactions inside the zeolites [32-34].

Figure 5 shows how the Mg^{2+} mobility upon the adsorption of water in MgX zeolites can be studied by combining the information given by infrared spectroscopy (A and B) with those obtained by molecular modelling (D). C illustrates a supercage of a faujasite-type zeolite and some sites where the cations should be located. Structural studies demonstrated that the occupancy of these sites is different in hydrated and in dehydrated forms of MgX [35], whereas the IR results shown in sections A and B demonstrated that water promotes the transfer of Mg^{2+} cations from sites I, inside the prismatic units, to sites I', in the sodalite (see the computer modelling representation in the section D). Particular evidence for this motion is the reversible appearance of a band at 570 cm^{-1}, assigned to deformation modes of double six rings units, and of bands in the 200-400 cm^{-1} range, assigned to translational modes of Mg^{2+} located in site I, upon adsorption-desorption of water [33].

2.3   FAST RECORDING SPECTRA

Many reactions can be followed by *in situ* IR measurements and the intermediates distinguished on the basis of their specific vibrations thanks to the possibility of recording spectra very rapidly (0.1 sec. between two successive spectra is probably the present limit). For example, when the interaction of propene with acidic OH groups in HZSM-5 (see Figure 6 and Scheme 1) was studied [36], it was possible to observe first the formation of H-bond complexes and then how they were converted into protonated species which successively react with other molecules to give longer hydrocarbon chains (oligomerisation).

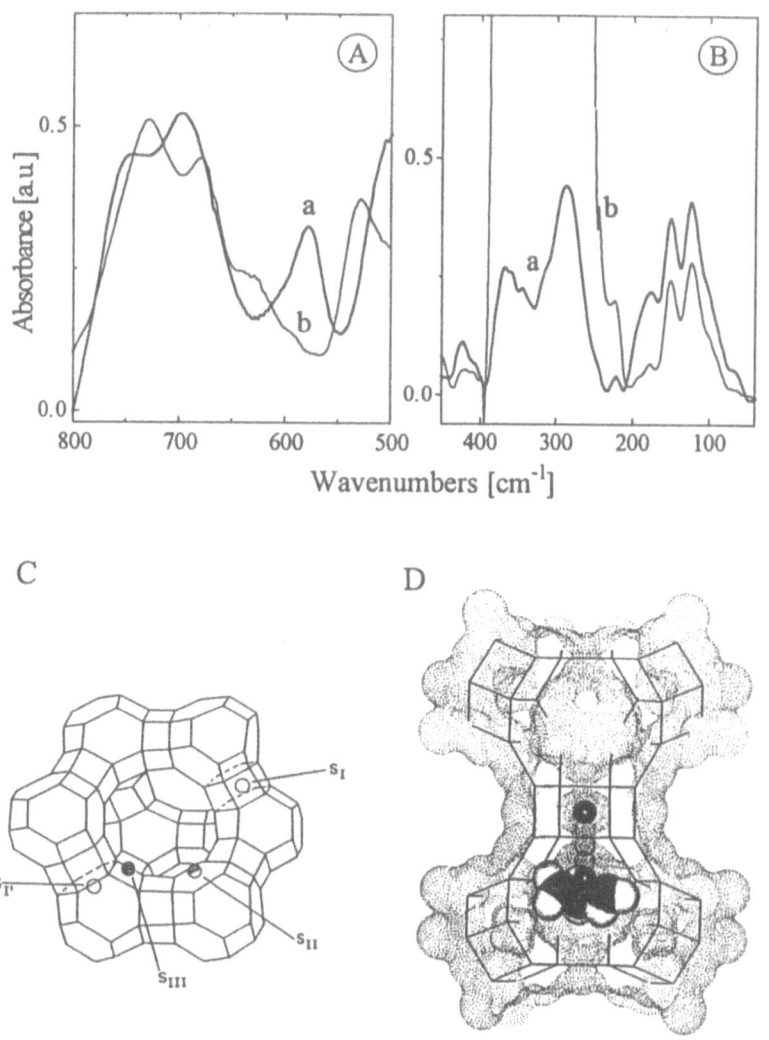

Figure 5. Low frequency region of the medium-IR (section A) and Far-IR (section B) spectra of hydrated (curves a) and dehydrated (curves b) forms of an MgX zeolite; section C: supercage of faujasite-type zeolites; section D: computer model representing the motion of Mg^{2+} cations from site I' to site I upon H_2O adsorption (reproduced from reference [33]. Copyright 1995 The Royal Society of Chemistry).

86

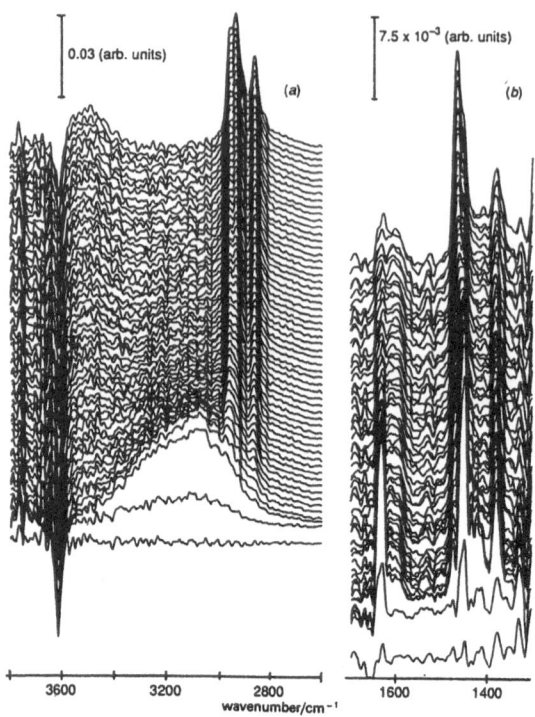

Figure 6: IR spectra of propene on H-ZSM-5. The spectra were taken every 1.5 sec. and each spectrum was the average of 10 interferograms. (Reprinted from reference [36]. Copyright 1994 The Royal Society of Chemistry).

2.4 TIME-RESOLVED (PICOSECONDS) VIBRATIONAL SPECTROSCOPY

Since the first detailed study of the relaxation of the OH stretching mode on a silica surface conducted by Heilweil *et al.* [37], vibrational lifetimes of adsorbates on the surface of various materials have been measured using the pump-probe method (see the references quoted in Hirose *et al.* [38] and Brugmans *et al.* [39]). Vibrational lifetimes of hydroxyls in acidic zeolites have been recently measured by using the experimental setup described in references [38-39] obtaining very interesting information on the local environment of these species. With conventional IR spectroscopy (Figure 7, top) two OH groups absorbing at 3637-3645 and 3378-3558 cm^{-1} are observed for zeolites H-Y. By using the time-resolved technique, it was found that the protons absorbing at higher wavenumbers (HF protons) have significantly longer lifetimes, 200-300 ps, than those absorbing at lower frequency (LF protons) which were found to be in the 35 to 100 ps range. The higher relaxation rates of LF protons were explained assuming that these OH groups are hydrogen-bonded to structural oxygens so that they lose the excess of vibrational energy through a stronger coupling to accepting modes (lattice vibrations) [39].

Further evidence that LF hydroxyls in faujasite-type zeolites and aluminophosphates are H-bonded to lattice oxygens was obtained by means of a combined 1H MAS NMR and FTIR study [40].

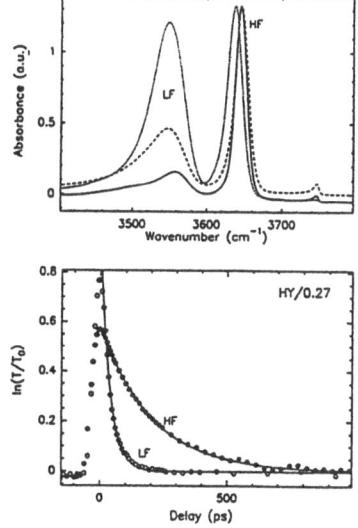

SCHEME 1: Early stages of propene oligomerisation

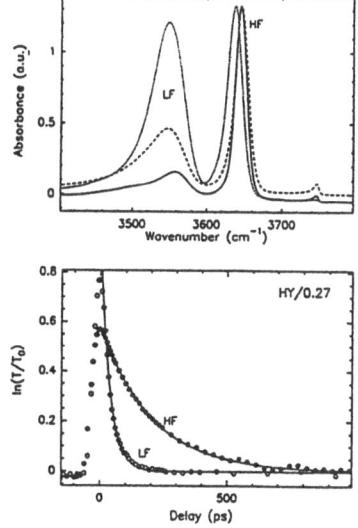

Figure 7. Upper: Conventional IR spectra of HY and Na/HY zeolites in the stretching OH region; the three curves are related to samples with various degrees of Na⁺/H⁺ exchange. Lower: Relative transmission of an infrared probe pulse as function of the delay between pump and probe pulses for zeolite HY/0.27; the laser frequency was tuned to the maxima of the LF and HF absorption bands (reproduced from reference [39]. Copyright 1994 Elsevier Science B.V.)

3.   The Nature of Surface Centres Studied by Using Molecular Probes

It has long been recognised that the complexes formed upon adsorption of molecular probes on the surface of a microcrystalline material give valuable information on the nature of the adsorbing centres (*e.g.* the oxidation state, the coordination number, the acidity strength of the surface cations) and it was demonstrated that the nature of surface complexes is strictly related to the nature of the surface adsorbing centres.

The methods of characterising acidic solids [14,18,21], basic catalysts [13,19], zeolites [14-16,19] and surface metal ions [10,20], by using different kind of molecular probes (*e.g.* NH₃, CO₂, CO, pyridine) have been recently reviewed. The importance of comparing the experimental and theoretical data of surface complexes formed by means of van der Walls interactions between adsorbate and adsorbent has been also emphasised [11]. In many of reviews it emerges clearly that CO is one of the prominent molecular probes for characterising surface centres in microcrystalline materials; but despite the fact that the electronic and vibrational spectra of the molecule in the gas phase are very well known, both the *adsorbate-adsorbate* and the *adsorbate-adsorbent* interactions occurring on the surfaces are still not completely understood because of their intrinsic complexity . This matter has been recently reviewed by studying the adsorption of CO and NO on well characterised oxide single microcrystals [20] and will be briefly summarised below.

3.1   CO AS MOLECULAR PROBE

The electronic charge of a probe molecule is invariably perturbed upon adsorption and when CO is adsorbed on cationic centres the frontier molecular orbitals 5σ and $2\pi^*$ (see Figure 8) are those mainly involved. However, it is still debated as to the degree of overlap, if any, between the orbitals of the molecule and those of the surface centres.

On low index faces of oxides, the CO molecule is bound to the surface cations through the carbon atom in an end-on configuration and forms a surface ad-layer of parallel oscillators. The $Me^{x+}\cdots CO$ bond (Me^{x+} = non transition metal ions) is mainly due to electrostatic and dispersive interactions leading to a polarisation of the molecule and to an increase in the frequency of the fundamental, ν_{CO}, transition higher relative to that of the free molecule (2143 cm⁻¹). The increase of the stretching frequency is roughly proportional to the polarising field [20, 41].

When CO is adsorbed on metallic surfaces, stronger bonds are established as, in this case, both a donation of electronic charge from the 5σ orbital of the CO molecule and back-donation of electronic charge from d orbitals of the metal atom to the $2\pi^*$ molecular orbital occur. The CO stretching in metal-complexes is at lower ν than that of the free molecule because the antibonding $2\pi^*$ orbital of the CO molecule is populated leading to a weakening of the C-O bond; its value depends on the extent of the orbital overlap, in particular on the degree of electronic donation compared to back-donation.

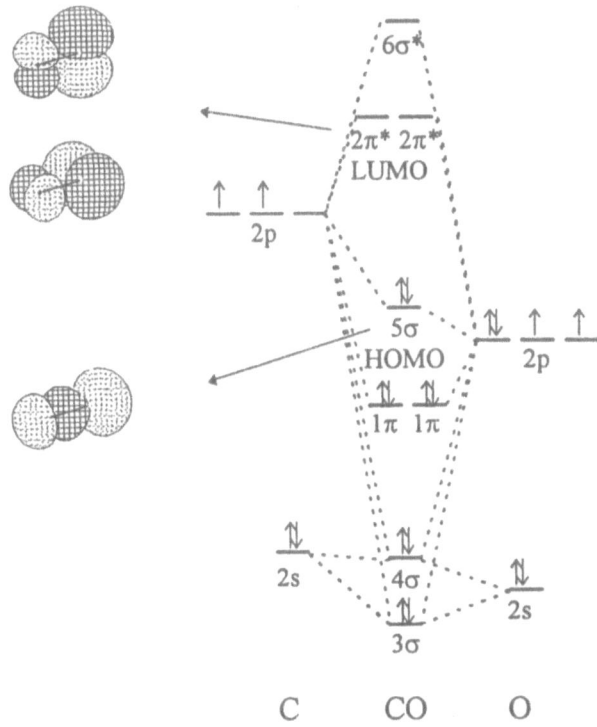

Figure 8: Molecular orbitals scheme of the CO molecule and representation of the HOMO (Highest Occupied Molecular Orbital) 5σ and LUMO (Lowest Unoccupied Molecular Orbital) $2\pi^*$ involved in the formation of surface complexes (1σ and 2σ are not included in the scheme).

The frequency shift and *band width* of adsorbed CO depend both on the surface heterogeneity, in that cations in different coordination adsorb (or polarise) the molecule differently, and on adsorbate-adsorbate interactions occurring among static and dynamic dipoles of the diatomic molecules in the ad-layer [7-9, 20 and references therein, 42-46]. The dynamic and static effects among the oscillators in overlayers formed by weakly adsorbed diatomic molecules (as on oxides) are essentially of the "through space" type; the dynamic interactions are due to coupling among parallel vibrating dipoles, whereas the static ones are due to reciprocal perturbation of the adsorbed molecules by "electrostatic" or "solvent" effects. In the case of the more strongly adsorbed CO on metal surfaces, the cooperative effects (the adsorbate-adsorbate interactions) determining the frequency and the width of the band occur also through the solid.

On well-defined surfaces of microcrystalline oxides CO forms regular arrays [20,46] and the static (Δv_{stat}) and dynamic (Δv_{dyn}) contributions to the overall v_{CO} shift can be measured by using $^{12}CO{:}^{13}CO$ isotopic mixtures (dilution limit). In practice, as ^{12}CO and ^{13}CO are not dynamically coupled because of the large difference in the stretching frequency, a 10:90 dilution is sufficient to keep ^{12}CO oscillators sufficiently well apart to be considered isolated. The importance in determining the dynamic coupling is in the fact that this contribution is related to the *vibrational polarisability*, α_v, of the adsorbed molecule through the modified Hammaker equation [43-44]:

$$(v / v_0)^2 = 1 + \frac{\alpha_v T_0}{1 + \alpha_e T_0},$$

where v is the monolayer frequency, v_0 the frequency of an isolated molecule, (single tone), α_e the electronic polarisability and T_0 the direct dipolar sum.

The vibrational polarisability, α_v, is related to $(\partial\mu/\partial Q)^2$, the square of the change of the dipole moment of the adsorbed molecule due to the change of the normal coordinate Q (in this case the elongation of the C-O bond). The relation between the molar extinction coefficient, ε, α_v and v, the frequency of the band, is given by $\varepsilon = 4\pi^3 \alpha_v v^2$ [47].

As α_v reflects the charge oscillations from the surface to the adsorbed molecule (and *vice versa*) occurring during the CO stretching, it depends very much upon the type of bond formed between CO and the surface. Table 1 summarises some spectroscopic features of CO adsorbed on various oxides and suggests that:

a) in the case of $Me^{x+}{\cdots}CO$ complexes with small values of Δv_{dyn} and α_v (*e.g.* for MgO, Al_2O_3 and TiO_2), electrostatic forces are mainly involved; the value of α_v is close to that of the free molecule [48]. The higher the polarising field, the higher the stretching frequency of the adsorbed molecule (the single tone, v_{singl}), whereas the intensity of the band does not vary significantly.

b) For ZnO, a substantial σ-bonding due to the overlapping between frontier orbitals occurs and the consequent electron transfer (from the 5σ orbital of the CO to the Zn^{2+} cations orbitals) plays an appreciable role in determining the α_v and v_{CO} values.

c) Higher α_v values have been found for $Me^{x+}{\cdots}CO$ complexes where transition metal cations are involved; in this case, a small contribution of d-π overlap forces greatly affects static and dynamic dipoles localised on adsorbed CO.

TABLE 1. Spectroscopic features of CO adsorbed on various metal oxides

Material	charge and coord. n°	Δv_{stat} (cm^{-1})	Δv_{dyn} (cm^{-1})	v_{singl} (cm^{-1})	α_v (Å^3)
MgO	Mg^{2+} (5c)	- 11.3	+ 3.3	2157	0.0306
Al_2O_3	Al^{3+} (3c,5c)	- 23.6	+ 3.6	2204	0.030
TiO_2	Ti^{4+} (5c)	- 16.5	+ 3.5	2179	0.025
ZnO	Zn^{2+} (3c)	- 28.0	+ 6.0	2190	0.0526
NiO	Ni^{2+} (5c)	- 33.0	+ 19.0	2152	0.256
α-Cr_2O_3	Cr^{3+} (5c)	- 27.5	+ 13.5	2167	0.101

Δv_{stat} = static shift; Δv_{dyn} = dynamic shift; v_{singl} = singletone frequency; α_v = dynamic polarizability.

Finally, we note that when d-π contributions are present in surface complexes, the stretching of CO causes negative charge to flow from the adsorbing centre to the molecule, whereas the compression of the CO bond causes the charge to flow from the molecule to the adsorbing centre. The result of this process is that the intensity of the CO vibration is greatly enhanced. Such an effect is very much operative in CO-metal complexes, and the intensity of the band is higher than in cation-CO complexes. However, the number of factors contributing to α_v and to the dipole coupling is larger when CO is adsorbed on metals [43-45].

Some examples of the use of CO as a molecular probe of both acidic and basic surface sites on MgO will now be illustrated. These case-studies show that the nature of the bond in the very weak surface complexes $Mg^{2+}\cdots CO$, as well as the reactivity between CO and the surface O^{2-}, are revealed by IR spectroscopy and are strictly dependent on the surface morphology of the microcrystals [20,46,49-50]. The formation of very strong CO-metal complexes will be also discussed examining the case of CO adsorbed on the Ni/MgO system.

4. Surface Morphology and Reactivity Towards CO of MgO Particles

MgO is often adopted as a model system in studies of dispersed oxides because it can be easily obtained in the form of high surface area powders with microcrystals showing a simple cubic structure predominantly bounded by (100) faces. Accordingly, it was determined by spectroscopic studies [51-54] that the sites exposed on the surface are cations and anions in low coordination states [Mg^{2+}_{LC} and O^{2-}_{LC}] and, more specifically, five-coordinated on the (100) facelets [Mg^{2+}_{5C} and O^{2-}_{5C}], four coordinated on edges [Mg^{2+}_{4C} and O^{2-}_{4C}] and three-coordinated on corners [Mg^{2+}_{3C} and O^{2-}_{3C}] as shown in Figure 9.

One further advantage of MgO as a model system is that it can be produced in two forms differing considerably in the morphology of the microparticles: a) MgO produced by thermal decomposition of the hydroxide (MgO-h) [51-54] and b) MgO smoke (MgO-s) produced by burning magnesium in air [55].

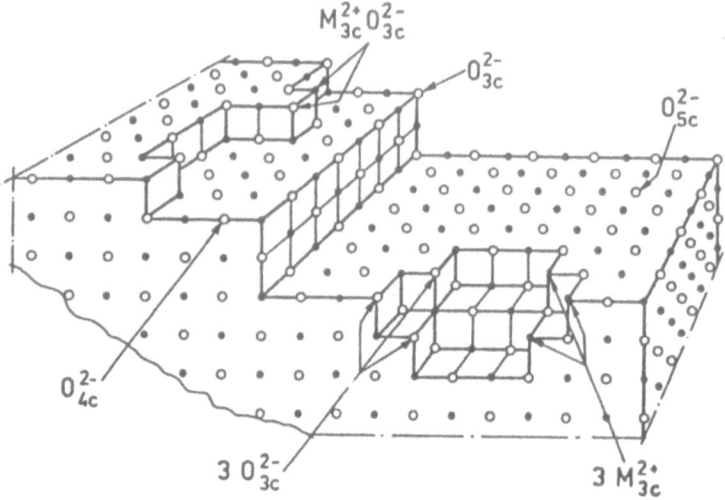

Figure 9: Ions in low coordination on the surface of MgO (reprinted from reference 59a. Copyright 1981 Kodansha Ltd).

Figure 10: High resolution electron micrograph of MgO ex-hydroxide.

MgO ex-hydroxide microcrystals have a plate-like structure characteristic of the starting material [56]: details of the rough, highly irregular surface of such particles are illustrated in the high magnification photograph in Figure 10. Steep successions of steps and small terraces and, consequently, high concentration of edges and corners are characteristic features of MgO-h microcrystals. Fringes 2.09 Å apart parallel to the edges confirm that (001) planes are overwhelmingly exposed. Heavily terraced segments simulate the presence of high index planes, but the actual surface are (001) faces, though, in most cases, of extremely reduced dimensions [49].

By contrast, *MgO smoke* microcrystals have a nicely shaped cubic habit with edge lengths in the 30 to 200 nm range (see Figure 11) [46,55,57]. The high magnification image in the inset of Figure 11 confirms that the surface smoothness is maintained down to the level of a few angstroms [46,58], as very few morphological defects such as steps and kinks, with unit cell dimensions may be observed; consequently, the proportion of sites on edges and corners as compared with sites on (001) planes is much lower. This material is particularly appropriate for studying the dynamic and static interactions occurring upon CO adsorption because it simulates closely the (001) surface of single crystals [46].

The relative population of the three families of surface sites with different coordination (3C, 4C, and 5C) are significantly different on the surfaces of the two samples, so that the activities towards hydrogen [59] and CO [60-63] of the various sites could be studied separately. It was found that only an exceedingly small fraction (<0.5%) of surface sites, *i.e.* O_{LC}^{2-} ions with the lowest coordination, on high surface area MgO are able to adsorb CO at room temperature (300K) and give complex anionic polymeric species.

However, acidic sites (that is Mg_{LC}^{2+} cations) were also revealed on MgO by adsorption of CO at low temperature [46,49,64]. In fact, CO coverage is very high at 77K and, in the case of MgO ex-hydroxide, a complex absorption produced by the overlap of several bands is observed in the 2220-2120 cm^{-1} range. A similar absorption, though much simpler, is observed with MgO smoke, and such differences allow each IR component to be related to a specific family of acidic Mg_{LC}^{2+} sites.

The very simple character of MgO has made this solid one of the most useful for testing the validity of different theoretical approaches for calculating the spectroscopic features of molecular probes in the field of the gas-solid interactions. For instance, quantum chemical molecular models [65] and quantum mechanical embedded cluster models [66,67] have been used to predict bonding and vibration of CO adsorbed on low-coordinated surface sites of MgO.

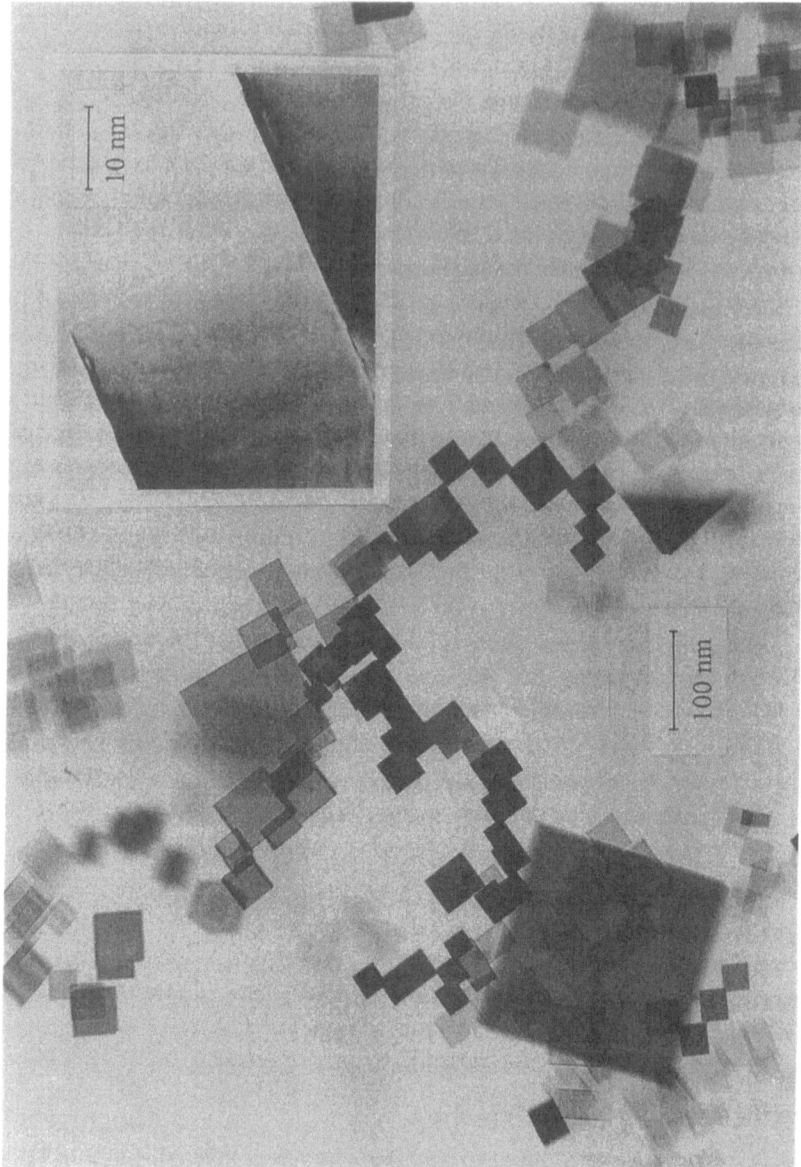

Figure 11: Low magnification electron micrograph of a collection of MgO smoke particles. Inset: High resolution image of a particle showing few defects at unit-cell dimension level and fringes of (100) planes (reprinted from reference 46. Copyright 1992 Elsevier Science Publishers).

4.1 DYNAMIC AND STATIC INTERACTIONS IN CO LAYERS ON MgO SMOKE (100) FACELETS

IR spectra obtained at different coverages of CO are shown in Figure 12. The spectrum at maximum coverage (p = 20 Torr) is dominated by a narrow peak (full width at half maximum (FWHM) = 4.5 cm^{-1}) and an intense one centred at 2148.1 cm^{-1}; other much weaker and sharper peaks belong to the ro-vibrational spectrum of CO in the gas phase.

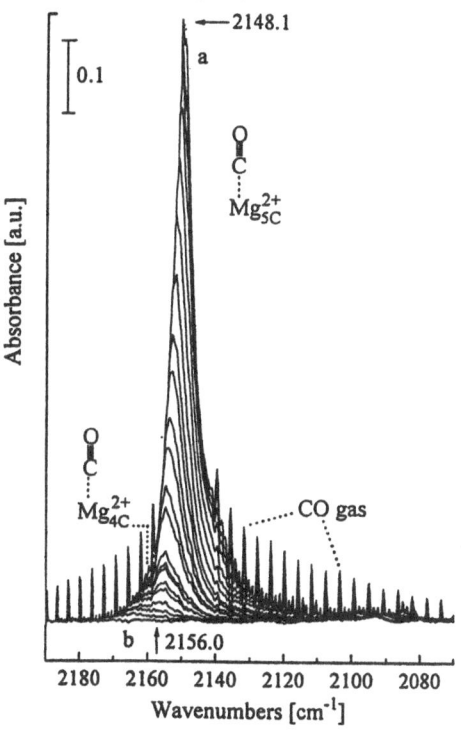

Figure 12: IR spectra of CO adsorbed under progressive reduced pressure from 20 Torr (curve a) to 0.001 Torr (curve b) on highly regular MgO smoke (100) facelets at 77K (reprinted from reference 46. Copyright 1992 Elsevier Science Publishers).

Moreover, two shoulders approximately located at 2160 and 2140 cm^{-1} are present on each side of the main peak. The shoulder at 2140 cm^{-1} was attributed to CO molecules in a liquid-like structure on the basis of its behaviour as a function of the pressure [46] and because liquid-like CO was reported to absorb at 2139 cm^{-1} [68].

The peak at 2148.1 cm^{-1} and its shoulder at 2160 cm^{-1} are at a frequency higher than the stretching mode of the free CO molecules (2143 cm^{-1}) and were assigned respectively to the stretching mode of CO adsorbed on pentacoordinated Mg^{2+}_{5C} sites exposed on the extended (001) microplanes and to CO adsorbed on tetracoordinated Mg_{4C}^{2+} sites on edges [46,49]. The proportion of the latter sites is small on the highly regular MgO smoke microcrystals, leaving the band due to CO adsorbed on (001) planes as the dominant and, in practice, the most significant band in the spectrum.

We note that the band-widths measured at 55 K in refs. [69-70] for CO adsorbed on NaCl (100) single crystal (ca. 0.4 cm^{-1}) is an order of magnitude lower than that measured for CO adsorbed on MgO (001) microcrystals (4.5 cm^{-1}). This broadening was interpreted in terms of both homogeneous and heterogeneous effects [46, 69-70], the latter being due to surface irregularities, such as steps of atomic dimension (see the high magnification image in Figure 11) which create flat surface regions differing in size and shape. A monolayer of adsorbed CO, in these conditions, must be considered as constituted of many distinct two-dimensional finite domains, each of them having slightly different spectroscopic features (bandshape and band centre).

Spectra in Figure 12 show that the dominant peak shifts to higher frequencies by decreasing the CO coverage. At maximum coverage (curve a) the peak is at 2148 cm^{-1} and tends to 2156.0 cm^{-1} as the pressure of CO and, consequently, the coverage tends to zero. The latter value is the stretching frequency of an isolated ^{12}CO oscillator (CO single tone). The difference (-7.9 cm^{-1}) between the two values at θ = max and $\theta \to 0$ measures the overall effects due to dynamic and static interactions experienced by the oscillators in a full monolayer (see above).

Dynamic and static contributions were determined by using a ^{12}CO-^{13}CO (10:90%) isotopic mixture (see Table 1).

4.2   CO ADSORBED ON HIGH SURFACE AREA MgO

Figure 13 (A) shows the spectra at decreasing coverages of CO adsorbed at 77 K on the high surface area MgO-h (ex-hydroxide) sample and, in section B, a comparison between the spectra of CO adsorbed, at the maximum coverage, on MgO-h and on MgO-s. The spectra of CO adsorbed on more defective MgO are more complex with both a larger number of components in the region of CO adsorbed on Mg_{LC}^{2+} acidic sites (at ν > 2143 cm^{-1}) and also a group of bands in the 2120-2080 cm^{-1} region which are absent in the case of MgO-s sample. These latter bands are related to anionic species formed by interaction of CO with the more basic O_{LC}^{2-} surface sites [50] following a mechanism which will be illustrated below.

4.2.1   Acidic Sites
CO stretching frequencies higher than 2143 cm^{-1} are typical of molecules adsorbed on positively charged sites with the carbon atom pointing towards

the cation, the shift increasing as the strength of the positive electric field increases (see above and references 20,64,66-67,71). Though the positive centres are all Mg_{LC}^{2+} ions, different coordinations generate cations with different polarising power. Indeed, the positive electric field associated with an Mg_{5C}^{2+} site on a plane, where it is surrounded by five next neighbours O^{2-}, must be lower than the field of Mg_{4C}^{2+} on an edge where the same charge is balanced by only four next neighbouring negative ions and, in turn, this must be lower than the field on Mg_{3C}^{2+} on a corner position, surrounded by only three O^{2-} anions (see scheme in Figure 9).

The weak band at the highest frequency (2200 cm^{-1}) was assigned to CO adsorbed on Mg_{3C}^{2+} [49-50,61]. CO molecules adsorbed on Mg_{4C}^{2+} and Mg_{5C}^{2+} are expected to absorb at progressively lower frequencies, and it was proposed that the two main bands observed at 2159 and 2152 cm^{-1} respectively are associated with such species. Note that only one band is observed in the case of MgO-s (Figure 13B), assigned to CO adsorbed on the predominant Mg_{5C}^{2+} sites [46].

As the extinction coefficient does not vary significantly in this range [20,48,72], the relative intensities support such an assignment. In fact, the 2152 cm^{-1} band associated with the abundant sites on (001) planes is the most intense, whereas the 2200 cm^{-1} band, associated with the much less numerous sites in the corner position, is the weakest. The reversibility also agrees with the assignment as: (a) species associated with the less acidic sites (5C) on planes and absorbing at 2152 cm^{-1} are the less stable and, consequently, the first to desorb (Figure 13A, a-c); (b) species stabilised by 4C sites (band at 2159 cm^{-1}) are more resistant (Figure 13A, c-g); and (c) CO adsorbed on the most acidic sites on corners (3C) is not reversible at 77K and are also present at 300K [49].

It is worth stressing that both binding and vibrational properties of CO adsorbed on different Mg_{LC}^{2+} site have been accurately predicted by means of various theoretical approaches [65-67].

Finally, the band at 2148 cm^{-1} in the spectra of CO adsorbed on MgO-h, which shows a frequency lower than that of CO on Mg_{5C}^{2+} but it is more resistant to desorption (curves c-h), was tentatively assigned to CO anchored both by the carbon and oxygen atoms to two cations at step defect sites [49].

4.2.2 Basic Sites

The adsorption of CO at room temperature on MgO has been unambiguously associated with the reactivity of basic O_{LC}^{2-} sites exposed with the lowest coordination (3C) onto the surface of MgO microcrystals [60]. Some conclusions have been reached on the structure of the resultant species, which are formed by CO polymerisation. The assignment of the IR bands and the reaction mechanisms have been repeatedly discussed, though the models are the subject of continuing revision [50,61-62,73-74].

The nature of the surface species formed during the first stages of the reaction can be deduced by means of the IR spectra of CO adsorbed at 77K (Figure 14) and represented in scheme 2.

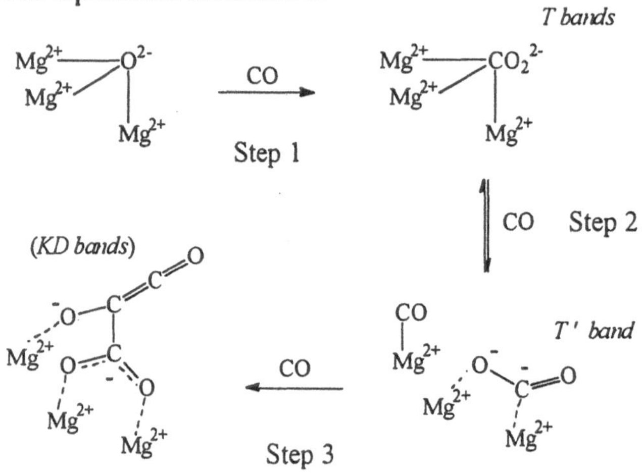

SCHEME 2: Early stages of the CO polymerisation leading to anionic molecules on defective MgO.

Step 1: transient CO_2^{2-} species (T bands in Figure 14a) are formed by nucleophilic attack of O_{LC}^{2-} on adsorbed CO (see Scheme 2); step 2: upon increasing the CO pressure (Figure 14b) the T species (multidentate CO_2^{2-}) are transformed into bidentate ones (T' band) characterised by a main peak at 1475 cm^{-1} and this transformation is reversible; step 3: the latter CO_2^{2-} transient species are precursors of trimeric $C_3O_4^{2-}$ species which are formed upon further addition of CO (KD bands in Figure 14b-c).

Some of the trimeric species may evolve, by addition of CO, into more complex polymers which may have cyclic [61] or, more probably, linear structures and these further evolve with time, by fragmentation, in oxidised (carbonate-like groups) and reduced counterparts [49-50].

The anionic species are not formed on MgO smoke samples which confirms that the oxygens mainly involved in the reaction are the most undercoordinated and, consequently, the most basic located in corner positions, which are very rare on the highly regular microcrystals of the smoke sample.

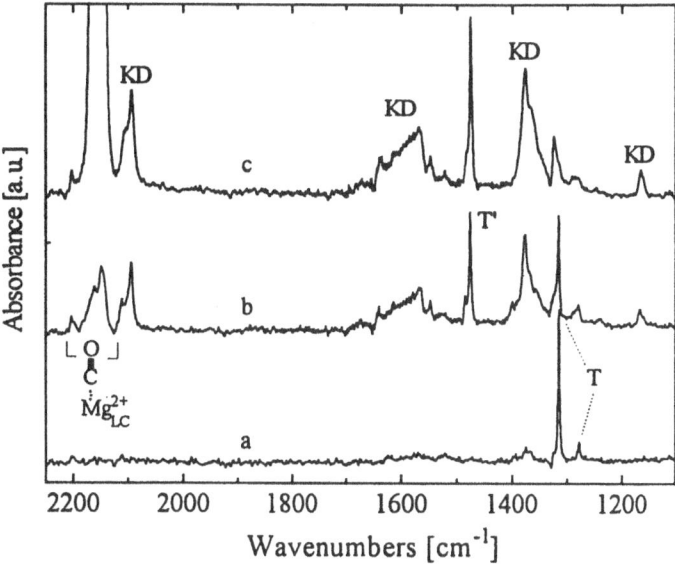

Figure 14: IR spectra of CO adsorbed at 77 K on MgO-h under increasing pressure from 0.01, curve (a), to 0.1 Torr, curve (c).

5. Surface Structure and Surface Reactivity Towards CO of Ni/MgO Catalysts

The activity of dispersed nickel catalysts in promoting reactions of great industrial interest stimulated extensive research work and, following the pioneering paper by Primet *et al.* [75], most characterisation studies were devoted to Ni dispersed on SiO_2 [76-78] and Al_2O_3 [78-79] acidic supports. Attention was also devoted to basic MgO matrices which produce bulk solid solutions with the precursor NiO and support reduced Ni on the surface of the particles [80-81].

Systematic studies showed that morphology, surface structure and reactivity of Ni/MgO catalysts strictly depend on the pretreatment conditions in that: a) calcination temperature controls the extent of solid solution and, thus, the overall reducibility; b) reduction temperature controls the extent of the metal phase and also size and shape of Ni particles [82-84]. After calcination at 673K and subsequent reduction at 1073K, Ni/MgO catalysts exhibit well dispersed Ni particles, with an average size 100-110Å [84-85] and rough and irregular shape. The high resolution electron image of Figure 15 shows a Ni particle embedded in the MgO matrix; its surface appears heavily stepped and with terraces of very small size.

The adsorption of successive doses of CO (Figure 16) produces two absorptions at *ca.* 2040 and *ca.* 1975 cm^{-1} assignable respectively to stretching vibrations of linear and bridged carbonyls bonded to Ni centres [76-78,80].

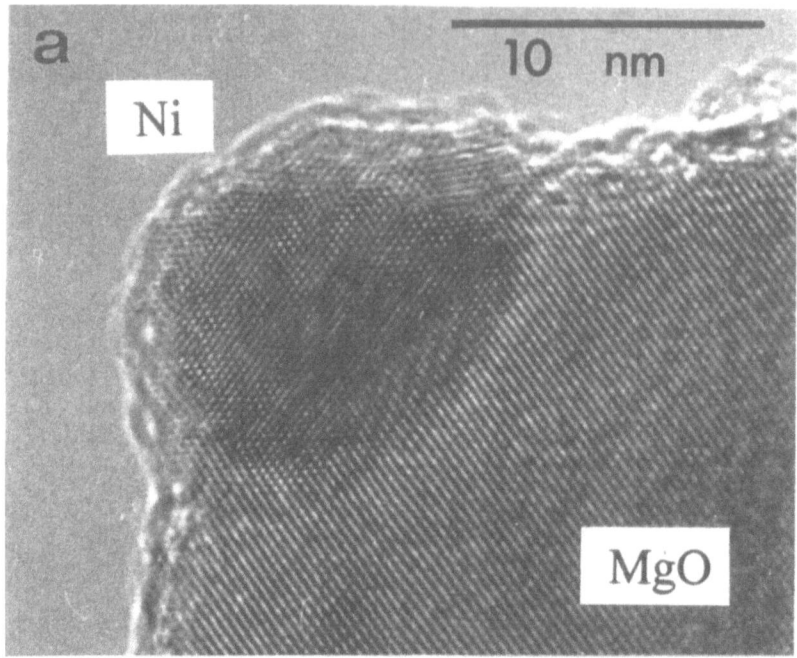

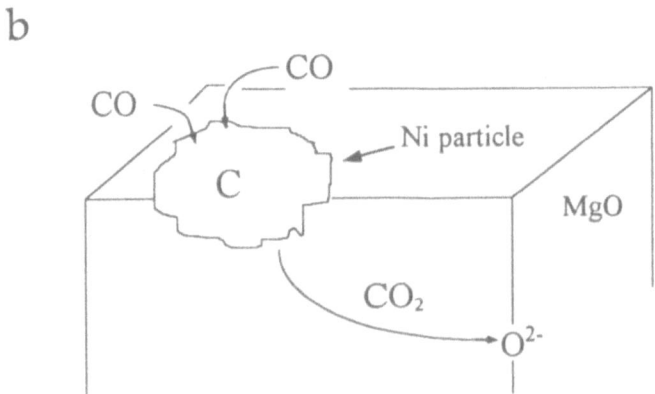

Figure 15: Section a: High resolution electron micrograph of a Ni/MgO catalyst showing two series of lattice fringes related to (110) and (111) planes of a Ni particle embedded in a MgO microcrystal. Fringes of (111) planes of the MgO are also clearly observed. Section b: schematic representation of the disproportionation reaction (2CO → CO$_2$ + C) on supported Ni particles.

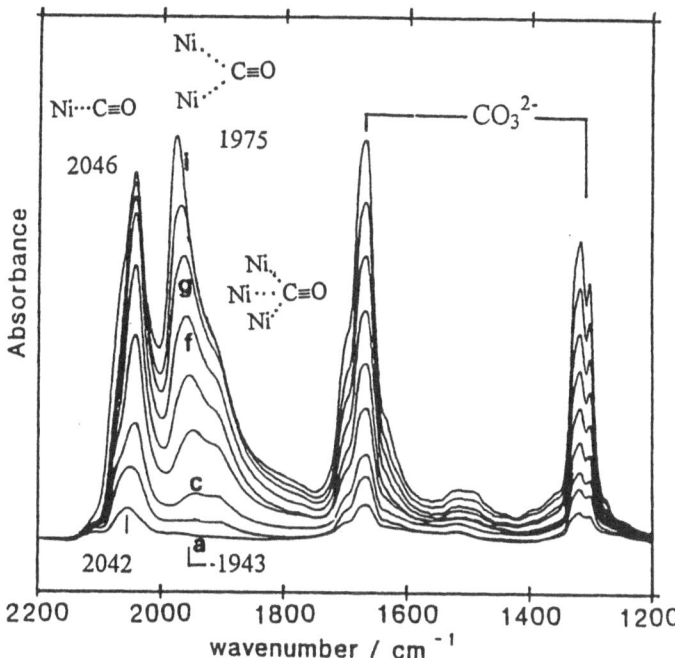

Figure 16: IR spectra of CO adsorbed at room temperature on a Ni/MgO catalyst: successive doses [(a) to (i)] up to a pressure of 0.5 Torr.

The adsorption of successive doses of CO (Figure 16) produces two absorptions at *ca.* 2040 and *ca.* 1975 cm^{-1} assignable respectively to stretching vibrations of linear and bridged carbonyls bonded to Ni centres [76-78,80]. These values are much lower than that of the free CO molecule due to the formation of strong chemical bonds between CO and the surface metal sites; in such bonds, d-π contributions play the dominant role in determining the electronic population of the $2\pi^*$ antibonding molecular orbitals of the adsorbed CO, its bond order and, in turn, its stretching frequency. This effect explains also the strength of the CO-Ni bond in such complexes: the carbon monoxide is desorbed from the Ni surface only at temperatures higher than 473K, whereas in the case of very weak CO-Mg^{2+} complexes, the CO is desorbed at 77K.

In the range of linear carbonyls, the first dose of CO (Figure 16a) produces a weak band at 2042 cm^{-1} which progressively increases in intensity and shifts upwards to 2046 cm^{-1}. An unresolved shoulder is always present on the high frequency side. In the range of bridged carbonyls, two overlapped bands are present at the lowest coverages (curves a-c), with maxima at 1943 and 1900 cm^{-1}. Upon adding CO, the former progressively grows and shifts to 1975 cm^{-1} shielding the other component (curves f-i).

This complex behaviour reflects both the surface heterogeneity of Ni particles exhibiting sites with different coordination on microfaces, edges, and corners and the dipole coupling effects which are strongly dependent on CO coverage.

Two structured absorptions are observed, with maxima at 1670 and 1315 cm^{-1}, in the 1800-1200 cm^{-1} range. Such bands are related to each other and can be assigned to bidentate carbonate-like groups formed on the surface of MgO (86). The presence of carbonate groups indicates [83-84] that *disproportionation* (Boudouard reaction) of CO occurs ($2CO \rightarrow CO_2 + C$), followed by the reaction of CO_2 with basic O^{2-} ions of the surface leading to CO_3^{2-} species, while carbon is stabilised on the metal particles [79,85,87]; such a mechanism is represented in Figure 15b.

The *disproportionation reaction is fully inhibited by the preadsorption of hydrogen*, as no bands due to carbonate groups were observed under these circumstances [85]. Systematic experiments of progressive cleaning of the surface confirmed that the activity towards the Boudouard reaction is restored only after the elimination of the last traces of adsorbed hydrogen by outgassing at 573K. The sites which stabilise hydrogen more strongly certainly comprise those on highly uncoordinated positions on corners and edges [88-90].

The role of *defects sites* in determining the activity of Ni catalysts was also shown by enhancing the smoothing of the originally irregular crystallites by means of a treatment in H_2 at 1173K. The expected smoothing was confirmed by HRTEM [85]. Significantly, IR. spectra of CO adsorbed on such catalysts indicated that the production of carbonates is inhibited by treatments at the highest temperature, and confirmed that the sites with the lowest coordination, typical of heavily stepped particles, promote the disproportionation reaction.

The sensitivity of CO to probe carbided surfaces of Ni particles was also checked [85] in properly designed experiments (Figure 17). After heating in CO at 473K, the intensity of carbonates bands increased drastically (as the Boudouard reaction proceeded much further than at room temperature) and the sample was outgassed at the same temperature to remove most carbonyl groups: a small fraction of adsorbed CO was left on the surface after this treatment (Figure 17a).

The spectra have a complex evolution when the coverage is gradually increased by adding doses of CO (Figure 17, b-i) in that a band at 1855 cm^{-1} is progressively depleted while another absorption at 1920 cm^{-1} (curve a) increases in intensity and progressively shifts to 1945 cm^{-1}. In addition, a new band appears at 2075 cm^{-1} upon the admission of the first dose of CO (curve b) and increases dramatically to become dominant, while shifting to 2085 cm^{-1} at the highest coverage (curve 1).

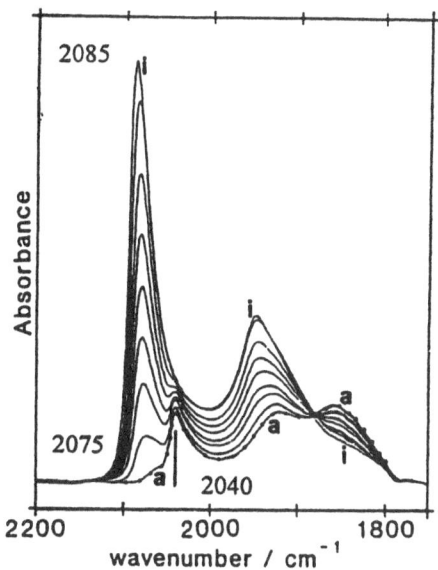

Figure 17: CO probing a carbided Ni/MgO catalysts. IR spectra of: (a) the sample carbided by heating in CO at 473K and then outgassed at the same temperature; (b-i) successive doses of CO admitted at RT on such treated sample.

The overall effects of the presence of carbon is shown by comparing the spectra at high coverage of Figures 16 and 17: the intensity of the adsorption due to bridged carbonyls (1950-1800 cm^{-1}) is much lower in the presence of carbon than in the case of clean Ni surfaces. Moreover, the band of linear carbonyls is found at higher frequency when carbon is present (2085 cm^{-1} for carbided Ni as compared to 2042 cm^{-1} for clean Ni).

Such phenomena were also observed for Ni dispersed on other supports [75-76,79] and for Ni single crystals [91]. Carbon is accommodated on faces where it can be stabilised by a number of Ni sites which explains the lower intensity of the absorption due to bridged carbonyls because they are produced on planes where multiple bonds to a number of metal sites are possible. Besides this competition for the same patches of sites [76,91], depletion of bridged carbonyls may be favoured by the distortion of the outmost nickel layers induced by the presence of carbon, causing significant changes of Ni-Ni distances and reconstruction of the surface [92].

As for the high frequency shift of the absorption of on-top carbonyls up to 2085 cm^{-1}, a number of factors (dilution effects, reconstruction, faceting) associated with the presence of carbon may play a role. Certainly, the electron withdrawing effect of carbon [91] and the consequent reduction of the backdonation into the $2\pi^*$ antibonding orbital of CO should contribute significantly.

The parallel decrease of the intensity of the band at 1855 cm^{-1}, due to three-fold bridged carbonyls, and increase of an adjacent component at 1920 cm^{-1}, due to two-fold carbonyls, upon increasing coverage of CO, was interpreted in terms of transformation of one species into another because of an increasing of competition for multiple sites [85].

The shift of the band of the linear carbonyls from 2075 to 2085 cm^{-1} and that of the bridged carbonyls from 1920 to 1945 cm^{-1} were attributed to coupling effects increasing with coverages [91].

Carbon deposits produced by running the Boudouard reaction at 473-673K on Ni/MgO catalysts were identified [85] by HRTEM studies (Figure 18).

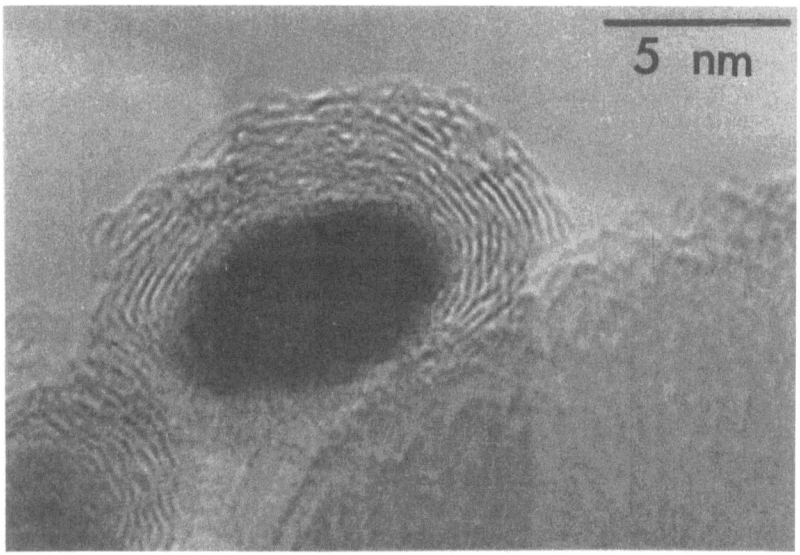

Figure 18: Electron micrograph showing carbonaceous residues on a Ni/MgO catalyst by heating in CO at 673K. The Ni particle appears surrounded by a shell of graphitic carbon with fringes 3.38Å apart.

ACKNOWLEDGEMENTS

The authors thank Dr. N. Damilano who has given valuable help in organising part of this contribution, Professor A. Zecchina for the stimulating discussions and all the undergraduated students of the Physical Chemistry group in Turin who significantly contributed in organising the Figures and Schemes.

REFERENCES

1. Terenin, A.N. (1957), in A.V. Kiselev (ed.), *Surface Chemical Compounds and Their Role in Adsorption Phenomena*, Moscow State University Press.

2. Eischens, R.P., Francis, S.A: and Pliskin, W.A. (1956) The effect of surface cove-rage on the spectra of chemisorbed CO, *J. Phys. Chem.* 60, 194-201.

3. Colthup, N.B., Daly, L.H. and Wiberley, S.E. (1990) *Introduction to Infrared and Raman Spectroscopy*, 3rd ed., Academic Press Limited, London.

4. Eischens, R.P. and Pliskin, W.A. (1958) The infrared spectra of adsorbed Molecu-les, *Adv. Catal.* 10, 1-56.

5. Little, L.H. (1966) *Infrared Spectra of Adsorbed Species*, Academic Press, New York.

6. Delgass, W.N., Haller, G.L., Kellermann R. and Lunsford, J.H. (1979) *Spectrosco-py in Heterogeneous Catalysis*, Academic Press, New York.

7. Bell, A.T. and Hair, M.L. (1980) *Vibrational Spectroscopies for adsorbed species*, American Chemical Society, Washington, D.C.

8. King, D.A., Richardson, N.V. and Holloway S. (1986) *Vibrations at surfaces*, Elsevier, Amsterdam.

9. Coluccia, S. and Marchese, L. (1988) Gas-solid interphase: characterization by infrared spectroscopies, in M. Schiavello (ed.), *Photocatalysis and Environment*, Kluwer Academic Publishers, pp. 205-222.

10. Davidov, A.A. (1990) *Infrared spectroscopy of adsorbed on the surface of transi-tion metal oxides*, John Wiley & Sons, New York.

11. Sauer, J., Ugliengo, P., Garrone, E. and Saunders, V.R. (1994) Theoretical study of van der Waals complexes at surface sites in comparison with the experiment, *Chem. Rev.* 94, 2095-2160.

12. Goodman, D.W. (1995) Model studies in catalysis using surface science probes, *Chem Rev.* 95, 523-536.

13. Hattori, H. (1995) Heterogeneous basic catalysis, *Chem. Rev.* 95, 537-558.

14. Corma, A. (1995) Inorganic solid acids and their use in acid-catalyzed hydrocarbon reactions, *Chem. Rev.* 95, 559-614.

15. Farneth, W.E. and Gorte, R.J. (1995) Methods for characterizing zeolite acidity, *Chem. Rev.* 95, 615-636.

16. van Santen, R.A. and Kramer, G.J. (1995) Reactivity theory of zeolitic Brönsted acidic sites, *Chem. Rev.* 95, 637-660.

17. Busca, G. (1996) Vibrational spectroscopy of adsorbed species on metal oxides: an introduction, *Catal. Today*, in press.

18. Lercher, J. (1996) IR studies of the surface acidity of oxides and zeolites using adsorbed probe molecules, *Catal. Today*, in press.

19. Lavalley, J.C. (1996) IR studies of the surface basicity of oxides and zeolites using adsorbed probe molecules, *Catal. Today*, in press.

20. Zecchina, A., Scarano, D., Bordiga, S., Ricchiardi, G., Spoto, G. and Geobaldo, F. (1996) IR studies of CO and NO adsorbed on well characterized oxide single microcrystals, *Catal. Today*, in press.

21. Morterra, C. and Magnacca, G. (1996) A case study: surface chemistry and surface structure of catalytic aluminas as studied by vibrational spectroscopies of adsorbed molecules, *Catal. Today*, in press.

22. Rosevcwaig, A. (1980) *Photoacustics and photoacustic spectroscopy*, John Wiley & Sons, New York.

23. Griffith, P.R. and de Haseth, J.A. (1986) *Fourier Transform Infrared Spectrome-try*, John Wiley & Sons, New York.

24. Marchese, L., Bordiga, S., Coluccia, S., Martra, G. and Zecchina, A. (1993) Structure of the surface sites of δ-Al_2O_3 as determined by high-resolution transmission electron microscopy, computer modelling and infrared spectroscopy of adsorbed CO, *J. Chem. Soc. Faraday Trans.* 89, 3483-3489.

106

25. Anderson, R.B. and Dawson, P.T. (1976) *Experimental Methods in Catalystic Research*, vol III, Academic Press, New York.

26. a) Tsiganenko, A.A. and Babaeva, M.A. (1983) Infrared spectrum of ammonia adsorbed by Si-OH groups on a silica surface, *Opt. Spektrosk.* 60, 1117-1120.

    b) Lokhov, Yu.A. and Bredikhin, M.N. (1990) Variable temperature high vacuum IR cells, *Bruker Optics News* 1, 12.

27. Hair, M.L. (1967) *Infrared spectroscopy in surface chemistry*, Marcel Dekker, Inc., New York.

28. Chen, J., Wright, P.A., Thomas, J.M., Natarajan, S., Marchese, L., Bradley, S.M., Sankar, G., Catlow, C.R.A., Gai-Boyes, P.L., Towsend, R.P. and Lok, C.M. (1994) SAPO-18 catalysts and their Brönsted acid sites, *J. Phys. Chem.* 98, 10216-10224.

29. Smith, L., Marchese, L., Cheetham, A.K., Thomas, J.M., Wright, P.A., Chen, J. and Morris, R.E. (1996) On the nature of water bound to a solid acid catalyst, Science, in press.

30. de Man, A.J.M. and van Santen, R.A. (1992) The relation between zeolite frame-work structure and vibrational spectra, *Zeolites* 12, 269-279.

31. van Santen, R.A: and Vogel, D.L. (1989) Lattice dynamics of zeolites, in C.R.A. Catlow (ed.), *Advances in solid-state chemistry*, J.A.I. Press Inc., London, pp. 151-223.

32. Jacobs, W.P.J.H., van Wolput, J.H.M.C. and van Santen, R.A. (1993) An in situ Fourier transform infrared study of zeolitic vibrations: dehydration, deammona-tion, and reammonation of ion-exchanged Y zeolites, *Zeolites* 13, 170-182

33. Martra, G., Damilano, N., Coluccia, S., Tsuji, H. and Hattori, H. (1995) Cationic mobility in MgX zeolite.A FTIR study, *J. Chem. Soc., Faraday Trans.* 91, 2961-2964.

34. Peuker, C. and Kunath, D. (1981) Far-infrared spectra of zeolites, *J. Chem. Soc., Faraday Trans. 1* 77, 2079-2085.

35. Anderson, A.A., Shelepev, Y.F. and Smolin, Y.I. (1990) Structural study of Mg-exchanged NaX and CaX zeolites in hydrated (25°C) and dehydrated (400°C) states, *Zeolites* 10, 32-37.

36. Spoto, G., Bordiga, S., Ricchiardi, G., Scarano, D., Zecchina, A. and Borello, E. (1994) IR study of ethene and propene ologomerization on HZSM-5: hydrogen-bonded precursor formation, initiation and propagation mechanisms and structure of the entrapped oligolmers, *J. Chem. Soc. Faraday Trans.* 90, 2827-2835.

37. Heilweil, E.J. Casassa, M.P. Cavanagh, R.R. and Stephenson, J.C. (1985) Vibra-tional deactivation of surface OH chemisorbed on SiO_2: Solvent effects, *J.Chem. Phys.* 82, 5216-5231.

38. Hirose, C., Goto, Y., Akomatsu, N., Kondo, J. and Domen, K. (1993) Measure-ment of vibrational energy relaxation of a surface hydroxyl group by ultrashort tunable infrared pulses, *Surf. Sci.* 283, 244-247.

39. Brugmans, M.J.P., Kleyn, A.W., Lagendijk, A., Jacobs, W.P.J.H. and van Santen, R.A. (1994) Hydrogen bonding in acidic zeolites observed by time-resolved vibrational spectroscopy, *Chem. Phys. Lett.* 217, 117-122.

40. Makarova, M.A., Ojo, A.F., Karim, K., Hunger, M. and Dwyer, J. (1994) FTIR study of weak hydrogen bonding of Brönsted hydroxyls in zeolites and alumino-phosphates, *J. Phys. Chem.* 98, 3619-3623.

41. Zaki, M.I. and Knözinger, H. (1989) An infrared specttroscopy study of carbon monoxide adsorption on α-chromia surfaces: probing oxidation states of coor-dinatively unsaturated surface cations, *J. of Catal.* 119, 311-321

42. Hammaker, R.M., Francis, S.A. and Eischens, R.P. (1965) Infrared study of inter-molecular interactions for carbon monoxide on platinum, *Spectrochim. Acta* 21, 1295-1309.

43. Mahan, G.D. and Lucas, A.A. (1978) Colettive vibrational modes of adsorbed CO, *J. Chem. Phys.* 68, 1344-1348.

44. Persson, B.N.J. and Ryberg, R. (1981) Vibrational interaction between molecules adsorbed on a metal surface: The dipole-dipole interaction, *Phys. Rev. B* 24, 6954-6970.

45. Hollins, P. and Pritchard, J. (1980) Intermolecular interactions and the infrared reflection-absorption spectra of chemisorbed carbon monoxide on copper, in R.F. Willis, *Vibrational Spectroscopy of Adsorbates*, Springer-Verlag, Berlin, pp.51-74.

46. Marchese, L., Coluccia, S., Martra, G. and Zecchina A. (1992) Dynamic and static interactions in CO layers adsorbed on MgO smoke (100) facelets: a FTIR and HRTEM study, *Surf. Sci.* **269/270**, 135-140.

47. Richardson, H.H., Chang, H.C., Noda, C. and Ewing, G.E. (1989) Infrared spec-troscopy of CO on NaCl (100). I Photometry, *Surf. Sci.* **216**, 93-104.

48. Bolis, V., Fubini, F., Garrone, E. and Morterra, C. (1989) Thermodynamic and vibrational characterization of CO adsorption on variously pretreated anatase, *J. Chem. Soc. Faraday Trans. I* **85**, 1383-1395.

49. Coluccia, S., Baricco, M., Marchese, L., Martra, G. and Zecchina A. (1993) Surface morphology and reactivity towards CO of MgO particles: FTIR and HRTEM studies, *Spectrochim. Acta* **49A**, 1289-1298.

50. Zecchina, A., Coluccia, S., Spoto, G., Scarano, D. and Marchese, L. (1990) Revi-siting MgO-CO surface chemistry: an IR investigation, *J. Chem. Soc. Faraday Trans.* **86**, 703-709.

51. Zecchina, A., Lofthouse, M.G. and Stone, F.S. (1975) Reflectance spectra of surfa-ce states in magnesium oxides and calcium oxides, *J.Chem. Soc. Faraday Trans. 1* **71** 1476-1490.

52. Garrone, E., Zecchina, A. and Stone, F.S. (1980) An experimental and theoretical evaluation of surface states in MgO and other alkaline earth oxides, *Phil. Mag.* **B42**, 683-703.

53. Nelson, R.L. and Hale, J.W. (1971) Electronic spectra of the surfaces of alkaline earth oxides, *Discuss. Faraday Soc.* **52**, 77-88.

54. Coluccia, S., Deane, A.M. and Tench, A.J. (1978) Photoluminescent spectra of surface states in alkaline earth oxides, *J. Chem. Soc. Faraday Trans. 1* **74**, 2913-2922.

55. Coluccia, S., Tench, A.J. and Segall, R.L. (1979) Surface structure and surface states in magnesium oxide powders, *J. Chem. Soc. Faraday Trans. 1* **75**, 1769-1779.

56. Moodie, A.F. and Warble, C.E. (1971) Electron microscopic investigations of magnesium oxide morphology and surfaces, *J. Cryst. Growth* **10**, 26-38.

57. Tanji, T. and Cowley, J.M. (1985) Interactions of electron beams with surfaces of MgO crystals, *Ultramicroscopy* **17**, 287-302.

58. Tanji, T., Masaoka, H, Ito, J., Yada, K. and Cowley, J.M. (1989) Charging effect on the HRTEM imaging of small MgO crystals, *Ultramicroscopy* **27**, 223-232.

59. a) Coluccia, S. and Tench, A. (1981) Spectroscopic studies of hydrogen adsorp-tion on highly dispersed MgO, Proceed. 7th ICC, Tokyo, Kodansha Ltd., Tokyo, pp. 1154-1169.

    b) Coluccia, S., Barton, A. and Tench, A.J. (1981) Reactivity of low-coordination sites on the surface of magnesium oxide, *J. Chem. Soc. Faraday Trans. 1* **77**, 2203-2207.

60. Zecchina, A. and Stone, F.S. (1974) Reflectance spectra of CO chemisorbed on MgO, and evidence for the formation of cyclic adsorbed species, *J. Chem. Soc., Chem. Commun.* 582-584.

61. Guglielminotti, E., Coluccia, Garrone, E., Cerruti, L. and Zecchina, A. (1979) Infrared study of CO adsorption on magnesium oxide, *J. Chem. Soc. Faraday Trans. 1* **75**, 96-108.

62. Babaeva, M.A., Bystrov, D.S., Kovalgin, A.,Y. and Tsyganenko, A.A. (1990) CO interaction with the surface of thermally activated CaO and MgO, *J. Catal.* **123**, 396-416.

63. Garrone, E., Zecchina, A. and Stone, F.S. (1988) CO adsorption on MgO and CaO, *J. Chem. Soc. Faraday Trans. 1* **84**, 2843-2854.

64. Platero, E.E., Scarano, D., Spoto, G. and Zecchina, A. (1985) Dipole coupling and chemical shifts of CO and NO adsorbed on oxides and halides with rock-salt structure, *Faraday Discuss. Chem. Soc.* **80**, 183-193.

65. Pelmenschikov, A.G, Morosi, G., Gamba, A. and Coluccia, S. (1995) A check of quantum chemical molecular models of adsorption on oxides against experi-mental infrared data, *J. Phys. Chem.* **99**, 15018-15022.

66. Pacchioni, G., Minerva, T. and Bagus, P.S. (1992) Chemisorption of CO on defects sites of MgO, *Surf. Sci.* **275**, 450-458.

67. Neyman, K.M. and Rösch (1993) Bonding and vibration of CO molecules adsorbed on low-coordinated surface sites of MgO: LCGTO-LDF cluster investigation, *Surf. Sci.* **297**, 223-234.

68. Ewing, G.E. (1962) Infrared spectra of liquid and solid carbon monoxide, *J. Chem. Phys.* **37**, 2250-2256.

69. Disselkamp, R., Chang, H.C. and Ewing, G.E. (1990)) Infrared spectroscopy of CO on NaCl(100). II. Bandshape analysis, *Surf. Sci.* **240**, 193-210.

108

70. Noda, C., Richardson, H.H. and Ewing, G.E. (1990) Infrared spectroscopy of CO on NaCl(100). II. Vibrational dephasing and band shapes, *J. Chem. Phys.* **92**, 2099-2105.

71. Hush, N.S. and Williams, M.L. (1974) Carbon monoxide bond length, force costant and infrared intensity variations in strong electric fields: Valence-shell calculations, with applications to properties of adsorbed and complexed CO, *J. Molec. Spectrosc.* **50**, 349-368.

72. Seanor, D.A. and Amberg, C.H. (1965) Infrared-band intensities of adsorbed carbon monoxide, *J. Phys. Chem.* **42**, 2967-2970.

73. Marchese, L., Coluccia, S., Martra, G., Giamello, E. and Zecchina, A. (1991) Novel dimeric species produced by CO interaction with surface F-type centres on Mg-doped MgO: an IR study, *Mater. Chem. Phys.* **29**, 437-445.

74. Giamello, E., Murphy, D., Marchese, L., Martra, G. and Zecchina, A. (1993) Electron paramagnetic resonance investigation of the interaction of CO with the surface of electron-rich magnesium oxide: evidence for the CO radical anion, *J. Chem. Soc. Faraday Trans.* **89**, 3715-3722.

75. Primet, M., Dalmon, J.A. and Martin, G.A. (1977) Adsorption of CO on well-defined Ni/SiO$_2$ catalysts in the 195-373 K range studied by infrared spectroscopy and magnetic methods, *J. Catal.* **46**, 25-36.

76. Wielers, A.F.H., Aaftink, G.J.M. and Geus, J.W. (1985) The adsorption of CO on clean and carbided Ni/SiO$_2$ catalysts, *Appl. Surf. Sci.* **20**, 564-580.

77. Blackmond, D.G. and Ko, E.I. (1985) Structural sensitivity of CO adsorption and H$_2$/CO coadsorption on Ni/SiO$_2$ catalysts, *J. Catal.* **96**, 210-221.

78. Bartholomew, C.H. and R.B. Pannell (1980) The stoichiometry of hydrogen and carbon monoxide chemisorption on alumina and silica-supported nickel, *J. Catal.* **65**, 390-402.

79. Galuszka, J., Chang, J.R. and Amenomiya, Y. (1981) Disproportionation of CO on supported Ni catalysts, *J. Catal.* **68**, 172-181.

80. Bond, G.C. and Sarsam, S.P. (1988) Reduction of nickel/magnesia catalysts, *Appl. Catal.* **38**, 365-377.

81. Zecchina, A., Spoto, G., Coluccia, S. and Guglielminotti E. (1984) Spectroscopic study of the adsorption of CO on solutions of NiO2 and magnesium oxide, *J. Chem. Soc., Faraday Trans. 1* **80**, 1891-1901.

82. Arena, F., Horrel, B.A., Cocke, D.L., Parmaliana, A. and Giordano, N. (1991) Magnesia-supported Ni catalysts, *J. Catal.* **132**, 58-67.

83. Parmaliana, A., Arena, F., Frusteri, F., Coluccia, S., Marchese, L., Martra, G. and Chuvilin, A.L. (1993) Magnesia-supported nickel catalysts. II. Surface properties and reactivity in methane steam reforming, *J. Catal.* **141**, 34-47.

84. Martra, G., Arena, F., Baricco, M., Coluccia, S., Marchese, L. and Parmaliana, A. (1994) High loading Ni/MgO catalysts. Surface characterization by IR spectra of adsorbed CO, *Catal. Today* **17**, 449-458.

85. Martra, G., Marchese, L., Arena, F., Parmaliana, A. and Coluccia, S. (1994) Surface structure of Ni/MgO catalysts: effects of carbon and hydrogen on the reactivity towards CO. HRTEM and FTIR studies, *Topics in Catal.* **1**, 63-73.

86. Busca, G. and Lorenzelli V. (1982) Infrared spectroscopic identification of species arising from reactive adsorption of carbon oxides on metal oxide surfaces, *Mater. Chem.* **7**, 89-126.

87. Kuijpers, E.G.M., Kock, A.J.H.M., Niewesteeg, M.W. and Geus, J.W. (1985) Disproportionation of CO on Ni/SiO2: kinetics and nature of the deposited carbon, *J. Catal.* **95**,13-20.

88. Canning, N.D.S. and Chesters, M.A. (1986) The co-adsorption of H2 and CO on Ni(110), *Surf. Sci.* **175**, L811-L816.

89. Martensson, A.S., Nyberg, C. and Andersson, S. (1988) Adsorption of hydrogen on a stepped nickel surface, *Surf. Sci.* **205**, 12-24.

90. Haq, S., Love, J.G. and King, D.A. (1992) The adsorption of CO on Ni(110) and its interaction with H: a RAIRS study, *Surf. Sci.* **275**, 170-184.

91. Bartolini, J.C. and Tardy, B. (1981) Vibrational EELS studied of CO chemisorbed on clean and carbided (111), (100) and (110) nickel surface, *Surf. Sci.* **102**, 131-150.

92. Onuferko, J.H., Woodruff, D.P. and Holland, B.W. (1979) LEED structure analysis of the Ni(100) (2x2)c(p4g) structure: a case of adsorbate-induced sub-strate distortion, *Surf. Sci.* 87, 357-374.

SOLID STATE NMR SPECTROSCOPY OF NON-INTEGER SPIN NUCLEI

CLARE P. GREY
SUNY Stony Brook,
Chemistry Department,
Stony Brook,
NY 11794-3400, U.S.A.

1. Introduction

Quadrupolar nuclei comprise the majority of the magnetically active nuclei, of which the non-integer spin nuclei are the most numerous. NMR spectroscopy of these nuclei differs from that of I=1/2 nuclei in that spectra of these nuclei are typically influenced by the interaction between the electric quadrupole moment of the nucleus and the electric field gradient at the nuclear site caused by the surrounding atoms. This quadrupolar interaction can sometimes be so large that the nucleus is extremely difficult to observe with NMR methods (e.g. for ^{35}Cl, ^{63}Cu, ^{127}I, etc.), and in this case, nuclear quadrupole resonance (NQR) spectroscopy may be a more appropriate technique with which to study these nuclei. For many nuclei (e.g. ^7Li, ^{11}B, ^{17}O, ^{23}Na, ^{27}Al, ^{51}V), however, the quadrupolar interaction is typically sufficiently small that many of the NMR experiments used to study I=1/2 nuclei can be applied. The central transition (m = 1/2 to - 1/2) for non-integer spin quadrupolar nuclei is not broadened to first order by the quadrupolar interaction, and relatively narrow resonances may be observed. The presence of the outer satellite transitions is often ignored, and the non-integer spin is treated as a 'fictitious spin-half system'. The validity of this assumption will depend on parameters such as the radio frequency (rf) field amplitude, ω_1, used to excite the nuclei, and on the size of the quadrupole coupling constant, QCC. There are a variety of consequences that result from the presence of these outer transitions, with different practical implications for obtaining quantitative spectra. This article aims to address this issue, point out when some of the assumptions sometimes made are no longer valid, and to discuss methods for obtaining quantitative high

C.R.A. Catlow and A. Cheetham (eds.), New Trends in Materials Chemistry, 111–139.

resolution NMR spectra. For more detailed reviews of the theory of NMR of non-integer spin nuclei, the reader is referred to recent reviews by Freude and Haase [1] and Vega [2]. The article concludes with a brief discussion of new double resonance NMR methods for studying non-integer spins, since this is an area where many new experiments are being developed and are currently being applied to problems in materials chemistry.

2. Theory

2.1 THE FIRST-ORDER QUADRUPOLAR INTERACTION

The Hamiltonian for a quadrupolar nucleus in a magnetic field, with quadrupole moment Q, is a sum of the Zeeman and quadrupolar terms:

$$H = \omega_L I_z + H_Q \qquad (1)$$

where (1) is written in frequency units. Standard symbols for frequencies are used throughout this article, where ω, ν and δ denote frequencies expressed in units of radians, Hz and ppm, respectively. The first-order quadrupolar interaction, H_Q', is given by:

$$H_Q' = (1/2)Q'(\theta,\phi)\{I_z^2 - I(I+1)/3\} \qquad (2)$$

where Q', the quadrupolar splitting, is a function of θ and ϕ, the polar angles that define the orientation of the principal axis system of the quadrupole tensor (V_{XX}, V_{YY}, V_{ZZ}) in the Zeeman field (or laboratory frame) (Figure 1) and the asymmetry parameter, η, (i.e. $(V_{XX}-V_{YY})/V_{ZZ}$):

$$Q'(\theta,\phi) = (\omega_Q/2)[(3\cos^2\theta-1)-\eta\sin^2\theta\cos2\phi] \qquad (3)$$

ω_Q, the quadrupole frequency, depends on the quadrupole coupling constant e^2qQ/h (or QCC):

$$\omega_Q = 3e^2qQ/[2I(2I - 1)\hbar] \qquad (4)$$

(where $eq = V_{ZZ}$). Note that the QCC is zero when the nucleus is located at a site with cubic, octahedral or tetrahedral point group symmetry. The energy levels of the quadrupolar nucleus, E_m, in the Zeeman field are given by:

$$E_m = m\omega_L + Q'(\theta,\phi)\{m_I^2 - I(I+1)/3\}/2 \tag{5}$$

The first-order quadrupole-split spectrum of a single crystal thus comprises 2I evenly spaced resonances with splittings given by Q', and intensities that are proportional to:

$$<m|I_x|m+1>^2 = \{I(I+1) - m(m+1)\}/4 \tag{6}$$

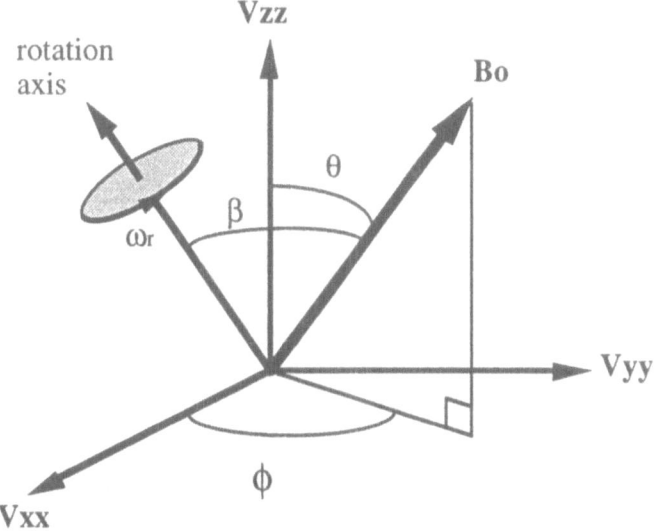

Figure 1. The angles that define the orientation of the quadrupole tensor (V_{XX}, V_{YY}, V_{ZZ}) and the rotor axis, with respect to the static magnetic field, B_0.

One of the most important consequences that follows from (5) is that for non-integer spins, the frequency of the m = 1/2 to m = -1/2, or central, transition does not depend on Q', and is therefore unaffected to first order by the quadrupolar interaction. The outer, or satellite transitions, occur at ±Q', ±2Q' etc., where Q' varies between ω_Q and $-\omega_Q/2$, depending on the orientation of the quadrupole tensor in the magnetic field. For a powder, characteristic lineshapes are observed, from which the QCC and η can be extracted. In practice, however, it is difficult to excite the whole spectrum with a single pulse for large QCCs. If necessary, spectra can be acquired at different resonance offsets, and the different spectra combined, taking into account effects such as the finite-pulse width and the band-width of the probe, which both result in a reduction of the signal at large frequency offsets. The first-order quadrupolar interaction can be averaged by Magic

Angle Spinning, and the broad resonances from the satellite transitions are split into evenly spaced sidebands, separated by the spinning speed, that can spread over many kHz or MHz (Figure 2). Quadupolar lineshapes for spinning at finite speeds have been calculated, and QCC and η can be extracted from the simulations of the experimental spectra [3-4]. The baseline of these spectra are typically severely distorted, since the first few points of the free induction decay (FID) cannot be collected. The baselines can be corrected with baseline correction software (e.g. by fitting the baseline to a cubic spline function) or by linear prediction [5] of the first few points of the FID. The isotropic resonance, or center band, is a sum of the central transition resonance, and the center bands of the (2I-1) satellite transitions.

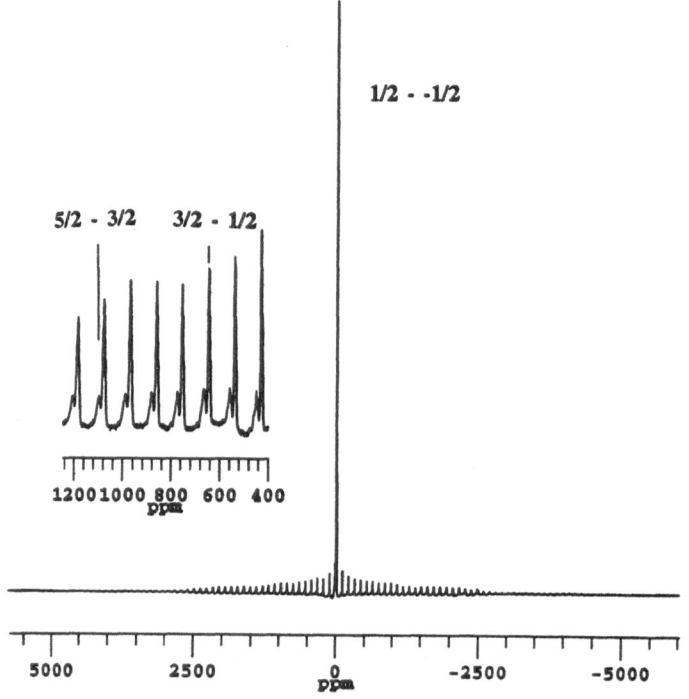

Figure 2. The ^{27}Al MAS NMR spectrum of aluminum acetylacetonate (spinning speed, $\nu_r = 10.2$ kHz, $\nu_L = 93.82$ MHz (i.e. 8.4 T)). The first 6 points of the FID were 'back-predicted' with linear prediction methods, prior to performing the Fourier transform. The $\pm 1/2 <\to> \pm 3/2$ and $\pm 3/2 <\to> \pm 5/2$ satellite transitions are clearly resolved in the enlargement of a region of the spectrum shown in the inset.

All the transitions depend on Q' for integer spins, hence the spectra of these nuclei, and the experimental methods used to acquire the spectra, differ considerably from those used for non-integer spin nuclei. Integer spin nuclei will not be discussed further.

2.2 THE SECOND-ORDER QUADRUPOLAR INTERACTION

A quadrupolar nucleus (with QCC > 0) does not align exactly along the static magnetic field (B_0), but along a field which is a combination of the static and quadrupolar fields (i.e., the quadrupolar nucleus is no longer quantized along B_0). Since the standard spin states |m> (|±3/2>, |±1/2> etc.) used in NMR are those for nuclei quantized along B_0, off-diagonal elements (i.e., terms that do not commute with I_z) appear in the Hamiltonian, H_Q. These terms can be ignored to first order and equation (2) results. For large QCCs, these terms have to be considered, and the full Hamiltonian is required:

$$H_Q = (1/6)\omega_Q\{3I_z^2 - I(I+1) + \eta(I_x^2 - I_y^2)\} \tag{7}$$

Second-order perturbation theory can be used to calculate the second-order correction to the energy levels, $\omega_Q^{(2)}$, which is proportional to ω_Q^2/ω_L. Hence $\omega_Q^{(2)}$ decreases at higher fields. The powder lineshapes that result from the second-order energy shifts have been calculated and are discussed in detail in many reviews and papers [1,6,7]. For example, for I=3/2, the second-order correction to the frequency of the central transition is given by:

$$\omega_Q^{(2)} = -(\omega_Q^2/16\omega_L)(I(I+1)-3/4)(1-\cos^2\theta)(9\cos^2\theta-1) \tag{8}$$

when $\eta = 0$.

2.2.1. *Magic Angle Spinning:*

Unlike the first-order term of the quadrupolar interaction, the second-order term is no longer a second rank tensor, and not averaged to zero by MAS. This can be seen by expressing $\omega_Q^{(2)}$ in terms of the zero, second- and fourth-order Legendre Polynomials $P_n(\cos\theta)$:

$$P_2(\cos\theta) = (3\cos^2\theta-1)$$
$$P_4(\cos\theta) = (35\cos^4\theta-30\cos^2\theta+3) \tag{9}$$

The averaged value for $\omega_Q^{(2)}$ under sample rotation, $<\omega_Q^{(2)}>_{rot}$ is given by:

$$\langle\omega_Q^{(2)}\rangle_{rot}= A_0 +A_2 P_2(\cos\beta)+ A_4 P_4(\cos\beta) \qquad (10)$$

where A_0 is the isotropic shift, and A_2 and A_4 are functions of ω_Q, ω_L, η, and the relative orientation of the quadrupolar tensor and rotor axis. β is the angle between the rotor axis and the static magnetic field (see Figure 1). $P_2(\cos\beta)$ and $P_4(\cos\beta)$ are averaged to zero and -7/18, respectively, for sample rotation at the Magic Angle. Hence MAS only reduces the linewidths of the resonances obtained from powdered samples by approximately 1/3, and for large QCCs significant second-order quadrupolar line broadening remains. Characteristic lineshapes are observed from which the QCC and η can be extracted (Figure 3) [1,2,8]. A shift in the center of gravity of the resonance to lower frequencies also occurs. This is called the second-order quadrupolar shift and is given by A_0 (or $\omega_{Qiso}^{(2)}$):

$$A_0 = \omega_{Qiso}^{(2)} = -(I(I+1) - 3/4)(1 + \eta^2/3)\omega_Q^2/30\omega_L \qquad (11)$$

$\omega_{Qiso}^{(2)}$ depends on the Larmor frequency, and it is sometimes necessary to acquire spectra at more than one magnetic field before the chemical shift, δ_{CS}, and $\omega_{Qiso}^{(2)}$ can be separated. The quadrupolar shift can shift resonances out of the typical chemical shift ranges observed for different environments. This is particularly important for ^{27}Al NMR, where a large QCC can, for example, shift the resonance of a tetrahedrally-coordinated aluminum atom into the chemical shift range typically observed for 5- and 6-coordinated aluminum atoms.

Considerable narrowing can be obtained by spinning at angles other than the magic angle. This technique is called variable-angle spinning or VAS, the optimum angle for spinning being determined by η [9]. This method will be successful in the absence of any significant contributions from other interactions such as the chemical shift anisotropy (CSA) or dipolar coupling, which will no longer be completely averaged, and can give rise to additional broadening.

The second-order quadrupolar interaction differs for each m to (m+1) transition, and the different satellite transitions can sometimes be resolved (see Figure 2). The ratios of $\omega_Q^{(2)}$ and $\omega_{Qiso}^{(2)}$ for the different transitions have been calculated by Samoson, and for I=5/2 nuclei are 1:0.292:-1.833 ($\omega_Q^{(2)}$) and 1:-0.125:-3.50 ($\omega_{Qiso}^{(2)}$) for (+1/2 <-> -1/2) : (±1/2 <-> ±3/2) : (±3/2 <-> ±5/2), where a negative sign denotes a reversed lineshape or shift [10]. The smaller value of $\omega_Q^{(2)}$ for the ±1/2 <-> ±3/2 transitions of I=5/2 nuclei is exploited in the technique SATRAS, or satellite-transition spectroscopy, where the QCC and η are extracted from analysis of the sidebands [10]. The resonances in the ^{27}Al spectra of amorphous materials that contain 4, 5, and 6 coordinated aluminum atoms are often overlapping in the central band, but can sometimes be resolved in the satellites. The technique has, for example, been applied to study Al_2O_3 and B_2O_3 containing glasses, and has been reviewed recently in [4].

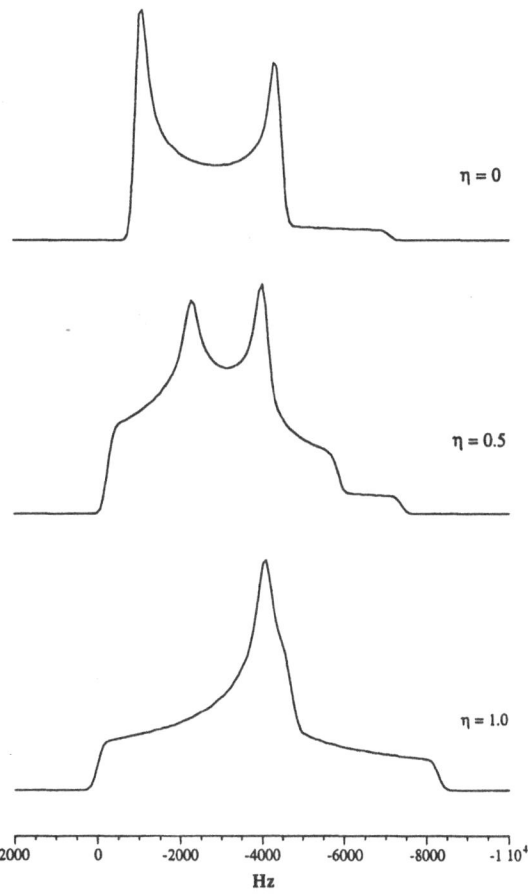

Figure 3. Simulated second-order quadrupolar lineshapes under MAS for I=5/2 nuclei. (QCC = 6.7 MHz, v_L = 93.818 MHz, δ_{CS} = 0).

The second-order broadening can sometimes be so large that the central transition resonances are no longer detected under conditions of MAS, or are only visible as broad humps in the baseline. These spins are often termed "invisible". It is often difficult to distinguish between these broad components and any baseline distortions that may be present. It is especially difficult when a range of QCCs is present, and no sharp discontinuities are visible in the second-order quadrupolar lineshapes; this is sometimes the case in the ^{27}Al NMR spectra of amorphous materials. Additionally, where narrower resonances are present, the sidebands from the narrower resonances may spread out over many ppm, and make quantification of the broad humps extremely difficult.

3. Removing the Second-Order Quadrupolar Broadening

There are now four major approaches to reducing or removing the second-order quadrupolar broadening. The most straightforward approach is to work at as high fields as possible, since the broadening is inversely proportional to ω_L. Two other approaches involve mechanical averaging of the second-order broadening (Dynamic Angle Spinning (DAS) [11], and Double Rotation (DOR) [12]). The most recent technique [13] makes use of the different second-order quadrupolar broadenings of the single and triple or 5-quantum transitions to accomplish the averaging. The latter three techniques are only likely to be successful for "visible" spins.

3.1 DOR

Averaging of the second-order quadrupolar broadening can be achieved by simultaneously spinning at two angles [11]. These angles, β_1 and β_2, are chosen such that both the second- and fourth-order Legendre polynomials are reduced to zero:

$$P_2(\cos\beta_1) = 0 \quad \beta_1 = \text{arc } \cos(1/3^{1/2}) \text{ (the Magic Angle)}$$
$$P_4(\cos\beta_2) = 0 \quad \beta_2 \approx 30.55 \text{ or } 70.12^{\circ} \tag{12}$$

In practice this achieved with a small rotor containing the sample (the inner rotor) which spins inside another rotor (the outer rotor). The axis of rotation of the inner rotor is inclined at angle of 30.55° to the axis of rotation of the outer rotor. The outer rotor is then spun at the Magic Angle. The isotropic shift, δ, observed under conditions of DOR is a sum of the chemical shift, δ_{CS}, and the second-order quadrupolar shift, $\delta_{Qiso}^{(2)}$, defined in (11), (where $\delta_{Qiso}^{(2)} = \omega_{Qiso}^{(2)}/\omega_L$), and is therefore field dependent. $\delta_{Qiso}^{(2)}$ must be determined independently (e.g. from studies at different fields) in order to obtain δ_{CS}. The field dependence of $\delta_{Qiso}^{(2)}$ can be exploited to separate resonances with similar values for δ_{CS} but with different QCCs. The technique has been combined with cross-polarization (CP) [14], and the CP dynamics under DOR have been studied.

A major limitation to the technique remains the spinning speed of the outer rotor: speeds of not more than 1.2 kHz are typically achieved, and the spectra often contain many overlapping resonances. Rotor synchronization will, however, eliminate half the spinning sidebands [15]. The poor filling factor of the coil also results in long acquisition times for low sensitivity nuclei such as ^{17}O. Zwaniger et al. have recently shown the power of a combined DOR and MAS study [16]. They used ^{17}O DOR to resolve three different oxygen sites in a B_2O_3 glass, and were able to maintain stable spinning for up to 20 hours. The ^{17}O DOR studies were performed at two fields, and δ_{CS}, and $\delta_{Qiso}^{(2)}$ were extracted for each of the sites from the field dependence of the isotropic shifts. The values for QCC and η cannot be separately determined from the DOR spectra, since $\delta_{Qiso}^{(2)}$ is proportional to $(1 + \eta^2/3)^{1/2}e^2qQ/h$ (from (11)). Hence, values for η were obtained from simulations of the DOR and MAS spectra. The large number of overlapping sidebands and the poor signal to noise of the spectra prevented accurate determination of the intensities of the resonances from the DOR spectra, and the populations of the three sites were extracted from the simulated ^{17}O MAS spectra.

3.2 DAS

Consider the VAS spectra obtained for spinning about two different orientations to the static magnetic field $\beta_1 = 37.88°$, $\beta_2 = 79.19°$ shown in Figure 4. The second-order quadrupolar broadening is of equal magnitude for the two angles, but opposite in sign. The DAS experiment works by spinning separately about these two DAS complementary angles [11, 17]. A 2-dimensional experiment is performed (Figure 5), where the spins are returned to the z-direction to preserve the magnetization while the flip between the two angles is implemented. The evolution due to $\omega_Q^{(2)}$ is refocussed when $t_1 = t_2$, and an echo forms. Data acquisition is started at the echo maximum at t_2. A Fourier transform along t_1 yields the isotropic resonance, and quadrupolar second-order lineshapes are obtained after a Fourier transform along t_2. Other DAS complementary angles can be found by finding solutions to the two simultaneous equations:

$$P_2(\cos\beta_1) + kP_2(\cos\beta_2) = 0$$
$$P_4(\cos\beta_1) + kP_4(\cos\beta_2) = 0 \qquad (13)$$

For example, when k = 5, $\beta_1 = 0^\circ$, and $\beta_2 = 63.32^\circ$. Hence spinning first about $\beta_1 = 0^\circ$ and then about $\beta_2 = 63.32^\circ$ will produce an echo at $5t_1 = t_2$. CP/DAS experiments are carried out with these set of angles, the CP being performed when spinning at β_1 [18].

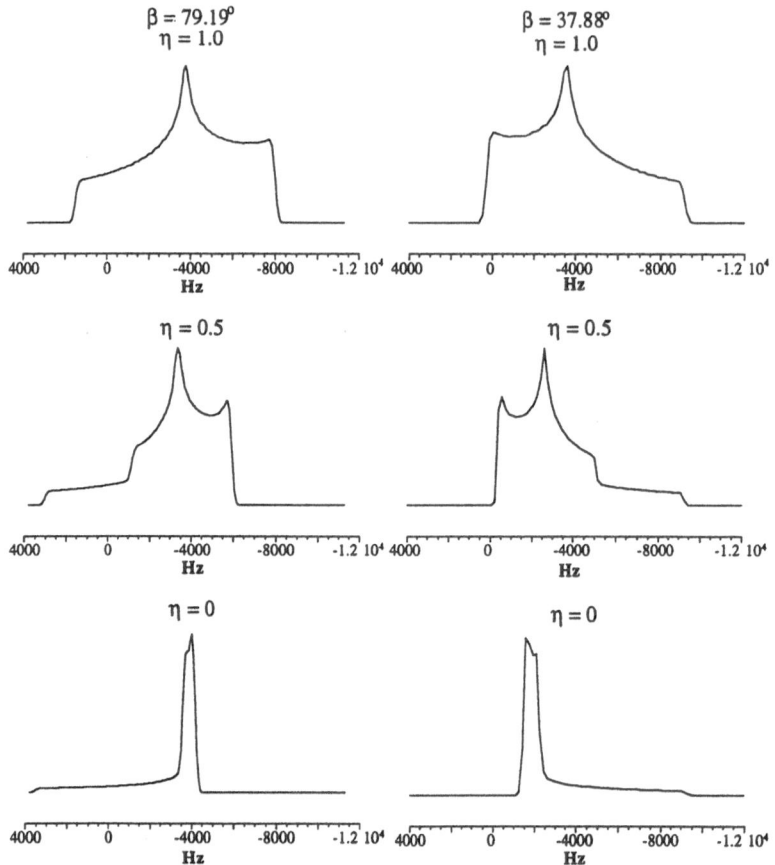

Figure 4. Simulations of the VAS spectra for spinning at one set of DAS complementary angles. (QCC = 6.7 MHz, ν_L = 93.818 MHz, $\delta_{CS} = 0$)

DAS is only effective for nuclei with sufficiently long spin-lattice relaxation times, so that significant magnetization is not lost during the time used to flip the rotor. In addition, DAS will not remove the homo-nuclear dipolar couplings. Spin exchange may occur during the flipping time which will result in broadening of the DAS spectra. Since this is a 2D experiment, experiment times are typically longer than for DOR. An additional $\pi/2$ pulse can be added so as to acquire pure-adsorption mode lineshapes with an improvement in the signal-to-noise (the shifted-echo approach) [19]. DAS has recently been combined with a HETCOR experiment to study $^{23}Na/^{31}P$ connectivities in glasses [20].

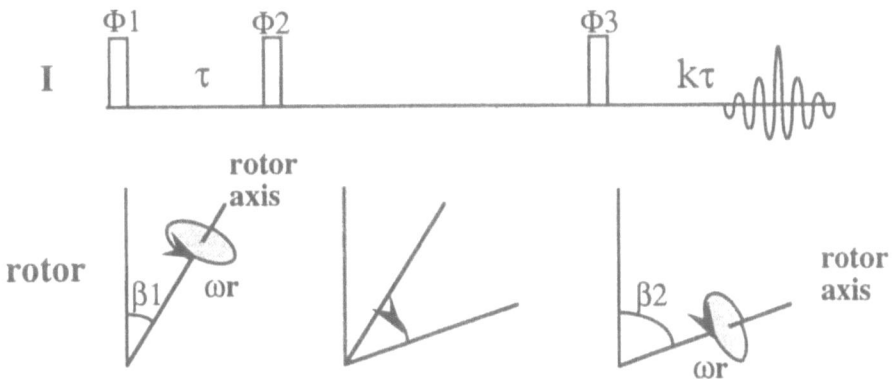

Figure 5. The DAS pulse sequence. The rotor flip from angle β_1 to β_2 is performed after the second pulse.

3.3. MULTIPLE QUANTUM NMR

In addition to the central transition, all odd-order multiple quantum (MQ) transitions of quadrupolar nuclei (i.e. 3Q, 5Q etc.) are unaffected by the first-order quadrupolar interaction. These MQ transitions are not directly observable, but can be observed indirectly if the MQ coherence is transferred back to the observable 1Q coherence [21]. The MQ transitions are affected to second-order by the quadrupolar interaction, $\omega_Q^{(2)}$, by an amount that depends on the order of coherence. Both the A_2 and A_4 terms defined in equation (10) depend on the order of coherence. For example:

$$A_{4(1)}/A_{4(3)} = -54/42 \qquad [14]$$

for I=3/2, where $A_{4(1)}$ and $A_{4(3)}$ are the A_4 terms for the 1Q and 3Q coherences, respectively. Recently Frydman et al. demonstrated that the dependence of $\omega_Q^{(2)}$ on coherence order could be exploited to average the second-order interaction [13].

The experiment is performed under conditions of MAS, so that the $A_2 P_2(\cos\beta)$ terms are averaged to zero. Averaging of the $A_4 P_4(\cos\beta)$ terms is then achieved by allowing the spins to evolve for different time periods in the single and multiple quantum coherences, t_{1Q} and t_{MQ}, respectively, such that the A_4 terms average. i.e. for I=3/2, $t_{1Q}/t_{3Q} = 42/54$. The experiment can be considered analogous to DAS, except that now the averaging is achieved by allowing the spins to evolve in two different MQ coherences; the experiment is also performed in a similar fashion. Initially the 3Q (or MQ) transition is excited and the spins are allowed to evolve for t_1. The 3Q coherence is then converted to the 1Q coherence (i.e. observable magnetization) and an echo forms at t_2. Echo formation occurs when the A_4 terms cancel. Performing the Fourier transform along the echo maximum, as a function of t_1, provides a spectrum free from second-order quadrupolar broadening. A Fourier transform performed in a direction perpendicular to the echo provides the second-order lineshape. A shearing transformation can be applied to the data, $S(t_1, t_2)$, that rotates the 2D FID, so that the isotropic and anisotropic spectra are observed in F1 and F2, respectively, after the Fourier transform.

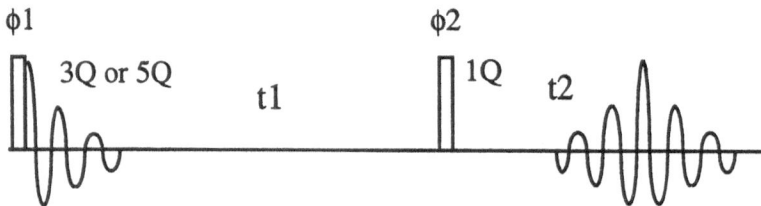

Figure 6. The MQ/1Q pulse sequence.

A single pulse, ϕ_1, is used to excite the MQ transition (Figure 6) [22-24], which results in a more efficient excitation of the MQ transition than the three pulses originally used [13]. Appropriate phase cycling of ϕ_1 and ϕ_2 is used to select the order of coherence. Time-proportional-phase-incrementation (TPPI) is used to acquire quadrature spectra in t_1 [23-24]. Alternatively, two data sets can be collected (where the real and imaginary components from t_1 are converted into the 1Q coherence in separate experiments); the data sets are then combined allowing a complex Fourier transform (FT) to be applied in both dimensions [22]. The shifted-echo approach has also been applied [22]. For moderate QCCs, 2D spectra have been acquired in relatively short times, whose second-order quadrupolar lineshapes are very close to the simulated ones, indicating a uniform excitation of the powder. A ^{23}Na 2D spectrum, acquired in only 50 minutes, is shown for a sample of Na_2HPO_4 (Figure 7). Three resonances can clearly be resolved in the isotropic dimension, for the three crystallographic sites, and the discontinuities in the second-order quadrupolar lineshapes can be discerned in the anisotropic dimension. Averaging of $\omega_Q^{(2)}$ with both the 3Q/1Q and 5Q/1Q experiments has been demonstrated for ^{27}Al [22-23].

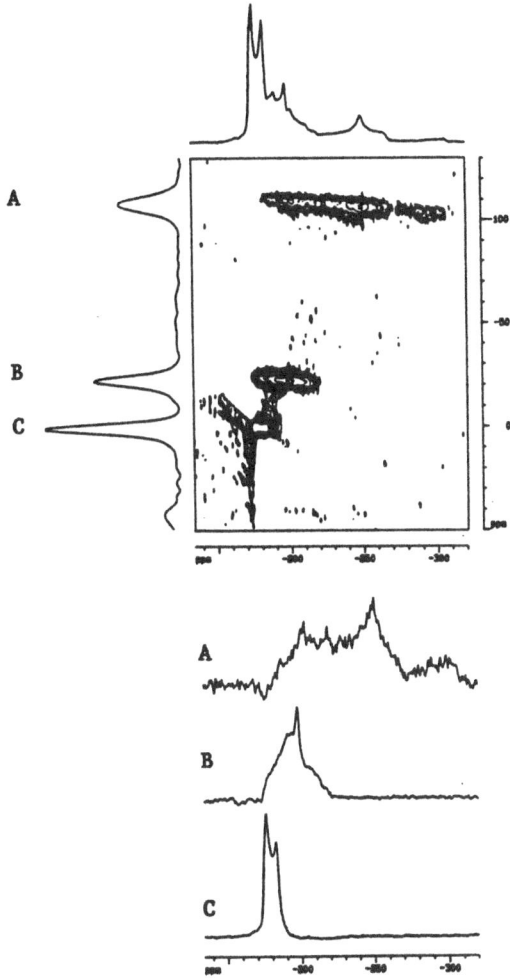

Figure 7. The ^{23}Na 3Q/1Q 2D spectrum of Na$_2$HPO$_4$. The F1 and F2 projections are shown; Slices through the three resonances in the F1 dimension (labeled A-C) are plotted below. A shearing transformation has been applied, prior to the 2D FT. (Data courtesy of S. Steuernagel, Bruker Analytische Messtechnik, GMBH).

The nutation frequency of the triple quantum excitation is inversely proportional to the quadrupolar splitting, Q', and is given by $2\omega_1^3/3Q'^2$. This has two consequences: firstly, the flip angle of the triple quantum excitation pulse will depend on Q', and will not be constant for the whole powder, resulting in non-uniform excitation of the sample. Secondly when $\omega_Q^{(2)}$ is much larger than $2\omega_1^3/3Q'^2$, the whole powder pattern will not be excited. Thus the method is expected to be less successful for large QCCs, especially at lower fields. However, given that the method is relatively straightforward to implement on a conventional MAS probe, and does not require any additional hardware, large numbers of applications for this exciting technique can be envisioned.

4. Acquiring Spectra

Two extreme cases can be distinguished, for single pulse excitation close to the Larmor frequency, which depend on the relative magnitudes of the QCC and ω_1. When $\omega_1 >> QCC$, the whole quadrupole spectrum is excited in a one pulse experiment. In contrast, when $QCC >> \omega_1$, only the central is affected by ω_1. The transition can then be treated as an isolated or so-called 'fictitious spin half' transition. Unfortunately, the nutation frequency of the spins (i.e. the frequency with which the spins precess around the applied rf field) is different for these two cases. The signal intensity for $I=3/2$ spins varies as $2\sin\omega_1 t$ and $\sin 2\omega_1 t$ for Q'=0 and $|Q'|>>\omega_1$, respectively, for a pulse of length t. The intensity of a resonance is thus dependent on Q' (Figure 8). In many cases, the QCC is of the same order as ω_1 (the intermediate regime). The orientation dependence of Q' gives a spread of nutation frequencies, which vary from ω_1 to $2\omega_1$ for $I=3/2$. The second-order quadrupolar lineshape is also distorted, since the powder is no longer uniformly excited. The situation is more complex under MAS since Q' is partially averaged during the pulse, and Q' appears smaller [25]. Thus the nutation frequency will also depend on the spinning speed. Since $\sin 2\theta \approx 2\sin\theta$ for small flip angles, θ, quantitative spectra and undistorted lineshapes can be obtained with short excitation pulses. This can be seen in Figure 8, where the signal intensity obtained for large QCCs and Q=0 can be seen to be very close if pulses of 15^0 or less are used. More generally , the nutation frequencies for all non-integer spins vary between $\sin(I+1/2)\omega_1 t$ and $(I+1/2)\sin\omega_1 t$ for $|Q'|<<\omega_1$ and $|Q'|>>\omega_1$, respectively, and quantitative spectra can similarly be obtained with short flip angles. The spread in nutation frequencies obtained in the intermediate regime can be used to determine the QCC and η; this is the basis of nutation spectroscopy [26].

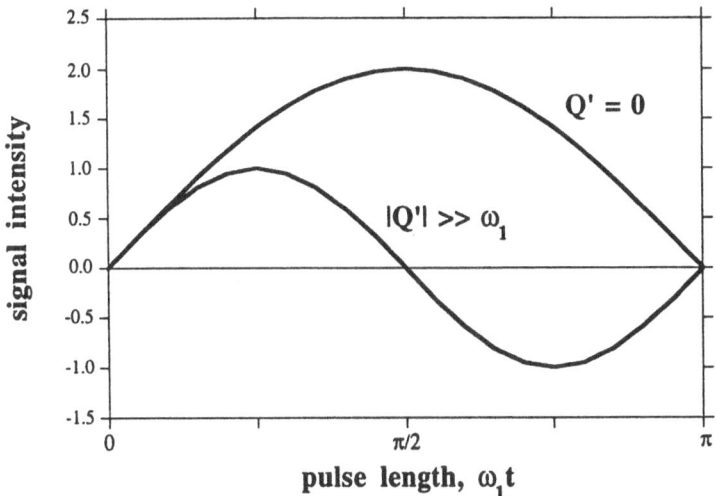

Figure 8. The signal intensity versus pulse length, t, for zero and a large quadrupolar spitting, Q'.

5. Spin Counting

Since it is sometimes difficult to observe all the spins in the sample, it is important to be able to estimate the number of spins that are actually observed. Without accurate spin counting, it is not possible to determine whether the species observed in the NMR spectra are representative of the whole sample, or whether they comprise a small subset of the spins that are present in the least distorted local environments. This is particularly important when obtaining ^{27}Al NMR spectra of high surface area materials, where a large fraction of the sample is often not detected. Careful spin counting has also been shown to be important in ^{23}Na NMR. For example, in the assignment of the [23]Na spectra of faujasite zeolites, the concentrations of the different extra-framework cation sites were shown to be very close to those obtained from diffraction data, after the intensities were scaled to account for the differences in the QCCs of the different sites [27].

The theoretical intensity of each of the transitions can be calculated from (6) and will depend on I. Those for the central transition are shown in Table 1. Unfortunately, the total number of spins that contribute to the center band will depend on the QCC. Three regimes can be identified:

(i) $\omega_Q = 0$

This will be the case, for example, in liquids, for nuclei in sites with cubic site symmetry, or for mobile species (e.g. hydrated cations on surfaces or in molecular sieves). In this case, all transitions will be excited, and the center band, in the absence of other anisotropic interactions, will contain intensity from 100% of the spins.

(ii) $0 < \omega_Q^2 < \omega_L \omega_r$

This is typically the case for ^{23}Na and ^{7}Li. Spinning sidebands from the satellite transitions are now visible, and the contribution of these satellites to the center band needs to be estimated. The spinning speed is greater than the static linewidth of the central transition (since $\omega_Q^2/\omega_L\omega_r < 1$), and the intensity of the central transition in the sidebands is very small and can often be ignored. Note that other interactions, such as the CSA or dipolar coupling, may result in intensity in the sidebands. Fast spinning will clearly increase the intensity of the central transition in the center band, but will also increase the contribution of the satellites to the center band. Both contributions can be conveniently estimated with the graphical method proposed by Massiot et al. (see below) [28]. Alternatively, the sideband intensities from the satellite transitions are often fairly constant close to the center band, allowing the contribution to the center band to be estimated from the intensity of other nearby sidebands. Having obtained the absolute intensity of the central transition resonance, the total number of spins in the sample can then be calculated from Table 1.

Table 1. The intensity of central transition as a percentage of the total intensity.

Spin	Intensity (%)
3/2	40
5/2	26
7/2	19
9/2	15

(iii) $\omega_r \ll \omega_Q$

The contribution of the outer satellites to the center band can be considered negligible when ω_Q is two or more orders of magnitude greater than ω_r. This is often the case for ^{27}Al NMR. In many cases the outer satellites are not even observed due to poor setting of the magic angle, insufficient signal-to-noise, or ionic motion of the nucleus (e.g. in some glasses). A significant concentration of the central transition, however, may now be present in the sidebands. This may be accounted for by integration of the whole resonance including sidebands (if interference from the outer satellites is not a problem). In cases where outer satellites are observed, it is often easier to use the graphical method of Massiot et al..

Massiot et al. [28] have plotted two graphs, the first showing the contribution of the satellite transitions to the intensity of the center band, as a function of $|1+m|\omega_Q/\omega_r$, for a transition from m to (m+1). The second graph shows the contribution of the central transition to the center band, as a function of $\omega_Q^2/\omega_L\omega_r$, for different values of η. Clearly this method requires that ω_Q be known. Often an estimate for ω_Q is sufficient, which can be obtained from the size of the second-order quadrupole broadening, or the total width of the satellite transition manifolds (for small ω_Q). This paper demonstrated the importance of working with as high MAS speeds as possible, for large QCCs, to concentrate the central transition intensity in the isotropic resonance.

Typically, spin counting is performed by comparing the intensity of the center band in a spectrum of a sample of known weight, with the intensity from a standard sample acquired under identical conditions. Often a liquid is used as the standard. In this case, $\omega_Q = 0$, and all the transitions are observed. The intensity of the standard needs to be appropriately scaled, so as to compare it to the solid where generally only the intensity of the central transition is obtained. Solid aluminum acetylacetonate (see Figure 2) can be used as an ^{27}Al standard, since, unlike many of the aluminum salts sometimes used for spin counting (e.g. aluminum sulfate), aluminum acetylacetonate is not hydroscopic, and the number of moles present in the sample can be determined accurately. The second-order quadrupolar lineshape of the central transition of aluminium acetylacetonate can be fitted giving values for QCC and η of 2.9 (±0.1) MHz and 0.12 (±0.02). All spectra should be acquired with short flip angles for reasons described earlier.

6. Cross-Polarization of Non-Integer Spin Nuclei

Cross-polarization from an abundant I=1/2 spin such as ^{1}H or ^{19}F to a quadrupolar nucleus, S, is accomplished by matching the nutation frequencies of the two spins I and S (the Hartmann-Hahn condition). As discussed earlier, the nutation frequency of non-integer spins depends on Q', and two different Hartmann-Hahn conditions apply for large and small values of Q':

$$(i) \ \omega_{1S} = \omega_{1I} \qquad\qquad (|Q'| \ll \omega_{1S})$$
$$(ii) \ (S+1/2) \ \omega_{1S} = \omega_{1I} \qquad (|Q'| \gg \omega_{1S}) \qquad\qquad (15)$$

As for single pulse excitation, case (i) applies when the whole quadrupole spectrum is excited, while case (ii) applies when the isolated fictitious spin-1/2 system is excited. In the intermediate regime, $|Q'| \approx \omega_{1S}$, and a range of nutation frequencies is observed; the Hartmann-Hahn condition is, therefore, poorly defined in this regime.

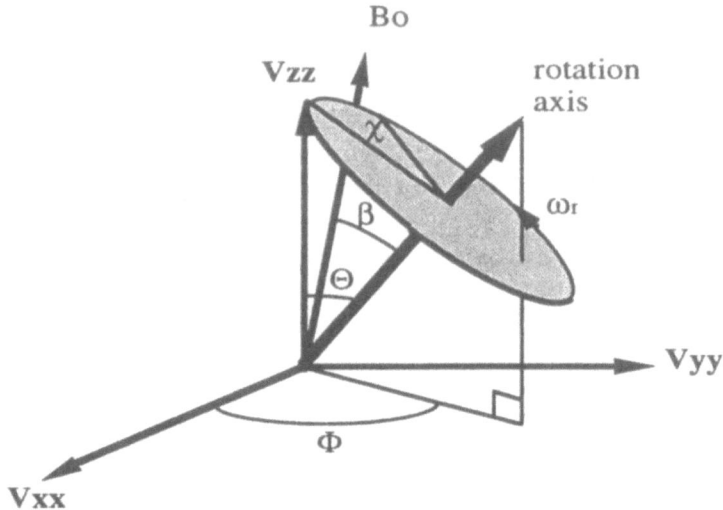

Figure 9. The angles that define the orientation of the quadrupole tensor (V_{XX}, V_{YY}, V_{ZZ}) and the static magnetic field, B_0, with respect to the rotor axis. χ is 0° when the rotation axis, B_0 and V_{ZZ} all lie in the same plane.

Additional complications arise under conditions of MAS, which may also result in inefficient CP, as was shown by Vega in [29-30]. These arise from the time dependence in Q' introduced by the sample spinning. $Q'(t)$ depends on the orientation of the quadrupolar tensor (V_{XX}, V_{YY}, V_{ZZ}) with respect to the static magnetic field (defined by the polar angles θ and ϕ). This orientation varies continuously under MAS and $Q'(t)$ oscillates between positive and negative values. In order to demonstrate this, it is convenient to rewrite Q' in terms the new angles Θ, Φ and χ, which describe the orientation of the quadrupole tensor with respect to the rotor axis (Figure 9). Only χ is time-dependent under MAS and is given by $(\omega_r t + \chi_0)$, where χ_0 is the initial value for χ at $t = 0$. Q' is then given by:

$$Q' = (\omega_Q/2)[2^{1/2}\sin2\Theta\cos(\chi_0 + \omega_r t) + \sin^2\Theta\cos2(\chi_0 + \omega_r t)] \quad (16)$$

where an axial quadrupole tensor has been assumed, and hence Φ is undefined.

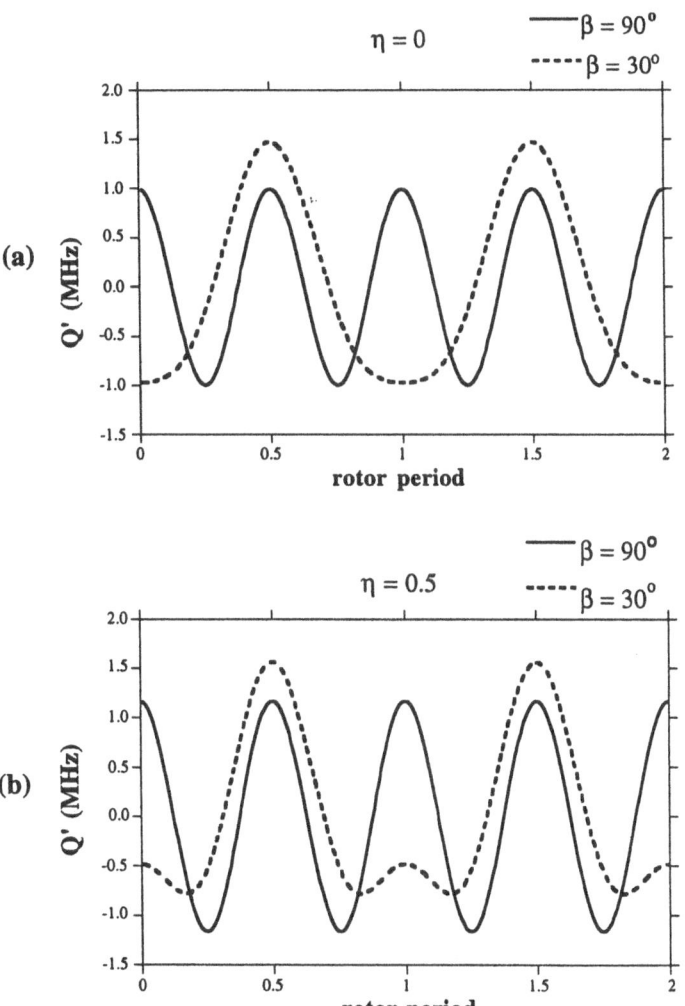

Figure 10. Plots of the variation of Q' as a function of time in the rotor period, for $\omega_Q = 1$ MHz, and different values of Θ; $\eta = 0$ and 0.5, in (a) and (b) respectively. $\chi_0 = 0^0$ in (a) and (b), and $\Phi = 0$ in (b).

Q' crosses through zero either two, or four times per rotor period (the zero-crossings), depending on the value of Θ. This can be seen in Figure 10 where Q' is plotted for $\Theta = 90$ and $30°$ as a function of the time in the rotor period, for $\eta = 0$ and 0.5. Note that the value of Q' is averaged to zero over the whole rotor period, since spinning is performed at the Magic Angle.

Continuous on- or close-to-resonance irradiation is applied to the quadrupolar nucleus during the contact time of the cross-polarization experiment (the spin-locking field). The outer satellite transitions for large QCCs (i.e. case (ii)) are unaffected by the rf field, ω_1, in the absence of MAS, for close-to-resonance irradiation, and magnetization build-up occurs along the direction of the spin-locking field for the central transition coherence only. Under MAS, however, zero crossings in Q' occur. At these times during the rotor period, ω_1 is greater than $|Q'|$, and the rf field induces transitions between *all* $|m\rangle$ Zeeman levels. A clearer understanding of the effect of MAS can be obtained by considering Figure 11, which shows a plot of the eigenvalues and eigenstates of an I=3/2 nucleus as a function of Q', for close-to-resonance irradiation. For large values of $|Q'|$ the eigenstates are given by $\{|1/2\rangle \pm |-1/2\rangle\}$ and $|\pm 3/2\rangle$, and spins present in these states are said to be spin-locked. Spins present in the $|+1/2\rangle$ and $|-1/2\rangle$ states are not spin-locked, and will start to nutate around the spin-locking field, oscillating between the $|1/2\rangle$ and $|-1/2\rangle$ states. Any magnetization present in these states will decay rapidly.

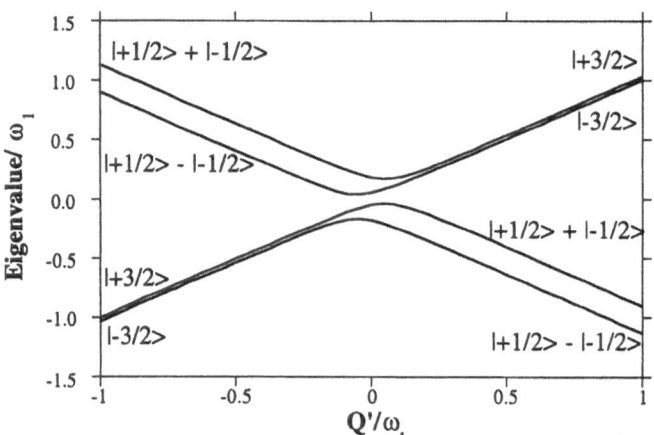

Figure 11. The eigenvalues for an I=3/2 nucleus as a function of Q', for close to on-resonance irradiation. The eigenstates for large Q' are marked.

Under MAS, Q' varies continuously, and a sweep from -Q' to +Q', for example, results in a smooth conversion of the central transition coherences

{|1/2> ± |-1/2>} to the outer Zeeman levels |±3/2>. If this sweep is performed sufficiently slowly (i.e. is adiabatic), all the spins that populate the central transition coherences {|1/2> ± |-1/2>} are transferred into the |±3/2> states. Similarly, the populations in the |±3/2> states are transferred into the central transition coherences. In contrast, a very fast sweep will leave the populations unchanged in their original states. A sweep performed at some intermediate rate will result in the transfer of populations into non-spin locked coherences. These non-spin locked coherences oscillate rapidly under the influence of the rf and quadrupolar field (i.e. they do not commute with the Hamiltonian), and any magnetization associated with these coherences decays rapidly. An adiabaticity parameter for the zero crossing, α', can be defined for a powder [29], which gives a measure of the efficiency of the population transfers:

$$\alpha' = \omega_1^2/\omega_r\omega_Q \qquad (17)$$

Fast passages, intermediate and adiabatic passages occur for $\alpha' \ll 1$, $\alpha' \approx 0.4$, and $\alpha' > 1$, respectively.

Magnetization builds up in one of the spin-locked coherences {|1/2> ± |-1/2>}, during the spin-locking period of the CP experiment, assuming the Hartmann-Hahn condition is adequately matched. Under MAS, however, the magnetization will not necessarily remain in this coherence. Very slow MAS (i.e. adiabatic passages) results in the transfer of the magnetization into the |±3/2> states at the zero-crossings for Q'. At the next zero-crossing, however, all the magnetization returns to the {|1/2> ± |-1/2>} and no magnetization is lost. This will be the case at the end of a rotor period, where an even number of crossings will have occurred. CP for values of α' in the intermediate regime will result in a rapid decay of the spin-locked magnetization, and inefficient or no CP. Fast MAS will leave the magnetization associated with the {|1/2> ± |-1/2>} coherence unaffected, and CP will again be efficient. The effect of MAS on the spin-locked magnetization has been demonstrated experimentally by Vega [29].

In conclusion, obtaining CP spectra of non-integer spin nuclei is not necessarily straightforward, even if the Hartmann-Hahn condition is matched. As a result, the inability to transfer magnetization from I=1/2 to quadrupolar S spins does not necessarily indicate that the S spins are not dipolar-coupled to the I spins, but the converse (i.e. the detection of S-spin magnetization) can be used to demonstrate the close proximity of S and I spins. It is, however, relatively straightforward to calculate α', and to determine the conditions required for efficient CP. The passages between Zeeman levels have also been shown to be important in the CP/DOR and

CP/DAS experiments, [14, 18] and double resonance experiments such as TRAPDOR and REAPDOR (see below) [31-33]. Adiabatic sweeps can be used to enhance the intensity of the central transition resonance, by transferring populations from the outer to the $|\pm1/2>$ Zeeman levels [34]. For a fuller description of the effect of adiabatic passages on spin-locking and CP, the reader is referred to the two papers by Vega [29-30].

7. Methods for Observing "Invisible" Spins

7.1 STEPPED ECHO METHOD

Extremely broad central transition resonances can be observed with a spin-echo even when these resonances are "invisible" in the MAS spectra, and spectra free from baseline distortions can be obtained. Ernst et al. were able to detect the ^{27}Al central transition of various dehydrated H-zeolites under non-spinning conditions, and to extract values of, for example, 2.7 MHz and 0.7 for the QCC and η, respectively, for zeolite HY [35]. Spinning will not result in increased resolution for these very large QCCs. The method may be extended to study even larger QCCs by varying the carrier frequency in the static experiment, and mapping out the echo intensity as a function of the offset-frequency. Stochastic NMR methods have recently been proposed as an alternative method for obtaining uniform excitation of the quadrupole spectrum in a single experiment [36]. A string of very short weak pulses is applied, a point from the FID being acquired in between each pulse. The sweep width of the spectrum is determined by the interval between the pulses. The use of weaker pulses results in shorter recovery times of the receiver and less distorted baselines. This method can be combined with MAS, and it may be possible, for example, to use this technique to determine whether there are any extremely broad components present under narrower components of the spectrum. The experiment is relatively simple to implement, but requires software to unscramble the effect of the pulse train of semi-random phases. A slight modification of the preamplifier circuit of the NMR spectrometer is also required. A typical solid state NMR spectrometer contains crossed-diodes at the end of the cable from the high power amplifier (the transmission line), and before the preamplifier circuit. These allow the high power pulse (typically 50-1000W) through and into the probe. The NMR signal that returns from the probe cannot pass through the crossed diodes, but passes through the appropriately chosen $\lambda/4$ cable and another crossed-diode (grounded at one end) into the preamplifier circuit. Since much lower power pulses are used in the stochastic NMR method, they are too weak to pass through the crossed-diodes at the end of the transmission line. The crossed diodes should be replaced by a switch that connects the probe to the transmitter (during the pulses) or the probe to the preamplifier (during acquisition of the data points).

7.2 TRAPDOR NMR

The $^1H/^{27}Al$ TRAPDOR (transfer of populations in double resonance NMR) NMR experiment can be used to detect the "invisible" spins if there is a suitable visible nucleus (typically I=1/2) in close proximity. [32] In this experiment, the effect of continuous irradiation of a quadrupolar nucleus, S, is monitored via the dipolar coupling to the nearby nucleus, I. For example, in $^1H/^{27}Al$ TRAPDOR NMR (Figure 12), ^{27}Al irradiation is applied during the evolution period of the ^1H spin echo, and the loss of intensity at the ^1H echo is detected.

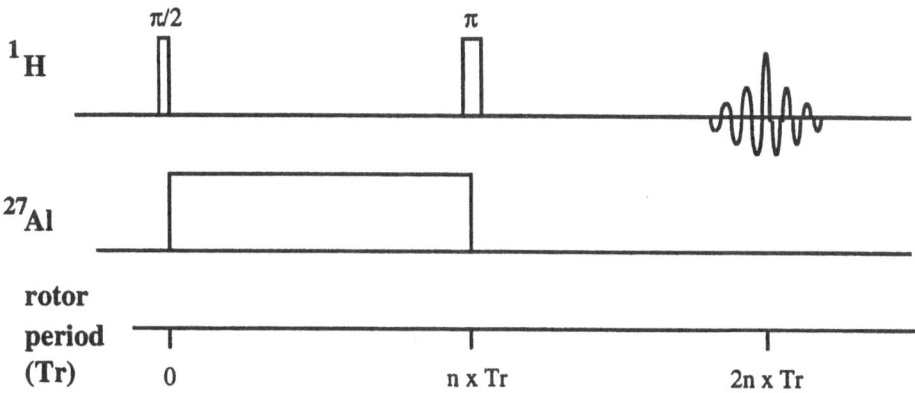

Figure 12. The $^1H/^{27}Al$ TRAPDOR NMR pulse sequence. The intensities (at the echo) are determined with (I) and without (I_0) ^{27}Al irradiation.

The $^1H/^{27}Al$ TRAPDOR experiment makes use of the passages that occur between the ^{27}Al Zeeman levels under conditions of slow MAS and continuous irradiation. These were discussed in detail for the CP experiment. The population transfers prevent the refocusing of the dipolar coupled I=1/2 spins at the rotor echo, causing a TRAPDOR effect. The experiment is similar to the Rotational Echo Double Resonance (REDOR) NMR experiment [37], except that in REDOR the refocusing of the dipolar coupled spins is prevented by applying π pulses to the S spins. The TRAPDOR fraction, defined as ($1-I/I_0$), will depend on the dipolar coupling between spins: the greater the dipolar coupling, the greater the dephasing of the I (^{1}H) spins, and thus the greater the TRAPDOR effect. Slower spinning ensures that the ^{27}Al passages are closer to adiabatic, resulting in more efficient population transfers and a larger TRAPDOR fraction. Values for $\alpha' > 1$ (where α' is defined in (17)) ensure that most of the ^{27}Al passages that occur for the whole powder sample are adiabatic, but even values for α' as low as 0.27 have been shown to give significant TRAPDOR dephasing [32]. A TRAPDOR effect can only be determined if the ^{27}Al irradiation lies within the ^{27}Al first-order quadrupole spectrum, and the largest TRAPDOR fraction is obtained for on-resonance ^{27}Al irradiation [32]. The size of the QCC for the quadrupolar nucleus, S, can be estimated by mapping out the intensity of the I echo, as a function of the S irradiation frequency offset, and determining where the TRAPDOR fraction drops to zero. For I=5/2 nuclei, the edge of the first-order quadrupole spectrum occurs at $\pm 2\nu_Q = (3/10)$ QCC. A semi-quantitative theoretical treatment of the experiment, which allows for the numerical prediction of TRAPDOR signal reduction, has been developed [32].

^{1}H/^{27}Al TRAPDOR NMR has recently been used to determine a value for the QCC of the ^{27}Al invisible spins in dehydrated zeolite HY of 15.3 MHz [32]. The Lewis acid sites in dehydroxylated HY can be observed indirectly via the ^{15}N NMR of the adsorbed base monomethylamine (MMA). ^{15}N/^{27}Al TRAPDOR NMR has demonstrated that MMA is coordinated directly to an aluminum atom, and has been used to determine a QCC of 8.5 MHz for this Lewis acid-MMA complex [38].

8. Double Resonance Experiments

Many new double resonance experiments for determining the proximity between spins have been used, in recent years, to study non-integer spins. Experiments such as REDOR [37], TEDOR [39] (transferred echo double resonance) and SEDOR [40] (spin echo double resonance) NMR, were originally developed for determining the dipolar coupling, and hence internuclear distances, between I=1/2 nuclei. These experiments can be applied to study coupling to non-integer spin nuclei, by detecting the central transition, and applying pulses to the dipolar-coupled I=1/2 nucleus. For example, SEDOR has been used to determine ^{27}Al-^{31}P distances in aluminophosphates molecular sieves [41]. REDOR and TEDOR have been applied to study Al/Si connectivities in zeolites [42], and REDOR has been used to study the structure of MMA bound to zeolite HY [43]. INEPT experiments, which expoit the J-coupling between heteronuclei, have also been used to probe Al/P and Si/Na connectivities [44].

TRAPDOR experiment (discussed above) can be applied to probe internuclear distances in two situations where the REDOR NMR experiment may prove difficult. Firstly, the TRAPDOR experiment can detect dipolar coupling to 'invisible' spins, and secondly, the TRAPDOR experiment can be used to study the structures of molecules adsorbed on, for example, catalyst surfaces, where it may be difficult to distinguish between the resonances from the surface and the bulk of the catalyst. In this case, a REDOR experiment designed to measure coupling between the surface species and the molecules, where π pulses are applied to the nuclei in the adsorbed molecules, will not show a very significant REDOR fraction. The dipolar coupling measured in the TRAPDOR experiment is larger than that measured in the REDOR experiment, because the TRAPDOR experiment probes the coupling to spins in all the Zeeman levels of the quadrupolar nucleus. Hence, the TRAPDOR experiment may be more sensitive to longer range dipolar couplings.

Finally, another double resonance experiment that is likely to prove extremely useful is REAPDOR (Rotational Echo and Adiabatic Passages Double Resonance) NMR [33]. This experiment, as the name implies, is a combination of the REDOR and TRAPDOR experiment. A string of $(16n+2)$ π pulses is applied (using the XY-8 pulse sequence) to the $I=1/2$ nucleus for $(8n+2)$ rotor periods. The π pulse at the end of $(4n+1)$ rotor periods is omitted, and instead, irradiation is applied to the S spin for a fraction of the rotor period (typically 1/3) [45]. The loss of I spin intensity at the echo (i.e. after $(8n+2)$ rotor periods) is determined with and without S spin irradiation, to determine the REAPDOR fraction. This experiment does not require multiple rotor periods of S-spin irradiation to detect weak I-S coupling (c.f. TRAPDOR), and consequently the REAPDOR fractions are simpler to calculate. Hence this experiment may result in a method for measuring more accurate internuclear distances than can be obtained from the TRAPDOR experiment. The method has been exploited to determine $^{13}C/^{17}O$ proximities in small peptides [46], and to study $^{29}Si/^{27}Al$ connectivities in zeolites [45]. The experiment could also be applied to determine the QCCs of the quadrupolar nuclei.

Acknowledgments

S. Steuernagel is thanked for allowing the author to publish his ^{23}Na 1Q/3Q spectra of Na$_2$HPO$_4$, and for advice on the MQ experiments. A.J. Vega is thanked for many stimulating discussions, and for donating the computer programs used to simulate the second-order quadrupolar lineshapes. The unpublished work discussed in this paper was performed under the support of the NSF (DMR 9458017). Acknowledgment is made to the Donors of the Petroleum Research Fund, administered by the American Chemical Society, for partial support of this research.

General Reference

C.A. Fyfe in 'Solid State Chemistry: Techniques (ed. A.K. Cheetham and P. Day), Oxford, 1987

References

1. Freude, D. and Haase, J. (1993) Quadrupole effects in solid-state nuclear magnetic resonance, in B. Blumich (ed.), *NMR Basic Principles and Progress*, Springer-Verlag, Berlin, **29**, 1.
2. Vega, A.J. (1995) Quadrupolar Nuclei in Solids, in D.M. Grant and R.K. Harris (eds.), *Encyclopedia of NMR*, to be published, and other related articles in this encyclopedia.
3. See for example: Jakobsen, H.J., Skibstedt, J., Bildsoe, H., and Nielsen, N.C. (1989) Magic-angle spinning NMR spectra of satellite transitions for quadrupolar nuclei in solids, *J. Magn. Reson.*, **85** 173. Skibstedt, J., Nielsen, N.C., Bildsoe, H., and Jakobsen, H.J., (1991) Satellite transitions in MAS NMR spectra of quadrupolar nuclei, *J. Magn. Reson.*, **95**, 88.
4. Jäger, C. (1994) Satellite transition spectroscopy of quadrupolar nuclei, in B. Blumich (ed.), *NMR Basic Principles and Progress*, Springer-Verlag, Berlin, **31**, 133.
5. Hoch, J.C. (1989) Modern spectrum analysis in nuclear magnetic resonance: alternatives to the Fourier transform, *Methods in Enzymology*, Academic Press, Inc., **176**, 216.
6. See for example: Gonzalez-Tovany, L., and Betrán-López (1990) Second-order powder pattern and simulated spectra of EPR transitions in orthorhombic symmetry or NMR with quadrupole interactions, *J. Magn. Reson.*, **89**, 227.
7. Narita, K., Umeda, J. and Kusumoto, H. (1966), Nuclear magnetic resonance powder patterns of the second-order nuclear quadrupole interactions in solids with asymmetric field gradients, *J. Chem. Phys.*, **44**. 2719.

8. Müller, D., (1982) Determination of chemical shifts of NMR frequencies of quadrupolar nuclei from the MAS-NMR spectra, *Ann. Phys. (Leipzig)* **39**, 451.

9. Ganapathy, Schramm, S., and Oldfield, E. (1982) Variable-angle sample-spinning high resolution NMR of solids, *J. Chem. Phys.*, **77**, 4360.

10. Samosen, A. (1985) Satellite transition high-resolution NMR of solids, *Chem. Phys. Lett.*, **119**, 29.

11. Mueller, K., Sun, B.Q., Chingas, G.C., Zwanziger, J.W., Terao, T., and Pines, A. (1990) Dynamic-angle spinning of quadrupolar nuclei, *J. Magn. Reson.*, **86**, 470.

12. Samoson, A., Lippmaa, E., and Pines, A. (1988) High resolution solid-state NMR averaging of second-order effects by means of a double-rotor, *Mol. Phys.*, **65**, 1013.

13. Frydman, L., and Harwood, J.S., (1995) Isotropic spectra of half-integer quadrupolar spins from bidimensional magic-angle spinning NMR, *J. Am. Chem. Soc.* **117**, 5367.

14. Wu, Y., Lewis, D., Frye, J.S., Palmer, A.R., and Wind, R.A., (1992) Cross-polarization double-rotation NMR, *J. Magn. Reson.*, **100**, 425.

15. Samoson, A., and Lippmaa, E. (1989) Synchronized double-rotation NMR spectroscopy, *J. Magn. Reson.* **84**, 410.

16. Youngman, R.E., Haubrich, S.T., Zwanziger, J.W., Janicke, M.T., and Chmelka, B.F., (1995) Short- and intermediate-range structural ordering in glassy boron oxide, *Science*, **269**, 141.

17. Llor, A. and Virlet, J. (1988) Towards high-resolution NMR of more nuclei in solids: sample spinning with time-dependent spinner axis angle, *Chem. Phys. Lett.*, **152**, 248.

18. Baltisberger, J.H., Gann, S.L., Grandinetti, P.J., and Pines, A. (1994) Cross-polarization dynamic-angle spinning nuclear magnetic resonance of quadrupolar nuclei, *Mol. Phys.*, **81**, 1109.

19. Grandinetti, P.J., Baltisberger, J.H., Llor, A., Lee, Y.K., Werner, U., Eastman, M.A. and Pines, A. (1993) Pure-adsorption-mode lineshapes and sensitivity in two-dimensional dynamic-angle spinning NMR, *J. Magn. Reson.*, **103**, 72.

20. Jarvie, T.P., Menslow, R.M., Mueller, K.T. (1995) High-resolution solid-state heteronuclear correlation NMR for quadrupolar and spin-1/2 nuclei, *J. Am. Chem. Soc.*, **117**, 570.

21. Vega, S., and Naor, Y. (1981) Triple quantum NMR on spin systems with I=3/2 in solids, *J. Chem. Phys.*, **75**, 75.

22. Massiot, D., Touzo, B., Trumeau, D., Coutures, J.P., Virlet, J., Florian, P., and Grandinetti, P.J. (1995) Two-dimensional Magic-Angle Spinning

Isotropic Reconstruction Sequences for Quadrupolar Nuclei, *Solid-State NMR*, in press.

23. Fernandez, C., and Amoureux, J.P. (1995) Triple-quantum MAS-NMR of quadrupolar nuclei, *Solid State NMR*, in press.

24. Fernandez, C., and Amoureux, J.P. (1995) 2D multiquantum MAS-NMR spectroscopy of ^{27}Al in aluminophosphate molecular sieves, *Chem. Phys. Lett.*, **242**, 449.

25. Nielsen, N.C., Bildsoe, H., and Jakobsen, H.J., (1992) Finite rf pulse excitation in MAS NMR of quadrupolar nuclei. Quantitative aspects and multiple-quantum excitation, *Chem. Phys. Lett.*, **191**, 205.

26. Samoson A., and Lippmaa, E. (1983) Central transition NMR excitation spectra of half-integer quadrupole nuclei, *Chem. Phys. Lett.*, **100**, 205.

27. Engelhardt, G., Koller, H., and Weitkamp, J. (1994), Characterization of sodium cations in dehydrated faujasites and zeolite EMT by ^{23}Na DOR, 2D nutation and MAS NMR, *Solid State NMR*, **2**, 111.

28. Massiot, D., Bessada, C., Coutures, J.P., and Taulelle, F. (1990) A quantitative study of ^{27}Al MAS NMR in crystalline YAG, *J. Magn. Reson.*, **90**, 231

29. Vega, A.J. (1992) MAS NMR spin locking of half-integer quadrupolar nuclei, *J. Magn . Reson.* **96**, 50.

30. Vega, A.J. (1992) CP/MAS of quadrupolar S=3/2 nuclei, *Solid-State NMR*, **1**, 17.

31. Grey, C.P., Vega, A.J., and Veeman, W.S. (1993) ^{14}N/^{13}C/^{1}H triple resonance solid state NMR; a probe of ^{13}C-^{14}N internuclear distances, *J. Chem. Phys.*, **98**, 7711.

32. Grey, C. P., and Vega, A.J. (1995) The determination of the quadrupole coupling constant of the invisible aluminum spins in zeolite HY with ^{1}H/^{27}Al TRAPDOR NMR, *J. Am. Chem. Soc.*, **117**, 8232.

33. Gullion, T. (1995) Measurement of dipolar interactions between spin-1/2 and quadrupolar nuclei by rotational-echo, adiabatic-passage, double-resonance NMR, *Chem. Phys. Lett.* in press.

34. Haase, J., and Conradi, M.S. (1993), Sensitivity enhancement for NMR of the central transition of quadrupolar nuclei, *Chem. Phys. Lett.* **209**, 287. Haase, J., Conradi, M.S., Grey C.P., and Vega, A.J., (1994) Population Transfers for NMR of Quadrupolar Spins in Solids, *J. Magn. Reson A.*, **109**, 90.

35. Ernst, H., Freude, D., and Wolf, I. (1993) Multinuclear solid-state NMR studies of Brønsted sites in zeolites, *Chem. Phys. Lett.*, **212**, 588.

36. Chew, B.G.M., Liao, M.-Y., and Zax, D.B. (1995) Solid-state NMR at cryogenic temperatures using stochastic excitation, *J. Magn. Reson. A*, **116**, 277.

37. Gullion, T., and Schaefer, J. (1989) Rotational-echo double-resonance NMR, *J. Magn. Reson.*, **81**, 196.

38. Kao, H.-M., and Grey, C.P. (1995) Probing the Brønsted and Lewis acidity of zeolite HY: a $^1H/^{27}Al$ and $^{15}N/^{27}Al$ TRAPDOR NMR study of monomethylamine adsorbed on HY, submitted.

39. Hing, A.W., Vega, S., and Schaefer, J (1992) Transferred-echo double-resonance NMR, *J. Magn. Reson.*, **96**, 205.

40. Herzog B., and Hahn, E.L. (1956) *Phys. Rev.*, **103**, 148.

41. van Eck, E.R.H., and Veeman, W.S. (1992) The determination of the average ^{27}Al-^{31}P distance in aluminophosphate molecular sieves with SEDOR NMR, *Solid St. NMR*, **1**, 1.

42. Fyfe, C.A., Wong-Moon, K.C., Huang, Y., Grondey, H., and Mueller, K.T. (1995) Dipolar-based $^{27}Al/^{29}Si$ solid-state NMR connectivity experiments in zeolite molecular sieve frameworks, *J. Phys. Chem.*, **99**, 8707.

43. Grey, C.P., and Kumar, B.S.A. (1995) A $^{15}N^{27}Al$ Double Resonance Study of Monomethylamine Adsorbed on Zeolite HY, *J. Am. Chem. Soc.*, **117**, 907.

44. Fyfe, C.A., Wong-Moon, K.C., Huang, Y., and Grondey, H. (1995) INEPT experiments in Solid-State NMR, *J. Am. Chem. Soc.*, **117**, 10397.

45. Grey, C.P., Ba, Y., and Gullion, T. (1995) unpublished results.

46. Gullion, T. (1995) Detecting ^{13}C-^{17}O dipolar interations by rotational-echo, adiabatic-passage, double-resonance NMR, *J. Magn. Reson.*, submitted.

COMPUTER MODELLING AS A TECHNIQUE IN MATERIALS CHEMISTRY

C.R.A. CATLOW

The Royal Institution of Great Britain, 21 Albemarle Street, London W1X 4BS,UK

1. Introduction

Atomistic computer modelling methods are now established tools in contemporary science[1]. They are used routinely in the study of proteins and pharmaceuticals and in the comformational analysis of organic molecules. Computational methodologies have, however, a growing and diverse rôle in the field of materials chemistry, especially in complex systems such as microporous catalysts and superconducting oxides which are the focus of much attention in this book.

In this chapter we will describe the current status of the methodologies and applications of modelling techniques in the field of solid state inorganic chemistry. The field has been extensively reviewed by the present author and others in recent years[2-6]. There is also a separate account in Chapter 8 of electronic structure techniques. In this chapter, we will give greater attention to interatomic potential methods and emphasise the rôle of modelling techniques as an ancillary tool to be used in conjunction with related experiment.

2. Methodologies

We will summarise first the classes of modelling techniques that are used in simulating materials. Firstly, we have the simple empirical methods based on observed bond lengths. Secondly, calculations of purely electrostatic energies have proved of value in the investigation of ionic and semi-ionic solids. The third, and by far the largest class of methods, is based on the use of *interatomic potentials* including both short and long range terms; these have an enormous current range of applications, using *lattice statics*, *molecular dynamics* and *Monte-Carlo* methods. (The latter may be extended to include quantum effects *via* path integral methods.) Our final category is quantum mechanical methods in which the Schrödinger equation is solved approximately for a cluster or a periodically repeated array of particles; and

C.R.A. Catlow and A. Cheetham (eds.), New Trends in Materials Chemistry, 141–194.
© *1997 Kluwer Academic Publishers.*

in this context we will discuss briefly the rôles of both Hartree-Fock and Local Density Functional methodologies. A much more detailed account is given in Chapter 8.

2.1   EMPIRICAL METHODS

These approaches are limited to making approximate predictions of crystal structures. The simplest are based on effective atomic and ionic radii of which there are several published tabulations (see *e.g.* Shannon and Prewitt[7]). Simple geometrical criteria *e.g.* the well-known radius ratio rules[8] may then be used to predict coordination numbers and hence crystal structures. A variant of the same principle uses measured bond-lengths which in many cases do not show strong variations from compound to compound: for example the length of the Si...O bond is generally close to its mean value of 1.60 Å. In a complex crystal structure the restraints provided by the rather narrow bands of acceptable bond lengths enables approximate predictions of structure to be made. Indeed the procedure has been automated in the distance least squares (DLS) technique developed by Meier and Villiger[9].

The method requires a starting set of atomic coordinates, which may be random although symmetry constraints are normally included. Atomic coordinates are then varied using least squares techniques in order to minimise the function:

$$R = \sum_j w_j^2 \left[D_j^p - D_j^c \right]^2, \qquad\qquad 1$$

where D_j^p and D_j^c are the prescribed and calculated bond lengths. The weights W_j are intended to reflect the fact that different bond lengths show different ranges. They may be made proportional to bond strengths[10] or force constants[11].

The method has enjoyed a fair measure of success especially in structural mineralogy, being used by, for example, Dollase and Baur[12] to obtain a structural model of meteoritic low tridymite, while Gramlich and Meier[13] have proposed that the technique be used in zeolite structural chemistry to eliminate hypothetical frameworks which cannot yield low values of R. Further discussion is given in reference 3.

2.2   ELECTROSTATIC ENERGY CALCULATIONS

In this approach, guidance as to crystal stabilities and site substitution energies is obtained by calculating the Coulomb energy, U_c, of the crystal, which may be written as

$$U_c = \frac{1}{2} \sum_{i=1}^{N} \sum_{j=1}^{\infty} \cdot \frac{q_i q_j}{r_{ij}}, \qquad\qquad 2$$

where the sum over i extends to all N ions in the unit cell. The sum over j goes to infinity but omits the term $i = j$ (hence the use of the prime superscript); q_i and q_j are the charges on the ions. In evaluating such sums, the periodicity of the lattice structure may be exploited; and rapid convergence may be achieved by a technique due originally to Ewald which involves a transformation into reciprocal space. Further discussion of this method is given in references 14,15 and 16. An alternative procedure, devised by Evjen, obtained Madelung potentials at a point by calculating the potentials due to successive shells surrounding that point. The method is in general less flexible than the Ewald procedure which is used in most modern simulation codes. We should stress that electrostatic energy calculations on crystals must be taken to infinity, owing to the long range nature of the Coulomb forces. Calculations on restricted regions or blocks of the structure may give misleading results.

It is commonly useful to decompose the total Coulomb energy into site energies U_c^n for all n atoms in the unit cell. These are written as:

$$U_c^n = \sum_i \frac{q_n q_i}{r_{in}}, \qquad\qquad 3$$

where the sum is again taken over all ions in the crystal ($i \neq n$). U_c may be considered as a sum over the site energies, *i.e.*

$$U_c = \frac{1}{2} \sum_n U_c^n. \qquad\qquad 4$$

Considerable use has been made of electrostatic energies by Jenkins and coworkers (see *e.g.* references 17,18 and 19), who have extensively applied these methods to problems in layer structured silicates. Detailed studies are reported on micas including an investigation of the energetics of expansion of the structures along the axis perpendicular to the layers, which are characteristic of these crystals; from these calculations estimates were obtained of the interlayer binding.

These methods are therefore simple and effective. Their range of application is, however, very limited as they omit any account of short-range repulsion, polarisation or covalence terms. Methods which attempt to include these effects are discussed in the next section.

2.3 METHODS BASED ON INTERATOMIC POTENTIALS

2.3.1 *Potential Functions and their Evaluation.*

The potential energy function $U(r_1 \cdots r_N)$ expresses the energy of an assembly of N atoms or ions as a function of the nuclear coordinates $r_1 \cdots r_N$. The Born-Oppenheimer approximation is, of course, implicit in the use of such functions but there is no explicit inclusion of the effects of the electronic structure of the system: such effects are subsumed into the potential function. The energy zero for such functions is normally taken to be that of the component atoms (or ions) at rest at infinity, *i.e.* the self energies (electron-nuclear) of the atoms (or ions) are not included in U.

The potential function is commonly expanded as follows:

$$U = \frac{1}{2}\sum_{i}^{N}\sum_{j}^{N}{}' \varphi_{ij}(r_i, r_j) + \frac{1}{3}\sum_{i}^{N}\sum_{j}^{\infty}\sum_{k}^{\infty}{}' \varphi_{ijk}(r_i, r_j, r_k) + \cdots \qquad 5$$

where the φ_{ij} are 'two-body' functions which depend only on the positions of pairs of atoms i and j; φ_{ijk} are the 'three-body' terms depending explicitly on the positions of atoms i, j and k and where the prime symbol again indicates that we only include terms $i \neq j \neq k$. The expansion extends to four-body terms φ_{ijkl} and higher order terms which are, however, rarely included. Indeed it has been common to approximate U by including only the two-body component φ_{ij}, which may be usefully decomposed into Coulombic and non-Coulombic terms:

$$\varphi_{ij}(r_i, r_j) = \frac{q_i q_j}{r_{ij}} + V_{ij}(r_{ij}), \qquad 6$$

where q_i is the charge on the atom or ion (and is normally treated as a point entitu). The Coulombic term which is purely two-body in nature is handled by the procedures described in the previous section. The non-Coulombic, V_{ij}, is then usually approximated by analytical functions which generally include both attractive and repulsive components; the latter describes the Pauli repulsion due to overlap of closed shell electron configurations, and the former attractive terms from dispersion (*i.e.* induced dipole-induced-dipole) and covalence effects. Several such functions are available, notably the Lennard-Jones potential:

$$V(r) = Ar^{-12} - Br^{-6}, \qquad 7$$

the Morse potential:

$$V(r) = D\left\{1 - \exp\left[-\beta(r - r_e)\right]\right\}^2 \qquad 8$$

and the Buckingham potential:

$$V(r) = A\exp(-r/\rho) - Cr^{-6}. \qquad 9$$

Lennard-Jones potentials have been used widely in modelling rare-gas and molecular crystals. Morse potentials become more appropriate when covalent systems are being studied; D may then be interpreted as the covalent bond-dissociation energy and r_e the equilibrium bond length. Buckingham potentials have been very widely used in the study of ionic and semi-ionic solids[20].

Inclusion of many-body effects may be achieved by the use of angle dependent forces. For example, Sanders *et al.*[21] employed simple 'bond-bending' terms of the type,

$$E(\theta) = \frac{1}{2} k_B (\theta - \theta_0)^2,$$

10

where q is the angle subtended by three atoms and θ_0 is an equilibrium value; k_B is the appropriate force constant. The use of such terms is most appropriate in systems with directional covalent bonding, *e.g.* silicates, where θ_0 will correspond to the tetrahedral angle subtended by O-Si-O bonds in the tetrahedral SiO_4 groups.

Alternative, three-body functions are provided by the 'triple-dipole' formation of Axilrod and Teller[22], which takes the form:

$$V_{ijk} = \frac{k_T \left(1 + 3\cos\theta_1 \cos\theta_2 \cos\theta_3\right)}{r_{ij}^3 r_{jk}^3 r_{ik}^3},$$

11

where the angles and distances relate to the triangle defined by the three atoms i, j, and k.

Such functions describe the three body contributions to the induced dipole attractive forces, and have been shown by Meath and co-workers[23] to make important contributions to the cohesive energies of rare gas crystals. Work of Baetzold *et al.*[24] has shown that such terms may be of value in constructing potential models for the silver halide crystals.

Finally we note that atomic and ionic polarisation in solids is essentially a many-body effect. Moreover in ionic materials polarisation energies may be large especially in the vicinity of charged defects and surfaces. Omission of polarisation, while obviously leading to the failure to describe high frequency dielectric properties, also results in marked inadequacies in calculated phonon dispersion curves. The simplest way of modelling polarisability is to use the point polarisable ion (PPI) model which assigns a linear constant of proportionality *i.e.* the polarisability, α, between the dipole moment μ and the field E acting on the atoms, that is:

$$\mu = \alpha E.$$

12

However, as the dipole is a point entity, it cannot describe the physical basis of polarisability which is the displacement of the valence shell electron density in response to the applied field. Consequently, important physical effects, in particular the coupling of polarisability and short range repulsion, are omitted by these models which fail badly in describing both dielectric and lattice dynamical properties of crystals, as discussed in greater detail in reference[25]. These deficiencies are largely overcome by the shell model originally proposed by Dick and Overhauser[26] which describes the development of a dipole moment in terms of the displacement of a massless shell of charge Y (representing the valence shell electrons) from a core (representing the nucleus and core electrons), the core and the shell being connected by an harmonic restoring force for which the spring constant is k; the free atom polarisability, α, is then given by $\alpha = Y^2/k$. Shell model potentials have been very widely used in studies of lattice dynamical and defect properties of ionic solids. Further discussion is given in references 5 and 25.

Having established the framework of the potential model to be used, it is next necessary to fix the variable parameters, *i.e.* those used in the description of the short range potential, V(r), the shell model constants Y and k and the effective atomic charges, q (although we note that in many modelling studies of ionic oxides, halides and even silicates, these have been fixed at integral fully ionic values).

2.3.2 *Parameterisation*

This is undertaken by two procedures: first, *empirical methods*, in which variable parameters are adjusted, generally *via* a least squares fitting procedure to observed crystal properties. The latter must include the crystal structure (and the procedure of 'fitting' to the structure has normally been achieved by minimising the calculated forces acting on the atoms at their observed positions in the unit cell). Elastic constants should, where available, be included; and dielectric properties are required to parameterise the shell model constants. Phonon dispersion curves provide valuable information on interatomic forces; and *force constant* models (in which the variable parameters are first and second derivatives of the potential) are commonly fitted to lattice dynamical data. This has been less common in the fitting of parameters in *potential* models which are our present concern as they are required for subsequent use in simulations. However, empirically derived potential models should always be tested against phonon dispersion curves when the latter are available.

An important aspect of empirical potential parameterisation is the question of transferability. Are, for example, models derived in the study of binary oxides, transferable to ternary oxides? Considerable attention has been paid to this problem by Cormack *et al.*[27] who have examined the use of potentials in spinel oxides, *e.g.* $MgAl_2O_4$, $NiCr_2O_4$ *etc.*; in addition Parker and Price[28,29] have made a very careful study of silicates especially Mg_2SiO_4. These studies conclude that transferability works well in many cases, although systematic modifications are needed when potentials are transferred to compounds with different coordination numbers; for example the correct modelling of $MgAl_2O_4$ requires that the potential developed for MgO, in which the magnesium has octahedral coordination, be modified in view of the tetrahedral coordination of Mg in the ternary oxide. The correction factor is based on the difference Δr^c between the effective ionic radii for the different coordination numbers. If an exponential, Born-Mayer, repulsive term is used, the pre-exponential factor is modified as follows:

$$A' = A \exp\left(\Delta r^c / \rho\right), \qquad\qquad 13$$

where the unprimed and primed pre-exponential factors refer to original and modified coordination numbers. Cormack *et al.*[27] have shown that such procedures yield potentials which are successful in modelling spinels.

There are now several sets of empirical potential parameters for a range of solids. Parameter sets are available for organic, molecular solids (see, for example, the work of Williams[30] and of Kiselev[31]). Ionic solids have been widely studied in the last twenty years, and parameters are available for ionic halides[32], oxides[33] and silicates[21]. Reviews are given in references 20 and 25.

Empirical procedures are still extensively used in developing potential parameters. They do, however, have obvious inadequacies: firstly, it is necessary to have a good range of accurate empirical data which may not be possible when new systems are being studied; secondly, there is no guarantee of the validity of the potential function when used outside the range of interatomic spacings employed in the parameterisation. For this reason there has been considerable incentive in recent years for the development of reliable theoretical procedures for calculating parameters. To date, such efforts have been directed largely towards parameterisation of effective charges and of the short range potential, V, and in a restricted number of cases, the three-body components of the potential; it is still very difficult to calculate shell model parameters. Calculations of interatomic potentials have been of two types.

2.3.3 *Electron Gas Methods*
These methods are based on the electron gas description of the atom which is used to calculate the interaction energies of pairs or in principle larger numbers of atoms (or ions) as a function of their nuclear coordinates. The method was established by Wedepohl[34] and by Gordon and Kim[35] and wide ranging and successful studies of ionic halides and oxides are reported by Mackrodt and co-workers[see for example references 36,37] (note also reference

25 and the compilation of parameters given in reference 38). More recently, modified electron gas procedures developed by Cohen and co-workers[39] have enjoyed considerable success when applied to oxides and silicates.

As the electron gas method is now very standard and has been reviewed by the present author and others (references 25 and 36-39), a more detailed discussion is not presented here; a summary of the essential features of the methodology is described in reference 40.

2.3.4 *Explicit Quantum Mechanical Methods*

These are methods in which the Schrödinger equation is solved approximately for a cluster or periodic array of atoms for a variety of internuclear geometries; the resulting variation in the total energy is then fitted to an interatomic potential function. *Ab initio* Hartree-Fock methods are being increasingly used in such studies; examples are the early work of Mackrodt and co-workers[41], and the more recent studies of Gale *et al.*[42] and Harrison *et al.*[43]. The study of silicate systems has been particularly active, with calculation on clusters ranging from $Si(OH)_4$ to $Si_{15}O_{16}H_{12}$[44,45,46]. Results of the recent survey[46] of the energies of the larger cluster yielded potential parameters showing an encouraging measure of agreement with those obtained from the empirical parameterisation of Sanders *et al.*[21].

In using such methods care must be taken in the choice of basis sets (*i.e.* the atomic centred functions from which the LCAO-MOs are constructed). Unless good sets are used, then the resulting interatomic potential function will be inadequate due to basis-set superposition errors. This subtle effect arises from the use by each atom in the many atom calculation of basis functions on *other* atoms to reduce its own energy — an effect which invariably arises when incomplete sets are used. Correction procedures are available[47]; but in general, high quality potentials need high quality basis sets. Secondly, it is important to note that wave functions in crystals differ from those of free ions. It is necessary therefore to include in such calculations a representation of the effects of the surrounding ions on the interacting atoms. A simple procedure commonly used in ionic solids is to employ point charges to reproduce the Madelung potential of the surrounding lattice. Thirdly, it should be recalled that the Hartree-Fock method does not include effects arising from electron correlation; more sophisticated and computationally expensive configurational interaction or perturbation methods are needed to describe correlation effects. Since, however, correlation is responsible for dispersive interactions, other methods are needed to calculate these terms. Alternative ab-initio methods based on the local density approximation (LDA) discussed in Section 2.4.2 do include directly certain of the effects of electron correlation, although they are still unable to model dispersive effects accurately. Perturbation theories following the original work of

Mayer[48] may be used; but empirical methods may be needed for some time to come.

Ab initio techniques are generally far more expensive computationally than is the case for the electron gas techniques, and the cpu and disk requirements increase rapidly with the total number of electrons in the calculation. The former methods are, however, preferable, if high quality basis sets may be used, as they allow a more accurate description of atomic interactions, in particular the redistribution of electron densities consequent upon interaction. Electron gas methods clearly, however, provide a cheap, useful and generally applicable technique for deriving short range potential parameters; and as such they will have a continuing rôle in this field.

Finally we should note a valuable approach developed by Pyper and coworkers[49] for ionic solids in which Hartree-Fock methods are used first to obtain a set of crystal orbitals, the interactions between which are calculated as a function of distance including a full explicit evaluation of the exchange term (unlike the local density approximation used in the electron gas method). Estimates of the dispersive energy are then added to the resulting interaction energy. This approach is particularly successful for strongly ionic halides and oxides.

Having developed and parameterised potential models, the final stage before their use in a simulation study should be their *evaluation*. Non-empirically derived potentials should be evaluated by reference to their ability to predict empirical crystal properties. For empirical potentials, it is clearly necessary to use data outside the range employed in the parameterisation. We have already referred to the use of lattice dynamical data. Comparison with the results of high pressure studies, in particular the variation of structural and elastic properties with pressure, is also of great value; and in this context we should note the work of Harding and Stoneham[50] who examined several potentials for simple ionic materials by comparison with 'Hugoniots' *i.e.* pressure-volume trajectories obtained from shock wave studies. A further useful and demanding test is provided by the ability of potential models to predict crystal structures of polymorphs of a compound other than that used in the potential parameterisation.

The development of improved potential models will continue to be a central feature of simulation studies. Progress will require the increased availability of accurate crystal data and refinements in theoretical procedures for calculating potentials.

2.3.5 Simulation Methodologies: Static Techniques
The characteristic features of these methods is that they do not include any explicit representation of thermal motions. Energies and entropies are calculated for the static lattice; the entropy calculations normally assume

the harmonic approximation. High temperature properties can be calculated using the quasi-harmonic approximation which simply evaluates energies and force constants for the high temperature lattice parameters; within this framework it is possible to predict lattice expansivity by calculating the free energy minimum as a function of lattice parameter, for a series of temperatures. Calculations may be performed on both perfect and defective lattice configurations; both will now be reviewed.

2.3.6 Perfect Lattice Calculations

The most basic quantity open to calculation is the lattice energy. If we omit the zero-point vibrational energy (and four-body) and higher order terms this may be written as:

$$E = U_c + \frac{1}{2}\sum_{i=1j=1}^{N}\sum_{}^{\infty}{}' V_{ij} r_{ij} + \frac{1}{3}\sum_{i=1j=1k=1}^{N}\sum_{}^{\infty}\sum_{}^{\infty}{}' \varphi_{ijk}\left(r_i, r_j, r_k\right), \qquad 14$$

where U_c is the electrostatic energy as defined in Equation 4 of Section 2.2. The summation conventions are as in Equations 2 and 5 and the terms V_{ij} and φ_{ijk} are as defined in Equation 5 and 6 of Section 2.3.1. We note that the great majority of lattice energy calculations only include the two-body contribution to the short-range energy. One important matter of definition is that the lattice energy gives the *energy of the crystal with respect to component ions at infinity*. If it is desired to express the energy with respect to atoms at infinity (for which the more appropriate term is then the *cohesive* energy) then the appropriate ionisation energies and electron affinities will be added.

Lattice energy calculations are now routine, and may be carried out for very large unit cells containing several hundred atoms. The codes METAPOCS[51], THBREL[52] and GULP[53] undertake lattice energy calculations including both two- and three-body terms, using both bond-bending and triple-dipole formalisms.

Table 1:   Relative Energies (per mole) of Microporous Siliceous Structures with respect to Quartz

Structure	Energy (kJ / mole)
Silicalite	11.2
Mordenite	20.52
Faujasite	21.4

Lattice energy calculations provide valuable insight into the structure and stabilities of ionic and semi-ionic solids. The technique is most powerful when combined with energy minimisation procedures, which generate the structure of minimum energy. These are discussed below after the calculation of entropies have been described. The results in Table 1 give a good illustration of the value of lattice energy studies. They are the energy minimum lattice energies calculated for a number of purely siliceous microporous zeolitic structures which are compared with the lattice energy of α-SiO_2. The latter has the lowest value as would indeed be expected since the more porous structures are known to be metastable with respect to the dense α-SiO_2 polymorph. Of greater interest is the observation that of the porous structures, silicalite has the greatest stability. This accords with the fact that this polymorph can only be prepared as a highly siliceous compound unlike the case with the other zeolitic structures which are normally synthesised with high aluminium contents. The calculations which are discussed in greater detail in reference 54, suggest that this behaviour has its origin at least in part in the thermodynamic stability of the compounds. We note that recently very similar results were obtained by Henson et al.[94] who also showed that the calculated values were in excellent agreement with experiment.

In addition to calculating energies, it is also possible to calculate routinely a range of crystal properties, including the lattice stability, the elastic and dielectric and piezoelectric constants, and the phonon dispersion curves. The techniques used which are quite standard require knowledge of both first and second derivatives of the energy with respect to the atomic coordinates. Indeed it is useful to describe two quantities: first the vector, g, whose components g_i^α are defined as:

$$g_i^\alpha = \left[\frac{\partial E}{\partial x_i^\alpha} \right], \qquad\qquad 15$$

i.e. the first derivative of the lattice energy with respect to a given Cartesian coordinate (α) of the ith atom. The second derivative matrix W has components $W_{ij}^{\alpha\beta}$; defined by:

$$W_{ij}^{\alpha\beta} = \left[\frac{\partial^2 E}{\partial x_i^\alpha \partial x_j^\beta} \right]. \qquad\qquad 16$$

The expressions used in calculating the properties referred to above from these derivatives are discussed in greater detail in reference 16. For more detailed discussions of the calculation of phonon dispersion curves from the second derivative or 'dynamical' matrix W, the reader should consult references 55 and 56. Finally, we note that by the term 'lattice stability' we refer to the equilibrium conditions both for the atoms within the unit cell, and for the unit cell as a whole. The former are available from the gradient vector g, while the latter are described in terms of the 6 components $\varepsilon_1 \cdots \varepsilon_1$ which define the strain matrix ε, where

$$\varepsilon = \begin{bmatrix} \varepsilon_1 & \frac{1}{2}\varepsilon_4 & \frac{1}{2}\varepsilon_5 \\ \frac{1}{2}\varepsilon_4 & \varepsilon_2 & \frac{1}{2}\varepsilon_6 \\ \frac{1}{2}\varepsilon_5 & \frac{1}{2}\varepsilon_6 & \varepsilon_3 \end{bmatrix}. \qquad 17$$

So when the unit cell as a whole is strained, we describe the modification of an arbitrary vector $\underline{r}$ in the unstrained matrix to a vector $\underline{r}'$ in the strained matrix, using the equation:

$$\underline{r}' = (1 + \varepsilon)\underline{r} \qquad 18$$

where 1 is the unit matrix. The six derivatives of energy with respect to strain, $\left[\dfrac{\partial E}{\partial \varepsilon_i}\right]$, therefore measure the forces acting on the unit-cell. The equilibrium condition for the crystal therefore requires that $g = 0$ and $\left[\dfrac{\partial E}{\partial \varepsilon_i}\right] = 0$ for all i.

2.3.7 Entropy Calculations

The entropy in a solid arises first from configuration terms which for a perfect solid are zero; while for a solid showing orientational or translational disorder configurational expressions based on the Boltzmann expression S = kln(W) may be used. In this section we shall pay more attention to the second term, which is due to the population of the vibrational degrees of freedom of the solid. Thus the entropy of a solid may be written as:

$$S_{vib} = k \int_0^Q dQ \sum_i \left\{ \frac{hv_i}{kT}\left[\exp\left[\frac{hv_i}{kT}\right] - 1\right]^{-1} - \ln\left[1 - \exp\left[\frac{-hv_i}{kT}\right]\right]\right\}, \qquad 19$$

where the sum is over all phonon frequencies and the integral is over the Brillouin zone. In practice the integral is normally evaluated by sampling over the zone for which a variety of techniques are available[57]. Vibrational terms also give a contribution to the lattice energy of the crystal:

$$E_{vib} = kT \int_0^Q dQ \sum_i \left\{ \frac{hv_i}{2kT} + \frac{hv_i}{kT}\left[\exp\frac{hv_i}{kT} - 1\right]^{-1}\right\}, \qquad 20$$

which results in the following expression for the crystal free energy with respect to ions at rest of infinity:

$$F = E + KT \int_0^Q dQ \sum_i \left\{ \frac{hv_i}{2KT} + \ln\left[1 - \exp\left[\frac{hv_i}{KT}\right]\right]\right\}, \qquad 21$$

where E is the lattice energy (omitting vibrational terms) given by Equation 14.

2.3.8 Energy Minimisation

Having evaluated energies and free energies of a crystal structure it is desirable to be able to use these in an energy (or free energy) minimisation procedure. Let us consider first the simple case of minimisation to constant volume (*i.e.* within fixed cell dimensions). We write the energy of the crystal as a Taylor expansion in the displacements of the atoms, d, from that current configuration giving:

$$E(\delta) = E_0 + g\delta + \frac{1}{2}\delta W\delta + ...$$
22

If we terminate this function at the second order term and minimise E with respect to d, we obtain for the energy minimum:

$$0 = g + W\delta \qquad i.e. \quad \delta = -gW^{-1}$$
23

Displacement of the coordinates by δ as given in Equation 22 will generate the energy minimum configuration. Of course, in practice, it will not be valid to truncate the summation at the quadratic term, except when very close to the minimum. However, Equation 23 provides the basis of an effective iterative procedure for attaining the minimum. Indeed this 'Newton Raphson' method is widely used in both perfect and defect lattice energy minimisation, as it is generally rapidly convergent. Its main disadvantage is that it requires the calculation, inversion and storage of the second derivative matrix, W. Recalculation and inversion each iteration may be avoided by use of updating procedures due to Fletcher and Powell[58] which we and others have discussed elsewhere[2,5,59]. The storage problem may become serious with very large structures owing to the high cpu memory requirements. Recourse may be made to gradient methods, *e.g.* the well known conjugate gradients technique, which make use only of first derivatives. Such methods are, however, more slowly converging. The increasing availability of very large cpu memories is, however, reducing the difficulties associated with the storage of the W matrix.

For evaluation of the energy minimum with respect to constant pressure (*i.e.* with variable cell dimensions), first we note that we can define the six components of the mechanical pressure acting on the solid, corresponding to the six strain components, defined in Equation 17, *i.e.*

$$P^{\varepsilon_i} = \frac{1}{V}\left[\frac{dU_i}{d\varepsilon_i}\right],$$
24

where V is the unit cell volume. The strains can then be evaluated, using Hooke's law,

$$\varepsilon = PC^{-1}$$
25

where C is the (6x6) elastic constant terms, which may be calculated from W. Substitution of these calculated strain components into Equation 25 then yields the new cell dimensions and atomic coordinates. Again, the procedure is iterative, as it is only strictly valid in the region of applicability of the harmonic approximation. With a sensible starting point, however, only a small number of iterations (typically 2-5) is required.

The treatment above assumes that the pressure and corresponding strains are entirely mechanical in origin. However, at finite temperatures there will be a 'kinetic pressure' arising from the changes in the vibrational free energy with volume. These may be written as:

$$P_{vib}^{\varepsilon_i} = \frac{1}{V} \frac{dF_{vib}}{d\varepsilon_i}, \qquad\qquad 26$$

where F_{vib} is the vibrational free energy. These kinetic pressures are most simply evaluated by applying small arbitrary strains to the structure and calculating the corresponding changes in F_{vib}. If P_{vib} is added to the mechanical pressure P in Equation 24, it enables us to carry out free energy minimisation. Parker and coworkers have written a general computer code, PAPAPOCS, for such calculations. Further discussion is given in reference[56] which also describes how the techniques may be used to calculate lattice expansivity, either directly or by calculating the cell dimension as a function of temperature or by calculation of the thermal Gruneisen parameter.

2.3.9 Defect Simulations

Two techniques are available for calculating the properties of defects: first we may set up a periodic array of defects; second, we may embed the defect in an infinite representation of the surrounding crystal.

2.3.10 Super Cell Methods

The simplest procedure for calculating the formation energies, entropies and hence free energies of defects is to set up a defect supercell, *i.e.* a large periodically repeating structure with the defect being at the centre of each unit cell. The techniques described in the previous section may then be employed. Thus to evaluate the defect energy we perform a calculation on the perfect lattice, which may be equilibrated either to constant volume or pressure. The lattice energy E_{PERF} is then compared with the value E_{DEF} obtained under the same conditions for the defective lattice. Thus the defect formation energy E_D is simply written as:

$$E_D = E_{DEF} - E_{PERF}. \qquad\qquad 27$$

A similar procedure is, of course, applicable for the entropies and hence free energies. The method is attractively simple, but has a number of drawbacks the principal one being the need for large supercells, especially if large complex defects are being considered. However, we note that supercells containing several hundred atoms are accessible to modern computers. A second drawback is that even with large supercells, the calculated defect energy will inevitably include a term arising from defect interaction as well as formation terms. In certain contexts, however, this may be an advantage, and indeed by calculating the defect energy for different sizes of the supercell, defect interactions may be calculated as a function of defect spacing. Such a study was performed by Cormack *et al.*[60] who investigated the interaction of the two-dimensional shear planes which form in non-stoichiometric oxides; their calculated interaction function is shown in Figure 1. Thirdly, there are problems in including charged defects. Either the defects must be neutralised by including a defect of opposite charge in the unit cell, in which case the isolated defect formation energy cannot be studied, or electroneutrality is achieved by adding a background neutralising charge or by making small adjustments to the charges of the other atoms in the unit cell.

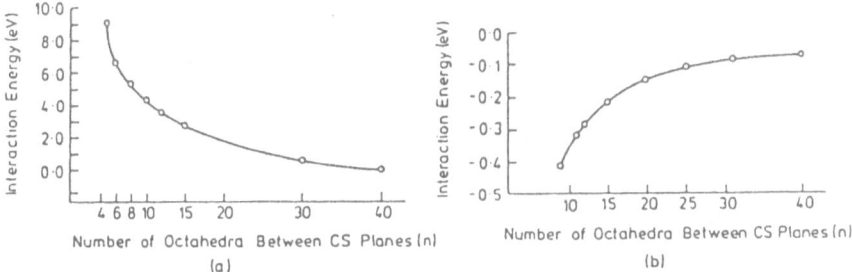

Figure 1: Interaction energy between shear planes in ReO_3 structured oxides: a) refers to relaxation at constant pressure and b) at constant volume

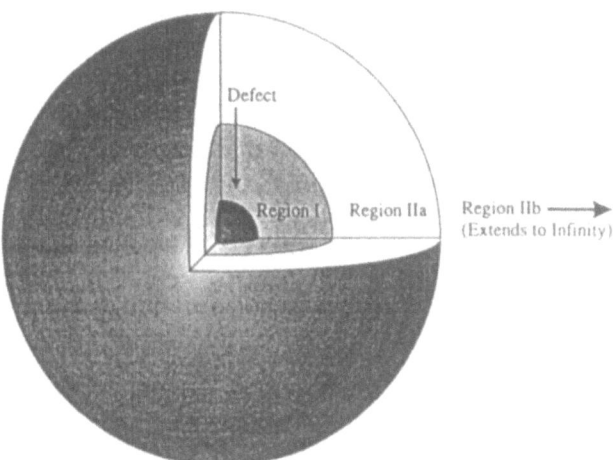

Figure 2: Schematic representation of the Mott-Littleton technique for treating relaxations around defects. (In practice an interface region IIa is included between the inner and outer regions.)

Despite these difficulties, supercell methods have found considerable application in defect studies[61,62,63], and indeed the results compare well with those obtained using the alternative techniques described below.

2.3.11 Mott-Littleton Methods

In these calculations, the isolated defect or defect cluster is embedded in the crystal which extends to infinity and the contrast between this approach and that used in the supercell methods is illustrated diagrammatically in Figure 2. The normal procedure in a Mott-Littleton calculation is to relax all the atoms in a region of crystal surrounding the defect, containing typically 100-300 atoms, until all the atoms are at zero force. Newton-Raphson minimisation methods are generally used. The relaxation of the remainder of the crystal is then described by more approximate methods in which the polarisation, P at a point r, is calculated for crystals which have dielectric isotropy, from the expression:

$$P = \frac{1}{4\pi} \frac{qr}{r^3}\left(1 - \varepsilon_o^{-1}\right), \qquad\qquad 28$$

where q is the charge on the defect, and ε_o is the static constant. More complex expressions are used for dielectrically anisotropic crystals. In practice, an interface region is normally used between the two regions described above; in the interface, displacements are calculated using formulae of the type given above, but are evaluated as the sum of the response to individual defects in the inner region, rather than as a response to the net charge of the defect configuration which is used for the remainder of the outer region. Short range interactions between ions in the inner region and those in the interface are also calculated explicitly.

The methods have been widely discussed and reviewed in the literature[2,5]. Several good automated computer codes are available: the HADES2 programme written by Norgett[64], which is confined to crystals of cubic symmetry; the HADES3 code[65] which extended the methodology to include crystals of any symmetry. Leslie[66] developed a program CASCADE, optimised for use on the CRAY vector processing super computers; the latest versions of this program permit the inclusion of many-body terms using bond-bending or triple-dipole formalisms. A large number of applications of these techniques are now reported and reviewed in the references given above. Indeed, work using these techniques has helped to establish the quantitative reliability of simulation techniques in solid state studies.

In addition to energies, defect entropies may also be calculated by related, although distinct, techniques. As with the case of the perfect lattice, defect entropies have two terms — configurational and vibrational — and the former can be calculated using simple combinatorial expressions. The latter arises from the perturbation of the vibrational frequency of the surrounding lattice atoms by the defect. The vibrational defect entropy S_{vib} is a function of the ratio of the perturbed to the unperturbed frequencies, *i.e.*

$$S_{vib} = -k \ln \frac{\prod\limits_{i=1}^{2N'} \omega_i'}{\prod\limits_{i=1}^{3N} \omega_i} + 3k(N'-N)\left[1 - \ln\left(\frac{\hbar}{KT}\right)\right]. \qquad 29$$

To evaluate the perturbed frequencies, Greens function techniques may be used. But Gillan and Jacobs[67] and Harding[68] have shown that embedded crystallite methods are more effective. Here the initial relaxation of the lattice surrounding the defect is undertaken as in the defect energy calculations; the perturbed force constant matrix for the relaxed region is then obtained. There are problems with the 'convergence' of the calculated entropy as a function of the size of the perturbed region; these may, however, be treated by the techniques discussed by Gillan and Jacobs[67].

Work of Harding and coworkers[68,69,70] has clearly established the value of defect entropy calculations using these methods. Moreover, we note that comparisons have been made[69] between entropies and energies calculated using supercell and embedded crystallite techniques. It is reassuring that the techniques yield the same defect parameters for large sizes of the supercell and of the crystallite.

Both energy and entropy calculations of the type described above are generally performed at constant volume, or more precisely constant lattice parameter. These may, however, be converted to constant pressure values using the following expressions:

$$h_p = u_v - V\beta T\left[\frac{df_v}{dV}\right], \qquad s_p = s_v - V\beta\left[\frac{df_v}{dV}\right], \qquad 30,31$$

where h_p and u_v are the constant pressure enthalpy and constant volume energy terms, s_p and s_v are the corresponding entropy terms, f_v is the constant volume free energy, the volume derivative of which may be calculated by studying the variation of u_v and s_v with lattice parameter; V is the unit cell volume and β is the expansivity of the solid. Discussion of these relationships is given in references 70 and 71. Work of Catlow *et al.*[72)] on AgCl and Jackson *et al.*[73] on UO_2 has shown their value in studying high temperature, defect properties.

In concluding this section we note that defect calculations may be used to study defect mobilities as well as defect formation and interaction processes. Assuming the validity of the hopping model of defect transport then the frequency of defect jumps n be written as:

$$v = v_0 \exp\left(\frac{-g_{ACT}}{kT}\right), \qquad\qquad 32$$

where g_{ACT} is the activation free energy for the defect migration, *i.e.* the difference between the free energies of the saddle point for the defect jump and that of its ground state. Saddle points may be identified by examining the potential energy surface for the migrating defect (although this may be difficult for crystals of low symmetry); once identified, the techniques discussed above may be used to calculate u_{ACT} and s_{ACT}.

Good discussions of the calculations of activation energies are given by Harding[70], and earlier work is reviewed in references 5 and 71. The validity of the hopping model, discussed in detail in reference 74, requires generally that $u_{ACT} \gg kT$ and that the 'jump time' for the migration process be far less than the 'residence time' of defects at individual sites. These conditions apply to the great majority of transport processes in solids. However, for materials with very high mobilities the conditions may break down and it becomes necessary to use the dynamical simulation methods discussed in the next section.

2.3.12 Molecular Dynamics Techniques

In contrast to the static methods discussed in the previous section, molecular dynamics (M.D.) includes thermal energies explicitly. The method is conceptually simple: an ensemble of particles represents the system simulated and periodic boundary conditions are normally applied to generate an infinite system. The particles are given positions and velocities, the latter being assigned in accordance with a target temperature. In simulations of crystalline solids the simulation box will normally be a super cell of the basic unit cell. The system is then allowed to evolve in time by solving the classical equations of motion using a specified time step, t (which is typically 10-15 - 10-14s). A variety of updating algorithms are available, as discussed in detail by Allen and Tildesley[75]. In the limit of an infinitesimal time step they all reduce to the simple classical equations of motion:

$$x(t + \tau) = x(t) + v(t)\tau, \qquad\qquad 33$$

$$v(t + \tau) = v(t) + \frac{f_i}{m_i}\tau, \qquad\qquad 34$$

where f_i is the force acting on the particle of mass m_i.

In the initial stages of the simulation, the ensemble equilibrates, *i.e.* it achieves an equilibrium distribution of velocities and equipartition between potential and kinetic energy. After this period, which may take several thousand time steps, the simulation is run for as long a period as is computationally feasible, and the trajectories of all particles are, if possible, stored.

M.D. simulations yield rich and detailed information on the system simulated. Structural information is available from radial distribution functions (r.d.f.s.), and diffusion coefficients may be calculated from the mean square displacement of the particle as a function of time, using the relationship:

$$D_\alpha = \frac{<r_\alpha^2>}{6t},$$

35

where D_α is the diffusion coefficient of the particle of type α. Typical results are shown in Figure 3 for the superionic compound Li$_3$N[76]. The increase of $<r_\alpha^2>$ with t clearly shows that diffusion is occurring; it is moreover evident that the simulation predicts that the rate of diffusion is greater perpendicular compared with parallel to the c axis, as is indeed observed experimentally. Calculation of the velocity auto-correlation function (v.a.f.) and van Hove self-correlation function yields more detailed dynamical information which may be compared with the results of inelastic neutron scattering studies; for further discussion we refer to our earlier reviews[71,77].

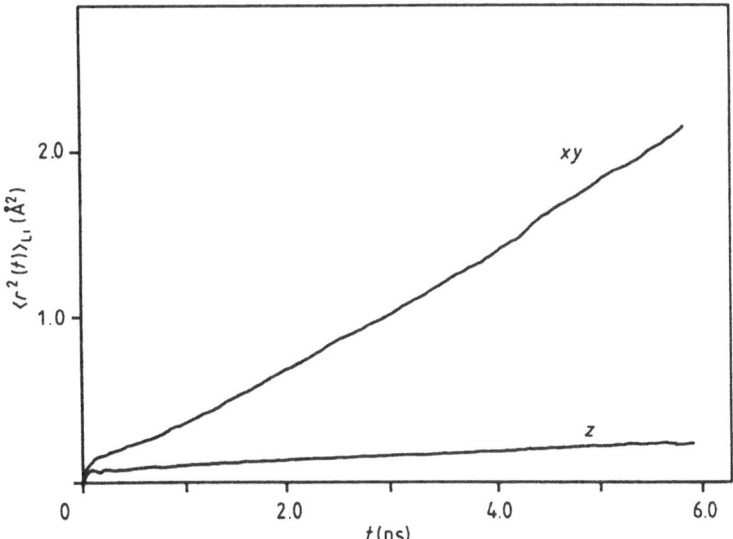

Figure 3: Mean square displacement vs time for Li$^+$ ions at 400K in Li$_3$N. Faster diffusion is apparent in the xy plane of this layer structured compound.

Despite its evident power, the technique has, however, serious limitations, the principal of which are as follows:

(i) Even with modern supercomputers it is rarely possible to run simulations for longer than 1ns. If the probability of the events of interest (*e.g.* a defect jump) are low in a period of this length, then the simulation may be of little value. Thus in practice, with currently available computer power, diffusion can only be studied for values of the diffusion coefficient $D > 10^{-8}$ cm^{2}sec^{-1}.

(ii) Inclusion of polarisability in the potential greatly increases the computational requirements for the simulation. For this reason most M.D. simulations performed do not include this term which may be a serious omission for ionic solids.

(iii) The calculation of thermodynamic properties is difficult using M.D. techniques. The case of defects is especially problematic and both formation and migration energies are far more effectively calculated using static simulation techniques.

(iv) In conventional M.D. employing periodic boundary conditions, there are no surface effects. Thus any process involving the generation or removal of species from the surface cannot be simulated. An example is provided by the generation of Schotty disorder in crystals, which requires displacement of ions to the surface. This cannot be simulated by conventional M.D. studies.

Despite these limitations, the scope of M.D. techniques is expanding rapidly mainly due to the rapid growth in computer power, especially the development of parallel architecture systems. It is now possible to carry out routinely, simulations on several thousand particles, for periods of ~100ps. Moreover, the scope of the method has been expanded by a number of more specialised developments. The two following examples are of particular importance in the study of inorganic materials.

2.3.12a   Constant Pressure M.D.

The conventional M.D. technique uses a fixed size for the simulation box, i.e. the calculation is performed under constant volume conditions. Using methods developed by Parrinello and Rahman[78] and by Berendsen and coworkers[79] it is now possible to undertake constant pressure simulations by allowing cell dimensions to vary during the simulation. Detailed discussions are given in reference[75]. The most obvious field of application of this technique is to the study of phase transitions, and useful applications have been reported to the study of melting and glass formation as discussed below.

2.3.12b   M.D. Simulations of Glasses

Increasing use is being made of M.D. techniques to generate structural models for amorphous solids. The procedure used is simple: a molten system is generated; the simulation is then rapidly cooled to below the glass transition temperature. R.d.f.s. and other properties of the resulting simulated glass may then be calculated. Pioneering work of Woodcock and Angell[80] and of Garafolini and coworkers[81] showed how such techniques could be used to simulate vitreous silica and demonstrated their success in reproducing r.d.f.s.

of the glassy material. And more recent work of, for example, Vessal *et al.*[82] has shown how the structural predictions can be refined by including bond-bending terms in the interatomic potential; further details will be given in Section 3. The main difficulty with the technique arises from the quenching procedure. The simulated quench rate is invariably several orders of magnitude faster than is achieved in any laboratory experiment . This may lead to structural models for the glasses which are significantly too disordered, although the extent and nature of the problem is still unclear.

2.3.13 Monte Carlo Techniques

Like M.D., Monte Carlo (M.C.) methods involve the generation of successive configurations of an ensemble of particles representing the system studied; but unlike M.D., there is no temporal connection between the different configurations. The aim of the technique is to generate a sufficient and representative number of configurations from which ensemble averages may then be calculated with acceptable accuracy.

A central feature of the procedure is the choice of a weighting scheme whereby configurations are included according to their probability. In particular it is desirable to weight the probability of including a configuration according to its Boltzmann factor:

$$W = \exp(-E / kT), \qquad\qquad 36$$

where E is the energy of the configuration. This may be achieved by by an ingenious and simple procedure, known as the Metropolis sampling method. Here the Boltzmann factor for a given configuration is calculated and compared with a random number P, which is generated in the range 0-1. If W < P, then the configuration is accepted, but if W > P it is rejected. It will be seen that, over a large number of configurations, the procedure weights the probability of their inclusion according to their Boltzmann factor. It necessarily involves the generation of configurations which, after evaluation of their energy, are rejected; and a number of procedures are available to increase the efficiency of the method by reducing the number of rejected configurations[75].

The manner in which the sequence of configurations is generated depends on the system studied. For example, in a collection of atoms, a given particle may be selected at random and displaced by a small amount. For molecules, displacement and rotation may be involved. Indeed, care must be exercised in generating configurations in order to maximise the sampling efficiency. One special type of system concerns vacancy disordered compounds in which we are interested in diffusional properties. Here the vacancies are selected at random as is their direction; the Boltzmann factor for the jump is calculated (or obtained from a 'look-up' table) and compared with the random number P following the normal Metropolis procedure. Such methods have been extensively used by Murch and co-workers to study diffusion in alloys[83] and an application to oxygen diffusion in the superionic oxygen conductor Y/CeO_2 was reported by Murray and coworkers[84].

We note that a Monte Carlo 'move' can consist of the insertion or deletion of a particle, to so-called Grand Canonical Monte Carlo technique, for detailed discussion of which and of other technical aspects of M.C. simulation we refer to reference 75.

We also note an important class of applications which uses the M.C. method to explore complex energy surfaces and to find low energy regions, from which minima can subsequently be located by standard minimisation procedures. *Simulated annealing* methods have been used with great effect in generating crystal structures from initial random configurations of atoms or ions. The 'energy term' in simulating annealing calculations may take a variety of forms. It is commonly a simple 'cost function' based on coordination numbers and connectivity. Lattice energies, calculated from Born model potentials, may of course be employed, but this procedure becomes computationally expensive. It should, however, be noted that simulated annealing procedures employing energies calculated by electronic structure techniques are becoming feasible; although in practice it would seem to be more computationally efficient to use simpler procedures to calculate the energy (or cost function), and to refine the approximate configuration generated by simulated annealing by electronic structure methods when the use of such methods is needed.

The 'temperature' used in a simulated annealing study is normally a parameter with no real physical significance. Higher temperatures will result in the acceptance of an increasing number of high energy configurations, allowing the exploration of a wider range of the energy surface but, of course, at increased computational cost. In practice, the simulation usually starts with a high 'temperature' which is then reduced as the lower energy regions of the surface are identified. Simulated annealing like M.C. methods may also be adapted to the analysis of experimental data. In this case the 'energy term' becomes the deviation between the calculated data (*e.g.* the calculated X-ray intensity *vs* scattering angle for a structural model) and experiment. Such techniques are described as 'Reverse Monte Carlo'[85].

2.4   ELECTRONIC STRUCTURE TECHNIQUES

Here we are concerned with methods which attempt to solve the Schrödinger equation at some level of approximation yielding information on electronic structure and bonding in the system studied. An extensive discussion of electronic structure calculations is beyond the scope of the article, and the reader should consult Chapter 8 for a detailed review. The present discussion will focus on those areas where quantum mechanical methods interact most strongly with the modelling techniques discussed elsewhere in the article. Our account will consider the two types of electronic structure techniques that are most widely used in contemporary studies, *i.e.* Hartree-Fock and Local Density Functional methods.

2.4.1 Hartree-Fock Methods

For accounts of the basic principles of this standard and well developed theory we refer the reader to, for example, reference 86. Applications to solids have now a long history, with earlier work being commonly based on the use of semi-empirical procedures (*e.g.* CNDO and MNDO) in which many of the integrals that must be evaluated in the Fock matrix are either set to zero or parameterised. Expansion of computer power has allowed modern work to make increasing use of *ab initio* techniques in which all integrals (above a certain threshold) are explicitly evaluated. Such calculations are of course subject to the inherent limitations of the Hartree-Fock methodology, in particular, as noted earlier, effects due to electron correlation are not included. Moreover, the accuracy may be limited by that of the basis sets used.

Hartree-Fock calculation in solids may be performed in two ways: the first are cluster calculations in which a small group of atoms is investigated. An important feature of such calculations concerns the termination of the cluster. Hydrogen atoms may be used to saturate unsatisfied valencies (or dangling bonds). The cluster may be embedded in point charges in order to reproduce the Madelung potential due to the surrounding lattice; this procedure is, in the opinion of the present author, particularly important when calculations are being performed on ionic or semi-ionic solids. A more sophisticated procedure is to surround the cluster by 'pseudo-potentials' *i.e.* functions on the surrounding atoms which minimise the effect of Pauli repulsion between the electrons on those atoms within the cluster. Most sophisticated are the 'Green function' techniques which attempt to map the wave function of the cluster on to that of the surrounding lattice atoms.

Hartree-Fock cluster calculations have been extensively used in studies of defects and of surface properties (including many studies of chemisorption as described in reference 87. In addition, many cluster calculations have been reported, especially in the field of silicate compounds[44,45,46,88], often with a view to elucidating the nature of the bonding in the corresponding solid. There is no doubt that such calculations will continue to play an important rôle in studies of solids, with expansion in computer power allowing large clusters to be examined at an increasingly high level and with improved embedding procedures.

Calculations on periodic systems using H.F. techniques have now acquired prominence owing to the pioneering work of Dovesi, Pisani, Causa and Saunders whose CRYSTAL code[89] has adapted H.F. methodologies to crystalline structure systems by building in translational symmetry into the wave function. Their method has been extensively used in studies of oxides, halides and silicates. Moreover, as noted earlier, it is finding applications in the parameterisation of interatomic potentials[42,43].

2.4.2. Local Density Functional Techniques

As discussed in Chapter 8, the starting point of this approach which has been widely used in the solid state physics community is the Hohenberg-Kohn theorem[90] which shows that the total energy of a system (molecule or solid) is a unique functional of the electron density ρ. The next step is the invocation of the local density approximation (L.D.A.) for the exchange and correlation energies which assumes that the exchange correlation potential V_{xc} at a point r has the value appropriate for a uniform electron gas with the same density as that at r. We should stress that the use of the L.D.A. term for the exchange and correlation energy is an approximation (as shown by the increasing use of non-local corrections). It is, however, an approximation that appears to work well in many systems and it is far more economical than direct calculation of the exchange terms by a series of many centre integrals as with Hartree-Fock methods. Moreover, unlike single determinantal H.F. methods, the method includes a representation of electron correlation.

Application of the variational principle to the expression for the total energy, results in a series of eigenvalue equations which must be solved self consistently as with Hartree-Fock methods. The electron density can be represented numerically or by using atomic basis functions. Pseudopotentials which represent the effect of core electrons without their explicit inclusion are commonly used in contemporary calculations.

An important development in L.D.A. theory was the proposal by Car and Parrinello[91] of a novel method incorporating an M.D.-like algorithm. Indeed this method opens up the possibility of *ab initio* dynamics, although in practice the method has often been used as an efficient procedure for calculating minimum energy geometries. Several elegant studies have been reported recently including a detailed study of vacancy defects in MgO[92], which give results that are in close agreement with experiment and with earlier Mott-Littleton studies. A good overview of recent applications is given by Payne[93] and further details are given in Chapter 8, which also discusses the use of more sophisticated density functionals, further reference to which will be made in the final section of this chapter.

Local Density Functional methods have unquestionably a major rôle to play in modelling inorganic materials and one that is complementary to the Hartree-Fock techniques discussed above.

3. Applications

As these are now so diverse, it is only possible within the compass of this article to give the reader a flavour of the types of problems and of systems that can now be investigated with modelling techniques. We have chosen to highlight six areas: first, modelling of structural and electronic properties of *crystalline solids*; second, the simulation of *defects* and of *atomic transport* processes in crystals; third, modelling of *surfaces* and *interfaces*; fourth, the prediction of the structures of *amorphous solids*; fifth, the simulation of the *diffusion* and *docking* of molecules within porous solids; and finally, and most challengingly, the modelling of *reaction pathways* in catalytic reactions.

3.1 CRYSTALLINE SOLIDS

Energy minimisation methods, together with interatomic potentials of the type discussed earlier in the article, can now accurately model the structures of even very complex inorganic solids including superconductors and zeolites. The accuracy of such calculations was highlighted by recent detailed comparison of experimental and energy minimised structures of high silica zeolites[94]. The need for a 'starting point' (*i.e.* an initial configuration) in such calculations means that they are essentially 'refinement' techniques rather than *ab initio* predictive methods. In this context, we should, however, note the recent work of Freeman[95] who showed that for the case of TiO_2 it was possible to use energy minimisation methods to generate an accurate crystal structure starting from a random set of atomic coordinates, given the constraints of the dimensions of the unit cell. Moreover, Newsam, Freeman and coworkers[96, 97] have shown how simulated annealing methods (which involve a Monte-Carlo like minimisation procedure) may be used to model zeolite structures with few assumptions and constraints.

Perhaps the most impressive recent examples of these simulation methods in structure prediction concern the modelling of low symmetry distortions in microporous structures. Earlier work of Catlow and Jackson[98] established the viability of the potential models that had previously been developed for α-SiO_2 (quartz) and α-Al_2O_3 (corundum) in modelling several classes of zeolites (microporous aluminosilicates). Bell *et al.*[99] then showed that the small monoclinic distortion in silicalite (a purely siliceous compound, *iso*-structural with the important catalyst H-ZSM-5 which contains a small component of Al) could be successfully reproduced by minimisation methods as shown in Table 2 and Figure 4. Recently, the structure of an unknown zeolite NU87[100] (synthesised and patented by ICI Plc.) was solved using simulations which again demonstrated the presence of a low symmetry distortion, as shown in Figure 5. The simulated low symmetry structure yielded a low 'R factor' for the high resolution powder diffraction pattern which it had not previously proved possible to refine.

Table 2: Comparison of Experimental and Simulated Silicalite*

	Experimental orthorhombic	Experimental monoclinic	Simulated
a/Å	20.07	20.17	19.986
b/Å	19.92	19.93	19.747
c/Å	13.42	13.42	13.324
a/°	90	90.64	90.803
b/°	90	90	90
g/°	90	90	90
Cell volume /Å	5365	5394	5298
Si-O Bond lengths /Å — range	1.52—1.67		1.595—1.608
Si-O Bond lengths /Å — average	1.59		1.601
O-Si-O Angles — range	97.5—129.0		106.3—115.0
Si-O-Si Angles — range	142.6—175.0		143.2—157.9
Si-O-Si Angles — average	155.0		148.8

* More details and reference to experimental work given in reference 99.

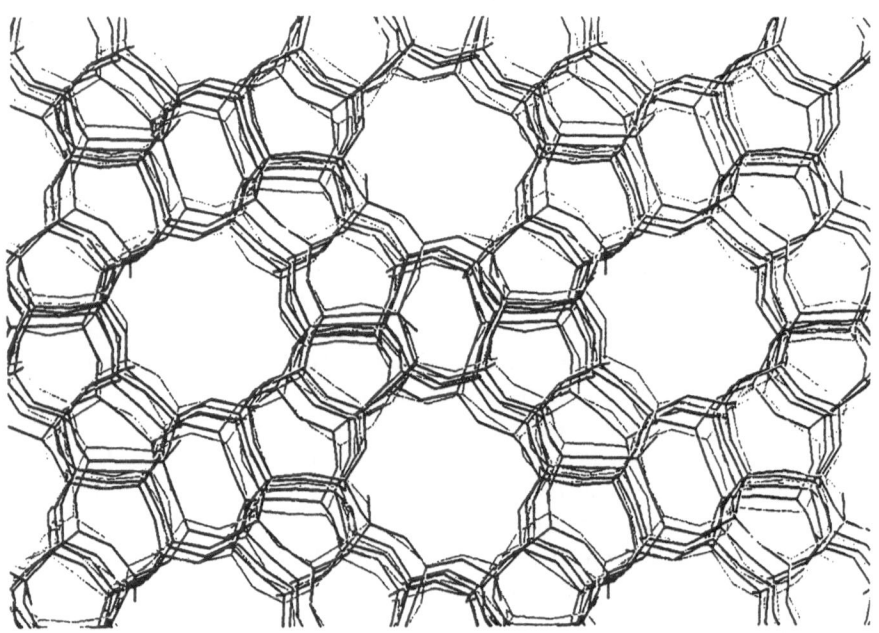

Figure 4: Simulated (darker) and experimental structure of the zeolite silicalite

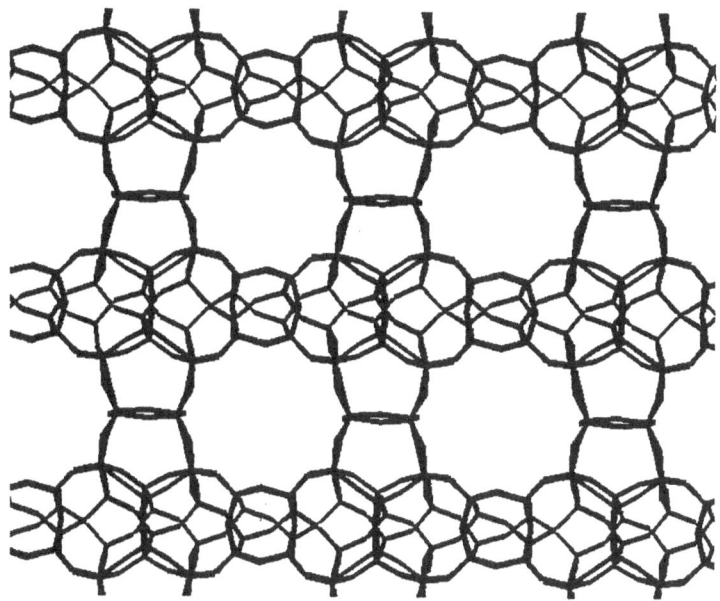

Figure 5: The energy minimised structure of zeolite Nu 87

Other recent applications include the identification of low symmetry structures of AlPO–5[101,102]

Modelling and refining of crystal structures by minimisation methods is clearly therefore now a routine matter — limited only by the availability of the necessary high quality interatomic potentials. The greatest challenge in the field is now the *prediction* of crystal structures, an exercise in global minimisation, which becomes, in practice, a question of identifying a suitable range of 'starting points' for subsequent energy minimisation. The challenge of generating plausible initial structures from random starting points is met by the use of the simulated annealing and genetic algorithm procedures. The latter approach uses evolutionary programming techniques to 'breed' successive structures whose probability of survival depends on the value of the cost function. The power of this approach can be illustrated by one example: the recent study of Bush *et al.*[103] who solved the structure of a complex ternary oxide, Li_3RuO_4, which had previously eluded structure determination. Their approach was based on a combination of genetic algorithm techniques (using a simple cost function related to Pauling's rule as discussed by Brown[104]) with energy minimisation techniques. The resulting structure leads to an acceptable Rietveld refinement of powder X-ray data. There is now a growing number of cases where unknown structures of this order of complexity have been solved by computer modelling techniques.

168

One of the most important recent areas of progress has concerned the development by Parker and coworkers[56] of free energy minimisation techniques which are coded in the PARAPOCS program. These allow simulations of the temperature and pressure variation of crystal structures. As an example of the former, we can cite the recent studies of the thermal expansivities of zeolites[105] where it was predicted that several structures (including zeolite X and L) have negative coefficients of expansion between 0K and 300K — a result which has been verified experimentally[105].

Recent high pressure studies of superconductors have provided a valuable stimulus to simulations, which have allowed the variations of critical bond length to be simulated[106,107] yielding results that are compatible with but amplify experimental data. Thus Figure 6 illustrates the calculated and experimental variations of the crucial Cu—O(2) bond (copper-apical oxygen bond in the Jahn Teller distorted octahedra) in the La_2CuO_4 superconductor. Correlations between the variations of such bond lengths and T_c are of great value in understanding the factors controlling superconductivity.

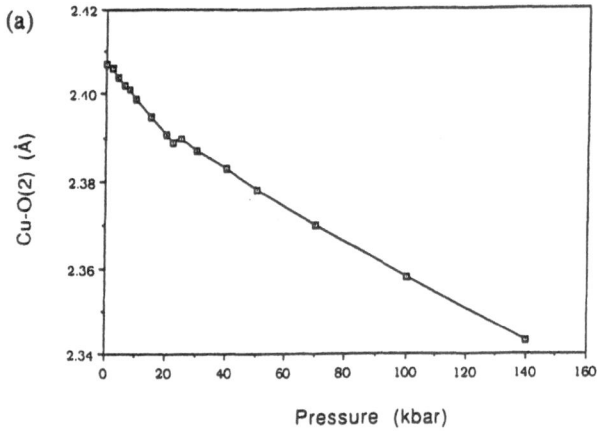

Figure 6: Calculated variation of the Cu—O(2) bond length with pressure in La_2CuO_4

We should also stress, as noted above, that substantial progress has been made in the modelling of the electronic properties of crystalline solids using both local density functional (LDF) and HF methods. As an example of the latter, we take the recent work of Nada et al.[108] who have studied SiO_2 in both the quartz and stishovite polymorphs. (Note that the latter is a high pressure phase, iso-structural with rutile, and in which Si is octahedrally coordinated.) The calculations accurately reproduced the structure of these two polymorphs and confirmed the semi-ionic nature of the bonding as shown by the electron density difference maps in Figure 7.

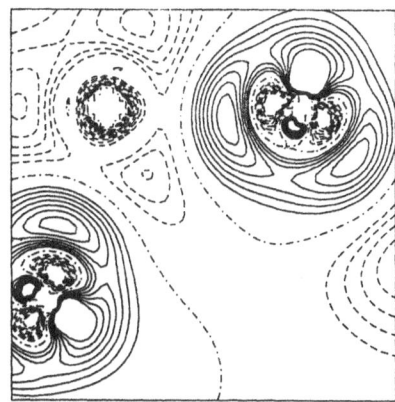

Figure 7: Calculated electron density difference map through a plane in the structure of a-quartz. Si is at left hand corner and oxygen along the x and y axes. Full lines indicate enhancement of electron density; dotted lines indicate reduction.

The ionicity is, as expected, significantly higher in the octahedrally coordinated stishovite than in the tetrahedrally coordinated α-quartz. Moreover, the calculations indicate that the d-orbitals on the Si play an important rôle in the bonding; in particular, they promote transfer of electron density from the non-bonding p-orbitals on the oxygen into the bonding Si....O s bonds. Further work with richer basis sets is, however, needed before this conclusion can be said to be definitive. Extensive work on silicates including $MgSiO_3$[109] has also been reported using these techniques. Several studies of transition metal oxides including TiO_2[110], and several of the series of rock-salt structured divalent oxides MO[111,112] (M = first series transition metal) have also recently been reported. The method is clearly having a major impact on our understanding of the electronic structure and bonding in inorganic solids. The methods are also having an impact on our understanding of more complex inorganic materials, *e.g.* the Jahn–Teller distorted structure, $KCuF_3$[113]. L.D.F. methods are also playing a major rôle in advancing our understanding of bonding in solids; further discussion will be found in reference 93 and in Chapter 8.

3.2 DEFECTS AND ATOMIC TRANSPORT

The last twenty years has seen a very extensive and successful range of applications of the basic Mott-Littleton methodology, coded in the HADES and CASCADE programs, to the formation and migration and aggregation of defects in the halide and oxide crystals. The work has firmly established the reliability of these techniques which can yield quantitative values of defect energies, as illustrated by the small selection of results collected in Table 3.

Table 3:   Some Calculated and Experimental Defect Energies*

Material	Process	Energies (eV)	
		Calculated	Experimental
NaCl	Schott Vacancy Pair Formation	2.4—2.6	2.3—2.7
NaCl	Cation Vacancy Migration	0.66	0.7—0.8
CaF_2	Anion Frankel Pair Formation	2.6—2.7	2.6—2.7
CaF_2	Anion Vacancy Migration	0.35	0.38—0.47
CaF_2	Anion Interstitial Migration	0.7	0.77—0.9
MgO	Schott Vacancy Pair Formation	7.5—7.7	6—7
MgO	Cation Vacancy Migration	2.0—2.4	2.0—2.3
NiO	Cation Vacancy Migration	1.8	1.7—2.0

* More details plus reference to calculated and experimental values given in reference 71.

One of the greatest challenges in contemporary defect physics and chemistry is provided by the high temperature superconducting materials. Defect processes indeed play a crucial rôle in most high T_C solids in which superconductivity requires doping by aliovalent ions (*i.e.* ions whose valence differs from that of the host cation which they replace). Doping (or deviation from stoichiometry) promotes hole and electron formation via appropriate redox reactions. An intriguing example of the importance and subtleties of the defect chemistry revealed by the applications of computer modelling techniques concerns the contrast between the behaviour of doped La_2CuO_4 which, on doping with divalent ions, becomes a 'hole' superconductor, and Nd_2CuO_4, in which doping with Ce promotes 'electron' based superconductivity. Calculations by Islam *et al.*[114] and Allan and Mackrodt[115,116,117] showed how oxidation of the former is energetically favoured, that is oxygen vacancies (created as charge compensators for low valence dopants) react exothermically with molecular oxygen according to the reaction:

$$V_O^{\bullet\bullet} + \frac{1}{2}O_2 \rightarrow O_O + 2h^{\bullet}. \qquad 37$$

In contrast, the reaction is endothermic with Nd_2CuO_4 as shown by the results summarised in Table 4. In contrast, Nd_2CuO_4 may be 'electron doped' by high valence dopants (which promote metal vacancy or oxygen interstitial compensation) owing to the exothermicity of reactions of the type:

$$V_M^{'''} + \frac{3}{2}O_O \rightarrow \frac{3}{4}O_2 + 2e'. \qquad 38$$

The corresponding reactions are endothermic for La_2CuO_4. The profound contrast between the hole and electron superconductivities of the two materials is thus fully explained by the differences in their defect/redox properties.

Table 4: Redox Reactions in High T_C Superconductors

Hole Superconductors (low valence doping)	$V_O^{\bullet\bullet} + \frac{1}{2}O_2 \rightarrow O_O + 2h^\bullet$
Material	Energy (eV)
La_2CuO_4	-0.25 [114] -1.7 [117]
Nd_2CuO_4	+0.1 [117]
Electron Superconductors (high valence doping)	$V_M''' + \frac{3}{2}O_O \rightarrow \frac{3}{4}O_2 + 2e'$
Material	Energy (eV)
La_2CuO_4	+8.6 [117]
Nd_2CuO_4	-2.5 [117]

A second recent example concerns the work of Zhang[118] who has applied molecular dynamics techniques to simulating the diffusion of oxygen in the YBCO superconductor. Standard microcanonical (NVE) techniques were applied to a system of composition $YBa_2Cu_3O_{6.91}$. A simulated Arrhenius plot (Figure 8) was obtained from which it is straightforward to obtain the parameters for a standard Arrhenius expression for the diffusion coefficient:

$$D = D_0 \exp(-E_{ACT}/kT). \qquad\qquad 39$$

The resulting values for D_0 and E_{ACT} show excellent agreement with experiment as shown in Table 5.

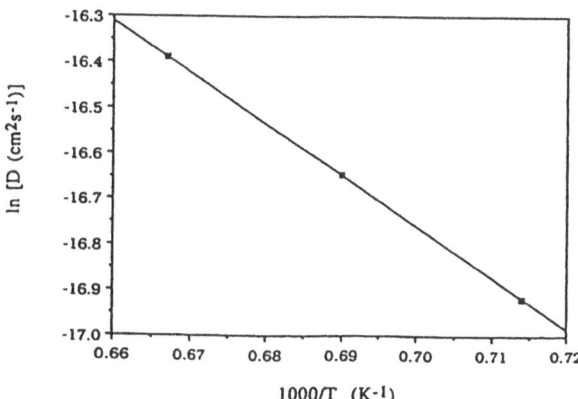

Figure 8: Calculated Arrhenius plot for oxygen diffusion in $YBa_2Cu_3O_{7-x}$. Calculated points are essentially superimposed on extrapolated experimental data.

Table 5: Comparison of Calculated and Experimental Diffusion Coefficients in $YBa_2Cu_3O_{6.91}$*

T (K)	Experimental D (x 10^{-8} cm^{2}s^{-1})	Calculated D (x 10^{-8} cm^{2}s^{-1})
1400	4.53	4.5±0.2
1450	5.97	5.6±0.6
1500	7.74	7.6±0.3

* See reference[118] for further details and references. Note that experimental data have been extrapolated to high temperatures at which MD calculations have been performed.

Analysis of the oxygen ion trajectories revealed that diffusion takes place predominantly by migration of oxygen between O(4) sites but *via* the O(1) and O(5) sites (see Figure 9). Propagation of 'interstitials' along O(5) sites which had been suggested in previous discussions of this system is not supported by these results. The power of modelling techniques both in directly simulating diffusion and in elucidating mechanisms is clearly demonstrated.

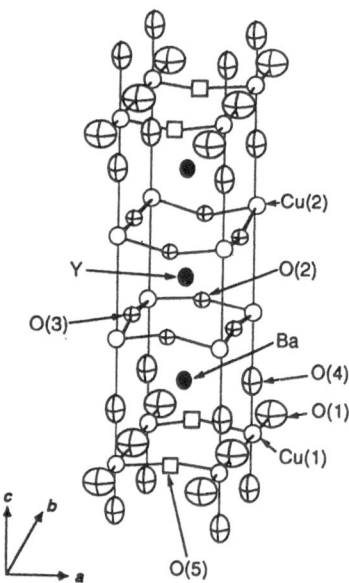

Figure 9: Crystal structure of $YBa_2Cu_3O_7$ indicating oxygen sites involved in diffusion processes.

The value of defect calculations in advancing understanding of ionic mobility in solids is also clearly shown by recent calculations on oxygen migration in perovskite structures oxides. Such materials may show high oxygen ion mobilities arising from mobile vacancies introduced by low valence cation dopants. Cherry *et al.*[119,120] used standard Mott–Littleton methods first to determine the vacancy migration mechanism, which they showed to proceed by a curved path to the migrating oxygen between the two lattice sites involved in the migration process. More significant was the demonstration that the migration energy is strongly dependent on the relative radii of the A and B metal ions in ABO_3 perovskite structured oxides. This result, which is in line with experimental data, provides guidance for the development of materials with optimised oxygen mobility.

3.3   SURFACES AND INTERFACES

Work of Tasker, Mackrodt, Colbourn and coworkers (see *e.g.* references 121, 122) established that both perfect and defect lattice simulations could be adapted to modelling surfaces and surface defects in ionic materials. Modelling of perfect surfaces involves equilibrating an infinite two dimensional slab, embedded on one side by an unrelaxed crystal which extends to infinity in the direction perpendicular to the slab as shown in Figure 10. Early calculations on simple materials demonstrated the importance of 'surface rumpling' in which cations and anions on the surface layer are displaced perpendicular to the surface in opposite directions. More complex surface relaxations take place in materials such as corundum as revealed by the work of Mackrodt and coworkers[122] whose simulated structure for the (0001) plane of this material is shown in Figure 11.

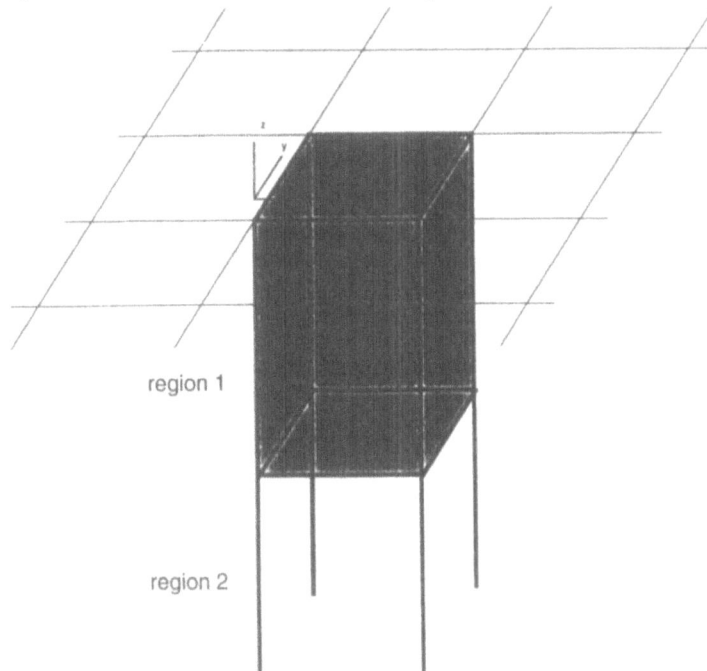

Figure 10: 'Two-Region' approach used in surface simulations

174

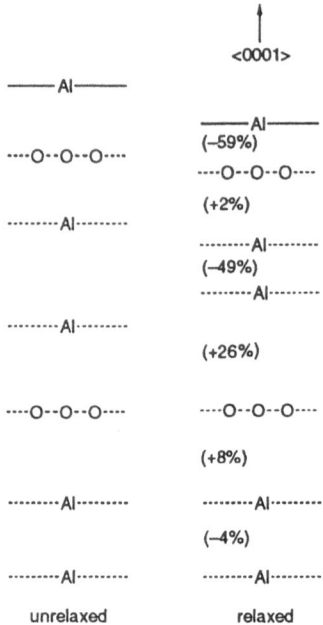

Figure 11: Comparison of the unrelaxed and classically relaxed stacking structure of the (0001) surface of α-Al₂O₃. (Figures in brackets are the percentage changes in the interplanar spacings) (kindly supplied by Dr W.C. Mackrodt)

Recent work of Gay, Rohl and coworkers has extended the complexity of systems that may be simulated by current surface simulation techniques. In particular, the MARVIN code[123], which uses the same basic approach as that employed in references 121 and 122, was written to allow routine generation and minimisation of the surfaces of complex crystal structures. Examples are provided by recent applications to the surfaces of quartz and of zeolites, illustrated diagramatically in Figures 12 and 13. The calculations showed firstly that stability of these surfaces required hydroxylation of terminal oxygen atoms. Secondly, the results again reveal significant surface relaxations. The calculations on the zeolitic systems represent the first attempt to construct atomistic models of the external surfaces of these materials.

Of even greater interest are the interactions between complex surfaces and sorbed molecules. A particularly topical case is provided by the interaction of growth inhibitor molecules with the surface of BaSO₄. BaSO₄ is, of course, highly insoluble and its precipitation in, for example, oil pipe lines can be a major technological problem. There is, therefore, considerable incentive for the design of effective inhibitors to prevent precipitation of this material. One strategy to achieve this end is to add 'growth inhibitors', of which a widely used class is the alkyl diphosphonates. Rohl *et al.*[124]

therefore investigated the interaction of such molecules with the surfaces of BaSO₄. They find strong surface inhibitor binding energies, which are, however, surface sensitive. Figure 14 shows the configuration of the docked molecules on the 100 surface. The strong binding of the inhibitor is clearly the basis of the growth inhibition mechanism.

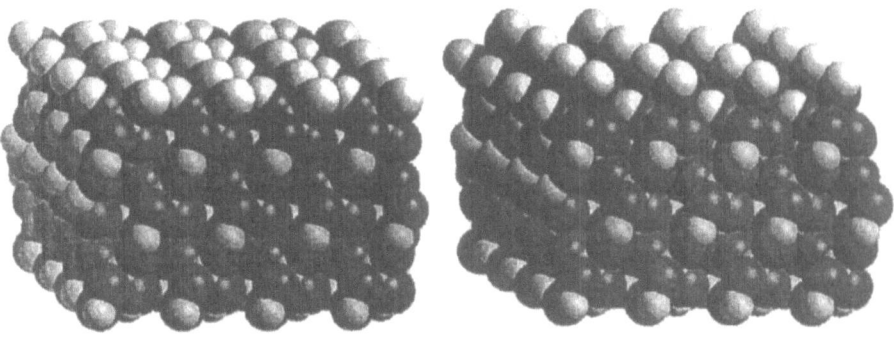

Figure 12: The simulated surface structure of α–quartz (lightest coloured spheres are hydrogen; darkest are oxygen). The left hand side diagram is the unrelaxed structure; the right hand side, the relaxed structure. Surface hydroxyl groups stabilise the surface.

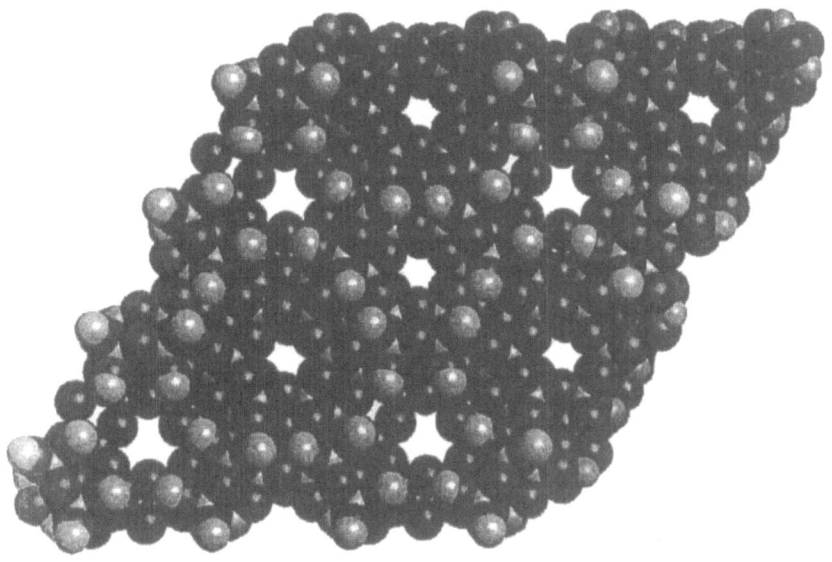

Figure 13: The simulated surface structure of the siliceous form of the zeolite faujasite. Surface hydroxyl groups are again present.

176

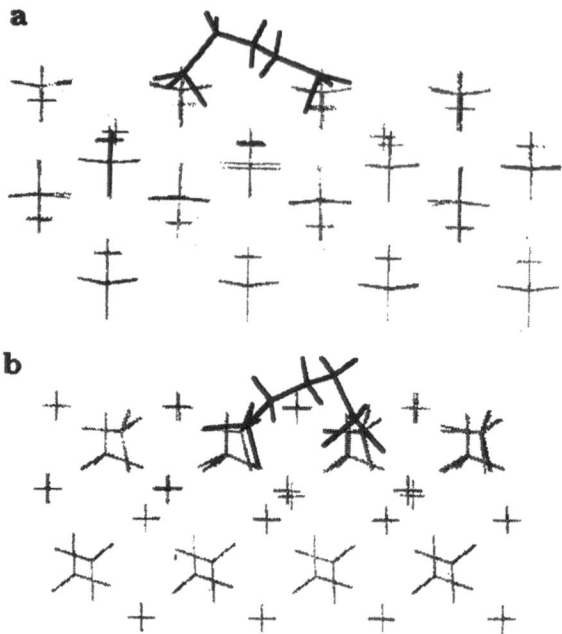

Figure 14: The two most stable docking sites for an alkyldiphosphonate on the surface of barite (a) on the 5.47Å site on the a cut of the (100) surface and (b) on the 5.66Å site on the (011) surface. The diphosphonate is coloured black and the surface barite ions grey.

Calculations on surface defects and impurities employ a modification of the bulk defect methodology described in Section 2.3.11, in which a hemispherical region of the lattice (centred on the defect) is relaxed to zero force, with the polarisation of the more distant regions being modelled employing a Mott-Littleton-like procedure. One of the most successful applications of such methods has been to the study of surface segregation whereby defects and impurities preferentially occupy surface sites (although the reverse behaviour, with depletion of surface occupancy is also known). The occurrence of segregation may drastically modify physical and chemical properties of surfaces. Detailed studies of ceramic materials have been reported by Mackrodt, Tasker and coworkers as reviewed by Mackrodt in references[122,125]. Here we highlight an interesting recent study of Lewis *et al.*[126] who have investigated the system Li doped MgO which as shown by Lunsford[127] can function as a partial oxidation catalyst, converting methane into C_2 species. As will be discussed in much greater detail in Section 3.6.1, it is generally considered that the catalytic activity of this material is promoted by hole states (which may be simply described as O^- species) which charge compensate the Li^+ substituting for Mg^{2+}. Calculations have revealed that both the Li^+ species and the lithium-hole (electroneutral) complex segregate to the surface. The latter is a plausible model for an active site of this catalyst. They found, moreover, that low coordinate sites, (corners and steps) are particularly effective traps for surface holes.

A further interesting development in our understanding of this system concerns the rôle of Cl^- ions on the catalytic chemistry. Introduction of surface Cl^- is known to enhance both the activity and selectivity of the catalyst[128]. Recent calculations of Lewis *et al.*[129] have highlighted the rôle of surface defect chemistry in this system. The lowest energy state for both bulk and surface holes is in a trapped state in which the hole occupies a site adjacent to a substitutional Li^+, to give the so-called Li^0 centre which we can represent as $(Li^+O^-)_S$ for the surface species. Lewis *et al.* showed that a surface Cl^- (substituting for an oxygen ion) could displace one of the hole states from the site adjacent to the surface Li^+ according to the reaction:

$$[LiO^-]_S + Cl_S^- \rightarrow [Li^+Cl^-]_S + O_S^-$$

The simulations also showed that, in an analogous manner, surface Cl^- ions could 'free' holes from low coordinate sites. The free surface O^- ions (O_S^-) may then provide much more active catalytic centres. Defect reactions clearly control catalytic chemistry. We return in Section 3.6.1 to a more detailed consideration of the reactivity of this system.

Calculations of the type we have described have been applied with success to the simulation of grain boundaries[130], while studies of the intriguing problems posed by metal-oxide interfaces are in progress. The use of modelling techniques to probe interfacial properties is expected to be a growth point in the field in the next few years.

3.4 AMORPHOUS STRUCTURES

Modelling of non-crystalline structures is one of the most difficult yet rewarding areas of application of computational techniques. The difficulties arise from the problems of developing reliable procedures for generating amorphous structures; the rewards, from the insights which computer generated models can give into the structures of non-crystalline networks.

The most widely used procedure in this field involves a simulated melt and quench, *i.e.* a simulation of the real procedure whereby glasses are made. Molecular dynamics techniques are used to melt a crystalline system, the temperature of which is reduced in steps to below the glass transition temperature. The problem with this procedure is of course the time scale: the simulated quench takes typically 50-500 ps; a real quench is ~1s or longer.

Despite this fundamental difficulty, the method has enjoyed considerable success. Pioneering work of Woodcock, Angell and Garafolini[80,81] of vitreous silica has been followed by several more recent studies[131-134,136-139]. Feusten and Garafolini[133] and Vashishta[134]

178

were able accurately to reproduce experimental radial distribution functions obtained from both X-ray and neutron diffraction studies. Recently Wright[135] has obtained very high quality neutron data on vitreous SiO_2 using the high energy pulsed neutron source ISIS. The resultant r.d.f. has been accurately reproduced by Vessal *et al.*[137] as shown in Figure 15. In an earlier study, Vessal *et al.*(82) were able, using constant pressure molecular dynamics, to simulate in detail the process of melting and glass formation in SiO_2. Figure 16 shows their plot of internal energy (E) *vs* temperature in a simulated heating/cooling cycle; both melting and the glass transition are clearly noted. The glass resulting from this simulation is, however, of considerably higher density than that normally observed, although Vessal *et al.*(82) argued that it provided a good model for glassy materials produced by shock waves, in which the conditions and timescales for glass production might be closer to those of the M.D. simulations.

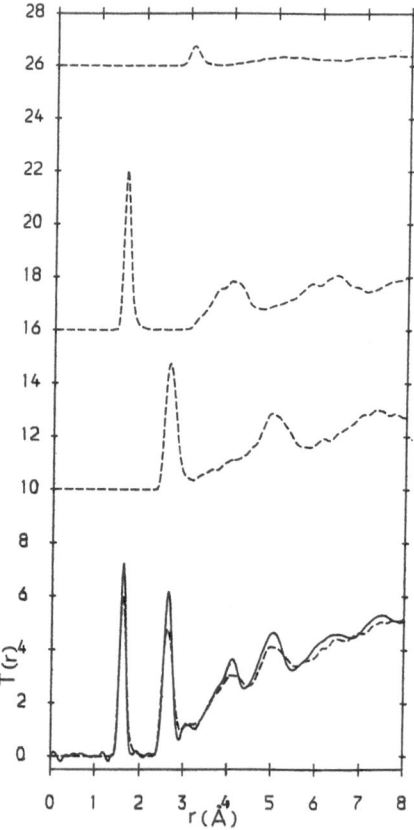

Figure 15: Calculated (dotted line) and experimental (full line) radial distribution functions (r.d.f.) for vitreous silica. The upper curves are partial r.d.f.s for respectively Si - -Si (top), Si - - O and O- - -O pairs.

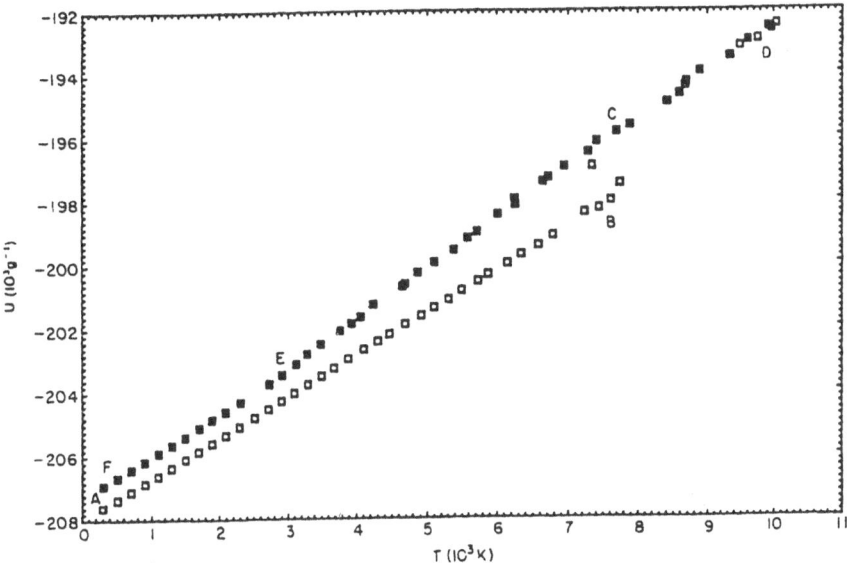

Figure 16: Plot of the total internal energy against temperature for silica. Open squares are points during the heating of the system; filled squares are points during cooling of the system.

Some of the most interesting recent studies of amorphous materials have concerned vitreous alkali silicates where work of both Cormack *et al.*[130] and Vessal *et al.*[139] has shown how alkali cations tend to cluster. Such structures support the modified random network models proposed by Greaves[140], one of the conceptual bases of which is that the network modifying cations considerably perturb the framework and group into channels. Such perturbations and channels are clear from the simulated structures.

The melt/quench procedure described above is, of course, not the only way of generating glass structures. Physical models have a long and successful history in this field, and will continue to play an important rôle — possibly reinforced and refined by computer simulations. Topological principles may also be used to guide the construction of models of amorphous systems as in the recent work of Marions and Hobbs[141]. There is no doubt, however, that computational methods have an important contribution to make to this difficult area of structural chemistry.

3.5 SIMULATION OF SORPTION

There is, of course, a long history of calculations on sorbed molecules on surfaces. Here we will focus on the study of sorption in the microporous solids, that is structures with enormous internal surfaces. Again, the problem has been extensively studied, with pioneering work of Kiselev and coworkers(31) whose investigations of hydrocarbon sorption in zeolites provided a valuable parameterisation for the interatomic potentials to be used in such studies. A landmark in the development of this field was the work of Wright *et al.*[142] who solved the problem of the location of pyridine in zeolite L by a combination of computer modelling and high resolution neutron powder diffraction techniques. The modelling calculations used simple energy minimisation methods within a rigid zeolite framework.

More recently Parker *et al.*[143] extended the technique by including full framework relaxation around the sorbed molecules in a study of C1-C8 hydrocarbons in the zeolites silicalite, ZSM-5 and H-faujasite. They obtained sorption energies in good agreement with experiment and showed, moreover, that the effect of including framework flexibility was significant for the larger molecules.

The above calculations employed conventional energy minimisation procedures whose use becomes increasingly problematic with the complexity of both the molecule and the host, owing to the increasing uncertainties caused by local minima. For this reason, Freeman *et al.*[144] have developed a more automated procedure based on a blend of energy minimisation, Monte Carlo and molecular dynamics methods. The latter is used in the first step of the calculations to generate for the sorbant a library of conformational states, each of which is then introduced into the zeolite by a crude M.C. procedure, *i.e.* random configurations are generated for the molecule within the solid, and only those are accepted whose energy falls below a specified threshold value; this in effect eliminates configurations which overlap too strongly with the walls of the zeolite cages. Each accepted configuration is then subjected to energy minimisation in which the framework relaxation should be included if computer time permits.

The method was successfully applied to the study of butene isomers in the zeolite H ZSM-5 which catalytically converts a mixture of such isomers predominately to iso-butene. The results summarised in Table 6 show that iso-butene is the least strongly bound of the isomers which is consistent with it being the dominant product in the isomerisation reaction. The energy minimised configuration (Figure 17) reveals that even in this structure, there is no satisfactory registry between the carbon atoms of the molecule and the oxygen of the framework, which may be identified as the cause of the low binding since the C···O van der Waals term is a dominant source of the molecule-host interaction.

Table 6: Butene isomer binding energies in silicalite (from reference 144)

Isomer	calculated minimum emergy kJ mol^{-1}	calculated average emergy kJ mol^{-1}	relative minimum emergy kJ mol^{-1}	relative average emergy kJ mol^{-1}
2–methyl propene	–38.58	–29.09	0	0
but–1–ene	–54.30	–44.41	–15.72	–16.30
cis–but–2–ene	–45.45	–31.64	–6.87	–3.55
trans–but–2–ene	–48.76	–38.44	–10.18	–10.35

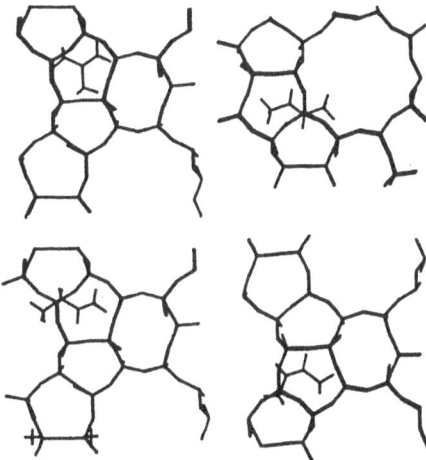

Figure 17: Typical minimum energy conformation for four isomers in silicalite

Calculations of the above type may be carried out cheaply and straightforwardly, and as such can contribute considerably to the study of sorption. They do, of course, tell us nothing about the dynamics of the sorbed molecules. For this reason, M.D. techniques are being used increasingly in this field. The viability of the technique in investigating the diffusion of sorbates within zeolite pores was demonstrated, for example, in the work of Pickett *et al.*[145] on Xe diffusion in a variety of zeolite systems. Work of Demontis *et al.*[146,147], June *et al.*[148,149] and Goodbody *et al.*[150] has applied the techniques to hydrocarbon diffusion principally in the purely siliceous zeolite silicalite. An example of the success of this approach is the work of Kawano *et al.*[151] who investigated methane diffusion in silicalite. Their simulated Arrhenius plot, obtained by running the M.D. simulation at a variety of temperatures and calculating the diffusion coefficient at each, is shown in Figure 18 which also shows experimental data. The agreement between the simulated and experimental results is seen to be excellent.

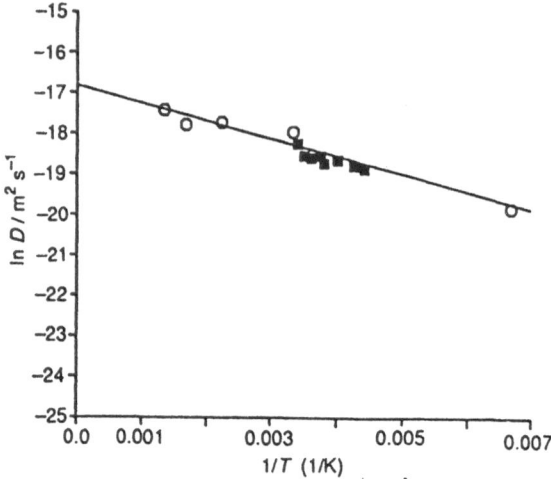

Figure 18: Calculated (circles) and experimental (squares) diffusion coefficients for CH_4 in silicalite represented as an Arrhenius plot

A more recent study of Hernandez and Catlow[152] explored the diffusion of butane and hexane in the pores of silicalite. The simulation found the expected decrease in diffusion coefficients with loading and achieved satisfactory agreement with experimental diffusion data. Of particular interest was the analysis of the migration mechanism, which was shown to proceed by a jump diffusion process, with jumps across the channel intersections of the silicalite structure. The study highlights the detailed nature of the information that is available from MD studies of sorbed molecules within microporous materials.

3.6 MODELLING OF REACTION MECHANISMS

One of the greatest challenges in theoretical solid state chemistry is to model the mechanism of reactions (especially those of catalytic importance) on the surface of solids and within their pores. Considerable effort has been devoted to chemisorption and simple reactions on surfaces — both metallic and ionic — of which a good review of earlier work is available from Colbourn[87]. To illustrate the present state of the art we highlight two of our own recent studies of key catalytic processes.

3.6.1 *CH-Bond Activation on Catalytic Oxides*

Oxidative coupling of methane (OCM) is known to be catalysed by a variety of oxidic materials[127,153]. Among these, Li/MgO (lithium doped magnesium oxide) may serve as the prototype, the defect chemistry of which was considered in Section 3.3. It has attracted a considerable number of investigations, both experimental[127,153,154] and theoretical[155-161] over the past decades. But although these studies created a large amount of data, many questions concerning the detailed nature of the physical and chemical processes involved are still to be clarified.

As discussed earlier, the catalytic activation of the process leading from methane to methyl radicals ($\cdot CH_3$, observed experimentally[162,163]) probably involves some kind of surface defect site. Evidence from e.p.r. measurements indicates a correlation between catalytic activity and the presence of O^- surface species[164,165]. In particular, as noted earlier, the $[Li]^0$ centre has been proposed as an active site for OCM catalysis on Li/MgO[154]. The importance of this type of defect, consisting of an electron hole localised on an oxygen ion and stabilised by an adjacent Li (dopant) ion, has been demonstrated[166], but has also recently been questioned by the work of Lunsford and coworkers[154]. In several theoretical approaches $[Li]^0$ has served as a model of the active surface site[155-161]. Other defects known to be present on MgO surfaces (like dislocations and charged or neutral vacancies) may also be considered (see *e.g.* references 160, 167, 168, 169).

Theoretical modelling of reactions at defective MgO surfaces has been undertaken by several workers employing different methods (ranging from EHT to MCSCF)[155-161, 167-169]. More than 20 years ago, *ab initio* calculations were used to investigate H_2 bond breaking at a cation vacancy site[159] and computational studies of the role of MgO surface defects in catalysis have continued to be performed[168,169,170,171].

The main challenge lies in choosing an appropriate description of both (a) the surface defect and (b) the reaction. While (a) can be achieved through atomistic modelling (see, for example, reference 172 and the discussion in Section 3.3), (b) requires a quantum mechanical approach. As we have found, however, the level of theory can affect results even qualitatively.

Since the long–range nature of the Madelung field demands an accurate treatment, to satisfy (a) one has to go beyond the simplest (*i.e.* most localised) approximation. This can be done by using large cluster models (allowing substrate relaxation), embedding of some kind or a (pseudo) two-dimensional representation of the system (slab). Ideally a model should fulfil both (a) and (b) as well as possible, but unfortunately the prohibitively high computational cost involved still restricts the degree to which this goal may be reached. We summarise below our recent work on this system, a more detailed account of which is given in reference 173.

The process here envisaged, $CH_4 + [Li^+O^-]_s \rightarrow \cdot CH_3 + [Li^+(OH)^-]_s$ (index s indicates surface species) involves the homolytic breaking of a CH bond; in the same process an OH bond is formed. We note that open shell systems are present in both reactants and products. Therefore, an electronic structure calculation (and subsequent geometry optimisation) ought to be carried out at a sufficiently high level of theory. Starting from Hartree Fock (HF) SCF level this also includes taking into account electron correlation to some extent. Although computationally demanding, a way to do this systematically is by applying post-HF schemes (like MP2, MP4 etc.). Alternatively, and at much lower cost, one can use an approach based upon gradient corrected density functional theory (DFT). Recent improvements in the gradient corrections applied (in this work we use the B-LYP correction[174,175]) have rendered this method very well capable of determining geometries and energies with a high accuracy[176,177]. It seems therefore to be ideally suited for the requirements outlined above.

In previous studies[156,157] very simple models have been used to describe the interaction of methane with a $[Li]^0$ centre. An efficient, though still crude way of improving the quality of the model is by using a cluster embedded in an array of point charges. In our model (Figure 19) only lithium, oxygen and methane are treated explicitly; the MgO crystal is represented by about 200 point charges (q = ±2.0 e) at their bulk positions. The CH and OH distances are varied and for each combination of d(CH) and d(OH) the remaining parameters are optimised. This is important for two reasons: (1) the oxygen ion accepting the hydrogen moves during this transfer process by several tenths of an Angstrøm and (2) the methyl radical generated in the process tends towards a planar structure gaining an energy of about 0.3 eV.

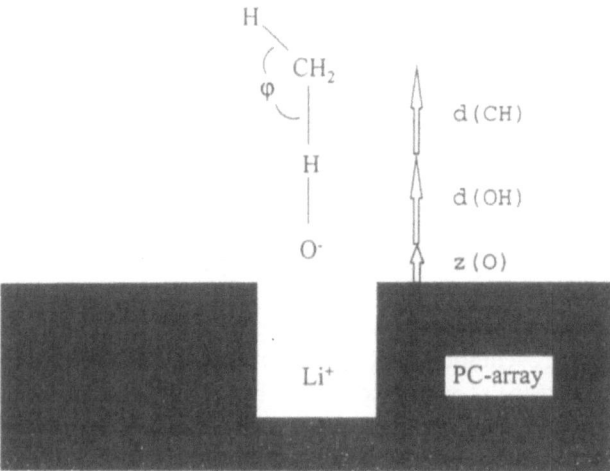

Figure 19: The $[Li^+O^-]$-H-CH$_3$ model system. Its geometry is determined by the bond distances d(CH) and d(OH), the position of oxygen relative to the surface z(O) and the HCH angle . Li is kept fixed at 2.11 Å below the surface.

The resulting energy surfaces (Figure 20) display several important features and some significant differences. From HF-calculations (Figure 20a) we find the classic shape of reactants valley (R), transition state (TS) and product valley (P). In contrast to the results of Zicovich-Wilson, however, this is not a late TS[156]; the saddle point is found in a rather symmetric position (at about 0.8 eV above the reactants level). This may be caused by the fact that the adoption of an increasingly planar structure, accompanying the release of ·CH₃ is only very crudely taken into account in reference 156; their model differs from the one used in this study in the position the dopant atom assumes, but this is not likely to be the cause of such a pronounced, qualitative effect; furthermore, in omitting the ionic crystal field entirely, their description differs from the present one considerably.

The surface resulting from B-LYP-DFT calculations is depicted in Figure 20b. Here no TS is observed. In fact, the transition of hydrogen from carbon to oxygen occurs without activation barrier. However, some energy is needed for the release of the methyl radical fragment. The minimum found at a CH distance of 1.7Å lies 0.43 eV below the product level (this value is reduced by a factor of two, when substrate relaxation is included[173]). Another difference seen in comparing Figure 20a and 20b is a slightly shorter OH distance at the large d(CH) observed in the HF results. This may be a compensation effect for a slightly smaller z(O) in the DFT calculation, a (very moderate) overbinding effect.

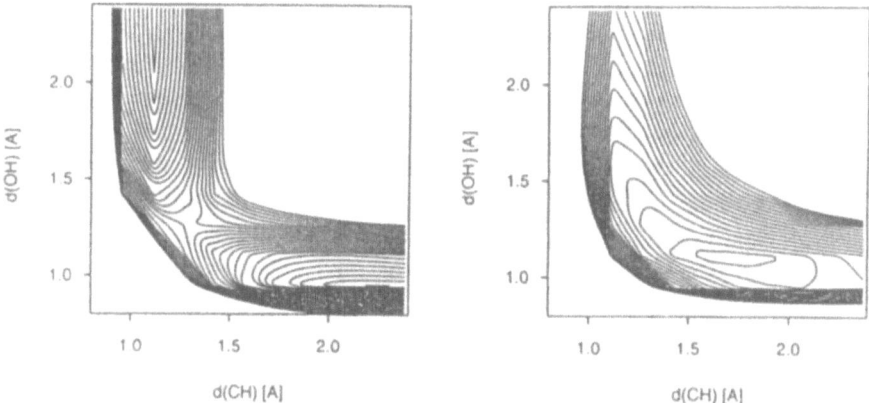

Figure 20: The total energy of [Li⁺O⁻]-H-CH₃ as a function of the CH and OH distance calculated using (a) HF and (b) B-LYP-DFT. The contour lines are spaced at 54 meV.

The results presented here clearly demonstrate, that a complete neglect of electron correlation (pure HF) leads to a qualitatively different description of the methane dissociation process than is achieved by using an advanced (gradient corrected) DFT scheme. Thus, in further studies employing large clusters it will be essential to use methods that include electron correlation. Considering the computational cost involved, DFT appears to be ideally suited for this challenge.

3.6.2 *Reactivity of Methanol at Alumino-Silicate Bronsted Acid Site*

Computational chemistry is contributing increasingly to our understanding of reactivity in zeolites. Of particular interest has been the prototypical methanol-to-gasoline (MTG) reaction[178] which, even after 20 years of experimental study (and more recently theoretical study), is still poorly understood.

Some years ago, Hutchings and Hunter suggested a mechanism for the MTG process in acidic zeolites that satisfied all of the accumulated experimental data[179]. They were also able to show that a number of popular mechanisms suggested in the literature were unlikely. Their mechanism, the methyl oxonium ion mechanism, proceeds schematically as follows:

Acid site   +   methylating agent   -------->   methylated acid site   (i)
(ZOH) CH_3OH, CH_3OCH_3 *etc.*   (ZOCH$_3$)

          methylated acid site   --------> surface stabilised carbene (ii)
               (ZOCH$_3$)   (ZO$^-$..CH$_2$)

          surface stabilised carbene   -------->   first C-C bond   (iii)
               (ZO$^-$..CH$_2$)   *e.g.*, $CH_2=CH_2$

Formation of the methylated acid site is the subject of the present section. The ongoing debate concerning the nature of the initial complex formed on methanol adsorption at an alumino-silicate Brønsted acid site[180-193] will not be discussed except to state that our current work suggests that at normal temperatures, the dominant adsorbed methanol species in acidic zeolites is a physisorbed species[193]. All reactions reported in the current work involving adsorbed methanol will be with respect to this physisorbed species.

A number of pathways for formation of a methylated acid site are conceivable. The simplest, an S_N2 reaction of a single CH_3OH molecule at a Brønsted acid site, [AlO(H)Si], has been studied by Blaskowskii and van Santen[187] and by Zicovich-Wilson *et al.*[194]. Due to different methodologies, the former reported an activation barrier of 184 kJ mol^{-1} whilst the latter reported a barrier of 217 or 171 kJ mol^{-1} depending on which oxygen site within their model cluster was methylated.

Another possibility for formation of the surface methoxyl species is that shown in Figure 21, *i.e.*, a pathway involving a single CH_3OH molecule but with more S_N1 character[193]. As expected on the basis of cluster calculations which omit long range electrostatic effects, no stable, charge separated S_N1 intermediate was observed. The calculated activation barrier for this process of around 230 kJ mol^{-1} at the TZVP/DFT(BLYP) level of theory indicates that this pathway is only likely to play a minor rôle in formation of the methylated acid site. We note however, that due to the polar nature of the mechanism, inclusion of long range electrostatic effects into the model is likely to lower the activation barrier.

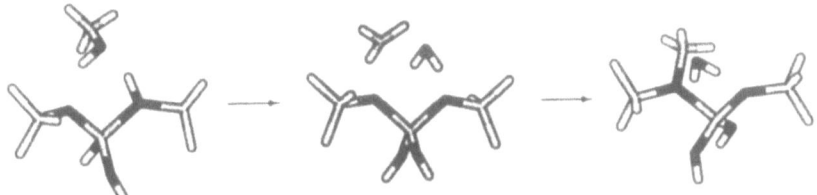

Figure 21: TZVP//DFT(BLYP) optimised structures for the methylation of a model Brønsted acid site by one methanol molecule. The acid site is modelled by a $H_3SiOAl(OH)_2O(H)SiH_3$ cluster and the central structure is the transition state.

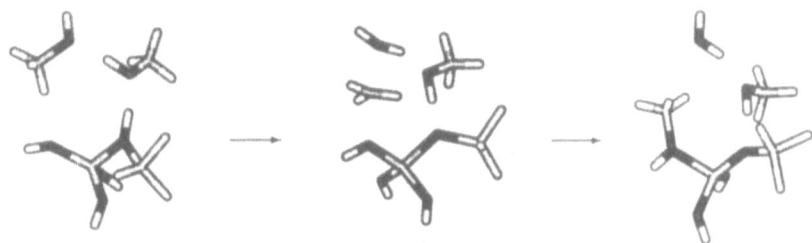

Figure 22: HF/3-21G optimised structures for the methylation of a model Brønsted acid site by two methanol molecules. The acid site is modelled by a $(HO)_2AlO(H)SiH_3$ cluster and the central structure is the transition state.

A third possibility for the mechanism of reaction (i) is *via* interaction of two methanol molecules with the acidic Brønsted site, Figure 22. We have been able to show that such a pathway proceeds with an activation barrier in the region of 130-160 kJ mol^{-1} at the MP2/6-31G**//HF/3-21G level of theory and using a range of model clusters[195]. The use of different model clusters also enabled us to demonstrate the significant effect of the difference in proton affinities of the two oxygen sites involved on the magnitude of the barrier. This effect was also discussed by Kramer *et al.* [196] who used it to rationalise the different reactivities of methane in MFI and faujasite type zeolites.

Finally, it is possible that dimethyl ether is the primary methylating agent in the initial stages of the MTG process. It is known that dimethyl ether forms almost immediately upon introduction of CH_3OH into acidic zeolites and well before the onset of hydrocarbon formation[197]. However, it is unclear as to whether dimethyl ether forms from the reaction of methanol with a surface bound methoxyl group ($ZOCH_3$)[198,199,200] or whether it forms from conventional condensation of methanol in the acidic environment of the zeolite pores[201,202].

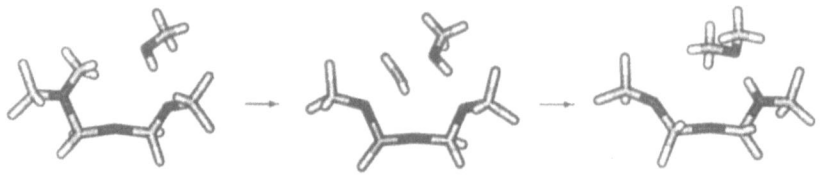

Figure 23: HF/6-31G** optimised structures for formation of dimethyl ether from a methylated acid site and adsorbed methanol. The methylated acid site is modelled by a $H_3SiO(CH_3)Al(H)_2OSi(H)_2OSiH_3$ cluster and the central structure is the transition state.

HF/6-31G** calculations of the formation of $(CH_3)_2OH^+$ from CH_3OH and $ZOCH_3$ (Figure 23) gave an activation barrier of 150 kJ mol^{-1} (desorption of the product was less activated). Assuming that the $ZOCH_3$ species was formed from adsorbed methanol (as above), the effective activation energy for formation of $(CH_3)_2OH^+$ (and from it, CH_3OCH_3) from methanol at a Brønsted acid site *via* a surface methoxyl species is estimated to be about 120 kJ mol^{-1}, *i.e.*, the activation barrier of 150 kJ mol^{-1} plus the energy to form ZOMe ($\sim$–5 kJ mol^{-1}) plus the energy of adsorption of CH_3OH at the methylated acid site ($\sim$–25 kJ mol^{-1}). Estimates of the effect of electron correlation (at the MP2 level) suggest that the barrier should probably be lower at about 100 kJ mol^{-1}. This value compares with the gas phase activation barrier of ~35 kJ mol^{-1} for formation of $(CH_3)_2OH^+$ and H_2O from $CH_3OH_2^+$ and CH_3OH[196]. It is thus clear that CH_3OCH_3 probably forms from condensation of CH_3OH in the zeolite pores. CH_3OCH_3 is therefore a good candidate for the methylating agent responsible for formation of methylated acid sites. The reaction of CH_3OCH_3 at an alumino-silicate Brønsted acid sites is currently under study, with initial results suggesting that the activation barrier for formation of $ZOCH_3$ from adsorbed dimethyl ether is indeed lower than that for its formation from methanol. Further studies of this problem have been reported recently by van Santen[203].

Thus although the mechanism of the MTG process is far from being understood, computational quantum chemistry methods are clearly bridging the gap in experimental knowledge.

4. Conclusions

Computer modelling techniques are now having a substantial impact on a broad range of problems in materials chemistry. They have developed well beyond the stage of 'demonstration' calculations which simply reproduce known experimental data. They are now able to provide insight as to structures and mechanisms at the atomic level, and they are developing an increasingly predictive rôle. Advances in techniques, algorithms and computer power offer an exciting future with a great expansion in the predictive and explanatory capabilities of the methods.

Acknowledgements

I am grateful to P.E. Sinclair and L. Ackermann for permission to quote their unpublished results. I would like to thank Professors Sir John Meurig Thomas, A.M. Stoneham and A.K. Cheetham for many useful discussions relating to the work discussed in this review.

References

1. C.R.A. Catlow, S.C. Parker and N.M.P. Allen (eds), 'Computer Modelling of Fluids, Polymers and Solids', NATO Series, Kluwer Academic Publishers, Dordrecht, 1989, Vol. 293.

2. C.R.A. Catlow and W.C. Mackrodt (eds), 'Computer Simulation of Solids', Lecture Notes in Physics, Springer, Berlin, 1982, Vol. 166.

3. C.R.A. Catlow and A.N. Cormack, (1987) Int. Rev. Phy. Chem., 6, 227.

4. W.C. Mackrodt, 'Transport in Non-Stoichiometric Compounds', eds. G. Petot-Ervas, Hj Matzke and C. Monty, North-Holland, Amsterdam, 1984.

5. (1989) J. Chem. Soc., Faraday Trans. 2 , Vol. 85.

6. C.R.A. Catlow and G.D. Price, (1990) Nature, 347, 243.

7. P.D. Shannon and C.T. Prewitt, (1970) Acta. Cryst., B26, 1046.

8. G. Wells, 'Structural Inorganic Chemistry', Oxford University Press, 1990.

9. W.M. Meier and H. Villiger, (1969) Z. Kristallogr. Kristalgeom., 129, 411.

10. M.J. Dempsey and R.G.J. Strens, 'The Physics and Chemistry of Minerals and Rocks' eds. R.G.J. Strens and J. Wiley, London, 1976, p.443.

11. D.L. Bish and C.W. Barnham, (1984) Amer. Miner., 69, 1102.

12. W.A. Dollase and W.H. Bauer, (1976) Am. Miner., 61, 971.

13. V. Gramlich and W.M. Meier, (1971) 2 Kristallogr. Kristallgoem., 133, 134.

14. P.P. Ewald, (1921) Ann. Physik, 64, 253

15. M.P. Tosi, 'Solid State Physics', eds. F. Seitze and K. Turnbull, Vol. 16, New York, Academic Press, 1964, p.1.

16. C.R.A. Catlow and M.J. Norgett, UKAEA Report, 1976, AERE-M2936.

17. H.D.B. Jenkins and P. Hartman, Phil.Trans.Roy.Soc., 1982, A304, 397

18. H.D.B. Jenkins, 'Computer Simulation of Solids', eds. C.R.A. Catlow and M.J. Norgett, Lecture Notes in Physics, Vol. 166, 1982, chap. 16.

19. H.D.B. Jenkins and P. Hartman, (1979) Phil. Trans. Roy. Soc., A293, 169.

20. A.M. Stoneham and J.H. Harding, (1986) Ann. Rev. Phys. Chem., 37, 53.

21. M.J. Sanders, M. Leslie and C.R.A. Catlow, (1984) J. Chem. Soc. Chem. Comm., 1271.

22. B.M. Axilrod and E. Teller, (1943) J. Chem. Phys., 11, 299.

23. W.J. Meath and R.A. Aziz, (1984) Mol. Phys., 52, 225.

24. R.C. Baetzold, C.R.A. Catlow, J. Corish, F.M. Healy, P.W.M. Jacobs, M. Leslie and Y. Tan, (1989) J. Phys. Chem. Solids, 791.

25. C.R.A. Catlow in 'Computer Modelling of Solids', eds. C.R.A. Catlow and W.C. Mackrodt, Lecture Notes in Physics, Vol. 166, Springer, Berlin, 1982.

26. B.G. Dick and A.W. Overhauser, (1958) *Phys. Rev.*, 112, 90.

27. A.N. Cormack, G.V. Lewis, S.C. Parker and C.R.A. Catlow, (1988) *J. Phys. Chem. Solids*, 49, 53.

28. S.C. Parker and G.D. Price, (1984) *Phys. Chem. Miner.*, 10, 209.

29. G.D. Price, A. Wall and S.C. Parker, (1989) *Phil. Trans. Roy. Soc. Lond.*, A328, 391.

30. D.E. Williams, (1981) *Top. Curr. Phys.*, 26, 3.

31. A.K. Kiselev, A.A. Lopatkin and A.A. Shulga, (1985) *Zeolites*, 5, 261.

32. C.R.A. Catlow, K.M. Diller and M.J. Norgett, (1977) *J. Phys. C.*, 10(9), 1395.

33. G.V. Lewis and C.R.A. Catlow, (1985) *J. Phys. C.*, 18(6), 1149.

34. P.T. Wedepohl (1967) *Proc. Phys. Soc.*, 92, 79.

35. R.G. Gordon and Y.S. Kim, (1972) *J. Chem. Phys.*, 56, 3122.

36. W.C. Mackrodt and R.F. Stewart, (1979) *J. Phys. C: Condensed Matter*, 12, 431.

37. W.C. Mackrodt and R.F. Stewart, (1979) *J. Phys. C: Condensed Matter*, 12, 5015.

38. A.M. Stoneham, UKAEA Report, 1981, AERE-R9598.

39. R.E. Cohen, L.L. Boyer and M.J. Mehl, (1987) *Phys. Rev.*, B35, 5749.

40. C.R.A. Catlow, in 'Solid State Chemistry: Techniques', eds A.K. Cheetham and P. Day, Oxford University Press, 1986.

41. W.C. Mackrodt, R.F. Stewart, J.C. Cambell and I.H. Hillier, (1980) *J. Phys (Paris)*, 41, C7:64.

42. J.D. Gale, C.R.A. Catlow and W.C. Mackrodt, (1992) *Modelling Simul. Mater. Sci. Eng*, 1(1), 73.

43. N. Harrison, M. Leslie, (1992) *Mol. Sim.*, 9, 171.

44. A.C. Lasaga and G.V. Gibbs, (1987) *Phys. Chem. Miner.*, 14, 107.

45. B.W.H. van Beest, G.J. Kramer and R.A. van Santen, (1990) *Phys. Rev. Lett.*, 60, 1955.

46. J. Purton, R. Jones, M. Heggie, S. Öberg and C.R.A. Catlow (1992) *Phys. Chem. Minerals*, 18, 389.

47. B. Johnson and B. Nelendar, (1977) *Chem. Phys.*, 25, 263.

48. J.E. Mayer, (1933) *J. Chem. Phys.*, 1, 270.

49. N.C. Pyper, in 'Advances in Solid State Chemistry', ed. C.R.A. Catlow, Vol II, JAI Press, 1992.

50. J.H. Harding and A.M. Stoneham (1984) *J. Phys. C: Condensed Matter*, 17, 3401.

51. S.C. Parker, C.R.A. Catlow and A.N. Cormack (1984) *Acta Crystallogr. Sect. B: Struct. Sci.*, B40(3), 200.

52. M. Leslie — Daresbury Laboratory, Warrington, WA4 4AD

53. J.D. Gale — *J. Chem. Soc. ,Faraday Trans.*, — in press

54. G. Ooms, R.A. van Santen, C.J.J. den Ouden, R.A. Jackson and C.R.A. Catlow (1988) *J. Phys. C.*, 92(15), 4462.

55. W. Cochran, (1971) *Crit. Rev. Solid Sci.*, 2, 1.

56. S.C. Parker and G.D. Price, in 'Advances in Solid State Chemistry' ed. C.R.A. Catlow, Vol I, JAI Press, 1990.

57. G. Filippini, C.M. Gramacciolli, M. Simonetta and G.B. Suffritti, (1976) *Acta Cryst.*, A32, 259.

58. R. Fletcher and M.J.D. Powell (1963) *Computer J.*, 6, 16.

59. M.J. Norgett and R. Fletcher, (1970) *J. Phys. C.: Condensed Matter*, 3, L190.

60. A.N. Cormack, Rachel M. Jones, P.W. Tasker and C.R.A. Catlow, (1982) *J. Solid State Chem.*, 44(2), 174.

61. M. Leslie and M.J. Gillan, (1985) *J. Phys.C: Condensed Matter*, 18, 973.

62. N.L. Allan, W.C. Mackrodt and M. Leslie, in 'Advances in Ceramics', ed. C.R.A. Catlow and W.C. Mackrodt, 1987, 23, 4257.

63. R.A. Jackson, J.E. Huntington and R.G.J. Ball (1991) *J. Mater. Chem.*, 1, 1079.

64. M.J. Norgett, UKAEA Report, 1974, R7650.

65. C.R.A. Catlow, R. James and W.C. Mackrodt (1982) *Phys. Rev. B: Condensed Matter*, 25(2), 1006.

66. M. Leslie, SERC Daresbury Laboratory Report, 1982, Rep. DL-SCI-TM31T.

67. M.J. Gillan and P.W.M. Jacobs (1983) *Phys. Rev.*, B28, 759.

68. J.H. Harding (1985) *Physica*, B131, 13.

69. J.H. Harding and A.M. Stoneham (1981) *Phil. Mag.*, B43, 705.

70. J.H. Harding (1990) *Rep. Prog. Phys.*, 53, 1403.

71. C.R.A. Catlow (1986) *Ann. Rev. Mater. Sci.*, 16, 517.

72. C.R.A. Catlow, J. Corish, P.W.M. Jacobs and A.B. Lidiard (1981) *J. Phys. C.*, 14(6), L121.

73. R.A. Jackson, A.D. Murray, J.H. Harding and C.R.A. Catlow (1986) *Philos. Mag. A.*, 53(1), 27

74. C.R.A. Catlow, *Solid State Ionics*, 1983, 8(2), 89.

75. M.P. Allen and D.J. Tildesley, 'Computer Simulation of Liquids', Oxford University Press, 1987.

76. M.L. Wolf and C.R.A. Catlow (1984) *J. Phys. C: Solid State Physics*, 17, 6635.

77. C.R.A. Catlow (1992) *Solid State Ionics*, 53, 955.

78. M. Parrinello and A. Rahman (1984) *J. Chem. Phys.*, 80, 860.

79. H.J.C. Berendsen, J.P.M. Postma, W.F. van Gunsteren, A.D. Nola and J.R. Haak (1984) *J. Chem Phys.*, 81, 3684.

80. L.V. Woodcock, C.A. Angell and P.A. Cheeseman (1976) *J. Chem. Phys.*, 64, 1564.

81. S.H. Garafolini (1982) *J. Chem. Phys.*, 76, 3189.

82. B. Vessal, M. Amini, D. Fincham and C.R.A. Catlow (1989) *Phil. Mag. B.*, 60(6), 753.

83. G.E. Murch (1982) *Phil. Mag.*, A46, 575.

84. A.D. Murray, G.E. Murch and C.R.A. Catlow (1986) *Solid State Ionics*, 18-19, 196.

85. D.A. Keen, R.L. McGreevy, (1990) *Nature*, 344, 423

86. A. Szabo and N.S. Ostland, 'Modern Quantum Chemistry', Macmillan, London, 1984.

87. E.A. Colbourn, in 'Advances in Solid State Chemistry', ed. C.R.A. Catlow, Vol I, JAI Press, 1989, 1.

88. F. Haase and J. Sauer (1995), *J. Am. Chem. Soc.*, 117, 3780.

89. C. Pisani, R. Dovesi and C. Roetti, 'Lecture Notes in Chemistry', Springer, Heidelberg, 1988, Vol 48.

90. P. Hohenberg and W. Kohn (1964) *Phys. Rev.*, B136, 864.

91. R. Car and M. Parrinello (1985) *Phys. Rev. Lett.*, 55, 2471

92. A. De Vita, M.J. Gillan, J.S. Lin, M.C. Payne, I. Stich and L.J. Clarke (1992) *Phys. Rev. Lett.*, 68, 3319.

93. M.C. Payne, M.P. Teter and D.C. Allan (1990) *J. Chem. Soc. Faraday Trans.*, 86, 1221.

94. N.J. Henson, A.K. Cheetham and J.D. Gale, (1994), *Chem. Mater.*, 6, 1647.

95. C.M. Freeman, S. Levine, J.M. Newsam and C.R.A. Catlow (1993) *J. Mater. Chem.*, *J.Mater.Chem.*, 3(5) 531-535.

96. C.M. Freeman, A.M. Gorman and J.M. Newsam in "Computer Modelling in Inorganic Crystallography" (1997) (ed. C.R.A. Catlow) p.116.

97. M.W. Deem and J.M. Newsam (1989) *Nature*, 342, 260.

98. R.A. Jackson and C.R.A. Catlow (1988) *Molecular Simulation*, 1, 207.

99. R.G. Bell, R.A. Jackson and C.R.A. Catlow (1990) *J. Chem. Soc., Chem. Comm.*, 782.

100. M.D. Shannon, J.L. Cusci, P.A. Cox and S.J. Andrews (1991) *Nature*, 353, 417.

101. N.J. Henson, A.K. Cheetham and J.D. Gale, (1996) *Chem. Mater.*, 8, 664.

102. A.R. Ruiz-Salvador, G. Sastre, D.W. Lewis, C.R.A. Catlow (1996) *J. Mater. Chem.*, 6(11), 1837.

103. T.S. Bush, C.R.A. Catlow, P.D. Battle, (1995) *J.Mats.Chem.*, 5(8) 1269.

104. I.D. Brown in Computer Modelling in Inorganic Crystallography, Academic Press, London, 1996

105. J. Couves, R.H. Jones, P. Tschaufeser, S.C. Parker and C.R.A. Catlow (1993) *J.Phys.: Condens. Matter*, **5**, L329-L332.

106. X. Zhang, C.R.A. Catlow, S.C. Parker and A. Wall (1992) *J. Phys. Chem. Solids*, 53, 761.

107. X. Zhang and C.R.A. Catlow (1992) *Physica C*, 323.

108. R. Nada, C.R.A. Catlow, R. Dovesi and C. Pisani (1990) *Phys. Chem. Minerals*, 17, 353.

109. R. Nada, C.R.A. Catlow, R. Dovesi and V.R. Saunders (1992) *Proc. R. Soc. Lond. A*, 436, 499.

110. B. Silvi, N. Fourati, R. Nada and C.R.A. Catlow (1991) *J. Phys. Chem. Solids*, 52, No.8, 1005.

111. W.C. Mackrodt, N.M. Harrison, V.R. Saunders, N.L. Allan, M.D. Towler, E. Aprà and R. Dovesi (1993) *Phil. Mag. A*, **68**, 653

112. M.D. Towler, N.L. Allan, N.M. Harrison, V.R. Saunders, W.C. Mackrodt and E. Aprà (1994) *Phys. Rev. B*, **50**, 5041.

113. M.D. Towler, R. Dovesi and V.R. Saunders (1995) *Phys. Rev. B*, **52**, 10150.

114. M.S. Islam et al. (1989) *J.Phys.C.*, 21, L109.

115. N.L. Allan and W.C. Mackrodt (1989) *J. Chem. Soc. Faraday Trans. 2*, 85, 385.

116. N.L. Allan and W.C. Mackrodt, in 'Advances in Solid State Chemistry', ed. C.R.A. Catlow, JAI Press, 1993, Vol 3.

117. N.L. Allan and W.C. Mackrodt, *J.Chem.Soc.Faraday Trans.*, 86, 1227, 1989

118. X. Zhang and C.R.A. Catlow (1993) *Phys. Rev. B.* 47(9), 5315-5319.

119. M. Cherry, M.S. Islam and C.R.A. Catlow (1995) *J. Solid State Chem.*, **118**, 125.

120. M.S. Islam, M. Cherry and C.R.A. Catlow (1996) *Journal of Solid State Chemistry*, 124, 230.

121. P.W. Tasker (1979) *Surf. Sci.*, 87, 315.

122. W.C. Mackrodt (1989) *J. Chem. Soc. Faraday Trans. 2*, 85, 541.

123. D.H. Gay and A.L. Rohl (1995) *J. Chem. Soc., Faraday Trans.*, **91**, 935.

124. A.L. Rohl, D.H. Gay, R.J. Davey, C.R.A. Catlow (1996) *J. Amer. Chem. Soc.*, 118, 642-648.

125. W.C. Mackrodt, in ' Advances in Ceramics', Vol 23, eds. C.R.A. Catlow and W.C. Mackrodt, 1987, 293.

126. D.W.Lewis, R.W. Grimes, C.R.A. Catlow (1995) *J. Molecular Catalysis A: Chemical*, 100, 103-114.

127. T. Ito and J.H. Lunsford (1985) *Nature*, 315, 721.

128. P.G. Hinson, A. Clearfield and J.H. Lunsford (1991) *J. Chem. Soc., Chem. Comm.*, 1430.

129. D.W. Lewis, C.R.A. Catlow (1994) *Topics in Catalysis* 1, 111

130. D.M. Duffy and P.W. Tasker (1985) *Physica*, 131B, 46.

131. C.A. Angell, P.A. Cheeseman and C.C. Phifer, Materials Research Society Symposium Proceedings, Vol 63, Materials Research Society, Pittsburgh, Pennsylvania, 1985, 85.

132. S.K. Mitra and J.M. Parker (1984) *Phys. Chem. Glasses*, 25, 95.

133. B.P. Feuston and S.H. Garafolini (1988) *J. Chem. Phys.*, 89, 5818.

134. P. Vashishta, R.K. Kalia and J.P. Rino (1990) *Phys. Rev.*, B41, 1297.

135. A.C. Wright, B. Bachra, T,M. Brunner, R.N. Sinclair, L.F. Gladden and R.L. Portsmouth (1992) *J. Non-Cryst. Solids*, **150**, 69.

136. B. Vessal, M. Leslie and C.R.A. Catlow (1989) *Molecular Simulation*, 3, 123.

137. B. Vessal, M. Amini and C.R.A. Catlow (1993) *J.Non-Crystalline Solids*, 159, 184.

138. A.N. Cormack and Y. Cao (1996) *Molecular Engineering*, 6, 183.

139. B. Vessal, G.N. Greaves, P.T. Marten, A.V. Chadwick, R. Mole and S. Houde-Walter (1992) *Nature*, 356, 504.

140. G.N. Greaves (1985) *J. Non-Cryst. Solids*, 71, 203.

141. C.S. Marions and L.W. Hobbs (1990) *J. Non Cryst. Solids*, **24**, 242; see also L.W. Hobbs (1995) *J. Non Cryst. Solids*, **192 and 193**, 79.

142. P.A. Wright, J.M. Thomas, A.K. Cheetham and A.K. Nowak (1985) *Nature*, 318, 611.

143. J.O Titiloye, S.C. Parker, F.S. Stone and C.R.A. Catlow (1991) *J. Phys. Chem.*, 95, 4038.

144. C.M. Freeman, C.R.A. Catlow, J.M. Thomas and S. Brode (1991) *Chem. Phys. Letts.*, 186(2,3), 137.

145. S.D. Pickett, A.K. Nowak, J.M. Thomas, B.K. Peterson J.P.F. Swift, A.K. Cheetham, C.J.J. den Ouden, B. Smit and M.F.M. Post (1990) *J. Phys. Chem.*, 94, 1233.

146. P. Demontis, G.B. Suffritti, S. Quarkieri, E.S. Fois and A. Gamba (1988) *J. Phys. Chem.*, 92, 867.

147. P. Demontis, G.B. Suffritti, S. Quarkieri, E.S. Fois and A. Gamba (1987) *Zeolites*, 7, 522.

148. R.L. June, A.T. Bell and D.N. Theodorou (1990) *J. Phys. Chem.*, 94, 4329.

149. R.L. June, A.T. Bell and D.N. Theodorou (1990) *J. Phys. Chem.*, 94, 8232.

150. S.J. Goodbody, K. Watanabe, D. MacGowan, J.R.P.B. Walton and N. Quirke (1991) *J. Chem. Soc. Faraday Trans.*, 87, 1951.

151. M. Kawano, B. Vessal and C.R.A. Catlow (1992) *J. Chem. Soc., Chem. Comm.*, Issue 2, 879.

152. E. Hernandez and C.R.A. Catlow, (1995) *Proc. Roy. Soc. Lond. A*, 448, 143-160.

153. G. J. Hutchings, M. S. Scurrell, and J. R. Woodhouse, (1989) *Chem Soc. Rev.*, 18, 251.

154. J. H. Lunsford, (1995) *Angew. Chem. Int. Ed. Engl.*, 34, 970.

155. S. P. Mehandru, A. B. Anderson, and J. F. Brazdil, (1988) *J. Am. Chem. Soc.*, 110, 1715.

156. C.M. Zicovich-Wilson, R. González-Luque, and P. M. Viruela-Martín, (1990) *J. Mol. Sruct. (THEOCHEM)*, 208, 153.

157. P.M. Viruela-Martín, R. Viruela-Martín, C.M. Zicovich-Wilson, and F. Tomás-Vert, (1991) *J. Mol. Catal.*, 64,, 191.

158. K. J. Børve, and L. G. M. Pettersson, (1991) *J. Phys. Chem.*, 95, 3214.

159. K. J. Børve, and L. G. M. Pettersson, (1991) *J. Phys. Chem.*, 95, 7401.

160. K. J. Børve, (1991) *J. Chem. Phys.*, 95, 4626.

161. J. L. Anchell, K. Morokuma, and A. C. Hess, (1993) *J. Chem. Phys.*, 99, 6004.

162. K. D. Campbell, E. Morales, and J. H. Lunsford, (1987) *J. Am. Chem. Soc.*, 109, 7900.

163. Y. Feng, and D. Gutman, (1991)*J. Phys. Chem.*, 95, 6556; Y. Feng, J. Niiranen and D. Gutman, *ibid*, 6564.

164. H. S. Zhang, J. X. Wang, D. J. Driscoll, and J. H. Lunsford, (1988) *J. Catal.*, 112, 366.

165. C. H. Lin, J. X. Wang, and J. H. Lunsford, (1988) *J. Catal.*, 111, 302.

166. D. J. Driscoll, W. Martir, J. X. Wang, and J. H. Lunsford, (1985) *J. Am. Chem. Soc.*, 107, 58.

167. E. G. Derouane, J. G. Fripiat and J. M. André, (1974) *Chem. Phys. Lett.*, 28, 445.

168. S. A. Pope, M. F. Guest, I. H. Hillier, E. A. Colbourn, W. C. Mackrodt and J. Kendrick, (1983) *Phys. Rev. B*, 28, 2191.

169. H. Kobayashi, D.R. Salahub, and T. Ito, (1994) *J. Phys. Chem.*, 98, 5487.

170. G.M. Zhidomirov, V.I. Avdeev, N.U. Zhanpeisov, I.I. Zakharov and I.Y. Yudanov, (1995) *Calalysis Today*, 24, 383.

171. R. Orlando, F. Corà, R. Millini, G. Perego and R. Dovesi, (1996) *submitted for publication*.

172. C.R.A. Catlow, R.A. Jackson and J.M. Thomas, (1990) *J. Phys. Chem.*, 94, 7889.

173. L. Ackermann, J.D. Gale and C.R.A. Catlow, — to be published

174. A. D. Becke, (1988) *J. Chem. Phys.*, 88, 2547.

175. C. Lee, W. Yang and R. G. Parr, (1988) *Phys. Rev. B* 37, 786

176. T. Ziegler, (1991) *Chem. Rev.*, 91, 651.

177. P. Politzer, J. M. Seminario (Eds.), *Density Functional Theory: A Tool for Chemistry*, Elsevier 1995.

178. S.L. Meisel, J.P. McCullogh, C.H. Lechthaler and P.B. Weisz, (1976) *Chemtech*, 6,86.

179. G.J. Hutchings and R. Hunter, (1990) *Catalysis Today*, 6, 279.

180. R. Vetrivel, C.R.A. Catlow and E.A. Colbourn, (1989) *J. Phys. Chem*, 89, 4594.

181. J.D. Gale, C.R.A. Catlow and A.K. Cheetham, (1991) *J. Chem. Soc, Chem. Comm*, 178.

182. J. Sauer, C. Kölmel, F. Haase and R. Ahlrichs, (1992) *Proc. 9th Int Zeo. Conf*, Montreal, 679.

183. J.D. Gale, C.R.A. Catlow and J.R. Carruthers, (1993) *Chem.Phys.Lett.*, 216, 155.

194

184. F. Haase and J. Sauer, (1994) *J.Phys.Chem.* **98**, 3083.

185. S. Bates and J. Dwyer, (1994) *J. Mol. Struct. (Theochem)*, **306**, 57.

186. F. Haase and J. Sauer, (1995) *J. Am. Chem. Soc.*, **117**, 3780.

187. S.R. Blaskowskii and R.A. van Santen, (1995) *J. Phys. Chem.*, **99**, 11728.

188. S.P. Greatbanks, Ph.D. thesis University of Manchester (1995)

189. J.D. Gale, (1996)*Topics in Catalysis*, 3(1,2), 169.

190. R. Shah, M.C. Payne, M.-H. Lee and J.D. Gale, (1996) *Science*, **271**, 1395.

191. J. Limtrakul, (1995) *Chem. Phys.* **193**, 79.

192. E. Nusterer, P.E. Blöchl and K. Shwarz, (1996) *Angew. Chem. Int. Ed. Engl*, **35**, 175.

193. P.E. Sinclair and C.R.A. Catlow, (1996), **92**(12), 2099-2105.

194. C.M. Zicovich-Wilson, P. Viruela and A. Corma, (1995) *J. Phys. Chem.*, **99**, 13224.

195. P.E. Sinclair, and C.R.A. Catlow, (1996) *J. Chem. Soc., Faraday Trans.*, **92**(12), 2099-2105.

196. G.J. Kramer, R.A. van Santen, C.A. Emeis and A.K. Nowak, (1993) *Nature*, **363**, 529.

197. C.D. Changs and A.J. Silvestri, (1977) *J. Catal*, **47**, 249.

198. L. Kubelkova, J. Novakova, and K. Nedomova, (1991) *J. Catal.*, **124**, 441.

199. J. Bandiera and C. Naccache, (1991) *Appl. Catal.*, **97**, 10732.

200. C.E. Bronnimann and G.E. Maciel, (1986) *J. Am. Chem. Soc.*, **108**, 7154.

201. M.W. Anderson and J. Klinowski, (1990) *J. Am. Chem. Soc.*, **112**, 10.

202. E.P. Grimsrud and P. Kebarle, (1973) *J. Am. Chem. Soc.*, 95, 7939.

203. R.A. van Santen, *J. Molecular Catalysis* — in press.

ELECTRONIC STRUCTURE METHODS

E. WIMMER

Biosym/Molecular Simulations
Parc Club Orsay Université, 20 Rue Jean Rostand, 91893 Orsay, France

This contribution continues the discussion of the rôle of theory and computation in materials chemistry by providing an overview of the major electronic structure methods for materials chemistry. The first part highlights and compares the fundamental concepts of the two dominant electronic structure theories, namely the Hartree-Fock approach and density functional theory. the discussion of specific computational methods such as all-electron localized basis approaches, pseudopotential plane wave methods, and approaches based on augmentation constitutes the central part of this article. The capabilities of these approaches are illustrated by examples including the adsorption of Ag atoms on a MgO(001) surface, the expansion of graphite upon Li intercalation, the relaxation around an oxygen defect in silica, the optical properties of ruby, and ferromagnetic *vx.* anti-ferromagnetic ordering in artificially layered Co/Cu structures. The chapter concludes with a perspective on current development efforts which aim at higher accuracy, the ability to calculate larger systems, and the simulation of time-dependent phenomena such as chemical reactions.

1. Introduction

The richness of materials chemistry finds its fundamental explanation in the fascinating interplay between the motions of the nuclei and the distribution of the electrons. Therefore, the electronic structure plays a key role in the understanding and prediction of a large number of phenomena in materials chemistry such as the crystallographic structure of a solid, the distortion of a crystal lattice due to defects, the binding sites of metal atoms on a surface, but also the electro-optical properties of semiconductors and the characteristics of magnetic materials. Quantum mechanics as formulated in the 1920s captures this variety and complexity of phenomena by an elegant mathematical framework. However, for many decades the solution of the fundamental equation of quantum mechanics, namely Schrödinger's equation, seemed far too complicated for any but the simplest systems such as the hydrogen atom. It is most remarkable that present

C.R.A. Catlow and A. Cheetham (eds.), New Trends in Materials Chemistry, 195–238.

theoretical and computational approaches together with the unprecedented capabilities of computer hardware have made it possible to solve the quantum mechanical equations with sufficient accuracy to allow quantitative predictions of materials properties such as crystallographic structures, electron densities, and magnetic moments for fairly complex solids. It is the aim of this contribution to elucidate the physical concepts and the mathematical approaches underlying such calculations and to demonstrate the current capabilities by illustrative examples.

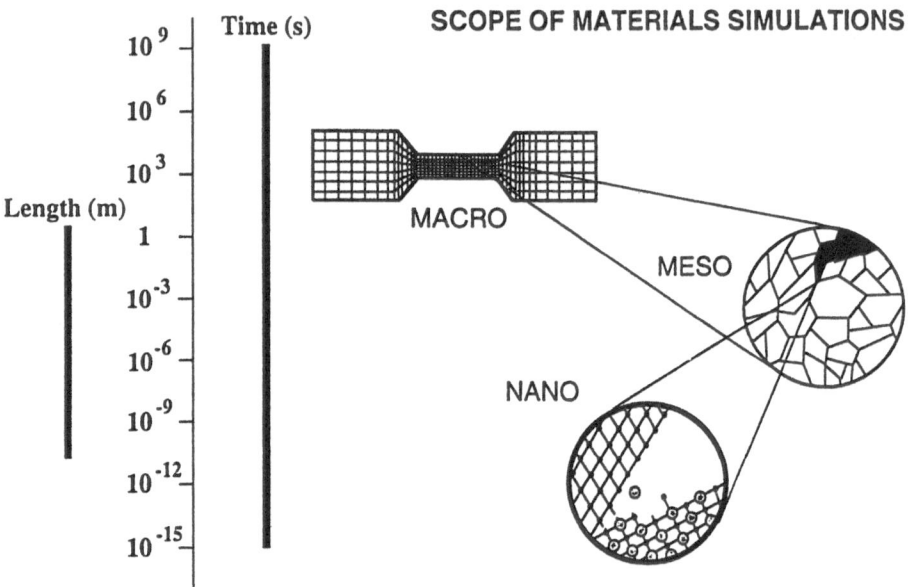

Figure 1. Length and time scales of atomistic and macroscopic phenomena. The macroscopic and mesoscopic domains can be described by classical continuum mechanics whereas quantum mechanics provides the theoretical framework for atomistic phenomena. Statistical mechanics links atomistic processes with macroscopic properties.

Chemistry is a macroscopic science, but its foundation rests in the atomistic scale. Bridging the gap between the atomistic and macroscopic scales thus represents a fundamental challenge of materials science. The typical length scale of atomistic phenomena is in the nano-meter range (1 nm = 10^{-9} m = 10 Å). For example, most interatomic bond distances are between about 0.1 - 0.4 nm. The smallest times relevant to atomistic processes are in the range of femto-seconds. For example, a stretching mode in the vibration of a C-H bond has a wave number of about $1/\lambda = 3000$ cm^{-1}. This means that a full cycle of this vibration takes about 10 femto-seconds. Thus, about nine orders of magnitude in the length scale and over fifteen in the time scale separate the atomistic domain from the macroscopic world, as illustrated in Fig. 1. As we discuss electronic structure theory and thus focus on ensembles of about ten to one hundred atoms per molecule, cluster, or periodic repeat unit, one should always keep in mind how this information is related to the macroscopic world.

The understanding and quantitative prediction of the electronic structure takes a central and fundamental role in today's concept of materials chemistry. The quality and reliability of any electronic structure theory of solids, surfaces, and molecules hinges on the ability to describe the many-body interactions accurately enough to allow quantitative predictions of physical properties. On the other hand, a useful theory also has to allow practical calculations with a reasonable computational effort on systems which are large enough to represent realistic models. The balance between accuracy and speed is intimately linked to the theoretical approach as well as to the computational implementation.

Since the formulation of quantum mechanics in the 1920's, two major theoretical approaches have emerged, namely Hartree-Fock (HF) theory [1,2] and density functional theory (DFT) [3,4]. A third approach, quantum Monte Carlo (QMC) [5], is promising but, so far, has remained limited to rather small systems. Because of its applicability to a wide range of systems including metallic, semiconducting, and insulating materials and its good balance between accuracy and computational efficiency, density functional theory has become the dominant approach for electronic structure calculations of solids and surfaces. Therefore, this lecture focuses on density functional methods and their applications to solid state problems. However, the reader should be aware that for organic molecules Hartree-Fock based approaches have been very successful in describing the electronic structures, binding energies, vibrational frequencies and other molecular properties [6]. Dovesi et al. [7] have developed a Hartree-Fock program for periodic systems that can also be used to describe surfaces. This approach has been applied to the study of oxides and related materials. In fact, one of the intriguing aspects of a solid state Hartree-Fock program is the ability to treat organic molecules on the same level of theory in the form of isolated molecules in the gas phase and adsorbed in the cages of a zeolite. In practice, however, solid state Hartree-Fock methods are very compute intensive and thus are limited to relatively small systems. Furthermore, major problems can arise if one tries to use Hartree-Fock theory for metallic systems. In this case, both fundamental issues as well as computational problems can limit the usefulness of this method.

Irrespective of a particular quantum mechanical approach, one should keep in mind that electronic structure calculations on solids are currently limited to a few hundred or perhaps one thousand atoms. The study of phenomena involving hundreds of thousands of atoms and time-scales of micro-seconds and longer require radically simpler theoretical and computational approaches such as empirical potential functions or force fields. Some of these approaches are covered in other contributions to this volume, especially in the chapter of Catlow.

2. Concepts of Hartree-Fock and density functional theory

2.1 GENERAL ASPECTS

The time-independent Schrödinger equation can be written in the form

$$H\Psi = E\Psi \tag{1}$$

The Hamiltonian operator, H, specifies a particular chemical system by defining the number, types, and positions of all atoms as well as the number of electrons. For most cases encountered in solid state chemistry, one can use the so-called Born-Oppenheimer approximation which exploits the fact that the mass of the nuclei is much larger than that of the electrons. Therefore, the electrons are assumed to adjust instantly to any changes in the positions of the atoms. This decouples the motions of the electrons from those of the nuclei and one has to solve Schrödinger's equation only for the electrons assuming fixed positions of the atomic nuclei. The N electrons of a system are represented by the many-electron wave function

$$\Psi = \Psi(1, 2, \dots N) \tag{2}$$

where the arguments 1,2,...N denote the Cartesian coordinates and the spin coordinate of each electron. A direct solution for a realistic atomistic model containing hundreds of electrons is not practical and approximations have to be introduced. As stated above, there are presently two major approaches which are compared below.

In Hartree-Fock (HF) theory, one uses the original Hamiltonian operator of Schrödinger's equation and seeks an approximation for the many-electron wave function. The simplest ansatz for such a wave function is a product of one-electron wave functions, ψ_i. However, such a wave function would violate the Pauli principle which requires that the total wave function of a system of Fermions such as electrons is antisymmetric, i.e. the wave function changes its sign if the coordinates of two electrons are exchanged. This requirement can be fulfilled with the product wave functions by creating a linear combination of products with alternating signs corresponding to all possible exchanges of coordinates. For convenience, one can write such a wave function in the form of a so-called Slater determinant indicated by eq. (3a). Now one has to find equations which allow the actual determination of the one-electron wave functions. This is done by using the variational principle which states that the expectation value of the total energy (5a) using any approximate many-electron wave function, such as a Slater determinant, is an upper bound for the exact total energy. Therefore, by varying each one-electron wave function such that it minimizes the total energy (6a), one obtains conditions for each wave function in the form of one-electron wave functions, which are known as Hartree-Fock equations (7a).

Hartree-Fock (1928, 1930)	Density Functional (1964,1965)	
$\Psi(1,2,...N) \approx \psi_1(1)\cdot\psi_2(2)\cdot...\cdot\psi_N(N)+...$	$\Psi^*\Psi = \rho(r) = \sum_i \psi_i^*\psi_i$	(3ab)
$E = E[\Psi]$	$E = E[\rho]$	(4ab)
$E[\Psi] = \dfrac{\int\Psi^*H\Psi d\tau}{\int\Psi^*\Psi d\tau}$	$E[\rho] = T_0[\rho] + U[\rho] + E_{xc}[\rho]$	(5ab)
$\dfrac{\delta E}{\delta\psi_i} = 0$ $\Downarrow$	$\dfrac{\delta E}{\delta\psi_i} = 0$ $\Downarrow$	(6ab)
$\left[-\tfrac{1}{2}\nabla^2 + V_C + \mu_x^i\right]\psi_i = \varepsilon_i\psi_i$	$\left[-\tfrac{1}{2}\nabla^2 + V_C + \mu_{xc}\right]\psi_i = \varepsilon_i\psi_i$	(7ab)
Hartree-Fock equations	Kohn-Sham equations	

The Hartree-Fock equations are formally similar to Schrödinger's equation. They consist of an operator acting on a wave function to yield the same wave function multiplied by a constant, i.e. mathematically the Hartree-Fock equations are an eigenvalue problem. The operator contains a one-electron kinetic energy term, a Coulomb potential term and a so-called exchange operator. The Coulomb potential contains the electrostatic interactions arising from all charged particles in the system, i.e. nuclei and electrons. The exchange-operator, μ_x, is a direct consequence of the choice of the Slater determinant. Its physical interpretation is as follows. The Pauli principle requires that two Fermions cannot have the same quantum numbers. If they could, then all electrons of any atom would collapse into the $1s$ state which is obviously not what we observe in nature. In real space the Pauli principle implies that if we take a reference electron with a given spin, then any other electron with the same spin does not come as close to the reference electron as it would without this exclusion principle. The net effect of the exchange operator is an effective reduction of the electrostatic repulsion between electrons of the same spin. In other words, each electron with a given spin is effectively surrounded by a positive "exchange hole". In HF theory, this exchange hole is constructed by using all one-electron wave functions of the Slater determinant. One can write the effective exchange operator of the HF equations in the following form [8]

$$\mu_x^i(\mathbf{r}) = -\sum_j \delta\left(\sigma_i, \sigma_j\right) n_j \frac{\int \psi_i^*(\mathbf{r}) \psi_j^*(\mathbf{r}') \frac{1}{\mathbf{r} - \mathbf{r}'} \psi_j(\mathbf{r}) \psi_i(\mathbf{r}') d\mathbf{r}'}{\psi_i^*(\mathbf{r}) \psi_i(\mathbf{r})} \tag{8}$$

This operator is different for each eigenstate i. The negative sign in eq. (8) indicates that the exchange operator acts as an effective attractive potential for the electrons, i. e. without this term the corresponding eigenvalue would be less negative. Through the sum over all states j and the integration $d\mathbf{r}'$ over the entire space of the system, this operator has a non-local character. It includes information from all other electrons throughout the entire system provided their wave functions overlap with the reference wave function ψ_i. The Kronecker δ insures that only electrons with the same spin σ_i are taken into account. The term for j=i cancels exactly a corresponding term in the Coulomb operator which simply means that an electron does not repel itself. Note that the occupation n_j enters into expression (8). The HF exchange operator is rather small for any unoccupied state i with $n_i = 0$. This has an important consequence for the interpretation of one-particle eigenvalues which will be discussed later.

In contrast to Hartree-Fock theory which uses the many-electron wave function as fundamental quantity, density functional theory is based on a rather remarkable theorem which states that the total energy of a system such as a bulk solid or a surface depends only on the electron density of its ground state. In other words, one can express the total energy of an atomistic system as a functional of its electron density as given by eqs. (4b) and (5b).

The idea of using the electron density as the fundamental entity of a quantum mechanical theory was first suggested by Thomas [9] and Fermi [10] in the early days of quantum mechanics. However, in the subsequent decades, the Hartree-Fock approach was developed and first applied to small molecular systems rather than the Thomas-Fermi approach. Calculations on realistic solid state systems were then out of reach. In 1951 Slater [11] used ideas from the electron gas with the intention to simplify Hartree-Fock theory to a point where electronic structure calculations on solids became feasible. Slater's work, which led to the so-called $X\alpha$ method [8], has contributed tremendously to the development of electronic structure calculations. Today, Slater's $X\alpha$ method can be seen as an early, simplified form of density functional theory. The $X\alpha$ method is hardly used in present electronic structure calculations and therefore will not be pursued here further and we return now to the explanation of density functional theory.

Eq. (4b) indicates the fundamental role of the electron density and the total energy. A critical aspect of the Kohn-Sham formulation of density functional theory is the decomposition of the total energy into three terms, as given in eq. (5b). The first term of eq. (5b) corresponds to the kinetic energy

$$T_o[\rho] = \sum_i n_i \int \psi_i^*(\mathbf{r})[-\frac{\hbar^2}{2m}\nabla^2]\psi_i(\mathbf{r})d\mathbf{r} \tag{9}$$

of non-interacting, effective electrons which are defined such that their corresponding one-particle wave functions generate the exact density of the interacting many-electron system

$$\rho(\mathbf{r}) = \sum_i n_i \psi_i^*(\mathbf{r})\psi_i(\mathbf{r}) \tag{10}$$

The second term on the right hand side of eq. (5b) represents the Coulomb energy. This term is purely classical and contains the electrostatic energy arising from the Coulombic attraction between electrons and nuclei, the repulsion between all electronic charges, and the repulsion between nuclei

$$U[\rho] = U_{en} + U_{ee} + U_{nn}$$

$$= -e^2 \sum_\alpha Z_\alpha \int \frac{\rho(\mathbf{r})}{|\mathbf{r} - \mathbf{R}_\alpha|} d\mathbf{r} + e^2 \iint \frac{\rho(\mathbf{r})\rho(\mathbf{r}')}{|\mathbf{r} - \mathbf{r}'|} d\mathbf{r}d\mathbf{r}' + e^2 \sum_{\alpha\alpha'} \frac{Z_\alpha Z_{\alpha'}}{|\mathbf{R}_\alpha - \mathbf{R}_{\alpha'}|} \tag{11}$$

where e denotes the elementary charge of a proton and Z_α is the atomic number of atom α at position $\mathbf{R}_\alpha$. The summations extend over all atoms and the integrations over all space. Once the electron density and the atomic numbers and positions of all atoms are known, expression (11) can be evaluated by using the techniques of classical electrostatics.

The third term of eq. (5b) includes all remaining complicated electronic contributions to the total energy and is called exchange-correlation energy, E_{xc}. The most important of these contributions is the exchange term. As discussed above in the context of Hartree-Fock theory, this term amounts to an effective attractive potential which lowers the one-particle energies (eigenvalues). In quantum mechanical approaches based on Hartree-Fock theory, electron correlation effects are described by improving the quality of the many-electron wave function beyond a single Slater determinant. For example, on can introduce perturbation theory or one can expand the many-electron wave function in a series of Slater determinants, each representing a different electronic configuration. This is called configuration interaction (CI) expansion. This approach is very satisfying from a formal standpoint since it allows, in principle, to approach the exact solution of the many-electron Schrödinger equation to any desired degree of accuracy. However, the extremely slow convergence of such expansions and the steep rise in computational effort with an increasing number of electrons makes this approach impractical for most cases.

As will be discussed below, there exist good approximations to the exchange-correlation energy which makes density functional calculations feasible even for fairly large systems. Assuming that one has an expression for the exchange-correlation energy in eq. (5b), one can apply the variational principle of Kohn-Sham theory which states that the total energy as a functional of the electron density is minimized by the exact electron density. Since the electron density is expressed by one-particle wave functions, this variational property of the total energy can be exploited to define conditions for the one-particle wave functions, as indicated by eq. (6b). This leads to a set of effective one-particle equations, called Kohn-Sham equations (7b). They are formally almost identical with the Hartree-Fock equations. In fact, as written in eqs. (7a) and (7b), the only difference is in the operator μ, which in Hartree-Fock theory contains only exchange effects and is different for each eigenstate whereas in density functional theory this operator includes all many-body exchange and correlation effects and is the same for all eigenstates. Despite this similarity between Hartree-Fock and Kohn-Sham equations one should keep in mind that the two theories are fundamentally different, one being based on two-electron interactions and the other resting on a collective description of the total electron system. It should be noted that eq. (7b) is a formally exact representation of the many-electron problem and no approximations have been introduced up to that point. In contrast, the corresponding HF equations contain already the severe approximation of using a single Slater determinant instead of the full many-electron wave function.

The magnitude of the various terms in eq. (7b) is illustrated by the electronic structure of single atoms. The total energy (i.e. the energy required to remove all electrons) of an isolated C atom is approximately -1000 eV, that of a Si atom -8000 eV and that of a W atom -44000 eV. The kinetic energy and the Coulomb energy terms are of similar magnitude but of opposite sign. The exchange-correlation term is about 10% of the Coulomb term and attractive for electrons (because the exchange-hole is positive). The correlation energy is smaller than the exchange energy, but plays an important role in determining the details in the length and strength of interatomic bonds. In fact, compared with the total energy, the binding energy of an atom in a solid or on a surface is quite small and lies in the range of about 1 to 8 eV. Energies involved in changes of the position of atoms on a surface can be even smaller. For example, only about 0.03 eV are required to flip an asymmetric Si-dimer on a reconstructed Si(001) surface from one conformation into another where the role of the upper and lower Si atom are reversed. It is a tremendous challenge for any theory to cope with such a range of energies. Density functional theory, as it turns out, comes amazingly close to this goal. A key to successful density functional calculations is a practical approximation for the exchange-correlation energy.

2.2 THE LOCAL DENSITY APPROXIMATION

As a simple and, as it turns out, surprisingly good approximation one can assume that the exchange-correlation energy depends only on the local electron density around each volume element dr. This is called the local density approximation (LDA)

$$E_{xc}[\rho] \approx \int \rho(\mathbf{r})\, \varepsilon_{xc}[\rho(\mathbf{r})]\, d\mathbf{r} \qquad (12)$$

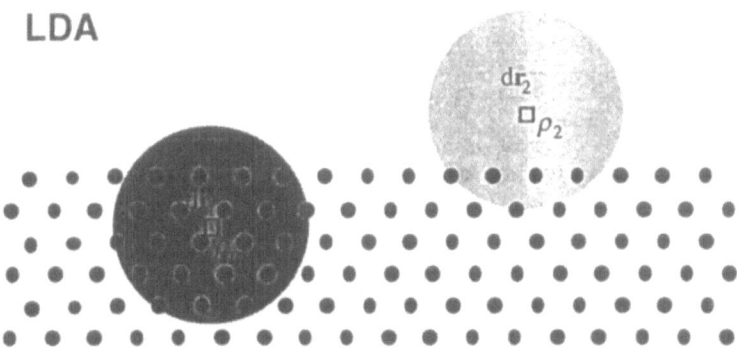

Figure 2. Illustration of the local density approximation (LDA). The solid dots represent positions of atomic nuclei, ρ_1 and ρ_2 denote the electron density in volume elements dr_1 and dr_2, respectively. In the LDA it is assumed that for the evaluation of the exchange-correlation effects, the real electron density surrounding each volume element can be replaced by a constant electron density of the same value as at the reference point. Note that this constant electron density is different for each point in space.

Fig. 2 illustrates the basic idea of the LDA. In any atomic arrangement such as a crystal, a surface, or a molecule, there is a certain electron density $\rho(\mathbf{r})$ at each point $\mathbf{r}$ in space. The LDA then rests on two basic assumptions: (i) the exchange and correlation effects come predominantly from the immediate vicinity of a point $\mathbf{r}$ and (ii) these exchange and correlation effects do not depend strongly on the variations of the electron density in the vicinity of $\mathbf{r}$. If conditions (i) and (ii) are reasonably well fulfilled, then the contribution from volume element dr would be the same as if this volume element were surrounded by a constant electron density of the same value as within dr (cf. Fig.

2). This is an excellent approximation for metallic systems, but represents quite a severe simplification in systems with strongly varying electron one careful assessment of its validity. This will be illustrated in the context of specific examples which are discussed later.

TABLE 1. Explicit form of the local density exchange as introduced by Gáspár [12] and Kohn and Sham, [4]. The correlation terms are those given by Hedin and Lundqvist [13] for a non-spin polarized system. Exchange and correlation energies per electron are denoted by ε and the corresponding potentials by μ. Both quantities are given in Hartree atomic units (1 Hartree = 2 Rydberg = 27.21165 eV). The units for the electron density are number of electrons / (Bohr radius)^3.

	Energy	Potential
	$\varepsilon_{xc} = \varepsilon_x + \varepsilon_c$	$\mu = \dfrac{\partial(\rho\varepsilon)}{\partial\rho}$
Exchange	$\varepsilon_x = -\dfrac{3}{2}\left(\dfrac{3}{\pi}\rho\right)^{\frac{1}{3}}$	$\mu_x = -2\left(\dfrac{3}{\pi}\rho\right)^{\frac{1}{3}}$
Correlation	$\varepsilon_c = -c\left[(1+x^3)\ln\left(1+\dfrac{1}{x}\right)+\dfrac{x}{2}-x^3-\dfrac{1}{3}\right]$	$\mu_c = -c\ln\left(1+\dfrac{1}{x}\right)$
	$c = 0.0225 \qquad x = \dfrac{r_s}{21} \qquad r_s = \left(\dfrac{3}{4\pi\rho}\right)^{\frac{1}{3}}$	

A system of interacting electrons with a constant density is called a homogeneous electron gas. Substantial theoretical efforts have been made to understand and characterize such an idealized system. In particular, the exchange-correlation energy per electron of a homogeneous electron gas, $\varepsilon_{xc}[\rho]$, has been calculated by several approaches such as many-body perturbation theory [13] and quantum Monte-Carlo methods [14]. As a result, $\varepsilon_{xc}[\rho]$ is quite accurately known for all densities of interest in solid state chemistry. For practical calculations, $\varepsilon_{xc}[\rho]$ is expressed as an analytical function of the electron density. There are different analytical forms with different coefficients in their representation of the exchange-correlation terms. These coefficients are not adjustable parameters, but rather they are determined through first-principles theory. Hence, the LDA is a first-principles approach in the sense that the quantum mechanical problem is solved without any adjustable, arbitrary, or system depended parameters. Table 1 shows an example of such local exchange-correlation terms. Note that there are two types of exchange-correlation terms, one for the energy and one for the potential. The energy, ε_{xc}, is needed to evaluate the total energy and the potential term, μ_{xc}, is required for the Kohn-Sham equations. The two terms are related by

$$\mu_{xc} = \frac{\partial[\rho\varepsilon_{xc}(\rho)]}{\partial\rho} \tag{13}$$

Using the explicit formulas given in Table 1, one can evaluate the exchange-correlation potential for any electron density $\rho(\mathbf{r})$. Thus, all terms of the effective one-particle operator in the Kohn-Sham equations are defined and one can proceed with a computational implementation. Before we present the various computational methods to solve the Kohn-Sham equations, we will discuss the interpretation of the one-particle eigenvalues of these equations.

2.3 INTERPRETATION OF ONE-PARTICLE ENERGIES

The fundamental quantities in density functional theory are the electron density and the corresponding total energy, but not the one-particle eigenvalues. However, the one-electron picture is so useful in chemistry that one seeks to exploit the Kohn-Sham eigenvalues and one-particle wave functions as much as possible. The Kohn-Sham equations have the form of an eigenvalue problem in which each wave function has an associated eigenvalue ε_i with an occupation number of n_i. Janak's theorem [15] provides the following relationship between the total energy and these eigenvalues.

$$\varepsilon_i = \frac{\partial E}{\partial n_i} \tag{14}$$

An eigenvalue ε_i equals the change of the total energy with respect to the occupation number of level i.

However, it is desirable to seek a more direct physical interpretation of the eigenvalues. Already before the formulation of present density functional theory, the one-electron picture has become widely used in solid state physics. For example, the distinction between a metal and an insulator is based on the analysis of the energy bands (energy bands are one-electron energies in periodic solids); the characteristics of semiconductors and semiconductor/metal junctions are explained in terms of energy band structures; photoemission experiments are conveniently interpreted by a one-electron picture, often with quite reasonable quantitative agreement between theory and experiment. Furthermore, the analysis of the s, p, and d character of partial densities of states has become an extremely useful tool in the understanding of chemical bonding in solids. While the direct interpretation of the Kohn-Sham eigenvalues as excitation energies often gives quantitative agreement with experimental photoemission spectra. there are significant differences in quantities such as energy band gaps in semiconductors. In fact, discrepancies of over a factor of two can be found between measured values and the LDA eigenvalues.

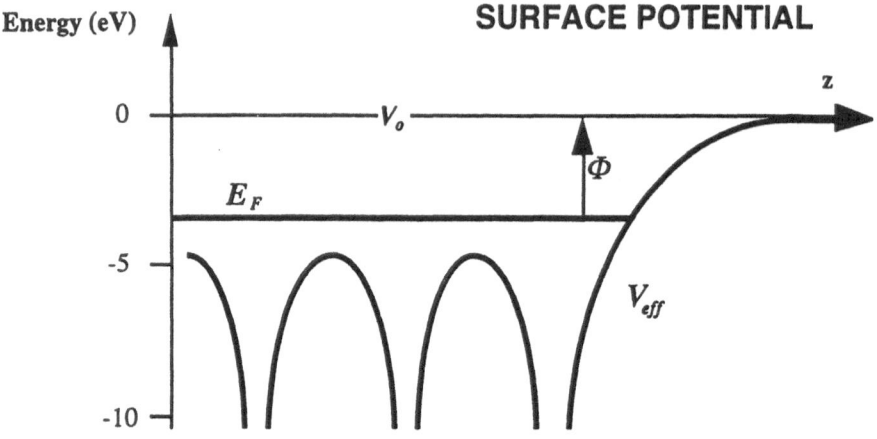

Figure 3. Schematic representation of the effective one-electron potential in a cross section of a surface. The singularities of the effective potential mark the positions of atomic nuclei. Typical values for work functions are in the range from 1 - 5 eV.

Such discrepancies between experimental excitation energies and differences between LDA eigenvalues are not necessarily a failure of the LDA, but rather an inappropriate interpretation of theoretical results. In the derivation of the Kohn-Sham equations given above, the effective one-particle eigenvalues were never required to be excitation energies! Only the total electron density and the corresponding total energy have rigorous meaning. It is possible, though, to use the results of density functional calculations as input into rigorous evaluations of excitation energies, as shown, for example, by Hybertsen and Louie [16].

The highest occupied electronic level in a metallic system is called the Fermi energy or Fermi level, E_F. The nature of the electronic states at E_F play a crucial role in determining materials properties such as electrical conductivity, magnetism, and superconductivity. On surfaces, the energy difference between E_F and the electrostatic potential in the vacuum region, V_o, above the surface is the work function, Φ (cf. Fig. 3). While in general the Kohn-Sham eigenvalues are not excitation energies, it can be shown [17] that for a metallic system the highest occupied Kohn-Sham eigenvalue can be directly interpreted as the work function as shown in Fig. 3. Thus, the agreement between experimental and calculated work functions provides a good test for the quality of actual calculations. With present LDA approaches, the calculated values are typically within 0.1-0.2 eV of the experimental results. Work functions depend strongly on the chemical nature and the geometry of the surface.

2.4 SPIN-POLARIZED AND MAGNETIC SYSTEMS

So far, the discussion of density functional theory was restricted to non-spin-polarized cases. However, many systems such as magnetic compounds containing Cr, Mn, Fe, Co, Ni, or rare earth elements such as Sm, or cases with molecular radicals involve unpaired electrons and thus require a spin-polarized method. In such systems, the number of electrons with "spin-up" can be different from that with "spin-down". Density functional theory has been generalized to accommodate spin-polarized systems which resulted in spin density functional theory with the local spin density (LSD) approximation [18,19].

In the local spin density functional (LSDF) theory, the fundamental quantities are both the electron density, $\rho(\mathbf{r})$, and the spin density, $\sigma(\mathbf{r})$. The spin density is defined as the difference between the density of the spin-up electrons and the density of the spin-down electrons

$$\sigma(\mathbf{r}) = \rho_\uparrow(\mathbf{r}) - \rho_\downarrow(\mathbf{r}) \qquad (15)$$

with the total electron density

$$\rho(\mathbf{r}) = \rho_\uparrow(\mathbf{r}) + \rho_\downarrow(\mathbf{r}) \qquad (16)$$

In LSDF theory, the exchange-correlation potential for spin-up electrons is in general different from that for spin-down electrons. Consequently, the exchange-correlation potential becomes dependent on the spin and thus the spin-polarized Kohn-Sham equations can be written as

$$\left\{ -\frac{\hbar^2}{2m} \nabla^2 + V_C + \mu_{xc}^\sigma [\rho(\mathbf{r}), \sigma(\mathbf{r})] \right\} \psi_i^\sigma(\mathbf{r}) = \varepsilon_i^\sigma \psi_i^\sigma(\mathbf{r}), \qquad \sigma = \uparrow \text{ or } \downarrow \qquad (17)$$

The exchange-correlation potential in LSDF theory depends on both the electron density and the spin density, as written in eq.(17). There are two sets of single-particle wave functions, one for spin-up electrons and one for spin-down electrons, each with their corresponding one-electron eigenvalues. For the case of equal spin-up and spin-down densities, the spin density is zero throughout space and LSDF theory becomes identical with the LDF approach. Note that in spin-polarized calculations, the occupation of single-particle states is 1 or 0, but there is still only one Fermi energy.

TABLE 2. Explicit form of local spin density exchange-correlation terms after von Barth and Hedin [18]. Energies and potentials are given in Hartree atomic units; the units for the electron and spin densities are number of electrons / (Bohr radius)^3. The electron gas density parameter r_s is defined in Table 1.

von Barth-Hedin

Exchange-correlation
energy

$$\varepsilon_{xc}(\rho,\zeta) = \varepsilon_{xc}^p(\rho) + \left[\varepsilon_{xc}^f(\rho) - \varepsilon_{xc}^p(\rho)\right]f(\zeta)$$

$$\varepsilon_{xc}^p = \varepsilon_x^p + \varepsilon_c^p \qquad \varepsilon_{xc}^f = \varepsilon_x^f + \varepsilon_c^f$$

$$\varepsilon_x^p = -\frac{3}{2}\left(\frac{3}{\pi}\rho\right)^{1/3} \qquad \varepsilon_x^f = 2^{1/3}\varepsilon_x^p$$

$$\varepsilon_c^p = -0.0225\ F\left(\frac{r_s}{21}\right) \qquad \varepsilon_c^f = -0.01125\ F\left(\frac{r_s}{53}\right)$$

$$F(x) = (1+x^3)\ln\left(1+\frac{1}{x}\right) + \frac{x}{2} - x^3 - \frac{1}{3}$$

$$f(\zeta) = \frac{(1+\zeta)^{4/3} + (1-\zeta)^{4/3} - 2}{2^{4/3} - 2} \qquad \zeta = \frac{\rho_\uparrow - \rho_\downarrow}{\rho}$$

Exchange-correlation
potential

$$\mu_{xc}^\sigma = A(\rho)\left(\frac{2\rho_\sigma}{\rho}\right)^{1/3} + B(\rho), \qquad \sigma = \uparrow \text{ or } \downarrow$$

$$A(\rho) = \mu_x^p(r_s) + v_c(r_s) \qquad B(\rho) = \mu_c^p(r_s) - v_c(r_s)$$

$$\mu_c^p(r_s) = -0.0225\ \ln\left(1 + \frac{21}{r_s}\right)$$

$$v_c = -\frac{4}{3}\frac{1}{2^{1/3}-1}\left[0.01125\ F\left(\frac{r_s}{53}\right) - 0.0225\ F\left(\frac{r_s}{21}\right)\right]$$

In magnetic systems, the spin-up and spin-down electrons are often referred to as "majority" and "minority" spin systems. It is also possible to use spin density functional theory to describe molecular systems with open shell singlet configurations. In this case the total number of spin up and spin down electrons is the same, but the spin density is non-zero at least in certain regions of space. In antiferromagnetic solid state systems, for example in bulk chromium, the total number of spin up and spin down electrons is equal, but there are large spin densities of alternating spin up and spin down character centered around each atom. Local spin density functional theory is able to describe these situations correctly.

Table 2 gives an example of a commonly used local spin density exchange-correlation formula introduced by von Barth and Hedin [18].

2.5 BEYOND THE LOCAL DENSITY APPROXIMATION

A large number of total energy calculations have shown that the LDA gives interatomic bond lengths within ±0.05 Å of experiment or better for a great variety of solids, surfaces and molecules. However, two systematic trends have been found: (i) weak bonds are too short; this includes cases such as the Ni-C bond in the Ni carbonyl $Ni(CO)_4$, the bond between two magnesium atoms (which are closed shell systems), and the length of hydrogen bonds such as that in the water dimer H-O-H$\cdots$$OH_2$; (ii) the binding energies calculated with the LDA are typically too large, sometimes by as much as 50% in strong bonds [20] and even more in weak bonds.

Gradient-corrected density functionals as suggested by Perdew in 1986 [21] and Becke in 1988 [22] seem to offer a remedy. The basic idea in these schemes is the inclusion of terms in the exchange-correlation expressions that depend on the gradient of the electron density and not only on its value at each point in space. Therefore, these corrections are also sometimes referred to as "non-local" potentials. As example, Table 3 gives the form suggested by Becke [22] for the exchange part and Perdew [21] for the correlation.

While dissociation energies calculated with these corrections rival in accuracy very good post-Hartree-Fock quantum chemistry methods, gradient corrected density functional calculations are computationally much less demanding and more general. At present, gradient corrected density functionals have been studied mostly for molecular systems as reported, for example in Ref. [23]. The results are very encouraging and this approach could turn out to be of great value in providing quantitative thermochemical data for bulk solids as well as for surface reactions, for example in catalytic and electrochemical processes.

Table 3. Correction to the total energy for exchange [22] and correlation [21] using a generalized gradient approximation (GGA). Energies are given in Hartree atomic units; the units for the electron and spin densities are number of electrons / (Bohr radius)^3. The constant b in Becke's formula is a parameter fitted to the exchange energy of inert gases. The explicit form of the functions f and g in Perdew's expression for the correlation energy is given in the original paper [21].

$$E_{GGA} = E_{LSD} + E_x^G + E_c^G$$

Becke (1988) Gradient-corrected exchange	$$E_x^G = b \sum_\sigma \int \frac{\rho_\sigma x_\sigma^2}{1 + 6bx_\sigma \sinh^{-1} x_\sigma}\, dr$$ $$x_\sigma = \frac{|\nabla \rho|}{\rho_\sigma^{4/3}} \qquad \sigma = \uparrow \text{ or } \downarrow$$
Perdew (1986) Gradient-corrected correlation	$$E_c^G = \int f(\rho_\uparrow, \rho_\downarrow)\, e^{-g(\rho)|\nabla \rho|}\, |\nabla \rho|^2\, dr$$

The one-particle eigenvalues obtained from gradient-corrected exchange-correlation potentials are not significantly different from the LDA eigenvalues. Therefore, these potentials do not (and are not intended to) remove the discrepancy between calculated and measured energy band gaps as discussed earlier.

3. Solution of the Kohn-Sham equations

In the previous section, the theoretical background of density functional theory has been presented without actually showing how one solves the Kohn-Sham equations in practice. These algorithmic and practical aspects form the topic of the current section.

The Kohn-Sham equations (7b) have the form of one-particle eigenvalue equations

$$H \psi_i(\mathbf{r}) = \varepsilon_i \psi_i(\mathbf{r}) \tag{18}$$

with the effective one-particle Hamilton operator

$$H \equiv -\frac{\hbar^2}{2m} \nabla^2 + V_C(\mathbf{r}) + \mu_{xc}(\mathbf{r}) \tag{19}$$

Following standard mathematical techniques for solving eigenvalue problems, one can expand the unknown solutions $\psi_i(\mathbf{r})$ in a set of known functions, $\phi_j(\mathbf{r})$, with unknown linear coefficients, c_{ij}.

$$\psi_i(\mathbf{r}) = \sum_j c_{ij}\phi_j(\mathbf{r}) \tag{20}$$

These coefficients are determined through a variational procedure which leads to the solution of the following matrix problem

$$\left(\underline{\mathbf{H}} - \varepsilon\underline{\mathbf{S}}\right)\mathbf{c} = 0 \tag{21}$$

$\underline{\mathbf{H}}$ and $\underline{\mathbf{S}}$ are the so-called Hamiltonian and overlap matrices with the following matrix elements

$$H_{ij} = \int \phi_i^*(\mathbf{r})\left[-\frac{\hbar^2}{2m}\nabla^2 + V_C(\mathbf{r}) + \mu_{xc}(\mathbf{r})\right]\phi_j(\mathbf{r})\,d\mathbf{r} \tag{22}$$

$$S_{ij} = \int \phi_i^*(\mathbf{r})\,\phi_j(\mathbf{r})\,d\mathbf{r} \tag{23}$$

ε represents an eigenvalue and $\mathbf{c}$ are the coefficients of a solution denoted as a column vector. In standard density functional calculations one diagonalizes the matrix ($\underline{\mathbf{H}}$ - $\varepsilon\underline{\mathbf{S}}$) to find the eigenvalues and coefficients (eigenvectors). The dimension of the matrices is determined by the number of basis functions in the expansion (20). A direct diagonalization can be avoided if one uses a Car-Parrinello or a conjugent gradient scheme as it is done for pseudopotential plane wave methods.

Both the Coulomb potential and the exchange-correlation potential in the Kohn-Sham equations (18) depend on the charge density, which is constructed from the one-particle wave functions. In other words, in order to set up the Kohn-Sham equations one needs to know their solutions. This problem is solved by an iterative, self-consistent procedure as shown in Fig. 4.

Start geometries for density functional calculations are constructed either by using experimental data such as bulk lattice constants, results from simpler computational approaches such as force field or semi-empirical methods, or from intuition. The start densities are then constructed from a superposition of atomic densities (calculated for individual free atoms or ions) or one re-uses the results from previous density functional computations.

GEOMETRY AND SCF CYCLES

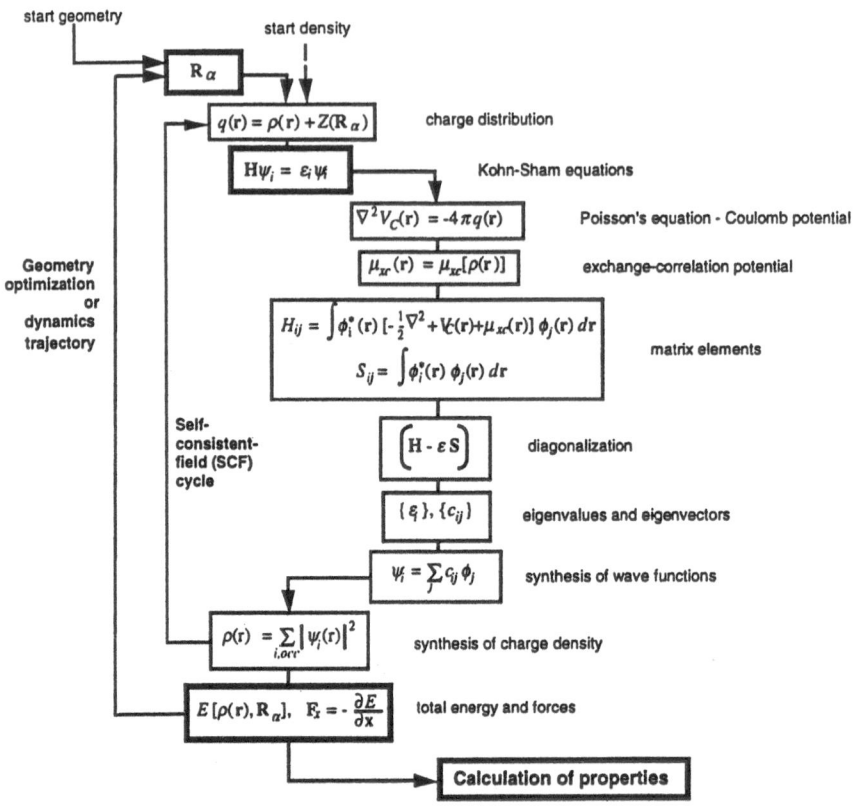

Figure 4. Scheme of typical electronic structure calculations. The outer cycle represents the geometry optimization or other manipulation of the geometry such as energy minimization, simulated annealing, dynamics trajectories or Monte Carlo procedures. The inner cycle is the self-consistency procedure to solve the Kohn-Sham equations. In methods such as the Car-Parrinello approach, both cycles are carried out simultaneously. The matrix diagonalization can be done either through explicit diagonalization or through iterative procedures. For simplicity, atomic units are used in the formulas.

The fundamental quantities of density functional theory are the electron density and, in the case of spin-polarized systems, also the spin density. With these quantities, and a given geometry, the Kohn-Sham equations are completely defined. Through the solution of Poisson's equation, the electrostatic Coulomb potential is obtained. Using an explicit form for the exchange-correlation potential as, for example, given in Table 1, the exchange-correlation potential operator is constructed. For a given variational basis set $\{\phi_j\}$, the Hamiltonian and overlap matrix elements can then be computed as indicated in Fig. 4. Subsequently, the matrix $\mathbf{H}$-$\varepsilon \mathbf{S}$ is diagonalized resulting in a set of one-particle eigenvalues and the variational expansion coefficients corresponding to each eigenvalue.

Using these coefficients, the single-particle wave functions are synthesized and a new ("output") electron density can be calculated. To this end, the occupation of each state is determined by using Fermi-Dirac statistics. This means that the eigenvalues are sorted by increasing energy and all states are filled until all electrons of the system are accommodated. In the case of systems with translational symmetry such as crystalline solids or surfaces, the synthesis of the charge density from one-particle wave functions involves the integration of the Brillouin zone in reciprocal space.

The new electron density is often referred to as "output" density indicating that each cycle in the self-consistency procedure essentially maps an "input" density into an "output" density. By definition, self-consistency is achieved when the output density equals the input density. In practice, one iterates until the residual difference between the input and output densities does not cause any significant errors in the total energy or other properties of interest.

While this self-consistency procedure is straightforward in principle, practical calculations especially for magnetic systems are often plagued by aggravating convergence problems. If one would feed the full output density back as input density, the self-consistency process would quickly diverge. Therefore, many computational schemes have been developed to enable and accelerate convergence. A straightforward method consists in mixing the output density with the input density. for example using 10% of the new output density, in order to construct the input density for the next iteration. In magnetic systems, feedback of more than 3% can sometimes already lead to divergence. More sophisticated convergence and extrapolation schemes have thus been developed and are implemented in various computer programs. The choice of a particular convergence scheme should not affect the final result unless for some reasons the scheme leads to a local minimum of the total energy as a functional of the electron density. In many non-magnetic systems, sufficient self-consistency can be reached within 10-20 cycles, but 50 steps or more may be necessary in more difficult cases. One possibility to accelerate or enforce convergence is the use of non-integer occupations of levels around the Fermi energy. Sometimes, this is referred to as "smearing". In this case, one actually moves away from the electronic ground state and possibly uncontrolled errors are introduced in the calculation.

Once self-consistency is achieved, the total energy and the forces on each atom can be calculated. Using this information, it is possible to optimize geometries, i.e. searching for minima or saddle points on the energy hypersurface. Geometries can be optimized by using only the total energy for various atomic arrangements. In practice, this is an extremely tedious process and only a few geometric parameters such as the lattice constant of a highly symmetric solid or the distance between an adsorbate atom and a surface can be optimized in this way. Knowledge of the forces on each atom greatly facilitates such geometry optimizations and makes it possible to relax many degrees of freedom simultaneously. In fact, quantum chemists have developed extensive experience with such methods and this experience can be readily applied to solid state calculations. Important optimization methods include steepest descent, conjugate-gradient, and the Broyden-Fletcher-Goldfarb-Shanno (BFGS) methods. However, none of these methods can guarantee that the global minimum has been found. Thus, as the complexity and degrees of freedom of a system increase, simulated annealing, molecular dynamics methods and Monte-Carlo searches are practical alternatives to identify families of low-energy structures.

To this end, an exciting and promising use of forces is the computation of dynamics trajectories. This enables the search for structural minima by simulated annealing techniques and, in principle, the computation of diffusion, phase transitions, and entire chemical reactions. However, the time-scales necessary to tackle these problems are in many cases far beyond the computational speed of present density functional calculations, but very encouraging first results have already been obtained. The final geometries are then used for the calculation of properties which can be derived from the charge density, the potential, and the one-electron wave functions.

There are shortcuts to this tedious self-consistency procedure. For example. Harris [24] has shown a formulation of the total energy as a functional of the density and the potential which does not require a self-consistency procedure. While this so-called Harris functional is approximate, it turns out to be sufficiently good to estimates structures. In fact, this approach together with a simulated annealing technique is a powerful method to get reasonable ground state structures even if one starts with completely arbitrary coordinates of the atoms. This procedure is particularly exciting if little information of a bonding topology is known or if one searches for new phases. Once a set of low-energy structures has been identified, it is always possible to switch to a self-consistent procedure to refine these structures and to obtain more accurate binding energies and other properties. Another possibility is the so-called Car-Parrinello method [25], in which the structural and electronic degrees of freedom are optimized or propagated simultaneously. Historically, the Car-Parrinello method is connected with the pseudopotential plane wave approach and thus will be discussed in that context.

4. Specific Electronic Structure Methods

The central task of electronic structure calculations is the solution of effective one-particle Schrödinger equations either in the form of Hartree-Fock or Kohn-Sham equations as discussed above. The overwhelming majority of Hartree-Fock electronic structure calculations is performed with Gaussian-type basis functions. The convenient mathematical properties of Gaussians allow the efficient computation of the difficult four-center integrals which occur in Hartree-Fock calculations. In density functional methods, these integrals can be avoided which opens the possibility to use other basis sets. Together with possible simplifications in the form of the potential, a variety of methods has emerged as summarized in Fig. 5.

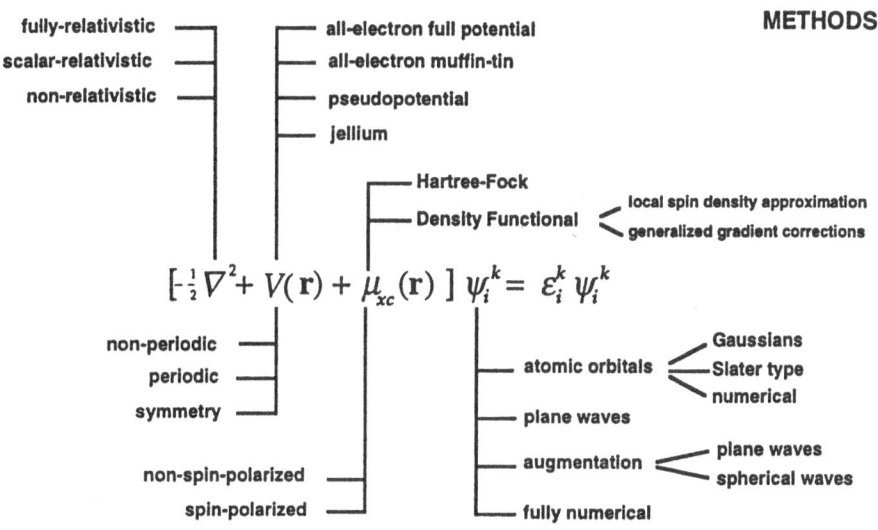

Figure 5. Overview of electronic structure methods for solving the Hartree-Fock and Kohn-Sham equations. Hartree-Fock methods are essentially restricted to Gaussian basis sets whereas the Kohn-Sham equations of density functional theory can be solved by any of the basis sets shown in this scheme. A detailed discussion of each of the choices is given in the text. After Ref. [26].

4.1 CHOICES FOR THE HAMILTONIAN

4.1.1 *Kinetic energy operator and relativistic effects*

The specification of a particular physical system, i.e. the type and number of atoms and their arrangement, is given in the Hamiltonian operator, which consists of a kinetic energy term, an electrostatic or Coulomb term, and the exchange-correlation operator. If a system contains elements with atomic numbers greater than about 40, a scalar-relativistic or a fully relativistic treatment is necessary because of the high kinetic energy of electrons close to the nuclei. Most electronic structure calculations are done either non-relativistically or on a scalar-relativistic level rather than using a fully relativistic approach. For heavier elements, the neglect of relativistic effects can be quite significant leading to an overestimation of bond distances of perhaps 0.1Å and incorrect ordering of energy bands.

4.1.2 Symmetry

Choices of specific geometric models and their space group symmetry are reflected in the potential. Symmetry such as an inversion center and mirror planes can greatly reduce the computational effort. In practical density functional implementations, the computational effort scales approximately with a third power or less in the number of atoms. If the number of inequivalent atoms of a system is reduced, for example, by a factor of 4, the computational time can thus be reduced by about 50 times.

A particular symmetry is the translational invariance of a perfectly periodic crystal. Certain translations map each atom of the crystal onto a completely symmetry-equivalent atom. One can determine the smallest number of atoms from which all other atoms of the crystal can be generated and group these atoms in a unit cell. The size and shape of the unit cell together with the type and position of all atoms in the unit cell represent a complete definition of the periodic solid. Mathematically, the translational symmetry is reflected in the one-electron wave functions by Bloch's theorem

$$\psi_i^k(r + T) = e^{ik \cdot T} \psi_i^k(r) \tag{24}$$

where T is a translation of the crystal mapping each atom into an equivalent atom in a neighboring unit cell. (Note that the subscript i counts the eigenvalues or bands whereas i in the exponent is the complex number $i^2 = -1$.) Upon such a translation, the electronic wave functions are not simply identical, but are multiplied by a factor which depends on the so-called k vector. A wave function with the property required by eq. (24) can be constructed by

$$\psi_i^k(r) = e^{ik \cdot r} u_i(r) \quad \text{with} \quad u_i(r + T) = u_i(r) \tag{25}$$

and is called a Bloch function.

The effective one-electron Schrödinger equations for a solid have to be solved not just once as it would be the case for a molecule, but for all k vectors. For molecular crystals or very large unit cells, only one k vector can be sufficient whereas for the calculation of optical properties of a metal thousands of k vectors may be needed. For structural properties and the understanding of the electronic structure of a solid, typically 10-100 k vectors are sufficient. For each k vector, a different set of discrete eigenvalues ε_i^k are obtained as indicated in Fig. 5. When plotted as a function of k, the well-known energy band structures are obtained. It is not within the scope of this article to go further into the discussion of the band theory of solids. The interested reader is referred to the many excellent text books such as that by Ashcroft and Mermin [27].

Electronic structure calculations can be carried out in a non-spin polarized or a spin-polarized form. In most cases, structural information such as equilibrium geometries of a molecule chemisorbed on a surface can be quite reliably obtained from non-spin polarized calculations. Also the characteristic features of the charge density distribution differ usually little between non-spin polarized and spin-polarized treatments. Therefore, one often first performs non-spin polarized calculations followed by spin-polarization.

4.1.3 Form of the effective potential

A characteristic feature of specific computational methods are the various simplifications to the form of the effective potential. In the most general case, the potential is constructed from all electrons and no additional approximation is made to its shape. In the late 1930's, the so-called muffin-tin approximation was introduced. In this approximation, the effective potential within spheres centered at each atom is taken to be spherically averaged, thus resembling a free atom. In the remaining interstitial region of a solid, the potential is taken to be constant. This approximation can be generalized to surface calculations by adding in the region above the surface a potential which depends only on the distance from the surface, obtained by averaging the potential in planes parallel to the surface. Muffin-tin potentials are reasonable for densely packed metallic systems, but lead to uncontrolled errors if applied to more open structures. Both the full-potential and the muffin-tin potential approach are used in all-electron calculations. One possibility to extend the range of muffin-tin potentials is the introduction of empty spheres, for example at lattice defect sites. Furthermore, the so-called atomic-sphere-approximation (ASA) has turned out to be an efficient approach. The basic idea of the ASA is an increase of the size of the spheres around each atom such that the sum of the volumina of the spheres equals the total volume of a unit cell. This can only be achieved if the atomic spheres overlap. For close-packed structures the relative overlap can be kept below 15% of the total volume which leads to acceptable errors. Larger overlap can be avoided by the introduction of empty spheres.

Pseudopotentials are frequently employed for calculations on semiconductors. In this approximation, the core electrons and the Coulomb singularity from the atomic nuclei are replaced by a smooth pseudopotential around each atom. At a certain cut-off radius away from each atomic position, the pseudopotentials becomes the actual effective potential. There are three major reasons for using pseudopotentials: (i) simple basis functions such as plane waves work well with pseudopotentials, (ii) there are fewer electrons thus simplifying the calculation, and (iii) relativistic effects which are mainly due to core electrons can be included in the pseudopotential. Together, these features lead to efficient and relatively simple computational implementations. The major drawbacks of pseudopotentials are (i) an additional approximation is introduced, (ii) pseudopotentials for transition metals, rare earths, and actinides are possible, but their accuracy is not very well established, and (iii) properties and effects involving the core electrons such as core level shifts, hyperfine fields, and spin polarization of the core electrons are not directly accessible from pseudopotential calculations. For these reasons, both all-electron and pseudopotential methods have their place in electronic structure theory. The jellium model is perhaps the simplest form of a potential. In this case, the discrete atomic nuclei are replaced by a constant positive background. This model has been used to describe metal surfaces yielding surprisingly good electron distributions and work functions [28].

4.1.4 *Exchange-correlation operator*

A central part of any electronic structure calculation is the treatment of the exchange and correlation effects. Within density functional theory, the appropriate choices have to be made in the selection of the exchange-correlation potential operator. Earlier, we have discussed the local density approximation and its generalization to the spin-polarized case. The local density approximation is uniquely defined by eq. (12). However, different ways, such as many-body perturbation theory and quantum Monte Carlo methods, to evaluate the exchange-correlation energy in the homogeneous interacting electron gas have led to slightly different explicit forms for the exchange-correlation potentials. The deviations between these different forms are typically smaller than the discrepancies between calculated and experimental values. Thus, one can consider these potentials as practically equivalent.

As discussed earlier, there are several ways to include effects beyond the local density approximation. While improvements such as the density-gradient corrections to exchange and correlation have become rapidly popular in quantum chemistry calculations, their advantages and benefits for solid state and surface calculations are still a matter of discussion and ongoing research. However, there is sufficient evidence that these corrections remove the problem of overbinding found in local density functional calculations and improve the description of weak bonding such as hydrogen bonds. However, gradient corrections may not be sufficient for a correct description of energy barriers in chemical reactions.

4.2 REPRESENTATION OF WAVE FUNCTIONS

As stated earlier, the energies involved in chemical processes or transformations of solids are extremely small compared with the total energy of atoms. Consequently, electronic structure calculations need to be very precise in order to be useful for solid state chemistry. The electronic wave functions play a critical role in achieving this accuracy. However, a high quality in the representation of the wave functions is usually associated with a high computational cost. Thus, the right balance between accuracy and speed rests on the right choice for the representation of the wave functions.

4.2.1 Linear Combination of Atomic Orbitals

In nature there is no sharp boundary between molecules, small particles, and extended solids. Therefore, electronic structure methods which are successful in describing molecules should also be applicable to solids. The vast majority of molecular electronic structure calculations are based on linear combinations of atomic orbitals,

$$\psi_i(\mathbf{r}) = \sum_\alpha \sum_j c_{ij}^\alpha \phi_j^\alpha(\mathbf{r}_\alpha) \tag{26}$$

with $\left\{\phi_j^\alpha(\mathbf{r}_\alpha)\right\}$ representing a basis set. The index i labels the one-electron wave functions. The first summation extends over all centers α, the index j identifies the different functions on center α, and $\mathbf{r}_\alpha$ denotes a point in space relative to atomic center α. The three-dimensional function $\phi_j^\alpha(\mathbf{r}_\alpha)$ is centered at position α and decomposed into a radial part and an angular-dependent part in the form

$$\phi_j^\alpha(\mathbf{r}_\alpha) = \sum_{\ell m} R_\ell^\alpha(r) Y_{\ell m}(\vartheta, \varphi) \tag{27}$$

The angular dependent part are the well-known spherical harmonics, which are fairly simple analytical functions. The radial function depends on the spherical part of the effective potential in a given sphere. The rationale for this expansion is the recognition that around each atomic center in a molecule the electronic wave functions resembles that of an isolated atom.

For an isolated molecule or cluster, one can use directly the form given by eq. (26). If one deals with a periodic system, then Bloch functions need to be constructed in order to fulfill the conditions imposed by the translational symmetry. To this end, one writes

$$\psi_i^k(\mathbf{r}) = e^{i\mathbf{k}\cdot\mathbf{r}} u_i(\mathbf{r}), \qquad u_i(\mathbf{r}) = \sum_T \sum_\alpha \sum_j c_{ij}^\alpha \phi_j^\alpha(\mathbf{r}_\alpha + \mathbf{T}) \tag{28}$$

Here $\mathbf{T}$ is a translational vector of the crystal lattice and the summation extends (formally) over the entire crystal. The index α runs over all atomic centers in the unit cell. (Note: the subscript i which labels the energy states should not be confused with the complex number i in the exponent.) The function $u_i(\mathbf{r})$ has the periodicity of the lattice and contains localized atomic functions ϕ_j^α. In principle, any functional form could be used. In practice, only three types of atomic orbitals have found widespread use, namely Gaussians, Slater-type orbitals, and numerical functions. These three classes are discussed next.

Gaussian basis sets. During the past forty years, Gaussian basis sets have become the standard in quantum chemistry. To a large extent, this choice was dictated by the occurrence of difficult four-center integrals in Hartree-Fock calculations, which require a mathematically and computationally efficient solution. Gaussian functions meet these requirements. For similar reasons, CRYSTAL [7] which is a Hartree-Fock program for periodic solids, is also based on Gaussians.

In density functional calculations, Gaussians have been successfully implemented for electronic structure calculations of solids and surfaces [29,30]. For molecular density functional calculations, the use of Gaussian-type orbitals has been pioneered by Sambe and Felton [31] and by Dunlap [32]. Further developments have lead to programs such as DeMon [33] and DGauss [23].

Gaussian functions have the basic functional form

$$R_j(r) = N_j e^{-a_j r^2} \tag{29}$$

for the radial part of the basis functions with N_j being a normalization factor. These functions do not have automatically the nodal structure of an atomic orbital. In fact, the association with a certain atomic-like function such as a C-$2s$ or Si-$3p$ function is lost. To overcome this deficiency, one combines several Gaussian functions with different sizes (as defined by the parameter a_j in the exponent) to construct a basis function with the correct radial behavior. This procedure is known as "basis set contraction". For light atoms such as C, N, and O there is extensive experience for basis set contractions. However, for heavier elements, for example Rh or W, the construction of Gaussian basis sets is less explored. To circumvent this problem, one can use pseudopotentials in conjunction with Gaussians [34,35]. Still, the construction of good Gaussian basis sets remains a tedious and delicate task [36] and one can never be sure to have reached full convergence.

Slater-type orbitals. While Hartree-Fock methods are practically limited to Gaussian-type orbitals, a variety of other basis functions can be employed for density functional calculations. One of the options are Slater-type orbitals (STO's). These functions are analytically defined as are Gaussians, but the shape of STO's is closer to actual atomic orbitals. The basic form of their radial part is

$$R_j(r) = N_j r^{n^*-1} e^{-\zeta_j r} \tag{30}$$

where N_j is a normalization factor and n^* is an effective principle quantum number. The parameter ζ controls the exponential decay at larger distances. The use of STO's in molecular density functional calculations was developed in the 1970's [37,38] and has led to programs such as the Amsterdam Density Functional (ADF) code applicable to systems with 1D, 2D, and 3D periodic boundary conditions [39].

Compared with a free atom, the one-particle wave functions in a molecule or the condensed phase are more compressed. The degree of the radial compression depends on the particular chemical environment. In order to provide the variational freedom to accommodate such a compression, one takes at least two radial functions for the representation of each atomic orbital such as an O-$2p$ function. In the case of STO's this requires two different values of ζ per orbital with quantum numbers $n\ell$ Hence this choice is called a double-zeta basis. In fact, this terminology is also being used in conjunction with Gaussian-type orbitals, since Gaussians can be used to mimic Slater-type orbitals. The concept of two radial functions for each $n\ell$ is quite general and occurs in a variety of methods as will be discussed below.

Despite the analytical form of STO's, current density functional programs such as the ADF code evaluate the Hamiltonian and overlap matrix elements by a numerical integration. This is in contrast to Gaussian-based methods, where all matrix elements except those of the density functional exchange-correlation potential operator are usually evaluated analytically.

Numerical atomic orbitals. In most density functional implementations the exchange-correlation terms are evaluated on a numerical grid. Therefore, it seems quite appealing to use numerical representations and numerical integrations also for all other terms. In particular, one can employ numerically defined atomic orbitals instead of analytical forms such as Gaussians and STO's. This approach has been pioneered by Averill and Ellis [40] and Delley and Ellis [41]. Subsequent developments by Delley [42,43] have lead to the DMol program and its generalization to periodic systems (DSolid).

Numerical atomic orbitals are generated by solving the Kohn-Sham equations for isolated atoms. The angular part of the atomic orbitals are spherical harmonics and the radial part is obtained by the solving the radial Schrödinger equation of an atom

$$\frac{1}{2r}\frac{d^2}{dr^2}\left[rR_{n\ell}(r)\right]+\left[-\frac{\ell(\ell+1)}{2r^2}+E_{n\ell}-V_{eff}(r)\right]R_{n\ell}(r)=0 \tag{31}$$

by an iterative numerical integration. This procedure yields functions which have automatically the correct nodal behavior close to the nucleus and an exponential decay at larger distances. In fact, to any desired degree of numerical accuracy these functions are the "exact" solutions to the atomic Kohn-Sham Hamiltonian.

As stated above, the wave functions in a condensed system tend to be more compressed than those of an isolated atom. To this end, one uses for each orbital with quantum numbers $n\ell$ at least two radial numerical functions, one corresponding to the isolated neutral atom and the other obtained from a positive ion. For example, to describe a Si-$3s$ function, on takes the $3s$ orbital from a neutral Si atom and a second $3s$ function from a Si^{2+} ion. The function from the positive ion is more compressed than the function from the neutral atom, thus giving the variational procedure the necessary freedom to find the correct radial extend of a molecular orbital by the appropriate linear combination of these two functions.

In addition to the functions which correspond to occupied atomic orbitals, so-called polarization functions are often required in order to account for the non-sphericity of the wave functions and charge density around each atom in a molecule or a solid. The concept of polarization functions is well known in quantum chemistry [6] and also applies to solid state and surface calculations. For example, $3d$ polarization functions are important to obtain an adequate description of the Si-Si bond length. The absence of polarization functions would cause the calculated bond length to be too long by about 0.1 Å or more.

Localized basis functions such as Gaussians, STO's or numerical functions are chemically appealing. They are well suited for the description of open structures such as zeolites and the study of surfaces interacting with molecules. However, localized atomic orbitals are not efficient to describe close-packed bulk systems such as metals and alloys. For these systems one would need extended atomic orbitals with slowly decaying tails to capture the metallic delocalized character of the valence electrons. This implies that each basis function would overlap with many surrounding atoms. One would need to orthogonalize such an extended basis functions to all core functions of each of the surrounding atoms, which can be computationally demanding. In other words, localized basis functions with slowly decaying tails are not a good choice for close-packed metals. In these cases plane waves are the "natural" representation. Plane waves are the correct solutions for the case of a constant potential, which is actually close to the effective potential experienced by the valence electrons of a metal such as Na.

4.2.2. *Plane waves*

Periodic functions can be expanded in Fourier series. The convergence of this expansion is controlled by a single parameter, namely the highest frequency at which the series is terminated. The solution of the Kohn-Sham equations for a periodic system can thus be expanded in plane waves, which amounts essentially to a three-dimensional Fourier series. The control of the basis set convergence by a single parameter is a very appealing feature, particularly when compared with the tedious task of basis set improvements with Gaussians or other localized functions. Mathematically, a plane wave expansion can be written as

$$\psi_i^k(\mathbf{r}) = \sum_{j=0}^{max} c_{ij} e^{i(k+G_j)\cdot r} \qquad (32)$$

The vector $\mathbf{G}$ is a reciprocal lattice vector. It defines the "frequency", i.e. the ability to describe spatial variations of a wave function. A large value for the cutoff G_{max} in a plane wave basis means the ability to resolve rapid variations of the wave functions and charge density. The sharp structures of wave functions near atomic nuclei are obviously a problem for a plane wave expansion. A radical way to eliminate this difficulty is the use of pseudopotentials. In fact, without pseudopotentials a plane wave expansion is so slowly converging that for most cases it would be useless. Therefore, plane waves are intimately linked to pseudopotentials. "Soft" pseudopotentials which are particularly smooth, require a smaller number of plane waves and thus are computationally more efficient.

In a plane wave expansion each basis function is no longer associated with a particular atom, but is defined over the entire unit cell. The number of plane waves required to reach a certain accuracy depends primarily on the size of the unit cell and not directly on the number of atoms. However, atoms like Si or S have softer pseudopotentials then, for example, C and O, and a larger number of plane waves is required if such atoms are present in the system under investigation. Periodic systems with a fairly uniform distribution of atoms are particularly well suited for a pseudopotential plane wave approach. On the other hand, the low coverage of a metal surface with chain-like molecules, for example, would represent a less favorable case for a plane wave basis.

As mentioned above, an important step in the variational solution of the Kohn-Sham equations is the diagonalization of the matrix $(\underline{H}-\varepsilon\underline{S})$. The computational effort for standard diagonalization algorithms increases by a third power with the size of the matrix. Since the number of basis functions determines the size of the matrix, large plane wave basis sets would be computationally inefficient.

To improve the efficiency, one can (i) use smoother pseudopotentials and thereby reduce the number of plane wave basis functions and (ii) avoid the direct diagonalization. In fact, both options are possible. To this end, for example, ultra soft pseudopotentials have been introduced [44]. On the other hand, the work by Car and Parrinello [25] has demonstrated that one can combine the iterations for obtaining a self-consistent solution of the Kohn-Sham equations with the iterations to find a minimum-energy geometric structure by using a generalized molecular dynamics scheme. Alternatively, a preconditioned conjugent gradient method has been implemented [45] which also avoids a direct diagonalization.

Pseudopotential plane wave methods have found particularly widespread use in the study of semiconductors such as Si, Ge, and GaAs. One of the reasons is the fact that very efficient and accurate pseudopotentials can be constructed for these main group elements. On the other hand, pseudopotential plane wave approaches for the systems with transition metals have only recently been investigated and the treatment of rare earth and actinides is questionable. In fact, the localized nature of d-states and especially f-electrons suggests the use of localized basis functions. On the other hand, one would like to keep the advantages of plane wave basis sets. Such considerations have lead to the concept of augmented plane waves, which is discussed below.

4.2.3. *Augmented plane waves*

The augmented plane wave (APW) method was introduced by Slater [46] in 1937. The basic idea of augmentation is the choice of the optimal form of the wave functions in each region of space. To this end, a solid or any other atomic assembly is partitioned into spherical regions around atoms, so-called muffin-tin or atomic spheres, and an interstitial region as shown in Fig. 6.

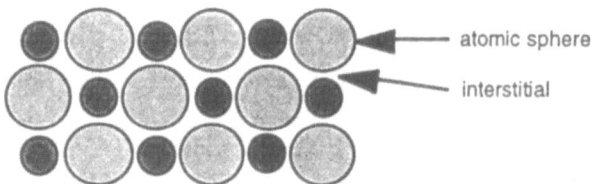

Figure 6. Partitioning of real space into atomic spheres and interstitial regions as used to construct augmented plane wave basis functions.

The dominant part of the effective potential inside the spheres has spherical symmetry. In the interstitial region the potential is fairly flat. Thus, one chooses as basis functions the solutions of Schrödinger's equations for these types of potentials, namely atomic-like functions, i.e. products of radial functions and spherical harmonics inside the atomic spheres and plane waves for the interstitial region. The muffin-tin approximation to the potential is used only to construct an efficient basis set. The actual charge density and the effective potential do not need to contain this shape approximation. This is analogous to LCAO methods where the basis functions are constructed by using just spherically symmetric atomic potentials of isolated atoms, yet the full spatial complexity of covalent bonds in molecules and solids can be described.

The basis functions in the linearized augmented plane wave (LAPW) method [47,48] have the following form.

$$\phi_j^k(\mathbf{r}) = \begin{cases} e^{i(\mathbf{k}+\mathbf{G}_j)\mathbf{r}} & \text{for } \mathbf{r} \in \text{interstitial} \\ \sum_{\ell m} R_\ell(r) Y_{\ell m}(\vartheta, \varphi) & \text{for } \mathbf{r} \in \text{sphere } \alpha \end{cases} \tag{33}$$

Eq. (33) shows the hybrid character of the LAPW method. In the interstitial region the basis functions are identical to those of the pseudopotential plane wave method whereas inside the spheres the basis functions are constructed in the same way as in a localized orbital method, namely as product of a radial function and a spherical harmonic centered at site α (the polar coordinates r, ϑ, φ are defined relative to center α). In the LAPW method the radial wave functions have the form

$$R_\ell^\alpha(r_\alpha) = A_{\ell m}^\alpha(\mathbf{k}+\mathbf{G}_j) R_\ell(E_\ell^\alpha, r_\alpha) + B_{\ell m}^\alpha(\mathbf{k}+\mathbf{G}_j) \dot{R}_\ell(E_\ell^\alpha, r_\alpha) \tag{34}$$

The radial part of the wave functions contains two basis functions per ℓ–value. As in the case of LCAO basis sets, these functions provide the necessary variational freedom to compress or expand the radial extent of the wave functions. In the LAPW method, the use of two radial functions per ℓ–value serves yet another purpose. Any acceptable wave function has to be continuous in value and derivative throughout all space and, in particular, across the boundaries between the spheres and the interstitial region. The two coefficients A and B shown in eq. (34) ensure that each plane wave with wave vector $\mathbf{k}+\mathbf{G}_j$ is joined smoothly to the atomic-like wave functions inside each sphere, thereby fulfilling the requirement of continuity of the total wave function. In the LAPW method, the radial wave functions in each atomic sphere are obtained by solving the following differential equation

$$\frac{1}{2r}\frac{d^2}{dr^2}[rR_{n\ell}(r)] + \left[-\frac{\ell(\ell+1)}{2r^2} + E_\ell - V_{eff}(r) + h_{rel}\right] R_{n\ell}(r) = 0 \tag{35}$$

Here $V_{eff}(r)$ is the spherically symmetric part of the current effective potential in each of the atomic spheres. In eq. (35) Hartree atomic units are used with $\hbar = 1, e = 1, m = 1$. The term h_{rel} is a semi-relativistic correction following Koelling and Harmon [49].

As one can see, the LAPW's are quite complicated compared with pure plane waves, especially in the case of a thin film geometry [50]. This is directly related to the fact that LAPW's contain detailed information about the actual potential whereas plane waves are "blind". In fact, the radial basis functions are recalculated in each iteration of the self-consistency cycle and thus are perfectly adapted to the actual shape of the potential in these regions. This is one of the reasons for the high accuracy of the full-potential linearized augmented plane wave (FLAPW) method [51-53]. Since each basis function contains detailed information about the particular system, significantly fewer basis functions per atom are required compared with a pure plane wave basis. On the other hand, the evaluation of the matrix elements with LAPW's is complicated whereas this step is very straightforward and computationally fast with pure plane waves.

In order to solve eq. (35) in each sphere, one needs values for the trial energies E_l. In practice, one chooses a value in the middle of the energy range of the corresponding band. For example, the d-band in metallic Cu is located between -6 and -1eV below the Fermi level and the corresponding energy parameter is chosen at about -3.5 eV.

The FLAPW method is an all-electron method. In each iteration of the self-consistency cycle, not only the radial parts of the valence functions are recalculated, but also the core wave functions are redetermined. This adds very little computationally overhead, yet it improves the value of the calculations significantly. For example, it allows the determination of core level shifts which can be used to interpret x-ray photoemission spectra. In the case of magnetic systems, the core electrons can show a significant polarization in response to the spin density of the valence states. This has direct consequences for the spin density at the nucleus which is needed to predict, for example, hyperfine splitting.

In summary, the FLAPW method is one of the most accurate but also algorithmically fairly complicated and compute-intensive approaches to determine the electronic structure of bulk solids [53] and surfaces [51]. This methods allows the evaluation of total energies [52] and forces [54] and thus can be used to determine equilibrium structures of bulk solids, interfaces, and surfaces for systems containing any atom type including transition metals, rare earth elements, and actinides.

4.2.4. *Augmented scattering functions (spherical waves)*

The essence of the LAPW method is the use of atomic-like functions around each atomic center. These functions are truncated at the sphere boundaries. Plane waves from the interstitial region link these functions by imposing the continuity conditions. In this way, information from one atomic center to all others is transmitted via plane waves and the matching coefficients A and B. This approach eliminates the problem of tails overlapping with neighboring atoms which can cause a computational bottleneck for closed-packed solids if one would use just atomic-like functions. Based on the concepts of the Korringa-Kohn-Rostoker (KKR) method [55,56], one can use scattering functions to describe the wave functions in the interstitial region. In its linearized form, this approach is know as the linearized-muffin-tin-orbital (LMTO) method [48,57]. Related to this method is the augmented-spherical-wave (ASW) approach [58]. These approaches are particularly well suited for the study of the electronic structure of close-packed solids containing transition metals and rare earths elements. These methods are fast, but in general not suited for geometry optimizations. This is related to the atomic-sphere approximations (ASA) used in typical LMTO or ASW calculations. The idea of the ASA is as follows. The atomic spheres around each atom are chosen such that the sum of the volumina of all spheres equals the total volume of the unit cell. Obviously, this leads to overlapping spheres. Experience has shown that this approximation is justifiable as long as the overlap region is not more than about 15% of the unit cell volume. In each of the spheres, only the spherically symmetric part of the effective potential is used for the calculation of the energy band structures and densities of states. Because of this simplification, LMTO and ASW calculations are about one order of magnitude faster compared with FLAPW calculations. The LMTO and ASW methods can be generalized to full-potential methods [59,60] at the cost of a substantial increase in computational effort.

4.2.5. *Fully numerical*

Given the Kohn-Sham equations (18), one could seek directly a numerical solution without any expansion in a variational basis set. In fact, this is common practice in the solution of the Kohn-Sham equations for isolated atoms. In this case, the spherical symmetry allows a separation of the solution in a radial part and spherical harmonics. The radial equations represent a one-dimensional problem which can be accurately solved by numerical techniques. To this end, a radial grid of about 500 mesh points is defined between $r=0$ and about $r=20$ Bohr radii. Since the radial wave functions show sharp nodal structures near the nucleus, but have a smooth exponential form at large radii, one uses a logarithmic mesh which is very dense for small radii and more widely spaced for larger radial values. The differential radial equation (31) is then discretized and integrated outward from the nucleus and inward from the last radial mesh points using a trial energy. In general, the solution from the outward integration and the inward integration are not continuous at their intersection (which is taken to be the radius where the trial energy equals the effective radial potential, i.e. the classical turning point). Using an iterative scheme, the trial energy is varied until the matching procedure results in a continuous and differentiable solution. At this stage, the trial energy is the eigenvalue and the numerical trial function is an acceptable radial wave function. This procedure is repeated for each eigenvalue. In the case of a molecule or solid, the direct numerical solution is challenging because of the high numerical precision required by electronic structure calculations. This is perhaps the main reason why fully numerical approaches have not yet found widespread use in the electronic structure theory of solids. However, with the continuing evolution of computer hardware combined with algorithmic improvements, it is possible that this type of approaches could gain more importance in the future.

5. Illustrative Examples

5.1 ADSORPTION OF Ag ATOMS ON A MgO(001) SURFACE

Magnesium oxide, MgO, with its simple rocksalt structure is a convenient system for the study of surface processes such as the formation of a metal/oxide interface. Upon deposition of a metal atom or any other adsorbate on a surface, the two immediate questions are: (i) where is the preferred adsorption site and (ii) what is the value of the binding energy. These aspects have recently been investigated by Spiess [61] for Ag atoms on a MgO(001) surface by using a first-principles local density functional approach. In these calculations, the surface was modeled by a finite cluster of 100 atoms. It was found that even such a relatively large cluster is not sufficient to describe the long-range electrostatic interactions in this ionic system. Therefore, an array of 194 point charges on MgO lattice sites surrounding the cluster was added to mimic the crystalline environment. Using Mulliken charges as guidelines, values of ±1.0 have been chosen for the point charges on the Mg and O sites, respectively.

The total energy local density functional calculations revealed that the most stable position of an adsorbed Ag atom on a MgO(001) surface is on top of an O atom with an Ag-O equilibrium distance of 2.36 Å and a binding energy of 0.66 eV. The four-fold hollow positions and the sites above the Mg atoms were found to be energetically less favorable by 0.13 and 0.30 eV, respectively. Furthermore, geometry optimizations of the clean MgO(001) surface show that the oxygen atoms in the surface layer are about 0.03Å above the ideal crystal plane whereas the Mg atoms relax towards the interior of the crystal by the same amount, thus giving rise to a small corrugation of the MgO(001) surface. Adsorption of Ag does not alter this surface corrugation within the accuracy of the calculation.

An important question in the formation of an overlayer is the actual growth mode. Bulk Ag crystallizes in a face-centered cubic lattice with an Ag-Ag distance which is about 3% smaller than the O-O distance on the MgO(001) surface. Therefore, an Ag overlayer on this surface could grow either epitaxially by retaining the on-top position of each Ag atom above an O atom or the overlayer could grow non-epitaxially if the Ag-Ag interactions are stronger than the Ag-O bonding. This intriguing question was addressed by Spiess [61] by considering an island of five Ag atoms on the 100-atom MgO cluster (a) in an epitaxial geometry and (b) in a non-epitaxial structure with the Ag-Ag distance fixed at the value of bulk Ag. The calculations show that the non-epitaxial geometry is slightly more stable, Furthermore, the Ag-O distance is found to increase from 2.36 Å (single Ag atom) to 2.47 Å (five Ag atoms) indicating a weakening of the adsorbate-substrate interaction with increasing Ag coverage.

This calculation clearly reveals the capabilities of present electronic structure methods as tool for the determination of structural and energetic properties of surfaces and adsorbates, thus giving insight and quantitative understanding which is difficult or impossible to gain from any other source. The Ag/MgO(001) system could also be studied by a thin film approach using two-dimensional periodic boundary conditions. In this case, the long-range Coulomb interactions would be automatically included. However, the periodic boundary conditions would restrict this approach to fairly high coverages with epitaxial registry or at least to adsorbate superstructures with a short periodicity. It would be desirable to carry out both cluster calculations as well as thin-film computations. A comparison of the two approaches would then allow an assessment of the errors introduced by the geometric model.

The binding energy reported by Spiess [61] is based on local density functional theory and thus is likely to be too large in absolute value. However, the relative energies between the different adsorption geometries is most likely to be significant and the conclusions drawn from the LDA results can be considered to be relevant.

5.2 INTERCALATION OF Li IN GRAPHITE

The Li-graphite system plays an important role in the development of rechargeable batteries. One of the problems in this type of batteries is the volume change associated with the operational cycle. Upon charging a graphite electrode with Li, the crystallites in the electrode expand. Upon discharge, the material shrinks. The repeated volume changes lead to fracture of the crystallites and eventually to a destruction of the electrode, thereby limiting the lifetime of the battery. A similar problem persists in the nickel-hydride batteries where the host material is derived from $LaNi_5$ and the battery cycles involve charging and discharging of H.

It would be desirable to develop materials which retain the necessary electrochemical properties, but show less volume changes during the operating cycles. In a preliminary study using LDA theory and numerical atomic basis functions [62], the volume changes of a Li-graphite model were studied by total energy electronic structure calculations. The graphite host was modeled by two small C_{30} sheets (see Fig. 7) in AA stacking mode. The dangling bonds on the borders of these graphitic sheets were saturated with H atoms.

The total energy as a function of the in-plane distance (see upper left panel of. Fig. 7) gives a sharp minimum at 1.41 Å which is very close to the experimental C-C distance of bulk graphite of 1.42 Å. The in-plane force constant is calculated to be 43 eV/Å^2 per C-C bond. For this model system, the calculated equilibrium interplanar distance is 3.53 Å, which is larger than the experimental value of 3.35 Å found in bulk graphite. In part, this discrepancy could be due to the fact that bulk graphite exists in an AB stacking whereas the model system has AA stacking. The force constant in the direction perpendicular to the graphitic sheets is 0.283 eV/Å^2 per C-C pair. Intercalation of one Li atom increases the interplanar equilibrium distance by about 5% to 3.71 Å while making the bond between the layers significantly stiffer. In fact, upon intercalation with Li, the interplanar force constant increases from 0.283 to 0.753 eV/Å^2 per C-C pair.

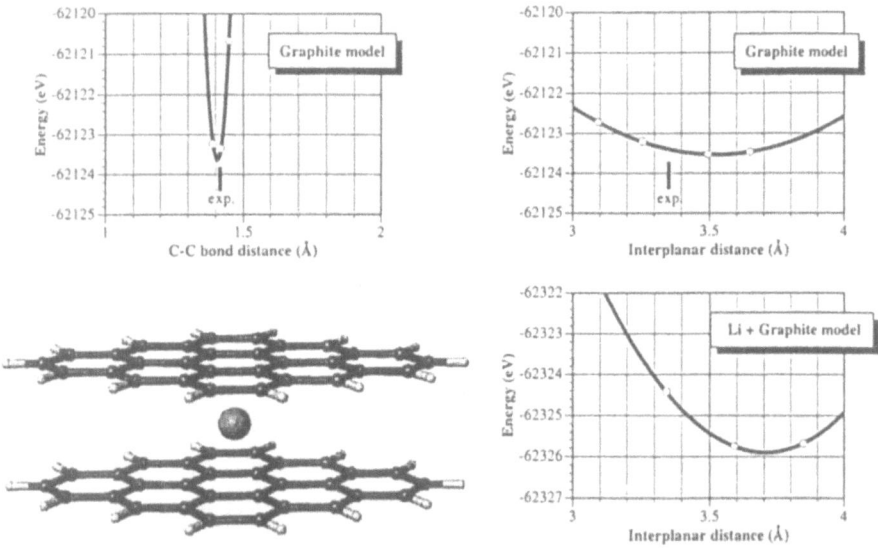

Figure 7. Models for graphite and Li intercalated in graphite. The three panels show the dependence of the total energy (a) as a function of the in-plane C-C distance, as a function of the interplanar distance of the pure graphite model and the interplanar distance with one Li atom intercalated between the two graphitic sheets. All calculations were performed on the local density functional level using a numerical atomic orbital basis as implemented in the DMol method [42,43].

The model study presented above describes the case of low Li concentration in a static picture using a reasonable, but still rather small cluster. The termination of the dangling bonds with H atoms on the edges is likely to be better than leaving unsaturated bonds on the boundary. In fact, this type of cluster termination is appropriate as long as no significant charge transfer occurs between hydrogen and the cluster atoms. As discussed above in the case of MgO, it would be desirable to complement the cluster model by thin film [63] or bulk calculations.

5.3 OXYGEN DEFECT IN SiO$_2$

The study of impurities in solids represents a fascinating research area with important technological consequences in areas such as microelectronics and optical data transmission. For example, in high-performance optical fibers, any impurity or defect is a potential source of scattering that can degrade the performance. A clear understanding of such defects greatly facilitates a rational approach to the improvements of these fibers. One of the defects in SiO$_2$ networks is caused by oxygen vacancies. Starting from the perfect SiO$_2$ network of α-quartz (see Fig. 8), the removal of an oxygen atom leaves two silicon atoms in the lattice with unsaturated bonds. It is not obvious how these atoms relax in order to minimize the total energy of the system. One possibility consists in the formation of an internal Si-Si dimer, thus closing the void created by the oxygen vacancy. Alternatively, the Si atoms adjacent to the oxygen defect could move away from the defect thereby enlarging the void.

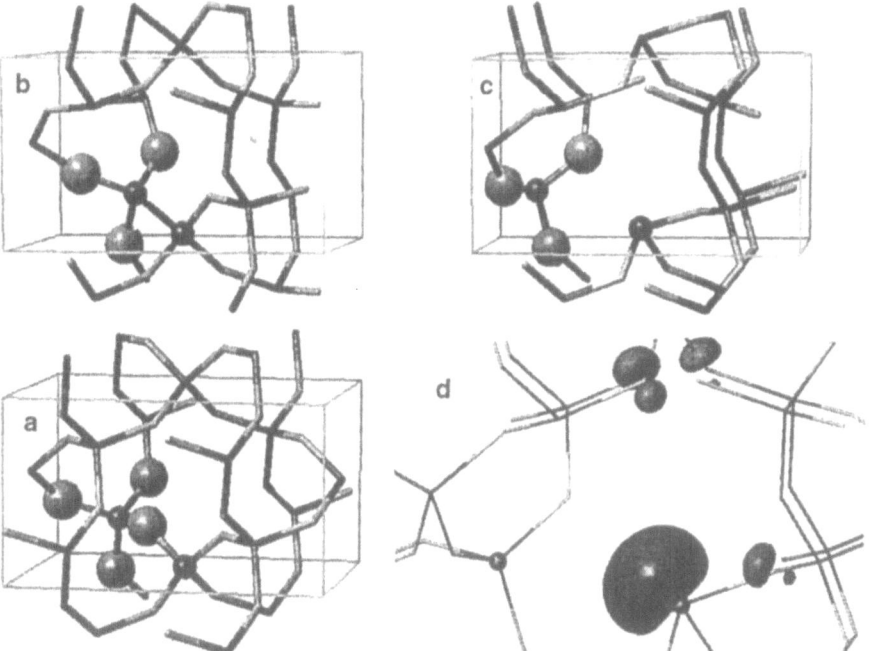

Figure 8. Unit cell of α-quartz of (a) the perfect crystal, (b) with an oxygen vacancy and a dimerization of the Si atoms adjacent to the defect and (c) with a retraction of one Si atom from the defect ("puckering"). Panel (d) shows the one-particle density of the highest occupied state of the puckered conformation [64]. Dark small spheres indicate Si atoms, light large spheres represent O atoms.

Local density functional pseudopotential plane wave calculations [64] reveal that upon removal of one oxygen atom the adjacent Si atoms spontaneously dimerize. However, a puckered conformation with an increased Si-Si distance is found as another possible metastable conformation (see Fig. 8). For this structure, the calculations give a highly localized electronic state as shown in Fig. 8. The energy of this state falls into the optical gap of the perfect silica and thus may be a cause for light absorption. A key finding, which is currently being further investigated, is the fact that the puckered structure is stabilized by the trapping of a positive charge. In fact, with this detailed knowledge of the nature of the oxygen defect and its sensitivity to the charge state, it is possible to design improved materials and processes that avoid this type of scattering centers, thus leading to better high-performance silica materials for applications such as optical fibers.

In this study [64], the random SiO_2 network of an actual glass is replaced by the ordered α-quartz structure and the oxygen defect is modeled by a supercell approach, i.e. the defect is periodically repeated. Both approximations introduce artifacts compared with the original problem. In the present case, the short-range nature of the atomic re-arrangement and the localized character of the defect state (cf. Fig. 8) are indications that the above model is probably relevant also for the random SiO_2 networks. However, these approximations need to be careful monitored. For example, one could determine the sensitivity of the results on the form and size of the supercell. At present, unit cells with up to about 100 atoms can be calculated with first-principles quantum. In the case of a pseudopotential plane wave approach, computer memory together with compute time become the limiting factors if one attempts to treat larger systems.

5.4 COLOR OF RUBY

Ruby crystals are appreciated as gem stones because of their beautiful red color, their great hardness and their chemical resistance. Technologically, ruby is exploited in lasers. The optical properties of ruby are directly related to its electronic structure and thus lend themselves to a computational study. Ruby crystals consist of an α-Al_2O_3 lattice in which some Al atoms are replaced by Cr atoms. In α-Al_2O_3 the oxygen atoms form a close-packed hexagonal structure with the Al cations occupying octahedral interstitial sites. Cr and Al ions have similar atomic radii, hence a substitution of Al by Cr does not cause significant distortions of the Al_2O_3 lattice. The Cr atoms in ruby are surrounded by six oxygen atoms in a slightly distorted octahedral coordination.

Fig. 9 shows a unit cell of α-Al_2O_3 in which one Al atom has been replaced by Cr. The electronic structure of this system was calculated [65] using local density functional theory and the augmented spherical wave (ASW) method with the atomic sphere approximation (ASA). The resulting energy band structure (cf. Fig. 9) shows a broad valence band which is derived from the O-$2p$ levels and a strongly dispersing unoccupied band which originates from the Al-$3s$ and Al-$3p$ states. Pure α-Al_2O_3 is an insulator which is reflected in the large separation of the valence band and the conduction band (see Fig. 9).

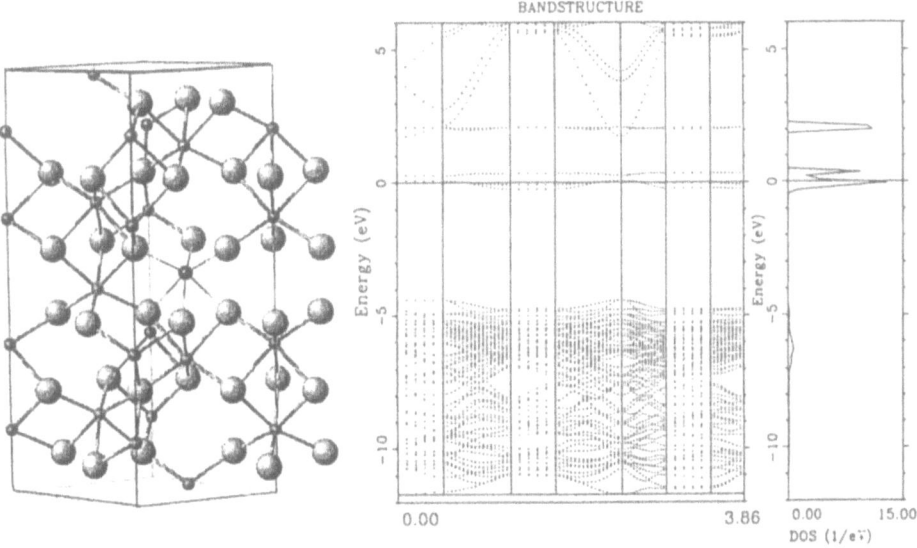

Figure 9. Crystal structure, energy bands, and Cr-d projected partial density of states of a model of ruby (Cr-doped Al_2O_3). The larger light spheres represent oxygen atoms and the smaller spheres are Al atoms. One Al atom in the α-Al_2O_3 unit cell is replaced by a Cr atom (shown as dark sphere). The electronic structure calculations were performed with the augmented spherical wave (ASW) method [58] using the atomic sphere approximation (ASA).

The replacement of an Al atom by a Cr atom introduces two flat energy bands separated by about 2 eV. The lower band is partially occupied and the higher band is empty. The partial density of states of Cr, shown in the panel to the right, identifies these flat bands as Cr-$3d$ states. In a perfectly octahedral crystal field, the d-states would be split into three t_{2g} levels and two e_g levels with the e_g at higher energies. The chromium atom in ruby has a formal charge of 3+, which amounts to an electronic configuration of s^0d^2. As can be seen from the Cr-d partial density of states (see Fig. 9), the lowest part of the Cr-d bands is occupied. Furthermore, due to the distortion of the octahedral coordination, the t_{2g}-derived states are split into two distinct peaks. The calculations give an energy difference between the partially occupied and the unoccupied Cr-d states of about 2 eV. This energy difference can be interpreted as a crystal field splitting. Since optical transitions between states of the same angular quantum number are symmetry forbidden, the life-time of the excited state is long and it is possible to pump electrons on many Cr atoms into these states. This effect allows the exploitation of ruby as material for lasers.

The energy of the optical transition is about 2 eV, which corresponds to the red part of the visible spectrum. As discussed earlier, energy differences between LDA one-particle energies should not be directly interpreted as excitation energies. However, in the case of a crystal field splitting, which has mostly an electrostatic origin, the LDA eigenvalues allow an almost quantitative interpretation of optical excitation energies.

5.5 MAGNETIC MULTILAYERS

Recording technology is one of the major driving forces in the search for novel magnetic materials. In this context, transition metal surfaces, atomic overlayers, sandwich structures, and artificially layered materials present particularly fascinating opportunities since their magnetic properties can be tailored through the control of the epitaxial process. First principles local spin density functional calculations have been highly successful in the prediction of magnetic properties of systems with reduced dimensionality [66]. The interest in magneto-optical recording and the discovery of the giant magnetoresistance effect have further intensified the interest in these systems and their investigation by electronic structure calculations [67].

An example of the present computational capabilities is the study of magnetic ordering in multilayers such as the Fe/Nb system [68]. When thin Fe films with a thickness of a few atomic layers are separated by a non-magnetic spacer material such as Nb, the Fe layers can couple either ferromagnetically or antiferromagnetically, depending on the thickness of the spacer. Sticht et al. [68] have carried out first-principles electronic structure calculations based on spin density functional theory to gain a detailed and quantitative understanding of this fascinating phenomenon.

Fe and Nb both crystallize in a body-centered-cubic lattice. In the bulk crystals, the Nb-Nb distance is larger than the Fe-Fe distance leading to a lattice mismatch of about 13%. In the calculations [68] an average lattice constant of 3.12 Å has been chosen to represent the multilayer system. Each repeat unit consisted of two Fe layers and a varying number of Nb layers, stacked in the [001] direction. Systems up to 2 Fe / 16 Nb have been studied [68]. The electronic structures of these systems were calculated with a semi-relativistic all-electron local spin density functional theory using the augmented spherical wave method within the atomic-sphere-approximation [58]. For each geometry, two self-consistent calculations were performed, one with ferromagnetic ordering and the other with antiferromagnetic ordering of the Fe layers. By using the same computational parameters such as the size of the atomic sphere radii and the number of k-points, the two resulting total energies for the ferromagnetic and antiferromagnetic ordering reveal which of the two spin arrangements is energetically more favorable. Given the large absolute values of the total energies of these systems, it is remarkable that numerically stable and relevant results can be obtained.

The calculations show that with increasing Nb spacer thickness, the coupling changes from ferromagnetic to antiferromagnetic at about 3 Å Nb thickness and back again at about 5 Å. This oscillation, which has a period of about 4.6 Å, continues up to fairly thick Nb layers of well over 20 Å thickness thus revealing the rather long-range character of the magnetic coupling between the Fe layers.

The analysis of the magnetic moments in a 2 Fe / 16 Nb multilayer system reveals that there is a small induced magnetic moment in the Nb layers which shows an oscillatory behavior of spin-up and spin-down spin density with a period of about three Nb layers. The chemical nature of the spacer material obviously plays an important role in the magnetic coupling between the iron layers, thus opening the possibility to tailor the material not only by modifying the composition of the active magnetic material, but also by optimizing the geometric structure and chemical nature of the originally non-magnetic spacer material.

6. Summary and Concluding Remarks

With the rigorous foundation of quantum mechanics and proven approximations to the many-electron problem, in particular Hartree-Fock-based approaches as well as density functional theory, electronic structure methods have become a central part of theoretical and computational solid state chemistry. Perhaps one of the most impressive successes of present electronic structure methods is their ability to predict geometric structures of any ensemble of atoms with an accuracy of a few hundreds of one Ångström for bond distances and a few degrees for bond angles and to evaluate the corresponding total energy. This is a most remarkable and useful capability especially in situation such as defects or surfaces, where experimental techniques are difficult, ambiguous, or simply impractical. The adsorption sites of Ag atoms on a MgO(001) surface, the relaxation of the Si atoms at an oxygen-defect in α-quartz, and the expansion of graphite upon Li intercalation as discussed in this chapter give a clear illustration of this capability.

The prediction of crystallographic structures with quantum mechanical methods has come to fruition only during the past 15 to 20 years. To a large part this development was due to the synergy between the enormous advances in computer hardware and the progress in computational methods, algorithms and computer programming. Earlier, electronic structure methods such as the pseudopotential approach for semiconductors and all-electron methods for transition metals and their compounds have mostly been used to calculate energy band structures, electronic densities of states and charge distributions in order to explain electrical, optical, and magnetic properties. These capabilities of electronic structure methods continue to be of unique value, as demonstrated here for the interpretation of the optical properties of ruby (i.e. Cr-doped α-Al_2O_3) and the prediction ferromagnetic vs. antiferromagnetic ordering of Fe layers separated by Nb.

A range of different electronic structure methods have been presented here including localized orbital methods, the pseudopotential plane wave approach and methods based on augmentation. In fact, each of these approaches has unique strengths and weaknesses so that a single method does not efficiently cover all types of solids, surfaces, and aggregates. For example, pseudopotential plane wave methods are very well suited for the study of three-dimensional solids such as semiconductors, but are much less efficient for very open structures (such as a large organic molecule interacting with a surface), where localized orbital methods are more appropriate. On the other hand, densely packed solids such as magnetic transition metal alloys or systems containing rare earths elements or actinides are better handled by all-electron methods.

Common to all accurate electronic structure methods is their need for substantial computational resources. However, during the past decade we have witnessed a dramatic development. Computations which ten years ago were only possible on supercomputers worth many millions of dollars can now be carried out on high-performance workstations which require a capital investment of less then $1/100^{th}$ of this amount. This has made electronic structure methods much more widely accessible and useful. The broader usage of these methods as tool in solid state physics, chemistry, and materials science is also exposing the present limitations very clearly, namely the limitations in system size, accuracy, and in the study of dynamic processes.

With present electronic structure methods and reasonable computing effort, systems containing up to about 100 atoms are accessible. This does not mean that computations on larger systems are impossible. In fact, fully quantum mechanical calculations have been reported for systems with thousands of atoms. However, these calculations represent special cases or special efforts involving, for example implementations on parallel machines. It seems that the two other limitations, namely accuracy and adequate description of dynamical processes represent perhaps the greater challenges. In general, the prediction of thermodynamic properties with current electronic structure methods does not yet reach chemical accuracy (i.e. $\pm$ 1 kcal/mol or less), perhaps with some exceptions in certain molecular systems. For example, it is currently not possible to predict melting points of compounds within a few degrees of experiment by using only first-principles electronic structure methods. The accurate predictions of reaction rates such as those occurring in chemical vapor deposition is still without reach. Probably qualitatively new developments on the fundamental theoretical level are needed to overcome this limitation.

The inspiring work of Car and Parrinello [25] has conceptually unified electronic structure methods and molecular dynamics. Today, ten years later, this and similar approaches have been used to follow the dynamical evolution of systems containing perhaps 100 atoms for times up to several pico-seconds. Compared with the time scale of actual chemical reactions there are more then 10 orders of magnitude yet to be bridged! And even then the accuracy to describe the energies of reaction barriers would still have to be improved substantially.

Obviously, electronic structure methods alone cannot be expected to provide complete solutions to the above mentioned challenges. Rather, as one approaches a problem of solid state chemistry and new materials with simulation methods, one needs to identify the relevant degrees of freedom and establish the required accuracy in order to solve a particular problem. In many cases, present electronic structure methods can offer at least qualitative insight which cannot be gained in any other way. As improved many-body theories are being developed and new, more efficient computational approaches and algorithms are being implemented, electronic structure methods can be expected to become an increasingly important research tool. The continuing progress in computer hardware, especially the increased usefulness and standardization of high-performance parallel machines together with advanced software environments will further fuel this exciting development.

Acknowledgments. It is a great pleasure to acknowledge the help and fruitful discussions with many colleagues and friends, especially Doug Allan, Jan Andzelm, Catalina Guerra, Bernard Delley, Art Freeman, John Harris, Dominic King-Smith, Jürgen Kübler, John Newsam, Jürgen Sticht, Michael Teter, and Art Williams.

7. References

1. Hartree, D. R. (1928) *Proc. Camb. Phil. Soc.* **24**, 89.
2. Fock, V. (1930) *Z. Phys.* **61**, 126.
3. Hohenberg, P. and Kohn, W. (1964) *Phys. Rev.* **136**, B864.
4. Kohn, W. and Sham, L. J. (1965) *Phys. Rev.* **140**, A1133.
5. Mitás, L. in *Computer Simulation Studies in Condensed-Matter Physics V*, Landau, D. P., Mon, K. K. and Schuttler, H. B. (1993) Springer, Berlin.
6. Hehre, W. J., Radom, L., Schleyer, P. and Pople, J. A. (1986) *Ab initio molecular orbital theory*, John Wiley & Sons, New York.
7. Dovesi, R., Saunders, V.R., and Roetti, C. (1992) *CRYSTAL 92, User Manual*, Univ. of Turin, Italy, and SERC Daresbury Laboratory, U.K.
8. Slater, J. C. (1974) *Quantum Theory of Molecules and Solids Vol. 4*, McGraw-Hill, New York.
9. Thomas, L.H. (1926) *Proc. Camb. Phil. Soc.* **23**, 542.
10. Fermi, E. (1928) *Z. Phys.* **48**, 73.
11. Slater, J. C. (1951) *Phys. Rev.* **81**, 385.
12. Gáspár, R. (1954) *Acta Phys. Acad. Sci. Hung.* **3**, 263.
13. Hedin, L. and Lundqvist, S.J. (1972) *J. Phys. (France)* **33**, C3-73.
14. Ceperley, D. M., and Alder, B.J. (1980) *Phys. Rev. Lett.* **45**, 566.

238

15. Janak, J. F. (1978) *Phys. Rev. B* **18**, 7165.
16. Hybertsen, M. S. and Louie, S. G. (1987) *Phys. Rev. Lett.* **58**, 1551.
17. Schulte, F. K. (1977) *Z. Physik B* **27**, 303.
18. von Barth, J. and Hedin, L. (1972) *J. Phys. C* **5**, 1629.
19. Gunnarsson, O., Lundqvist, B.I. and Lundqvist, S. (1972) *Solid State Commun.* **11**,149.
20. Weinert, M., Wimmer, E. and Freeman, A.J. (1982) *Phys. Rev. B* **26**, 4571.
21. Perdew, J.P. (1986) *Phys. Rev. B* **33**, 8822.
22. Becke, A. D. (1988) *Phys. Rev. A* **38**, 3098.
23. Andzelm, J. and Wimmer, E. (1992) *J. Chem. Phys.* **96**, 1280.
24. Harris, J. (1985) *Phys. Rev. B* **31**, 1770.
25. Car, R. and Parrinello, M. (1985) *Phys. Rev. Lett.* **55**, 2471.
26. Wimmer, E. (1993) *J. Comp.-Aided Mat. Design* **1**, 215.
27. Ashcroft, N. W. and Mermin, N. D. (1976) *Solid State Physics*, Holt, Rinehart and Winston, New York.
28. Lang, N.D. and Kohn W. (1970) *Phys. Rev. B* **1**, 4555.
29. Wang, C. S. and Callaway, J. (1974) *Phys. Rev. B* **8**, 4897.
30. Feibelman, P. J. (1985) *Phys. Rev. Lett.* **54**, 2627.
31. Sambe, H. and Felton, R. H. (1975) *J. Chem. Phys.* **62**, 1122.
32. Dunlap, B.I. (1986) *J. Phys. Chem.* **90**, 5524.
33. Salahub, D. R. (1987) in *Ab Initio Methods in Quantum Chemistry-II*, edited by Lawley, K. P., J. Wiley & Sons, New York, p. 447
34. Hay, P. J., Wadt, W. R. and Dunning, T. H. (1979) *Annu. Rev. Phys. Chem.* **30**, 311.
35. Huzinaga, S. (1985) *Comput. Phys. Rep.* **2**, 6.
36. Godbout, N., Salahub, D. R., Andzelm, J. and Wimmer, E. (1992) *Can. J. Chem.* **70**, 560.
37. Baerends, E.J., Ellis, D. E. and Ros, P. (1973) *Chem. Phys.* **2**, 41.
38. Rosen, A. and Ellis, D. E. (1976) *J. Chem. Phys.* **65**, 3629.
39. te Velde, G. and Baerends, E. J. (1991) *Phys. Rev. B* **44**, 7888.
40. Averill, F. W. and Ellis, D. E. (1973) *J. Chem. Phys.* **59**, 6412.
41. Delley, B. and Ellis, D. E. (1982) *J. Chem. Phys.* **76**, 1949.
42. Delley, B. (1990) *J. Chem. Phys.* **92**, 508.
43. Delley, B. (1991) *J. Chem. Phys.* **94**, 7245.
44. Vanderbilt, D. (1985) *Phys. Rev. B* **32**, 8412.
45. Payne, M. C., Teter, M. P., Allan, D.C. and Joannopoulos, J. D. (1992) *Rev. Mod. Phys.* **64**, 1045.
46. Slater, J. C. (1937) *Phys. Rev.* **51**, 846.
47. Koelling, D. D. and Arbman, G. O. (1975) *J. Phys. F* **5**, 2041.
48. Andersen, O. K. (1975) *Phys. Rev. B* **12**, 3060.
49. Koelling, D. D. and Harmon, B. N. (1977) *J. Phys. C* **10**, 3107.
50. Krakauer, H., Posternak, M. and Freeman, A.J. (1979) *Phys. Rev. B* **19**, 1706.
51. Wimmer, E., Krakauer, H. Weinert, M. and Freeman, A.J. (1981) *Phys. Rev. B* **24**, 864.
52. Weinert, M., Wimmer, E. and Freeman, A. J. (1982) *Phys. Rev. B* **26**, 4571.
53. Jansen H. J. F. and Freeman, A. J. (1984) *Phys. Rev. B* **30**, 561.
54. Soler, J.M. and Williams, A.R., 1989, Phys. Rev. B **40**, 1560.
55. Korringa, J. (1947) *Physica* **13**, 392.
56. Kohn, W. and Rostoker, N. (1954) *Phys. Rev.* **94**, 1111.
57. Skriver, H. L. (1984) *The LMTO method*, Springer Verlag.
58. Williams, A. R., Kübler, K. and Gelatt, J. R. (1979) *Phys. Rev. B* **19**, 6094.
59. Methfessel, M. (1988) *Phys. Rev. B* **38**, 1537.
60. Eyert V. (1991) *Thesis*, Technische Hochschule Darmstadt, Germany.
61. Spiess L. (1996) *Surf. Rev. Lett.* (submitted)
62. Guerra C. and Wimmer, E. (unpublished)
63. Posternak, M., Baldereschi, A., Freeman, A. J. and Wimmer, E. (1984) *Phys. Rev. Lett.* **52**, 863.
64. Allan, D. C. and Teter, M. P. (1990) *J. Am. Ceram. Soc.* **73**, 3247.
65. Sticht, J. (unpublished)
66. Wang, D.-S., Freeman, A. J. and Krakauer, H. (1982) *Phys. Rev. B* **26**, 1340.
67. Jansen, H. J. F. (1995) *Physics Today* **48**, 50.
68. Sticht, J., Herman, F., and Kübler, J. (1993) in *Physics of Transition Metals vol 1*, 456, World Scientific, Singapor; Oppeneer, P. and Kübler, J., editors

MIXED VALENT COPPER AND MANGANESE OXIDES: HIGH T_C SUPERCONDUCTORS AND COLOSSAL MAGNETO RESISTANCE (CMR) MATERIALS

B. RAVEAU

Laboratoire CRISMAT, URA 1318 associée au CNRS, ISMRA et Université de Caen, Bd du Maréchal Juin 14050, Caen cedex, France

1. Introduction

The mixed valence of transition elements tends to induce particular transport and magnetic properties in oxides. This is the case for the Cu(II)–Cu(III) mixed valence which is at the origin of superconductivity at high temperature, and for the Mn(III)–Mn(IV) mixed valence which was recently shown to induce colossal magnetoresistance effects (CMR) in manganites. The structures of both series of oxides, copper and manganese, derive from that of perovskite. Layered superconducting cuprates, involving bismuth, thallium and lead have been the subject of several reviews, and for this reason will be considered as well known. Thus the present paper will be devoted to two kinds of superconducting materials — the mercury based cuprates and the copper oxycarbonates — and to the manganese perovskites with CMR properties.

2. The mercury based cuprates

After the discovery of superconductivity in $HgBa_2CuO_{4+\delta}$[1], the complete series of cuprates $HgBa_2Ca_{m-1}Cu_mO_{2m+2+\delta}$[2-5] with a T_C up to 134K could be synthesised using pressures of several tens of kilobars. Nevertheless, it was rapidly realised that these oxides are difficult to synthesise as pure phases. For this reason, the possibility of investigating such materials by partially replacing mercury with foreign cations was investigated. This exploration was based on the fact that the $[HgO_\delta]_\infty$ layer exhibits a high oxygen deficiency which renders the structure metastable. Thus the introduction, on the mercury site, of an element with an oxidation state larger than that of Hg(II) should increase the oxygen content close to the mercury layers and consequently should stabilise the structure.

In this way, about fifty new superconductors were stabilised by introducing either rare earth cation, or bismuth, or lead, or a transition element into the structure.

C.R.A. Catlow and A. Cheetham (eds.), New Trends in Materials Chemistry, 239–283.
© *1997 Kluwer Academic Publishers.*

240

2.1   THE 1201 Hg–BASED CUPRATES

Seven series of "HgSr" cuprates with the "1201" structure (Figure 1a) have been synthesised. This result is of great interest if one takes into account that the cuprate $HgSr_2CuO_{4+\delta}$ could never be synthesised, even under high pressure. The list of these phases given in Table 1 shows that the bismuth and lead mercury cuprates exhibit a wide homogeneity range[6-7]: $0 \leq x \leq 0.75$ for $Hg_{0.5}Bi_{0.5}Sr_{2-x}La_xCuO_{4+\delta}$ and $0 \leq x \leq 0.65$ for $Hg_{0.3}Pb_{0.7}Sr_{2-x}La_xCuO_{4+d}$; in contrast, one observes a smaller homogeneity range for the praseodymium and cerium cuprates: $0 \leq x \leq 0.1$ for $Hg_{0.4}Pr_{0.6}Sr_{2-x}Pr_xCuO_{4+\delta}$[6] and $0.4 \leq x \leq 0.8$ for $Hg_{0.4}Ce_{0.5}Cu_{0.1}Sr_{2-x}La_xCuO_{4+\delta}$[8]. Finally, for stabilisation by transition elements — Cr, Mo and Re — there appears to be only one composition: $Hg_{0.7}Cr_{0.3}Sr_2CuO_{4+\delta}$[9], $Hg_{0.8}Mo_{0.2}Sr_2O_{4+d}$[10] and $Hg_{0.9}Re_{0.1}Sr_2CuO_{4+\delta}$[9] respectively.

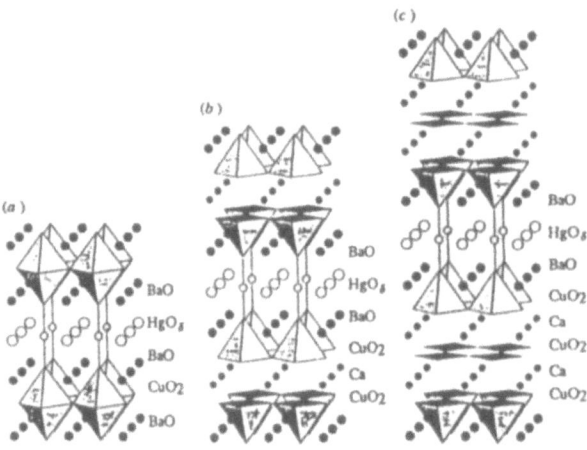

Figure 1: Structure of mercury–based superconductors: (a) $HgBa_2CuO_{4+\delta}$ ("1201"); (b) $HgBa_2CaCu_2O_{6+\delta}$ ("1212"); (c) $HgBa_2Ca_2Cu_3O_{8+\delta}$ ("1223"). The twofold coordination of mercury is evidenced, the oxygen sites at the level of mercury are partially occupied (open circles).

The HREM [010] images of the bismuth, lead, cerium and transition metal substituted mercury cuprates are typical of the 1201 structure (Figure 1a) as shown for $Hg_{0.5}Bi_{0.5}Sr_{1.5}La_{0.5}CuO_{4+\delta}$ (Figure 2). In contrast, the praseodymium phases differ from the classical tetragonal "1201" cuprates by their orthorhombic symmetry, characterised by a doubling of one parameter of the tetragonal subcell (a = 2a_p, b = a_p, c = c_{1201}). The [010] HREM image of this phase (Figure 3a) shows that this situation is due to an ordering of the Hg and Pr atoms in the mercury layers in such a way that one mercury row alternates with one praseodymium row along $\bar{a}$ (Figure 3b).

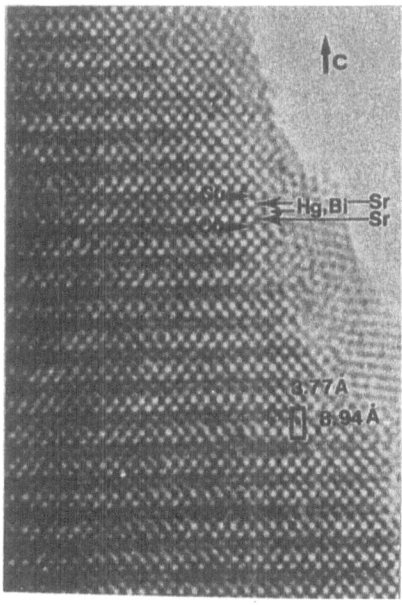

Figure 2: [010] image of $Hg_{0.5}Bi_{0.5}Sr_{1.5}La_{0.5}CuO_{4+\delta}$. The cationic positions are correlated with the bright dots. The nature of the layers stacked along $\bar{c}$ is indicated.

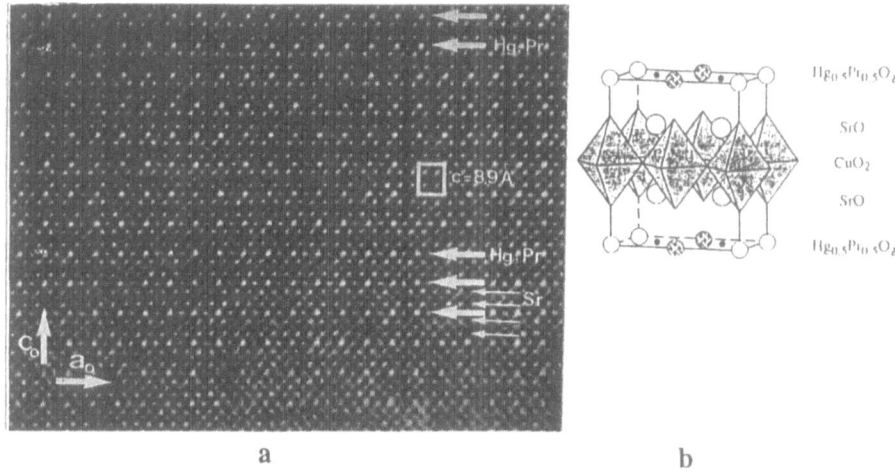

a   b

Figure 3: a) –[010] image of $Hg_{0.4}Pr_{0.6}Sr_2CuO_{4+\delta}$. The mixed [$Hg_{0.4}Pr_{0.6}O_\delta$.] layers are indicated by large white arrows and the [SrO] layers by small ones.
b) Structure of the cuprate $Hg_{0.4}Pr_{0.6}Sr_2CuO_{4+\delta}$. with an orthorhombic double cell.

Among these "Sr–Hg" cuprates, three phases — the bismuth, lead and chromium cuprates — exhibit bulk superconductivity with a maximum T_C of 27K, 40K and 60K respectively. In contrast, the praseodymium and cerium phases do not superconduct, whereas the molybdenum and rhenium phases, though they reach a T_C of 70K and 78K respectively, are characterised by a low superconducting volume fraction.

Table 1: Characteristics of the substituted mercury "1201" cuprates

Nominal Composition	Cell parameters			As synthesised			After annealing		
	a (Å) ±0.0003	b (Å) ±0.0002	c (Å) ±0.0001	T_{cmid}	T_{onset}	svf	T_{cmid}	T_{onset}	svf
$Ba_2CuO_{4+\delta}$	3.8830		9.536	84K	91K	50%			
$V_{0.2}Ba_2CuO_{4.3}$	3.8863		9.338	89K	96K	65%		D	
$Mo_{0.2}Ba_2CuO_{4.4}$	3.8819		9.378	66K	74K	60%		D	
$Mo_{0.1}Ba_2CuO_{4.2}$	3.8748		9.435	68K	75K	60%		D	
$W_{0.2}Ba_2CuO_{4.4}$	3.8713		9.416	37K	45K	30%		D or U	
$W_{0.1}Ba_2CuO_{4.2}$	3.8746		9.450	83K	89K	40%		D or U	
$Cr_{0.2}Ba_2CuO_{4.1}$	3.9261		9.306	Non–superconductor			57K	70K	55%
$Mn_{0.2}Ba_2CuO_{4.2}$	3.890		9.343	90K	93K	50%		U	
$Nb_{0.2}Ba_2CuO_{4.3}$	3.885		9.461	78K	83K	30%	65K	90K	80%
$Ru_{0.2}Ba_2CuO_{4.2}$	3.879		9.473	91K	93K	25%		D	
$Cr_{0.3}Sr_2CuO_{4.15}$	3.845		8.683		60K	100%			
$Mo_{0.2}Sr_2CuO_y$	3.797		8.818		78K	9%		72K	1%
$Re_{0.2}Sr_2CuO_y$	3.783		8.883		70K	2%			
$Bi_{0.5}Sr_{2-x}La_xCuO_{4+\delta}$ (0≤x≤0.75)	3.765– 3.778		8.94–8.84		27K				
$Pb_{0.7}Sr_{2-x}La_xCuO_{4+\delta}$ (0≤x≤0.65)	3.749– 3.768		8.87–8.99		40K				
$Pr_{0.6}Sr_{2-x}Pr_xCuO_{4+\delta}$ (0≤x≤0.1)	3.683	7.606	8.8	Non–superconductor					
$Ce_{0.5}Cu_{0.1}Sr_{2-x}La_xCuO_{4+\delta}$ (0.4≤x≤0.8)	3.733– 3.738	7.575– 7.750	17.946– 17.966	Non–superconductor					

svf: superconducting volume fraction; D: degradation; U: unchanged

In contrast to the "HgSr" 1201 cuprates, the "HgBa" 1201 cuprates can be stabilised for a large series of transition elements (Table 1): M = V, Mo, W, Cr, Mn, Nb and Ru [11–13], whereas one phase is also obtained for M = Bi [14]. Amongst these cuprates, the behaviour of the transition metal substituted "1201" cuprates is remarkable from the view point of their structural and magnetic properties. The evolution of their cell parameters, refined from powder XRD data and compared to those of $HgBa_2CuO_{4+\delta}$, shows indeed that the c parameter of the tetragonal cell is significantly decreased when Hg is partly replaced by the transition element, especially for M = V, Mo, Cr and Mn in agreement with the smaller size of the transition metal. This confirms that the transition element occupies the Hg sites and not the copper sites, in spite of its small size. In contrast, the "a" parameter is only slightly decreased (for Mo, Ru, W) or increased (for V, Nb, Mn) by the substitution, except for chromium which exhibits a significant increase of the "a" parameter. Clearly, the partial replacement of mercury by a transition element in the "HgBa" 1201 cuprates induces a remarkable variation of the structure anisotropy as shown from the c/a ratio.

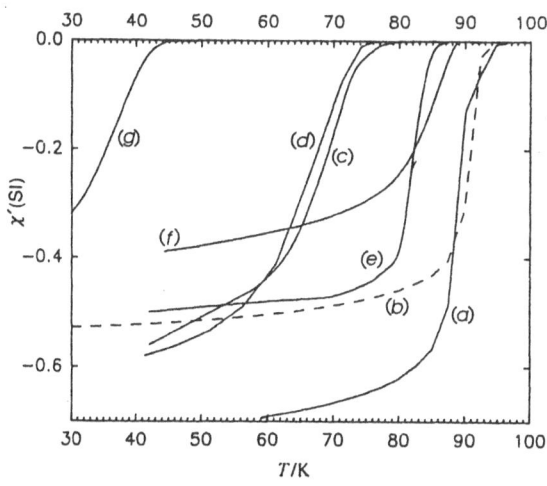

Figure 4. χ' (T) curves of $Hg_{0.8}V_{0.2}Ba_2CuO_{4+\delta}$ (a); $Hg_{0.8}Mn_{0.2}Ba_2CuO_{4+\delta}$ (b); $Hg_{0.75}Mo_{0.25}Ba_2CuO_{4+\delta}$ (c); $Hg_{0.9}Mo_{0.1}Ba_2CuO_{4+\delta}$ (d); $Hg_{0.9}Ba_2CuO_{4+\delta}$ (e); $Hg_{0.9}W_{0.1}Ba_2CuO_{4+\delta}$ (f); $Hg_{0.75}W_{0.25}Ba_2CuO_{4+\delta}$ (g).

The χ' (T) curves of several as-synthesised "HgBa" 1201 cuprate phases compared to that of as-synthesised $Hg_{0.9}Bi_2CuO_{4+\delta}$ are shown in Figure 4. The best characteristics are observed for the V phases (curve a) and Mn phases (curve b) with a T_c (midpoint) of 89–90K, whereas the molybdenomercury cuprates (curves c and d) exhibit a critical temperature T_c (midpoint) of 66–68K, *i.e.* smaller than that of $Hg_{0.9}Ba_2CuO_{4+\delta}$ of 81K (curve e). As for vanadium, the introduction of small amounts of tungsten on the mercury sites does not affect T_cs as shown for $Hg_{0.9}W_{0.1}Ba_2CuO_{4+\delta}$ (curve f) which corresponds to a T_c (midpoint) of 83K, very close to that of $Hg_{0.8}V_{0.2}Ba_2CuO_{4+\delta}$ (curve a). Nevertheless, the superconducting volume fraction observed for the tungsten phase (40%) is significantly smaller than that for vanadomercury and molybdenomercury cuprates. Moreover, the induction of higher W contents seems to decrease dramatically T_cs, as is shown for a sample $Hg_{0.75}W_{0.25}Ba_2CuO_{4+\delta}$ (curve g) that exhibits a T_c

(midpoint) of 37K. Curiously, the as–synthesised chromium phase does not superconduct. Subsequent annealing of these compounds, either in an argon flow or in an oxygen flow, tends to destroy superconductivity, except for the chromium phase whose T_C can be increased to 57K by annealing in an oxygen flow.

2.2 THE 1212 Hg–BASED CUPRATES

Like the "1201" cuprates, the Bi– and Pb–substituted "1212" "Hg–Sr" phases exhibit a wide homogeneity range (Table 2). Oxides with the general formula $Hg_{0.5}Bi_{0.5}Sr_2Ca_{1-x}R_xCuO_{6+\delta}$ where R = Nd, Y, Pr, are obtained for $0.35 \leq x \leq 1$ [15]. Their [010] HREM images (Figure 5) confirm that their structure consists of double pyramidal copper layers connected through $[Hg_{0.5}Bi_{0.5}Sr_2O_{1+\delta}]$ rocksalt layers (Figure 1b). The "1212" cuprate $Hg_{0.3}Pb_{0.7}Sr_2Ca_{1x}R_xCu_2O_{6+\delta}$ with $0.20 \leq x \leq 0.50$ (R = Y, Nd) [16, 17] is also very similar to the 1212" Bi–Hg" cuprate.

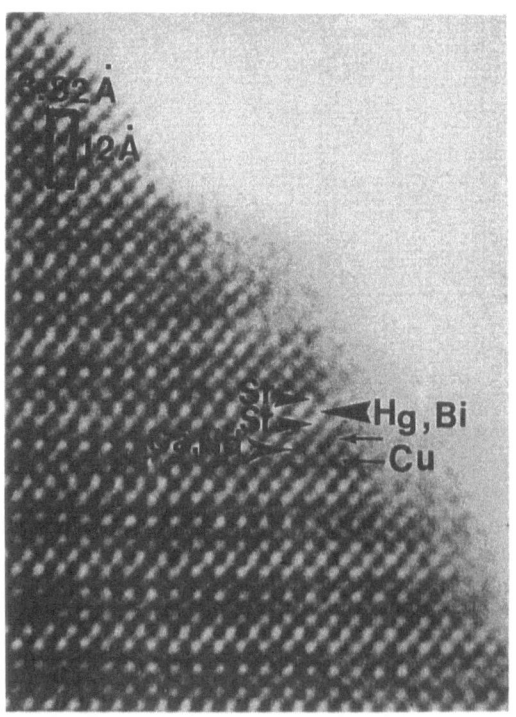

Figure 5. [010] HREM image of $Hg_{0.5}Bi_{0.5}Sr_2Ca_{0.5}Nd_{0.5}Cu_2O_{6+\delta}$. The cationic positions are correlated to the bright spots. The nature of the layers along $\bar{c}$ is indicated.

The lead and mercury cations form the oxygen deficient rocksalt layers that interconnect with the pyramidal copper layers. All these "Hg–Sr" substituted cuprates exhibit rather sharp transitions with T_Cs ranging from 90K to 94K. But the most remarkable relates to the fact that in contrast to the

"123" phase for which the substitution of Pr for Y between the copper layers kills superconductivity, the 1212–type Pr–phase superconducts and, moreover, exhibits the highest T_C.

Table 2: Characteristics of the substituted mercury "1212" cuprates.

Nominal composition	Cell Parameters			As synthesised			After annealing		
	a (Å) ±0.0003	b (Å) ±0.0002	c (Å) ±0.0001	T_{cmid}	T_{onset}	svf	T_{cmid}	T_{onset}	svf
$HgBa_2CaCuO_{6+\delta}$	3.8624		12.704	—	117K	—	—	127K	—
$Hg_{0.8}V_{0.2}Ba_2CaCuO_{6.3}$	3.8692		12.500	113K	119K	60%	124K	128K	
$Hg_{0.8}Mo_{0.2}Ba_2CaCu_2O_{6.4}$	3.8602		12.576	123K	127K	75%			75%
$Hg_{0.8}W_{0.2}Ba_2CaCuO_{6.4}$	3.8607		12.661	113K	118K	59%		D	
$Hg_{0.9}W_{0.1}Ba_2CaCu_2O_{6.2}$	3.8635		12.685	112K	118K	40%		D	
$Hg_{0.8}Ti_{0.2}Ba_2CaCu_2O_{6.2}$	3.8568		12.581	123K	127K	80%		D	
$Hg_{0.8}Cr_{0.2}Ba_2CaCu_2O_{6.1}$	3.8753		12.488	90K	110K	75%		D	
$Hg_{0.3}Pb_{0.2}Ba_2Ca_{0.7}Nd_{0.3}Cu_2O_7$	3.81		12.13		100K		106K	115K	98%
$Hg_{0.5}Bi_{0.5}Sr_2Ca_{1-x}Nd_xCu_2O_{6+\delta}$ $(0.35 \leq x \leq 1)$	3.809– 3.854		12.06– 12.09		94K 92K				
$Hg_{0.5}Bi_{0.5}Sr_2Ca_{0.65}Y_{0.35}Cu2O_{6+\delta}$	3.795		12.036		95K				
$Hg_{0.5}Bi_{0.5}Sr_2Ca_{0.65}Pr_{0.35}Cu2O_{6+\delta}$	3.805	3.819– 3.38	12.054		85K				
$Hg_{0.4}Pr_{0.6}Sr_2Sr_{1-x}Pr_xCu_2O_{6+\delta}$ $(0.20 \leq x \leq 0.70)$	3.838– 3.864	3.795– 3.802	12.22– 12.26		85K				
$Hg_{0.4}Pr_{0.6}Sr_2Ca_{1-x}Pr_xCu_2O_y$ $(0.20 \leq x \leq 0.70)$	3.833– 3.846		12.15		85K				
$Hg_{1-x}M_xBa_2Y_{0.6}Ca_{0.4}Cu_2O_{6+\delta}$ $(x= 0.40)$	3.870		12.537		90K				
$HgBa_2Nd_{1-x}Ca_xCu_2O_{6+\delta}$ $(0.40 \leq x \leq 0.60)$	3.877		12.607		110K				
$Hg_{0.4}Ce_{0.5}Sr_{2.5}Ca_{0.5}Cu_{2.1}O_7$	7.626	3.813	12.19		51K				
$Hg_{0.7}Mo_{0.3}Sr_2Ca_{1-y}Y_yCu_2O_{7-\delta}$	3.822		11.940		95K				
$Hg_{0.7}Re_{0.3}Sr_2Ca_{1-y}Y_yCu_2O_{7-\delta}$	3.818		11.879		95K				
$Hg_{0.7}Cr_{0.3}Sr_2Ca_{1-y}Y_yCu_2O_{7-\delta}$	3.8526		11.819		non-superconducting				

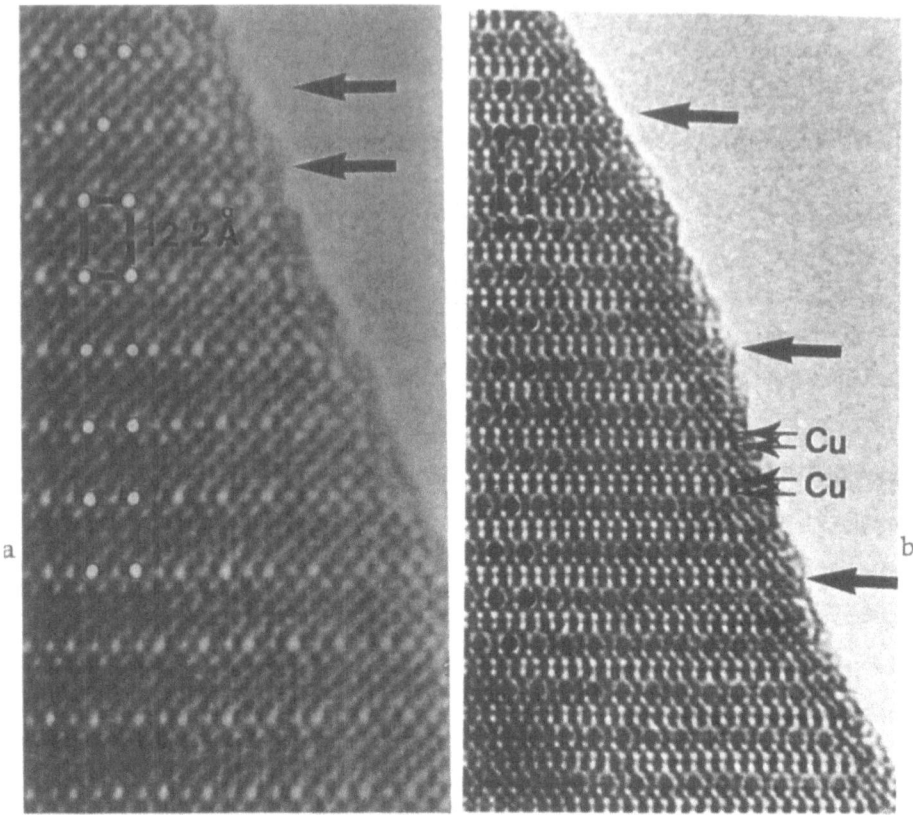

Figure 6: $Hg_{0.4}Pr_{0.6}Sr_{2.7}Pr_{0.3}Cu_2O_{6+\delta}$: [010] HREM image of the ordered crystals. One Hg row alternates with a Pr row along $\bar{a}$ a) orthorhombic I and b) orthorhombic II supercells.

The "1212" structure of "Hg–Sr" cuprates can also be stabilised by the introduction of praseodymium on the mercury sites (Table 2). In contrast to the Bi and Pb cuprates, the "1212" type oxides $(Hg_{0.4}Pr_{0.6})Sr_2Sr_{1-x}Pr_xCu_2O_{6+\delta}$ exhibit an orthorhombic symmetry, like the "1212" praseodymium phases. However, two kinds of crystals are observed for the same composition: the first is characterised by a doubling of the a parameter only (a = $2a_p$, b = a_p, c ≈ c_{1212}), whereas the second exhibits a doubling of the a and c parameters with respect to the tetragonal subcell (a = $2a_p$, b = a_p, c = $2c_{1212}$). The corresponding HREM images (Figure 6a and 6b) show clearly that this phenomenon corresponds to an ordering of the praseodymium and mercury ions within the mercury layers, similar to that observed for the "1201" "Pr–Hg" cuprate. In both kinds of crystals, one mercury row alternates with one praseodymium row along $\bar{a}$ leading to a doubling of this parameter; in the first kind of crystal (labelled O_1) two successive mercury rows are above each other along $\bar{c}$ (Figure 6a), so that c ≈ c1212, whereas such successive rows are shifted $\bar{a}/2$ in the second kind of crystals (Figure 6b), leading to a doubling of the c parameter. But the most spectacular feature deals with the superconducting properties of these phases which exhibit a critical temperature up to 85K as shown, for instance, for the phase $Hg_{0.4}Pr_{0.6}Sr_{2.7}Pr_{0.3}Cu_2O_{6+\delta}$ (Figure 7) for which a very high diamagnetic volume fraction (90%) and a sharp transition are observed.

This effect of praseodymium which enhances dramatically the superconducting properties of the "1212" phase, and increases especially T_{cs}, is fundamentally opposite to what is observed for other cuprates such as the "123" phase in which, as noted, the introduction of praseodymium kills superconductivity.

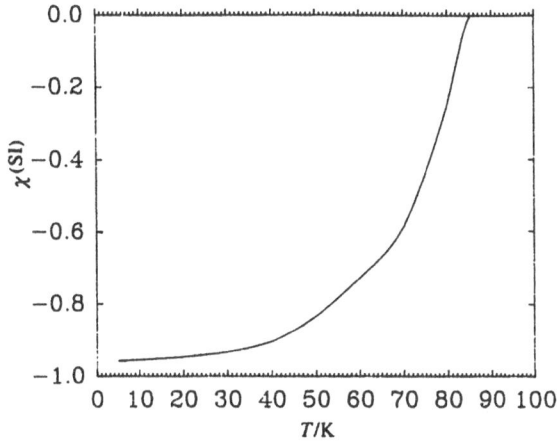

Figure 7: c (T) curves for $Hg_{0.4}Pr_{0.6}Sr_{2.7}Pr_{0.3}Cu_2O_{6+\delta}$

A systematic study of this phase will be of major importance for understanding the rôle of praseodymium in the appearance or disappearance of superconductivity in the layered cuprates, taking especially into consideration the oxidation state of praseodymium (III or IV or mixed) and its location within the rocksalt layers or in the fluorite type layers between the CuO_5 pyramids.

Like the "1201" cuprates, the "1212" cuprates can be stabilised by the introduction of transition elements on the Hg sites. Three cuprates, $Hg_{1-x}M_xSr_2Ca_{1-y}Y_yCu_2O_{6-\delta}$, have been synthesised for M = Mo, Re, Cr [9]: two of them (M = Mo, Re) are superconductors, whereas the third (M = Cr) does not superconduct (Table 2).

In contrast to strontium, the "1212" bismuth and lead substituted "Hg–Ba" phases have not been synthesised to date. All attempts to stabilise the "1212" phase (Hg, Bi or Pb) $Ba_2CaCu_2O_{6+\delta}$ failed. The "1212" phases could only be obtained by replacing simultaneously calcium by a rare earth cation according to the formula (Hg, Bi or Pb) $Ba_2(Ca, R)Cu_2O_{6+\delta}$ [13, 18]. The latter oxides exhibit a rather broad transition with a T_c(onset) of 128K.

Bismuth or lead is not necessary to stabilise the "1212" HgBa cuprates. Partial replacement of calcium by yttrium and neodymium in the ideal cuprate $HgBa_2CaCu_2O_6$ allows, indeed, two series of cuprates with the 1212 structure to be synthesised: $Y_{0.6}Ca_{0.4}Ba_2Cu_2Hg_{1-x}M_xCu_2O_{6+\delta}$ for M = Ca + Cu with $0.40 \leq x \leq 0.60$ [19] and $Nd_{1-x}Ca_xBa_2Hg_{1-x'}Cu_{2+x'}O_{6+\delta}$ with $0.40 \leq x \leq 0.60$ and $0 \leq x' \leq 0.20$ [20].

The superconducting properties of these materials are so far not optimised, showing very broad transitions that might be due to disordering of cations and also of oxygen and anionic vacancies. Nevertheless, the critical temperatures that can be reached, up to 110K for the neodymium compounds, are very promising.

The "1212", "Hg–Ba" cuprates can also be stabilised by introduction of transition elements on the mercury sites (Table 2). Six series of "1212" cuprates $Hg_{1-x}M_xBa_2CaCu_2O_{6+\delta}$ have been synthesised at normal pressure for M = V, Mo, W, Ti, Cr [11–13]. Nevertheless, these superconductors are more difficult to obtain as pure phases. For M = Cr, Mo and Ti, the "1212" phase was always obtained as a major phase from XRD but, for all x, secondary minor phases ($BaCuO_2$ and $CaHgO_2$) were detected. The amount of the "1212" phase is estimated, from the ED analysis, to more than 90% in the case of chromium and molybdenum cuprates. The titanomercury cuprates exhibit, in the same way, clean XRD patterns, but the electron microscopy investigations show that there exist two types of crystal. The first exhibit XRD patterns with sharp reflections and the HREM images confirm that the layers are regularly stacked along c according to the expected "1212" mode. In the second type of crystals, intergrowths of the "1212" and "1223" members are observed either in the form of defects or in the form of different regular sequences. The EDS analysis shows that for Mo and W the composition of the microcrystals is not very different from the nominal composition of the samples. On the contrary, the actual composition of the "1212"–Ti phase varies significantly from one crystal to the other, but this feature can be correlated with the HREM observations, whereas the "1212"–Cr phase exhibits a systematic deviation from the normal composition ($\Delta x \sim 0.2$). As for the "1201" cuprates, the c parameter of the "1212" HgBa cuprate phases is dramatically decreased by the introduction of the transition element, whereas the a parameter is not significantly modified except for the chromium phase. But the most remarkable feature deals with the fact that the evolution of a and c parameters is absolutely identical in both "1201" and "1212" series: $c_{Cr} < c_V < c_{Mo} < c_{Ti} < c_W < c_{Hg}$ and $a_{Ti} \leq a_{Mo} \sim a_W \leq a_{Hg} < a_V < a_{Cr}$.

The $\chi'(T)$ curves of the as-synthesised Mo substituted Hg–Ba samples (Figure 8) show that all these cuprates exhibit a rather sharp transition with high diamagnetic volume fractions ranging from 50% to 90%. One observes that the vanadomercury cuprate (curve a) has a $T_{c(midpoint)}$ much smaller than that of the molybdenomercury cuprate (curve b) whose T_c(midpoint) of 123K is close to the optimum. The critical temperature of the tungstomercury cuprates (curves c–d) is very similar to that observed for the as–synthesised vanadomercury cuprate (curve a), with a T_c(midpoint) ranging from 112 to 113K; thus in contrast to the "1201" tungstomercury cuprates, the T_cs are not modified by the tungsten content; nevertheless, the superconducting volume fraction is higher for $Hg_{0.9}W_{0.1}Ba_2CaCu_2O_{6+\delta}$ (68%) than for $Hg_{0.75}W_{0.25}Ba_2CaCu_2O_{6+\delta}$ (45%). Differently from the "1201" phase, the "1212" as–synthesised chromium phase (curve e) is superconducting; nevertheless, its critical temperature, T_c(midpoint) = 90K, remains smaller than those of other "1212" cuprates. The titanomercury cuprate (curve f) exhibits a high volume fraction (80%) and a high critical temperature, T_c(midpoint) = 123K; the presence of intergrowths in this phase will be confirmed by subsequent annealings.

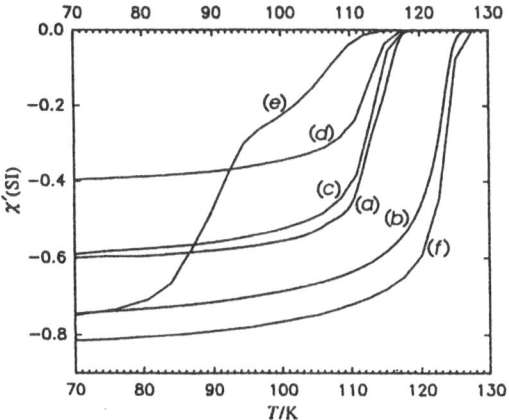

Figure 8: χ' (T) curves for samples $Hg_{0.8}V_{0.2}Ba_2CaCu_2O_{6+\delta}$ (a), $Hg_{0.75}Mo_{0.25}Ba_2CaCu_2O_{6+\delta}$ (b), $Hg_{0.75}W_{0.25}Ba_2CaCu_2O_{6+\delta}$(c), $(Hg_{0.9}W_{0.1}Ba_2CaCu_2O_{6+\delta}$ (d), $Hg_{0.6}Cr_{0.4}Ba_2CaCu_2O_{6+\delta}$ (e), and $Hg_{0.7}Ti_{0.3}Ba_2CaCu_2O_{6+\delta}$(f).

Subsequent annealings of the "1212 HgBa" molybdenomercury cuprate lead to a dramatic decrease of the critical temperature (Figure 9). Compared to the narrow transition of the as–synthesised phase at 123–127K (curve a), the argon annealed phase (curve b) exhibits a broad transition with a T_c(midpoint) of 113K and a much smaller superconducting volume fraction (55%); in the same way, the oxygen annealed phase is characterised by a broader transition (curve c) and a lower T_c(midpoint) = 103K. Thus the behaviour of the Mo–phase is different from that of the vanadium "1212" cuprate whose as–synthesised phase exhibits a T_c(midpoint) of 113K (curve d), and which can be optimised by annealing in an oxygen flow leading to a narrow transition (curve e) characterised by a high critical temperature, T_c(midpoint) = 124K, and a high superconducting volume fraction (75%).

As for vanadium, the superconducting properties of the chromium phase are improved by oxygen annealing (Figure 9). One indeed observes that starting from a T_c(midpoint) of 90K (Figure 9, curve f), a T_c(midpoint) of 106K can be obtained, keeping a high diamagnetic volume fraction (curve g). Nevertheless, the T_cs of the optimised "1212" chromium phase remains smaller than those of other cuprates, as for the "1201" chromium phase. The superconducting properties of the tungstomercury cuprate could not be improved, either by oxygen or by argon annealing, leading to a broadening of the transition. The χ'(T) curves of the titanomercury sample, $Hg_{0.8}Ti_{0.2}Ba_2CaCu_2O_{6+\delta}$ after annealing (Figure 10), are of interest. Starting from a narrow transition at 123–127K (curve a), one obtains a different shape of the χ'(T) curve (curve b) by oxygen annealing that indicates two transitions at about 105K and 131K respectively. The first transition can be assigned to the "1212" matrix and the second to the "1223" members or "1223–1212" intergrowths in agreement with the HREM observations. This means that in the as–synthesised sample, the "1212" phase is nearly optimised, whereas the "1223" member (or intergrowths) are not yet optimised, so that the narrow transition at 123K has all the characteristics of the "1212" matrix. Then, by annealing in an oxygen flow, the superconducting properties of the "1212" phase are degraded, leading to a lower T_c (105K) whereas those of the "1223" members are improved, leading to a higher T_c (131K). The fact that the "1212" and 1223" titanomercury cuprates require different redox conditions, *i.e.* different oxygen pressures, for their optimisation is confirmed further by the study of the 1223 titanomercury cuprate.

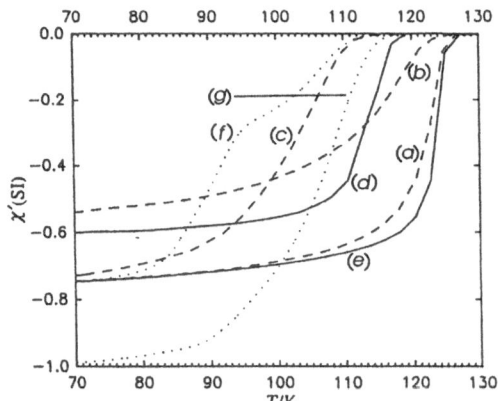

Figure 9: χ' (T) curves of
$Hg_{0.75}Mo_{0.25}Ba_2CaCu_2O_{6+\delta}$ as–synthesised (a), annealed in an Ar flow (b) or in an O_2 flow (c)
$Hg_{0.8}V_{0.2}Ba_2CaCu_2O_{6+\delta}$ as–synthesised (d) and annealed in an O_2 flow (e)
$Hg_{0.6}Cr_{0.4}Ba_2CaCu_2O_{6+\delta}$ as–synthesised (f) and annealed in an O_2 flow (g).

2.3 THE 1223 Hg–BASED CUPRATES

The difficulty in synthesising pure "1223" Hg cuprates is still greater, showing that the metastable character of these cuprates at normal pressure increases with the number of copper layers.

No "Hg–Sr" 1223 cuprate could be synthesised to date, whereas six "Hg–Ba" 1223 cuprates have been synthesised (Table 3) with the general formula $Hg_{1-x}M_xBa_2Ca_2Cu_3O_{8+\delta}$, for M = Bi and Pb [14, 18] and M = Cr, Mo, Ti, V [11–13]. Moreover, for transition metal substituted cuprates, secondary minor phases are always observed. For Cr, Mo and V, the purity level can be estimated to about 80% with $BaCuO_2$ and $CaHgO_2$ as main secondary phases, and only small proportions of the corresponding "1212". For the nominal composition $Hg_{0.9}Mo_{0.1}Ba_2CaCu_3O_{8.2}$ only $BaCuO_2$ and $CaHgO_2$ are (scarcely) visible on the XRD pattern. In the case of titanium, an extra reflection is observed at $2\Theta \approx 29.8°$, which could be attributed to the existence of the related phases built up from the intergrowth of the 1212 and 1223 members.

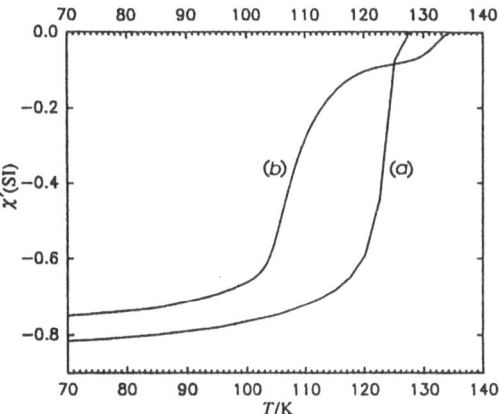

Figure 10: $\chi'(T)$ curves of $Hg_{0.7}Ti_{0.3}Ba_2CaCu_2O_{6+\delta}$, as synthesised (a) and annealed in an O_2 flow (b).

The electron microscopy characterisation of these phases shows that the different layers are very regularly stacked along the c axis, confirming the 1223 structure (Figure 1c). For the titanomercury cuprates, we observe a behaviour which can be compared to that of the 1212 member; most of the crystals, indeed, exhibit a regular sequence of the layers, whereas some others are characterised by the formation of complex intergrowths. The analogy of the evolution of the evolution of the "c" and "a" parameters of the tetragonal cell, with the two other series of "1201" and "1212" cuprates is remarkable (Table 3). The "c" parameter is indeed significantly decreased by introducing the transition element on the mercury site, whereas the a parameter is not noticeably modified.

Table 3: Characteristics of the substituted mercury "1223" cuprates.

Nominal composition	Cell Parameters		As synthesised			After annealing		
	a (Å) c (Å) ±0.0003±0.0001		T_{cmid}	T_{onset}	svf	T_{cmid}	T_{onset}	svf
$HgBa_2Ca_2Cu_3O_{8+\delta}$	3.8564	15.856	—	105K	—	—	134K	—
$Hg_{0.8}V_{0.2}Ba_2Ca_2Cu_3O_{8.3}$	3.8629	15.652	106K	120K	80%	122K	127K	80%
$Hg_{0.8}Mo_{0.2}Ba_2Ca_2Cu_3O_{8.4}$	3.8537	115.731	124K	129K	62%	131K	134K	67%
$Hg_{0.6}Ti_{0.4}Ba_2Ca_2Cu_3O_{8.4}$	3.8507	15.707	124K	128K	58%	132K	134K	60%
$Hg_{0.8}Cr_{0.2}Ba_2Ca_2Cu_3O_{8.1}$	3.8731	15.625	92K	95K	68%	104K	110K	76%
$Hg_{0.8}Bi_{0.2}Ba_2Ca_2Cu_3O_{8+\delta}$	3.850	15.813	130K					
$Hg_{0.66}Pb_{0.33}Ba_2Ca_2Cu_3O_{8+\delta}$	3.849	15.860	133K					

All these cuprates are characterised by a sharp transition and a high superconducting volume fraction (Figure 11). The as–synthesised cuprates exhibit a high T_C ranging from 106K for the V–phase (curve a) to 130K for the Bi–phase (curve e), the Mo and Ti cuprates (curves b and d) being characterised by a T_C of 124K. The lowest T_C is in fact observed for the Cr–phase: 92K (curve c).

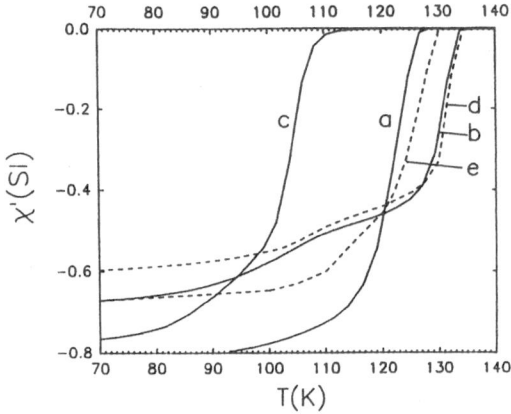

Figure 11: $\chi'(T)$ curves of $Hg_{0.7}V_{0.3}Ba_2Ca_2Cu_3O_{8+\delta}$ (a), $Hg_{0.75}Mo_{0.25}Ba_2Ca_2Cu_3O_{8+\delta}$ (b), $Hg_{0.6}Cr_{0.6}Ba_2Ca_2Cu_3O_{8+\delta}$ (c), $Hg_{0.7}Ti_{0.3}Ba_2Ca_2Cu_3O_{8+\delta}$ (d), $Hg_{0.8}Bi_{0.2}Ba_2Ca_2Cu_3O_{8+\delta}$ (e).

The superconducting properties of the transition element substituted 1223 cuprates can be significantly improved by annealing (Figure 12). The three cuprates of vanadium, (curve a), molybdenum (curve b) and titanium (curve c) are indeed optimised up to the maximum value of the critical temperature for these materials, i.e. $T_{c(midpoint)}$ = 131–132K, and $T_{c(onset)}$ = 134K. Here again, one observes that the chromium phase (curve c) has a lower critical temperature, even after annealing, *i.e.* $T_{c(midpoint)} = 104K$. Thus it seems that in this case, the thallium and mercury atoms occupy two different sites corresponding to thallium bilayers and mercury monolayers respectively.

A "Hg–Sr" solid solution with the "1212" structure has recently been synthesised. These oxides, with the formula $(Tl,Hg)_1Sr_{2+y}Nd_{1-y}Cu_2O_{7+d}$ [23], contain up to 20% mercury in the thallium layers, and exhibit critical temperatures up to 95K. Finally, the "TlSr" "1212" superconducting cuprate $Tl_{0.8}Hg_{0.2}Sr_{2.2}Ca_{0.7}Cu_2O_{7+\delta}$, with a $T_{c(midpoint)}$ of 65K and the 2212 phase $Tl_{1.5}Hg_{0.4}Sr_{2.2}Ca_{0.8}Cu_2O_{8+\delta}$, have recently been synthesised [24].

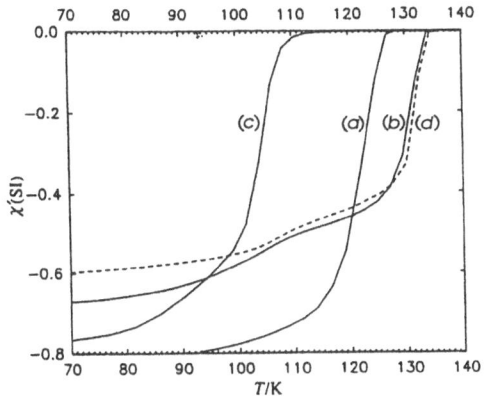

Figure 12: $\chi'(T)$ curves for samples annealed in O_2 flow: $Hg_{0.8}V_{0.2}Ba_2Ca_2Cu_3O_{8+\delta}$ (a), $Hg_{0.75}Mo_{0.25}Ba_2Ca_2Cu_3O_{8+\delta}$ (b), $Hg_{0.6}Cr_{0.6}Ba_2Ca_2Cu_3O_{8+\delta}$ (c), $Hg_{0.7}Ti_{0.3}Ba_2Ca_2Cu_3O_{8+\delta}$ (d).

During the investigation of the substitution of Bi for Hg in the "Ba–Hg" cuprate, a new "1212–1223" intergrowth was isolated in the composition $Hg_{2-x}Bi_xBa_4Ca_3Cu_5O_{14+\delta}$, [13] with $0.1<x<0.4$. This compound exhibits a very sharp transition with a high superconducting volume fraction. It is remarkable that its critical temperature, close to 130K, is similar to that of the "1223" phase; thus the presence of "1212" member as a regular intergrowth of the "1223" matrix does not affect significantly the critical temperature.

These results confirm that, in contrast to the 1201 and 1212 cuprates of d^0 transition elements, the 1223 corresponding cuprates are optimised by oxygen annealing. In contrast, the chromium phases are systematically improved by oxygen annealing.

2.4 OTHER MERCURY BASED SUPERCONDUCTING CUPRATES

The possible creation of new "Hg–Ba" cuprates is the theme of research into the mixed thallium–mercury superconductors. Substitution of mercury for thallium is limited in the oxides $Tl_{2-x}Hg_xBa_2Ca_{n-1}Cu_nO_{2n+4}$ prepared under normal oxygen pressure. This is, for instance, the case for the "2223" cuprate $Tl_{2-x}Hg_xBa_2Ca_2Cu_3O_{10-\delta}$ [21], that exhibits a maximum mercury content x = 0.4. Moreover, the T_C of this phase, 130K, is similar to that of the undoped material. Of particular interest is the synthesis of the 50K superconductor $HgTl_2Ba_4Cu_4O_{10+\delta}$ [22] whose structure consists of a regular intergrowth of the "2201" $Tl_2Ba_2CuO_6$ structure with the "1201" $HgBa_2CuO_4$ structure.

3. Creation of new copper oxycarbonates through shearing mechanisms

The $Tl_{0.5}Pb_{0.5}Sr_4Cu_2O_7CO_3$ structure [25] which consists of an intergrowth of the "1201" and $Tl_{1.5}Hg_{0.4}Sr_{2.2}Ca_{0.8}Cu_2O_{8+\delta}$ (Figure 13) has been obtained for several other oxycarbonates: $Tl_{1-x}M_xSr_4Cu_2O_7CO_3$ with M = Bi [26], Mo [27] and $Hg_{1-x}M_xSr_4Cu_2O_7CO_3$ with M = Pb [16], Bi [28], V [29], Mo and Cr [30]. All these oxides named $[1201]_1[S_2CC]_1$ intergrowth are superconductors with a critical temperature ranging from 37K to 76K. In the same manner, bismuth oxycarbonates $(Bi_2Sr_2CuO_6)_m(Sr_2CuO_2CO_3)_n$, intergrowths of the 2201 and $Sr_2CuO_2CO_3$ structures, can be synthesised as shown for the superconductor $Bi_2Sr_4Cu_2CO_3O_8$ [37-39] that exhibits a T_C of 40K. Starting from these two types of structures — $Tl_{0.5}Pb_{0.5}Sr_4Cu_2CO_3O_7$ and $Bi_2Sr_4Cu_2CO_3O_7$ — new structures can be generated by applying shearing mechanisms transversally to the copper layers.

3.1 THE (100)–COLLAPSED $[1201]_1[S_2CC]_1$ OXYCARBONS

Starting from the single intergrowth $[1201]_1[S_2CC]_1$ (Figure 13), a shearing mechanism can be applied along the (100) plane of the structure by introducing barium on the strontium sites, leading to a new structure that consists of slices of the $[1201]1[S2CC]1$ intergrowths, shifted by $^c/_2$ with respect to each other.

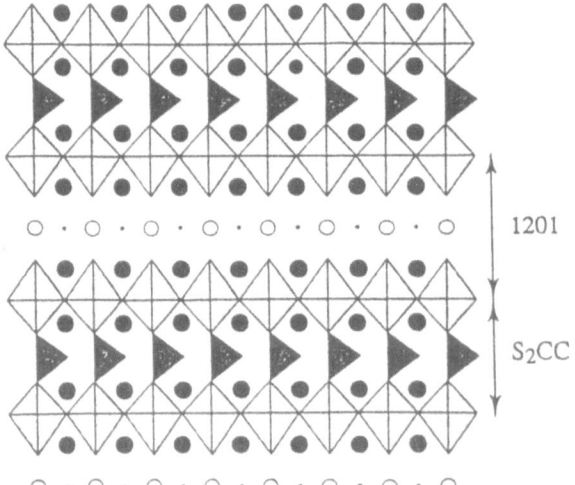

Figure 13: Schematic representation of the structure of the single intergrowth $[1201]_1[S_2CC]_1$. Triangles represent the carbonate groups.

The most recent member of this series is represented by the oxycarbonates $TlBa_2Sr_2Cu_2(CO_3)O_7$ [32], $Tl_{1-x}Hg_xBa_2Sr_2Cu_2(CO_3)O_7$ with x = 0.2 and 0.5 [32]. In this structure, two successive slices that are m = 4 CuO_6 octahedra thick are shifted $\bar{c}/2$ in the (100) plane, as schematised in Figure 14. From this shearing mechanism, it appears that the $[CuO_2]_\infty$ layers and the $[SrO]_\infty$ layers of each block remain unchanged and form infinite layers parallel to (001). This is not the case for the $[TlO]_\infty$ ribbons which are limited to four Tl atoms along b and are connected to ribbons of four CO_3 groups. Thus the $[TlO]_\infty$ and $[CO]_\infty$ layers of the structure $Tl_{0.5}Pb_{0.5}Sr_4Cu_2(CO_3)O_7$ are replaced by mixed layers $[(TlO)_4(CO)_4]_\infty$ characterised by a sequence of four thallium atoms and four carbonate groups along b.

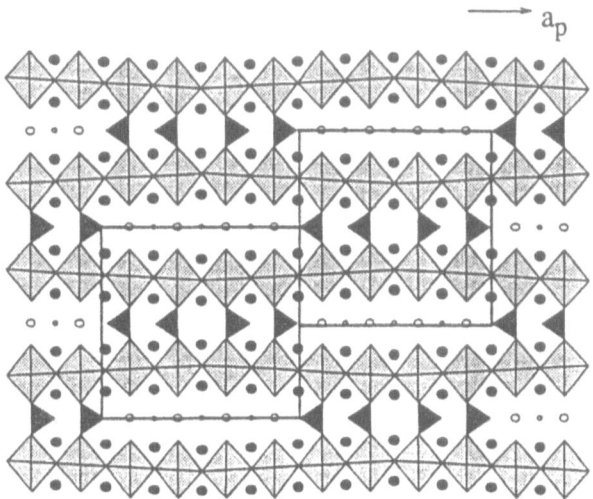

Figure 14: Idealised drawing of the structure of the collapsed oxycarbonate $TlBa_2Sr_2Cu_2(CO_3)O_7$. Two blocks of the single intergrowth $[1201]_1[S_2CC]_1$, four octahedra wide are evidenced.

This shearing phenomenon with respect to $Tl_{0.5}Pb_{0.5}Sr_4Cu_2(CO_3)O_7$ is evidenced by the [100] HREM images. The contrast can be described from rectangular areas (Figure 15) which correspond to half a cell "$b/2 \times c$". Considered separately, these rectangles exhibit exactly the same contrast as the oxycarbonate $Tl_{0.5}Pb_{0.5}Sr_4Cu_2(CO_3)O_7$; along b, each rectangle is four CuO_6 octahedra wide. Then the three groups of layers with different contrast stacked along c can be identified by their similarity with $Tl_{0.5}Pb_{0.5}Sr_4Cu_2(CO_3)O_7$.

The first group consists of three rows of alternated black dots which are correlated to the sequence of rows "(Ba,Sr)O–TlO–(Ba,Sr)O", also observed in some thallium cuprates characterised by thallium monolayers. The second corresponds to a single layer of grey dots which is correlated to a $[CuO_2]_\infty$ layer. The last group consists of three alternating rows of white dots, characteristic of the rows of CO_3 groups. The fundamental difference from the lead based oxycarbonate relates to the fact that two adjacent rectangles are generated one from the other by a translation of $b/2 + c/2$ in agreement with the A–type space group.

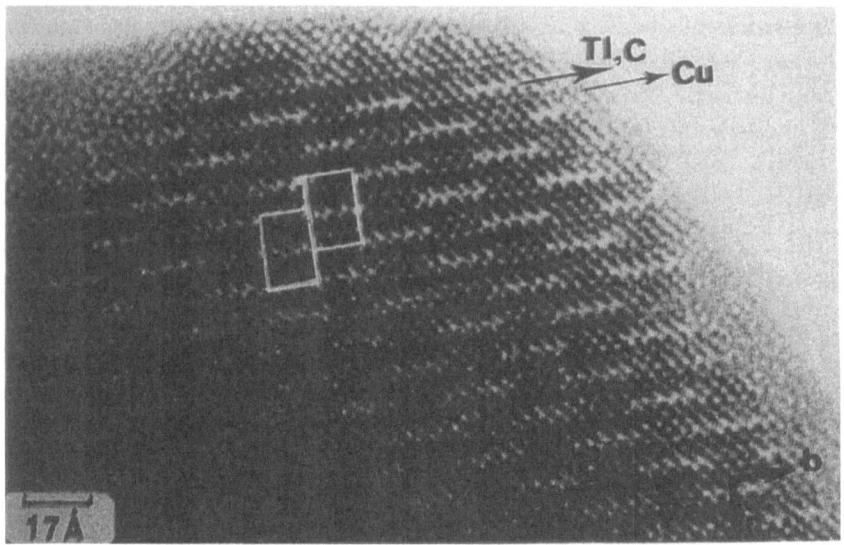

Figure 15: [100] HREM image of $TlBa_2Sr_2Cu_2(CO_3)O_7$.

Such a shearing phenomenon, which implies the coexistence of thallium and carbon within the same layer raises the issue of the adaptation of these elements owing to their large size differences, which imply Tl–O apical distances of about 2.00Å, much larger than the C–O distances along c (about 1.30Å). The answer to this question is given by changing the focus of the [100] HREM image. For some focus values (Figure 16), the atoms of the $[CuO_2]_\infty$ layers are highlighted: it can be seen that these layers undulate with a rather large amplitude. Thus the structure of $TlBa_2Sr_2Cu_2(CO_3)O_7$ (Figure 14) consists of single perovskite undulating layers involving the $[CuO_2]_\infty$ and the $[(Sr, Ba)O]_\infty$ sheets connected through $[(TlO)_4]_\infty$ layers. As a result, the cell is orthorhombic, and characterised by a superstructure along $\bar{b}$ with respect to the single intergrowth: $a\sim a_p\sim3.8$Å, $b\approx8 \times a_p$, $c\approx 17$Å.

In a more general way, the (100) plane is a crystallographic shear plane, along which a displacement may appear in every "m" octahedra, so that the (100)–collapsed oxycarbonates will be characterised by a super lattice "$a \approx a_p$, $b \approx 2m\, a_p$, $c \approx 17$Å"; the above phases correspond to m = 4.

Different shearings may appear that lead to various superstructures: (100)–shearing phenomena were observed in the system $TlSr_{4-x}Ba_xCu_2(CO_3)O_7$ [33] leading to the superlattice "$a_p \times 6a_p \times 17$Å" that corresponds to m = 3, and to the superlattice "$a_p \times 7a_p \times 17$Å" that corresponds to a (100) shearing phenomenon every three (m = 3) and four (m = 4) octahedra alternately. This shearing mechanism is sometimes more complex as shown, for instance, in the oxycarbonate $Tl_{0.3}Hg_{0.7}Ba_2Sr_2Cu_2(CO_3)O_7$ [32] that exhibits an incommensurate structure "$a_p \times 7.5a_p \times 17\text{Å}$", which implies that the distribution of the (100) shear planes is modulated along the b direction of the actual cell.

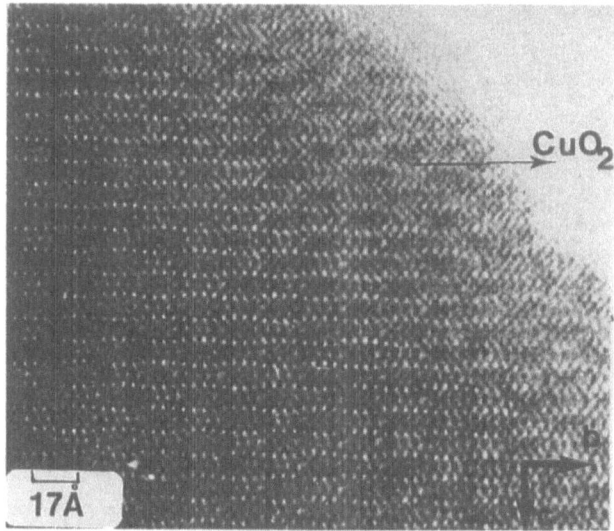

Figure 16: [100] HREM image of $TlBa_2Sr_2Cu_2O_7CO_3$: the [CuO$_2$] layers are highlighted.

It is remarkable that all these (100)–collapsed $[1201]_1[S_2CC]_1$ oxycarbonates are superconductors with critical temperatures ranging between 60K and 70K. This is in agreement with the fact that the [CuO2]∞ planes are not interrupted by the shearing phenomenon. Moreover this shows that the waving of the copper oxygen layers does not destroy the superconducting properties of these materials. At the present time, the rôle of barium in the (100)–shearing phenomenon is not explained. In particular, it is not known whether barium is absolutely necessary for the appearance of (100) shear planes.

3.2 THE (110)–COLLAPSED $[1201]_1[S_2CC]_1$ OXYCARBONATES

The introduction of transition elements on the thallium sites of the S_2CC–1201 oxycarbonate allows also a shearing phenomenon to be induced. This is the case for the oxycarbonates $Tl_{2/3}Cr_{1/3}Sr_4Cu_2(CO_3)O_7$ and $Tl_{0.8}V_{0.2}Sr_4Cu_2C(CO_3)O_7$ [30]. Nevertheless, in these oxycarbonates the shearing plane is no more the (100) plane but the $(110)_{S_2CC-1201}$ plane. In order to observe the atomic arrangements of these oxides, the crystals were viewed along [100]. This direction corresponds to $[110]_{S_2CC-1201}$ or $[110]_p$. Two images of the $Tl_{2/3}Cr_{1/3}Sr_4Cu_2(CO_3)O_7$ compound are given in Figures 17 and 18 illustrating the typical contrasts which have been registered; their interpretation, especially the identification of the layers, was made with the help of the [110]S2CC–1201 simulated images, calculated for an ideal unmodulated structure. In Figure 17 the heavy Tl(Cr or V) and Sr atoms appear as dark spots, for a focus value close to–50Å. The contrast of this image consists mainly of three rows of bright and grey spots: in the intermediate row, five bright spots correlated to the carbonate positions alternate with five dark spots correlated to the heavy ion positions (Tl, Cr), leading to a $5 \times a_p\sqrt{2}$ periodicity along $\bar{b}$; the position of the five bright dots is shifted by $5 \times a_p\sqrt{2.2}$ in the two adjacent equivalent rows, involving a centred image.

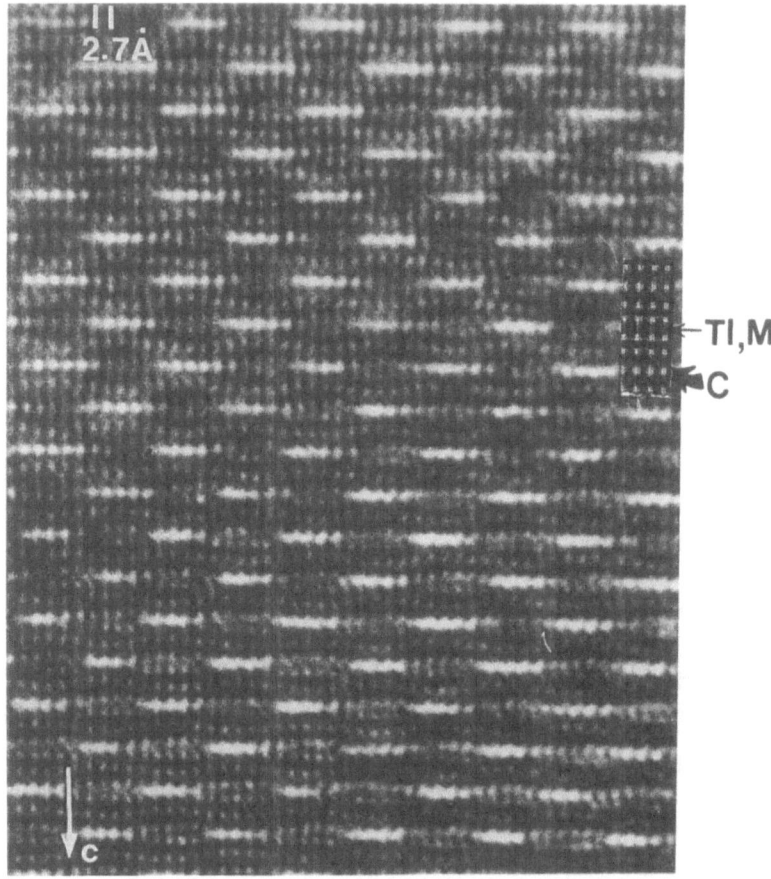

Figure 17: [010] HREM image of the (110)–collapsed $Tl_{2/3}Cr_{1/3}Sr_4Cu_2(CO_3)O_7$ oxycarbonate. The simulated image (t~31Å and focus value ~50Å) in insert has been calculated for the unmodulated subcell. The carbon segments appear as five successive dots.

The other two rows of bright spots correspond to the $[SrO]_\infty$ layers. The copper layers, which appear as rows of grey spots, are intercalated between the $[SrO]_\infty$ layers; they undulate with a rather large amplitude: they indeed sandwich the bright spots at a significantly smaller distance than the dark spots. In Figure 18, the contrast is different so that the thallium and strontium atoms, which appear as bright spots, are clearly shown; the three rows of highlighted spots are correlated to the SrO–Tl(Cr)O–SrO layers and the row of grey spots to the copper layers. for a focus value close to 200Å, the existence of mixed thallium(Cr)–carbon layers is easily imaged, as well. These images indicate, without any ambiguity, that the atomic arrangements of $Tl_{2/3}Cr_{1/3}Sr_4Cu_2(CO_3)O_7$ and $Tl_{0.8}V_{0.2}Sr_4Cu_2(CO_3)O_7$ are those of a (110) collapsed–S2CC–1201 structure, characterised by mixed intermediate layers where m = 5 (Tl + Cr) atoms alternate with m' = 5 carbon atoms along $\bar{b}$ (Figure 19a). The oxycarbonates $Hg_{0.5}Pb_{0.5}Ba_2Sr_2Cu_2(CO_3)O_{7-\delta}$ and $Hg_{0.5}Pb_{0.5}Sr_{3.2}Ba_{0.8}Cu_2(CO_3)O_7$ [35] exhibit also the (110) shearing plane. But in that case the 1201–S2CC structure is shifted by $\bar{a}/2$ every six and five CuO_6 octahedra alternately along [110]S2CC–1201 (Figure 19b) leading to a periodicity along [110] of $11a_p\sqrt{2}$. In the same way, the HREM image obtained for $HgBa_2Sr_2Cu_2(CO_3)O_7$ [34], that implies a periodicity = $9a_p\sqrt{2}$ along (110), can be interpreted by a (110) shearing of the [1201]1[S2CC]1 intergrowth with every five and four octahedra alternately.

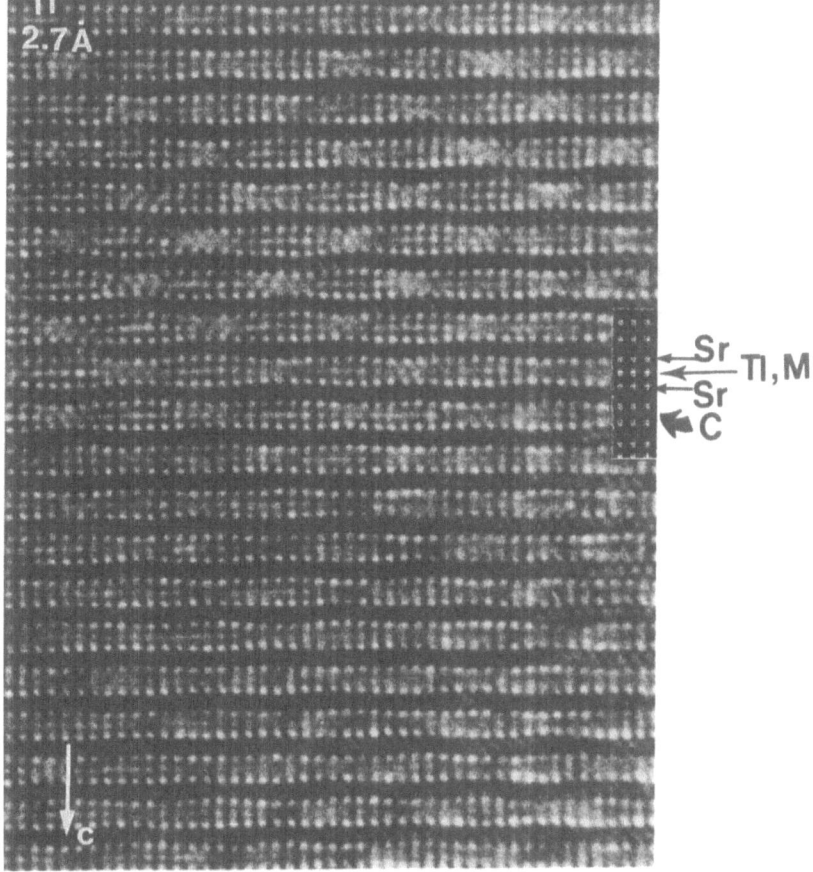

Figure 18: [010] HREM image for a focus value close to –550Å where the cation positions are correlated to the bright spots. The calculated image is in the insert.

The possibility of obtaining an incommensurate distribution of the (110) shear planes in these structures has also been observed. This is, for instance, the case of $Tl_{0.2}Hg_{0.8}Ba_2Sr_2Cu_2(CO_3)O_7$ [32] that exhibits 4.7 times incommensurate structure along the [110] direction.

As with the (100) "collapsed" oxycarbonates, the (110) collapsed oxycarbonates are all superconductors in agreement with the fact that the $[CuO_2]_\infty$ layers are not interrupted by the shearing mechanism. $HgBa_2Sr_2Cu_2(CO_3)O_{7-\delta}$ is a superconductor up to 65K, whereas critical temperatures ranging from 55K to 68K were observed for $Hg_{0.5}Pb_{0.5}Sr_{2+x}Ba_{2-x}Cu_2(CO_3)O_7$ (x = 0, 1.2). The critical temperature of the as–synthesised $Tl_{0.66}Cr_{0.333}Sr_4Cu_2CO_3O_{7-\delta}$ is significantly higher than that observed for other thallium oxycarbonates ($T_{C(onset)}$ = 72K; $T_{C(midpoint)}$ = 68K). Moreover successive annealings in an Ar/H_2 flow show that Tc is significantly increased ($T_{C(onset)}$ = 77K; $T_{C(midpoint)}$ = 74K).

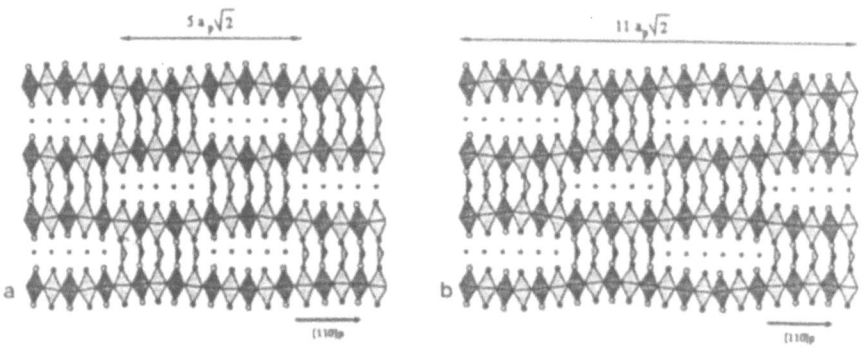

Figure 19: Idealised drawing of the (110)–collapsed structures: a) $(Tl, M)Sr_4Cu_2(CO_3)O_7$ and b) $Hg_{0.5}Pb_{0.5}Ba_2Sr_2(CO_3)O_{7-\delta}$.

3.3 THE COLLAPSED BISMUTH OXYCARBONATE $Bi_{15}Sr_{29}Cu_{12}(CO_3)_7O_{56}$

The shearing mechanisms can also be applied to the bismuth oxycarbonates. But in that case, the shearing phenomenon leads to an interruption of the $[CuO_2]_\infty$ layers due to the fact that the thickness of the $[SrBi_2O_3]_\infty$ layers is different from that of the octahedral layers.

Consequently, the corresponding collapsed structure does not superconduct. This is the case for the collapsed oxycarbonate [36], $Bi_{15}Sr_{29}Cu_{12}(CO_3)_7O_{56}$ that exhibits a monoclinic cell, with a = 18.796Å, b = 5.486Å, c = 39.506Å and β = 111.67°. as for the other bismuth based phases, the cationic distribution of this complex structure has been identified using HREM and the previous work on modulated collapsed and oxycarbonate phases.

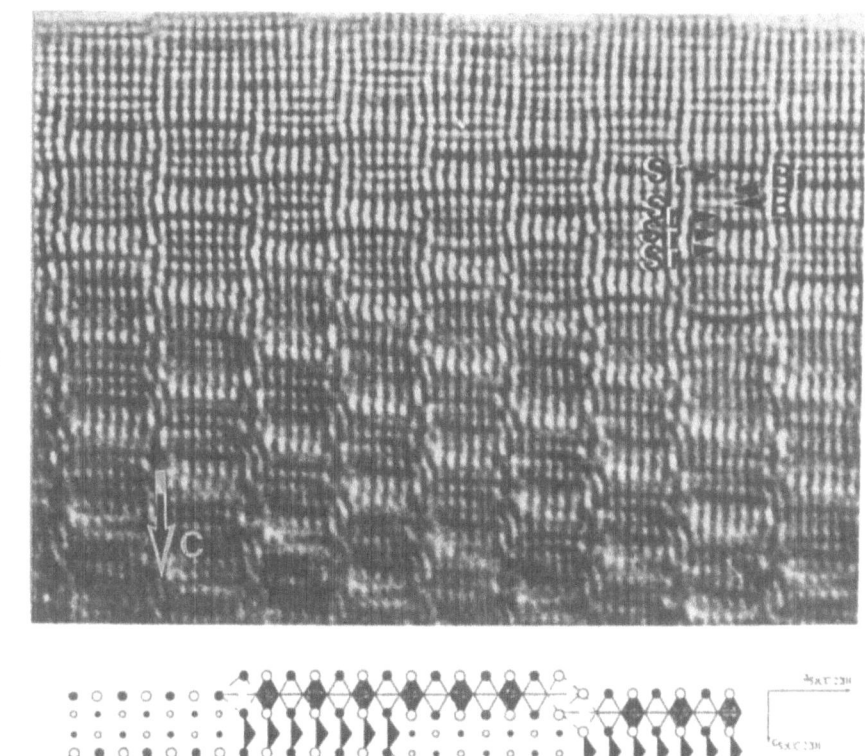

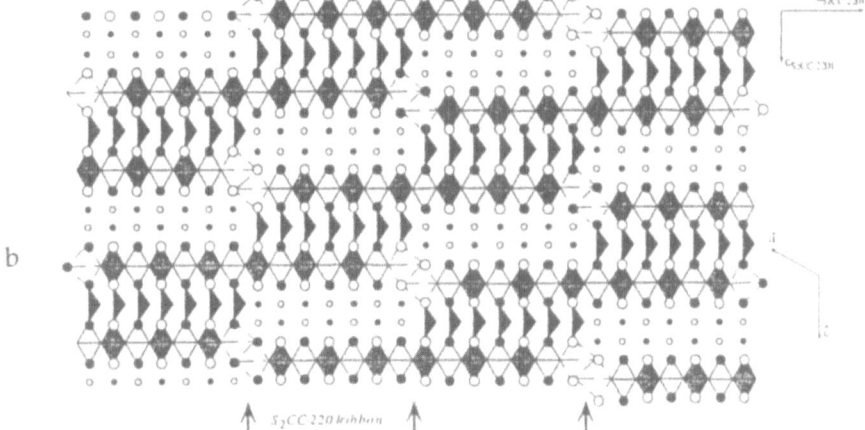

Figure 20: a) HREM image of $Bi_{15}Sr_{29}Cu_{12}(CO_3)_7O_{56}$ and b) idealised model of the collapsed oxycarbonate.

The existence of $Bi_2Sr_4Cu_2CO_3O_8$–type ribbons running along c is, here again, straightforward and the identification of the layer sequence within the 7 octahedra wide ribbons is the first step. In Figure 20a, the heavy atom positions appear as bright dots. The contrast can be described from two rows of bright dots: the first one is a group of four rows, oval shaped, correlated to the sequence $(SrO)_7$–$(BiO)_7$–$(BiO)_7$–$(SrO)_7$; the second is a group of two rows correlated to two $(SrO)_7$ ribbons. The layer sequence along c within the ribbons is that of the $m = 1$ and $n = 1$ oxycarbonate $Bi_2Sr_4Cu_2(CO_3)O_8$ [37–39].

The formation of ribbons results from a shearing operation: one shear plane is parallel to (010) plane of the orthorhombic oxycarbonate $Bi_2Sr_4Cu_2(CO_3)O_8$, *i.e.* to the $(010)_{2201}$ plane since the latter can be described from the intergrowth of 2201 and $Sr_2CuCO_3O_2$ units and exhibits a and c axes parallel to those of the 2201 structure. Two oval shaped "Bi" segments are translated for about 10Å along c.

The connection of the (010) tapes, through the shear planes, can be easily observed on such images. An idealised model of the collapsed oxycarbonate is shown in Figure 20b. Such a structure would correspond to the ideal formula $Bi_{14}Sr_{28}Cu_{14}O_{56}$; however, as mentioned for the 2212 collapsed cuprate, the contrast variation at the level of the shear planes suggests the existence of local cationic substitution, such as the replacement of copper by bismuth or strontium atoms at the extremities of the copper segments. This is in agreement with the nominal and actual composition which corresponds to the cationic ratio $Bi_{14}/Sr_{28}/Cu_{12}$ instead of $Bi_{14}/Sr_{28}/Cu_{14}$ This structure can thus be described as a $\{010\}_{2201}$ collapsed oxycarbonate which results from a shearing operation, every seven octahedra, in the $m = 1$ and $n = 1$ member of the Bi oxycarbonate $(Bi_2Sr_2CuO_6)_m(Sr_2CuCO_3O_2)_n$ [37–39].

4. Synthesis of metastable superconductors $(CaCuO_2)_m(Ba_2CuO_2CO_3)_n$ as thin films

After the discovery of high temperature superconductivity in 1987, laser ablation was revealed to be an exceptional tool for depositing thin films of cuprates. In fact, this method is of considerable interest, since it allows the realisation of new metastable frameworks, due to the fact that it involves a rather low deposition temperature on the substrate. Thus it can be considered as a new kind of soft chemistry, based on the quasi epitaxial character of the deposition on a monocrystalline substrate.

Based on these considerations, the system Ba–Ca–Cu–O, using laser ablation, was investigated, but introducing various partial pressures of carbon dioxide close to the substrate. By this method, a new superconducting oxycarbonate family $(CaCuO_2)_m(Ba_2CuO_2CO_3)_n$ is synthesised as a thin film [40]. The main phase that forms the matrix is $Ba_2Ca_3Cu_4CO_3O_8$ ($m = 3$; $n = 1$). The films are deposited on the (001) plane of a single crystal substrate $LaAlO_3$, using a pulsed KrF excimer laser Lambda Physik, working at $\lambda = 248nm$. A special preparation of the sintered target is necessary, the best results being obtained for the nominal composition $Ba_2Ca_3Cu_5O_{10}$. The O_2 and CO_2 pressures were carefully controlled during deposition, the optimisation of the film being reached for a gas mixture containing 6% CO_2.

The XRD patterns and the electron diffraction patterns show that this new phase is characterised by a pseudo–cubic subcell characteristic of the perovskite and exhibits, in fact, a tetragonal symmetry with an in–plane b parameter equal to the a parameter, perpendicular to the substrate. However, the [001] ED patterns recorded along a direction perpendicular to the substrate plane (Figure 21) show that the c parameter, in the substrate plane, varies and is a multiple of the subcell a_p parameter $c = (m + 2) \times a_p$. One indeed observes two important structural features. First, there exist streaks along two equivalent directions of the perovskite subcell (Figure 21). Second, from the reconstruction of the reciprocal lattice, it clearly appears that the streaks exist only along one direction, i.e. along $\bar{c}$, (the intense reflections corresponding to the perovskite subcell). Moreover, small nodes are observed which attest that the periodicity along $\bar{c}$ is, in fact, roughly a multiple of a_p, i.e. $c = (m + 2) \times a_p$.

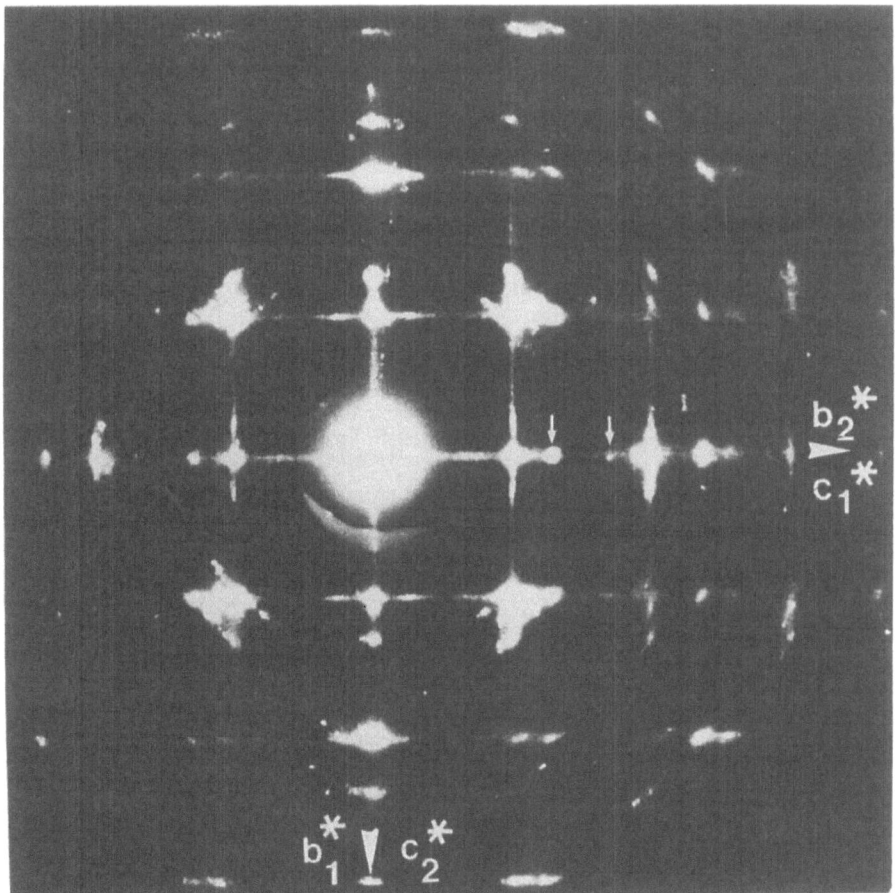

Figure 21: [100] ED pattern, recorded along a direction perpendicular to the substrate plane. The intense reflections are those of the perovskite subcell. Streaks are observed along two equivalent directions of the subcell, $c1^*$ and $c2^*$. Nodes, indicated by small arrows, are the 006 and 008 reflections of the m = 4 member of the series.

Figure 22: Enlarged [100] images: a) the cation positions are imaged as bright dots (Δf ~ –50nm); barium and copper layers are identified. The number of copper layers within one slice, *i.e.* m + 1, is indicated in the top part of the image. b) typical image (Δf ~ –15nm) where the carbon rows and surrounding oxygen atoms appear in the form of very bright dots (curved arrow).

Thus the periodicity along $\bar{c}$, which is supposed to result from an ordering phenomenon, is not well established so that the aleatory sequences can be expected with the ordering phenomena taking place along the two equivalent directions of the subcell, which would involve the formation of 90° oriented domains. These results suggest that the as–grown new tetragonal microphases, with $a = a_p$ and $c = (m + 2) \times a_p$ are a–axis orientated with respect to the substrate, *i.e.* with the a–axis perpendicular to the substrate plane.

The EDS analyses, performed on several zones of the films, lead to an average cationic composition of the new material (plotted to 2 for the barium content) of "$Ba_2Ca_{2.9}Cu_{4.15}$". The [100] high resolution electron microscopy images, recorded for two different focus values (Figure 22) clearly establish that the contrast is directly related to that observed in the copper based perovskites and oxycarbonates. In Figure 22a, the cation positions are imaged as bright dots ($\Delta f \approx -50nm$). Such an image allows the sequence of the layers to be identified. The characteristic features of the contrast consist in two rows of very bright dots separated by rows of slightly less bright dots. The very bright dots are correlated to the positions of the barium and the less bright ones to the copper positions. Rows of small grey dots are located between the copper layers; they are correlated to the calcium rows. In Figure 22b, the zones of light electron density are imaged as bright dots.

It can be seen that between the rows of dark dots which are now correlated with the cation positions, there exist two types of rows of bright dots. The first ones are located between the barium rows and exhibit a contrast which is typical of the rows of carbonate groups and the second ones, located between the rows of calcium atoms, are correlated with the copper rows and exhibit a contrast similar to that observed in 1223 or 1234 cuprates. The interlayer distance between two successive $[CuO_2]_\infty$ layers, close to 3.3Å, suggests that in the intermediate copper layers, located between two calcium layers, copper exhibits a square planar coordination, whereas calcium has a cubic coordination.

The number of copper layers between two successive carbonate layers can be easily obtained from these images (see white numbers in Figure 22a). It simply corresponds to the number of rows of bright dots intercalated between the two rows of very bright dots. One observes that it varies from 2 to 5, but is mainly 4. From these observations, a structural model can be proposed which describes the different members of the series (Figure 23) as derivatives of the infinite layer structure $CaCuO_2$ with intercalated layers of carbonate groups. At the level of each carbonate layer, calcium is replaced by barium, so that one carbonate layer is sandwiched by two BaO layers according to the sequence "$CuO_2–BaO–CO–BaO–CuO_2$". Such slices are then stacked with multiple $CaCuO_2$–type layers so that the general formula of these microphases can be written $(CaCuO_2)_m(Ba_2CuO_2CO_3)_n$. Most of the members observed in this film (Figure 23a) correspond to n = 1, whereas m varies from 1 to 4. The structure of the oxycarbonates $(CaCuO_2)_m(Ba_2CuO_2CO_3)_n$ consists of pyramidal copper layers stacked with layers of CuO_4 square planar groups and interconnected with layers of carbonate groups as shown, for example, in the first three members of this series (Figure 23a-b-c). Note that the m = 1 member, $Ba_2CaCu_2CO_3O_4$, (Figure 23a) is directly derived from the "123" structure by replacing rows of CuO_4 square planar groups by rows of carbonate groups. The m = 3 member, $Ba_2Ca_3Cu_4CO_3O_8$, (Figure 23c) which is the predominantly observed microphase (Figure 22a) consists of quadruple copper layers built up from double $[CuO_2]_\infty$ square planar layers sandwiched between two pyramidal copper layers and interconnected through carbonate layers. The n > 1 members are rarely observed; n = 2 members appear sometimes as a defect and correspond to the local microphase $CaBa_4Cu_3(CO_3)_2O_6$, (m = 1; n = 2). The corresponding structural model (Figure 23d) shows that the structure of the members $(CaCuO_2)_m,(Ba_2CuO_2CO_3)_n$, consists of octahedral and pyramidal copper layers interconnected through layers of carbonate groups. The AC susceptibility measurements of the films (Figure 24) show that the best T_c is obtained at 6% CO_2. DC resistance measurement of a sample grown with a 3% gas composition of CO_2 is shown in Figure 25. The T_c onset is greater than 100K and zero resistance is reached below 75K.

Above the transition, the signal is noisy, probably owing to the ill crystallised areas of the film, and the sample shows semiconducting behaviour. These new bulk oxycarbonates can be compared to those recently synthesised in the form of materials at high pressure [41][42]. Nevertheless, their structure "BaCaCu" is fundamentally different from those of high pressure carbonates such as $(Cu_{0.5}C_{0.5})Ba_2Ca_{n-1}Cu_nO_{2n+3}$ or $(Cu_{0.5}C_{0.5})Ba_3Ca_{n-1}Cu_nO_{2n+5})$ by the fact that in the films no "1–1" ordered carbonate/copper layers i.e. no ordering has been detected along b; thus the copper content of the carbonate layers of these films, if different from zero, can be assumed to be very low. In this respect, the laser ablation method seems to be more effective for carbonation than the high pressure method since it allows full carbonate layers to be stabilised.

However, the ideal formula of these phases and particularly of the main microphase $Ba_2Ca_3Cu_4CO_3O_8$ (m = 3; n = 1) does not suggest any mixed valent copper, i.e. it implies Cu(II) only and therefore presents the problem of discovering the origin of superconductivity in these compounds. The creation of hole carriers necessary for the appearance of superconductivity may result from the fact that CO_3 groups are statistically missing in the structure (Figure 26a), inducing mixed valence Cu(II) – Cu(III).

Another hypothesis proposes a partial replacement of the CO_3 groups by a small number of CuO_4 groups (Figure 26b) which would not be seen by HREM owing to the statistical distribution of the CuO_4 groups over the carbonate positions. The EDS analysis, indicating a slight excess of copper with respect to the ideal composition, supports this second hypothesis

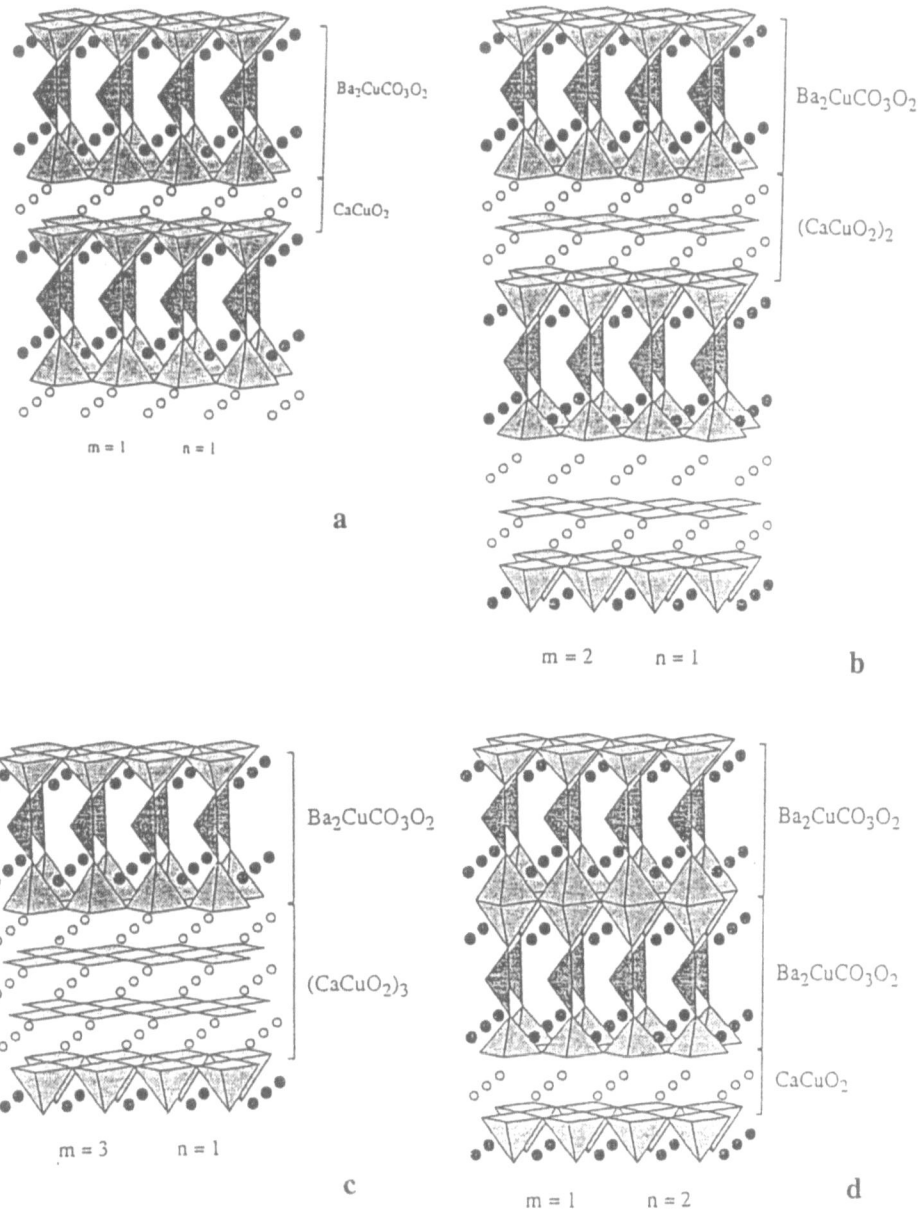

Figure 23: Idealised structural models of the first members of the family $(CaCuO_2)_m/(Ba_2CuO_2CO_3)_n$. a) m = 1, n = 1; b) m = 2, n = 1; c) m = 3, n = 1; d) m = 1, n = 2.

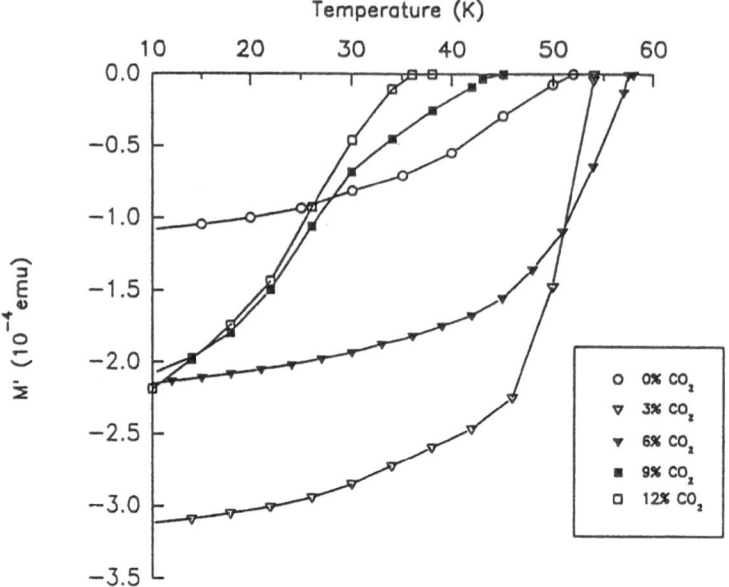

Figure 24: AC magnetic susceptibility measurements of thin films grown at 0, 3, 6, 9 and 12% gas composition of CO_2. Dimensions of the film grown under pure oxygen are 3mm x 2mm x 1500Å. Dimensions of the other films are 5mm x 4mm x 1500Å.

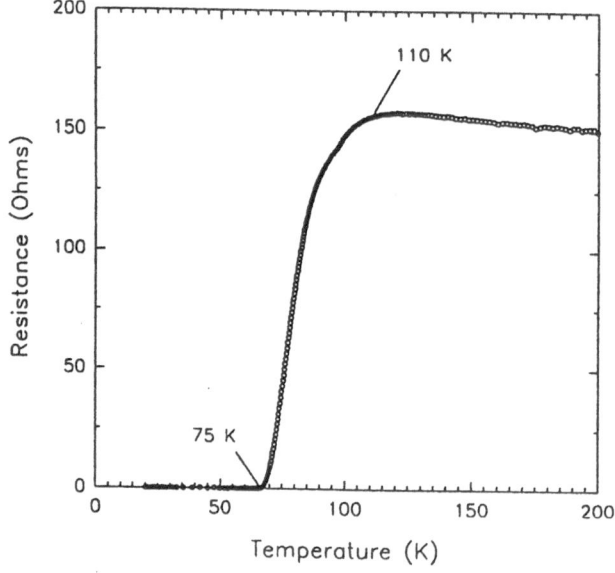

Figure 25: DC resistance measurement of a film grown with a 3% gas composition of CO_2.

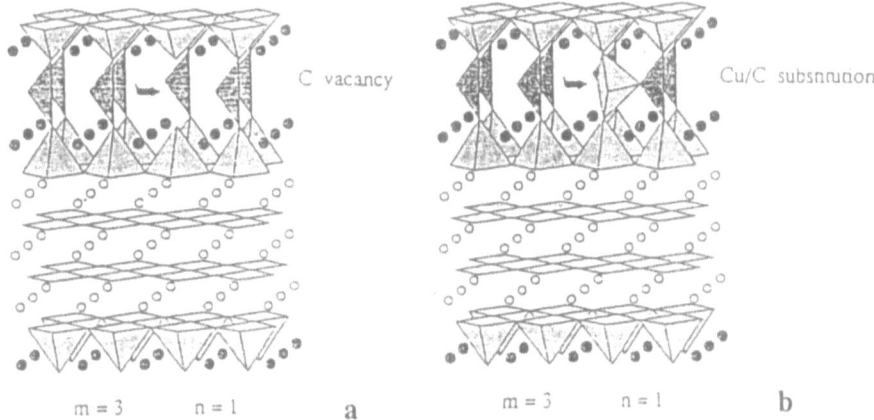

C vacancy

Cu/C subsntuuon

m = 3   n = 1   a

m = 3   n = 1   b

Figure 26: Idealised structural models illustrating a) the existence of carbonate group vacancies and b) the partial replacement of carbonate groups by copper polyhedra; both events would statistically occur in the carbonate layers.

The magnetic and electrical measurements that show critical temperatures ranging from 30 to 100K with rather broad transitions demonstrate that these materials are at present not optimised in agreement with the fact that several members are coherently intergrown in the same film. These results open a route to the synthesis of the different members and to the optimisation of their superconducting properties by controlling the target composition, the CO_2/O_2 atmosphere and the substrate temperature. There is no doubt that it will be possible to isolate thin films of single members, especially using MBE laser deposition.

5.   Colossal magnetoresistance properties of mixed valent perovskites

Much work has been performed on giant magnetoresistance (GMR) properties of metallic multilayers such as Fe–Cr systems, in which ferromagnetic layers are ordered antiferromagnetically and separated by a non magnetic layer. Such materials are of the highest interest for device applications since they offer the possibility of magnetic serving, and can be used for instance for magnetic information storage systems. More recently, the manganese oxides $La_{1-x}A_xMnO_3$ with A = Ca, Sr, Ba, Pb synthesised either as thin films [43–48] or as ceramics [49] were found to exhibit a similar behaviour. These oxides belong, like the superconducting cuprates, to the perovskite family. They exhibit the $GdFeO_3$–type structure (Figure 27) that corresponds to a monoclinic distortion of the ideal perovskite so that an orthorhombic cell is obtained with $a \approx b \approx a_p\sqrt{2}$ and $c \approx 2a_p$. This perovskite is characterised by a ferromagnetic ordering in the a–b plane and an antiferromagnetic ordering along the c–axis, below the Nél temperature (Figure 27). Thus one observes ferromagnetically ordered $[MnO_2]_\infty$ layers parallel to (001) separated by non magnetic $[(La, Ca)_1O]_\infty$ layers, two successive $[MnO_2]_\infty$ layers being antiferromagnetically ordered with respect to each other exactly as in the intermetallic Fe–Cr systems. Nevertheless, these oxides differ from the Fe–Cr films by the fact that the spin structure is at the scale of the crystallographic cell.

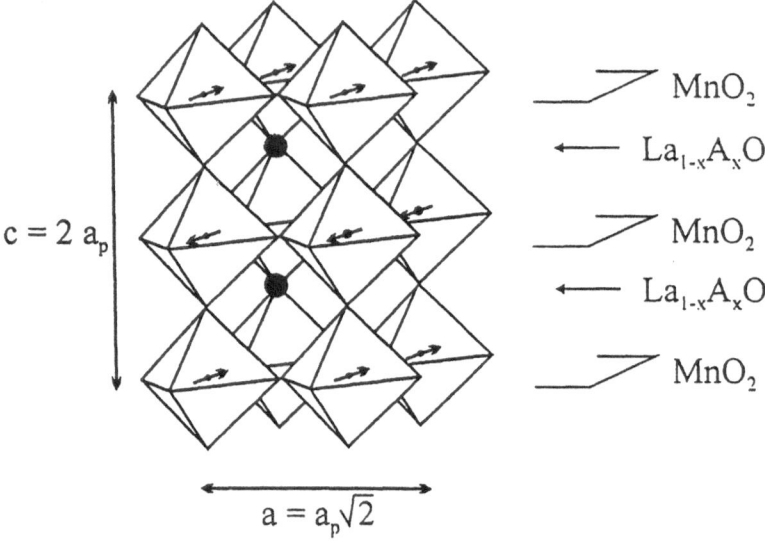

$c = 2 a_p$

$a = a_p\sqrt{2}$

MnO$_2$

La$_{1-x}$A$_x$O

MnO$_2$

La$_{1-x}$A$_x$O

MnO$_2$

Figure 27: The structure of La$_{1-x}$A$_x$MnO$_3$ manganate: arrangements of the spins in the [MnO$_2]_\infty$ layers at low temperature.

In these oxides, ferromagnetic properties and metallic conductivity appear simultaneously. Such a phenomenon can be explained by the presence of Mn(III) and Mn (IV) species that fluctuate due to electron transfer between them. Consequently, the antiferromagnetically ordered manganese spins cant, leading to both magnetisation and metallic conductivity, and the magnetisation increases as the canting angle increases. Although the interaction between spins and conduction electrons are not understood, it appears clearly that this phenomenon is at the origin of the magnetoresistance properties of these compounds. The effect of a magnetic field upon the resistivity of these materials is much higher than that observed for the Fe–Cr multilayers; we will see further that it can be several orders of magnitude lower; for this reason these properties will be called colossal magnetoresistance (CMR) effects. The amplitude of the CMR effect can be characterised by the R_0/R_H ratio where R_0 is the resistance of the material in a zero magnetic field and R_H is the resistance of the latter in an applied magnetic field $B = \mu_0 H$.

In the oxides $La_{1-x}A_xMnO_3$, R_0/R_H ratios ranging from 3 to 10^3 can be reached [43–48]. Lanthanum is not the only lanthanide that produces the CMR effect; the latter was indeed observed for the first time in the oxide $Nd_{0.5}Pb_{0.5}MnO_3$ [50]. More recently, a R_0/R_H ratio of 1.06×10^4 was obtained in thin films of $Nd_{0.7}Sr_{0.3}MnO_3$ [51] at 60K in a magnetic field of 8T, whereas a polycrystalline sample of $La_{0.6}Y_{0.07}Ca_{0.33}MnO_3$ [52] was shown to exhibit a R_0/R_H ratio of 100 at 140K with magnetic field of 6T.

Samarium and praseodymium manganites were also shown to exhibit similar properties: a R_0/R_H ratio of 500 at T = 92.5K with μ_0H = 5T was observed in the ceramic $Sm_{2/3}Sr_{1/3}MnO_3$ [53], whereas a R_0/R_H ratio of 7 was obtained for $Pr_{0.75}Sr_{0.25}MnO_3$ [54] at 210K in a field of 5T. But the record has recently been reached for praseodymium–calcium manganites doped with strontium, barium or lanthanum with the generic formulations $Pr_{0.7}Ca_{0.3}A_xMnO_{3-\delta}$ (A = Sr, Ba) and $Pr_{0.7-x}Ca_{0.3}MnO_{3-\delta}$ [55–58] for which R_0/R_H ratios ranging from $2.5.10^5$ to 10^{11} have been observed in a magnetic field of 5T, at temperatures between 30K and 85K.

At this point, it is important to determine the factors which govern the CMR properties of these materials. It can easily be understood that the magnetic and transport properties of these manganese perovskites depend at least on two parameters, the distances between the Mn neighbours which influences the overlapping of the Mn and O orbitals, and the hole carrier density. The first factor should be controlled by varying the cell parameters of the perovskite, *i.e.* by changing the mean size of the interpolated cations, so that the cations with various ionic radii ranging from 0.1nm to 0.136nm such as alkaline earth (Ca, Sr, Ba) and lanthanide (Ln), may influence dramatically the Mn–Mn distance and consequently the overlapping of the manganese–oxygen orbitals. The hole carrier density corresponds to the Mn(IV) content that is introduced with respect to the undoped Mn(III) phase $LnMnO_3$; this second factor is controlled by the charge balance of the interpolated cations that allow the Mn(III)/Mn(IV) ratio to be varied.

In order to understand these phenomena, three series of praseodymium manganites have been studied. Two of them, $Pr_{0.7}Ca_{0.3-x}Sr_xMnO_{3-\delta}$ and $Pr_{0.7-x}La_xCa_{0.3}MnO_{3-\delta}$, exhibit the same constant Mn(III)/Mn(IV) ratio of 0.7/0.3, whereas the third one, $Pr_{0.66}Ca_{0.34-x}Sr_xMnO_{3-\delta}$ exhibits a different Mn(III)/Mn(IV) ratio of 0.66/0.34. In each of this series, the mean size of the interpolated cation varies with x, the hole carrier density being kept constant.

5.1 THE PEROVSKITES $Pr_{0.7}Ca_{0.3-x}Sr_xMnO_{3-\delta}$ and $Pr_{0.66}Ca_{0.34-x}Sr_xMnO_{3-\delta}$

The magnetisation curves *versus* T of these two series are given in Figures 28 and 29. They are all characterised by a smooth transition from a ferromagnetic state to a paramagnetic state. One can in fact define two transition temperatures, a temperature T_1 (see Figure 28) which corresponds to the beginning of the transition and a temperature T_c (see Figure 28) that corresponds to the end of the transition, *i.e.* to the appearance of the paramagnetic state.

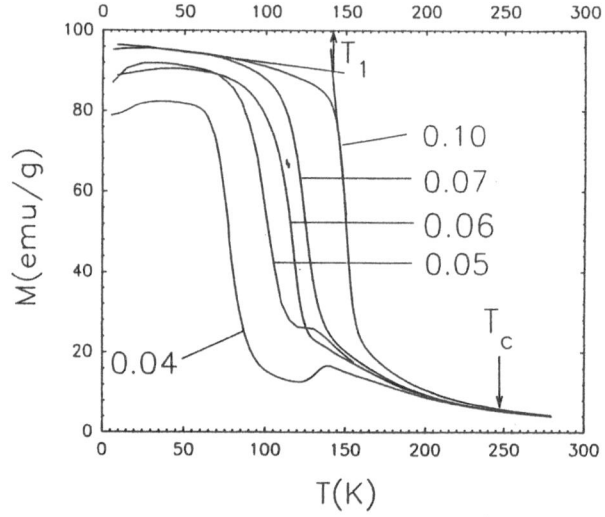

Figure 28:
T dependence of the magnetisation M(T) for samples of the series $Pr_{0.7}Ca_{0.3-x}Sr_xMnO_{3-\delta}$ with $0.04 \leq x \leq 0.10$ ($\mu_0H = 1.4T$). Above T_c, the compound is paramagnetic ($\chi \sim 1/(T - T_c)$).

One observes that T_1 increases dramatically as x increases, from 60K to 144K for $Pr_{0.7}Ca_{0.3-x}Sr_xMnO_{3-\delta}$ (Figure 28) and from 70K to 90K for $Pr_{0.66}Ca_{0.34-x}Sr_xMnO_{3-\delta}$ (Figure 29), whereas T_c is much higher (about 200K). This demonstrates that the transition temperature, T_1, increases dramatically with the mean size of the interpolated cation, *i.e.* with the Mn–Mn distance. Note that for the same x value, *i.e.* for an identical nature of the interpolated cations, the phases $Pr_{0.7}Ca_{0.3-x}Sr_xMnO_{3-\delta}$ (Figure 28) exhibit a transition temperature about 50K higher than that of the phases $Pr_{0.66}Ca_{0.34-x}Sr_xMnO_{3-\delta}$ (Figure 29).

The evolution of the resistance *versus* temperature for the manganites $Pr_{0.7}Ca_{0.3-x}Sr_xMnO_{3-\delta}$ (Figure 30) shows a semi–conducting behaviour for $x \leq 0.04$, whereas above $x = 0.04$ a metallic behaviour is observed at low temperature followed by a transition to a semi–conducting state, characterised by a resistance peak at a temperature T_{max} which coincides approximately with the transition temperature T_1.

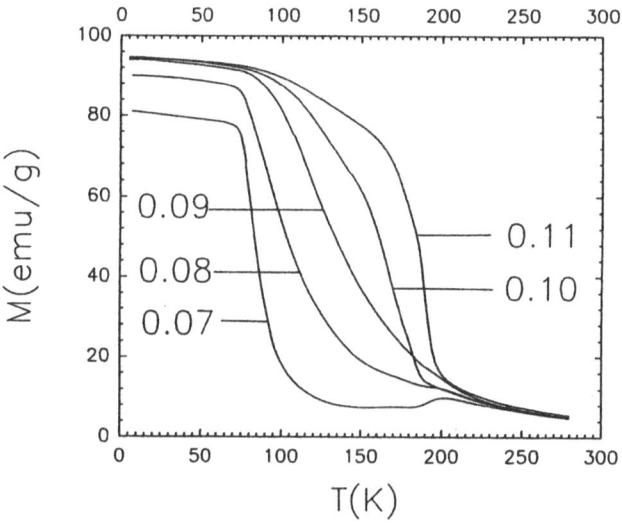

Figure 29: M(T) curves for the sample of the series $Pr_{0.66}Ca_{0.34-x}Sr_xMnO_{3-\delta}$ with $0.07 \leq x \leq 0.11$.

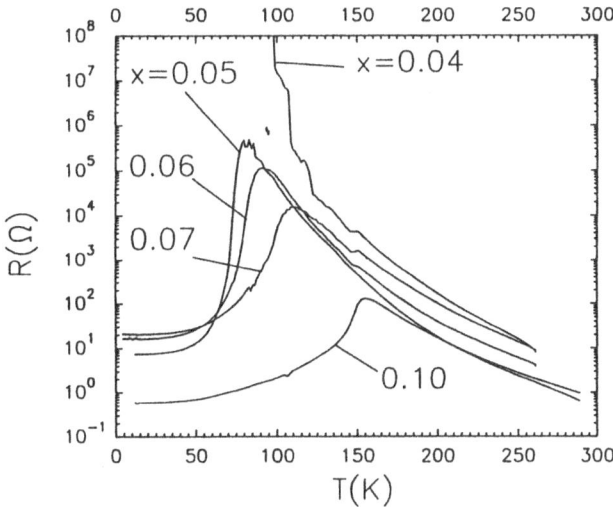

Figure 30: T dependence of the resistance R(T) at B = 0 for samples of the series $Pr_{0.7}Ca_{0.3-x}Sr_xMnO_{3-\delta}$ (0.04 ≤ x ≤ 0.10).

A very similar behaviour is observed for $Pr_{0.66}Ca_{0.34-x}Sr_xMnO_{3-\delta}$ (Figure 31) that is semi–conducting for $0 \leq x \leq 0.07$ and exhibits a metallic to a semi-conducting state transition characterised by a resistance peak for $x \geq 0.08$.

These results are of great interest since they show that a small doping by strontium changes dramatically the magnetic and transport properties of these phases.

But most important is the evolution of variation of the resistance ratio $R(T_{max})$ / R(10K) (RR) as x increases. For $Pr_{0.7}Ca_{0.3-x}Sr_xMnO_{3-\delta}$ (Figure 30) a large value of RR $\approx 3.10^5$ is observed for x = 0.05 with T_{max} = 88K, whereas the height of the peak decreases as x increases for x > 0.05, down to RR = 170 for x = 0.10. In the series $Pr_{0.66}Ca_{0.34-x}Sr_xMnO_{3-\delta}$ (Figure 31), the first peak appears at x = 0.08; it also corresponds to a resistance ratio RR $\approx 2.10^3$, but the corresponding temperature T_{max} = 50K is significantly lower.

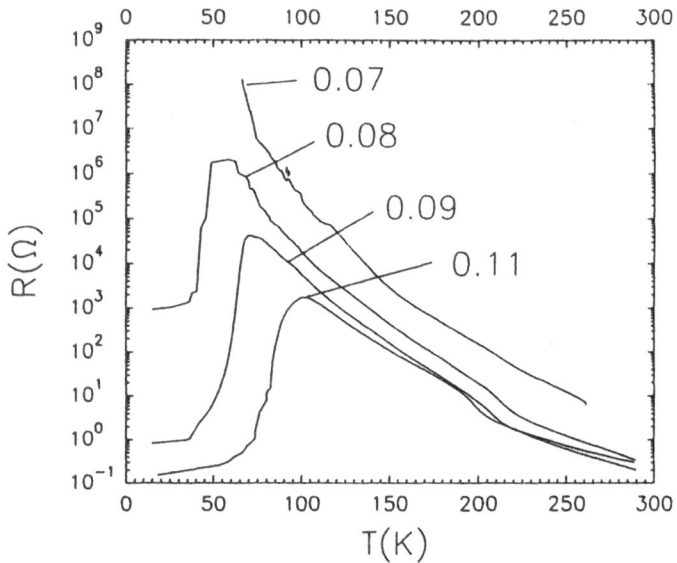

Figure 31: R(T) (B = 0) curves for samples of the series $Pr_{0.66}Ca_{0.34-x}Sr_xMnO_{3-\delta}$ (0.07 ≤ x ≤ 0.11).

The magnetoresistance curves of these oxides confirm their spectacular properties as shown for instance for two best performing compounds of these series, x = 0.05 in $Pr_{0.7}Ca_{0.3-x}Sr_xMnO_{3-\delta}$ (Figure 32a) and x = 0.08 in $Pr_{0.66}Ca_{0.34-x}Sr_xMnO_{3-\delta}$ (Figure 32 b).

These oxides exhibit indeed a peak for R_o/R_H (T) at T_{max} = 85K and 50K respectively, (μ_0H = 5T) corresponding to a variation of the resistance of more than five to seven orders of magnitude, *i.e.* 2.5×10^7% for $Pr_{0.7}Ca_{0.25}Sr_{0.05}MnO_{3-\delta}$ and 10^9% for $Pr_{0.66}La_{0.26}Sr_{0.08}MnO_{3-\delta}$. These CMR effects are confirmed by the evolution of the resistance measured at 88K and 50K respectively, showing for both oxides a spectacular drop in the range 0-2T (Figure 33a and b).

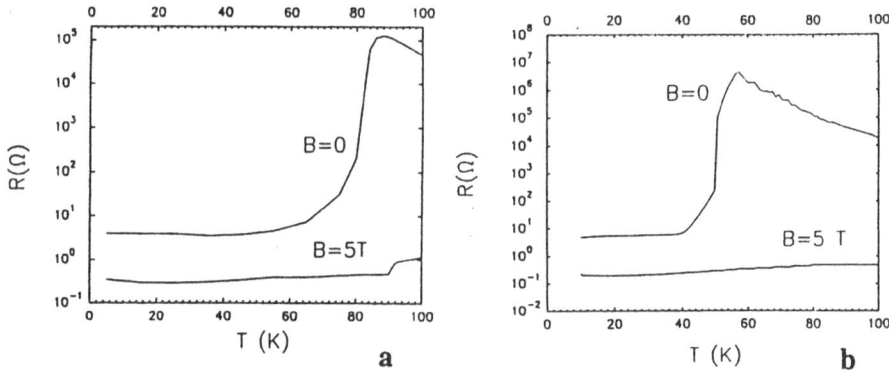

Figure 32: a) T dependence of the resistance at B=0 and at B=5T for $Pr_{0.7}Ca_{0.25}Sr_{0.05}MnO_{3-\delta}$
b) T dependence of the resistance at B=0 and at B=5T for $Pr_{0.66}Ca_{0.26}Sr_{0.08}MnO_{3-\delta}$.

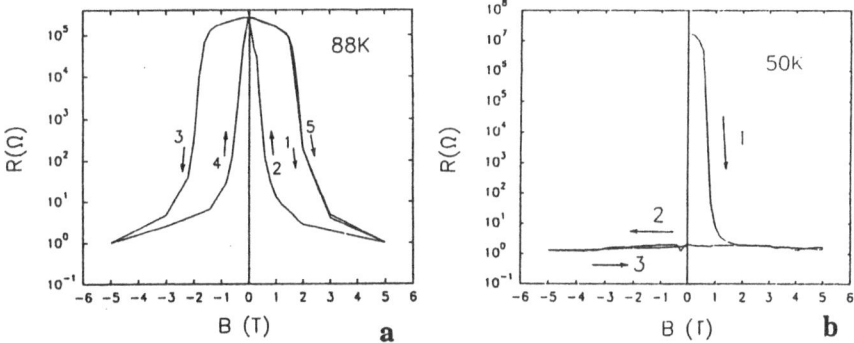

Figure 33: a) B dependence of the resistance registered after a zero field cooling process of $Pr_{0.7}Ca_{0.25}Sr_{0.05}MnO_{3-\delta}$. b) B dependence of the resistance registered after a zero field cooling process of $Pr_{0.7}Ca_{0.3-x}Sr_xMnO_{3-\delta}$ at 50K.

The influence of the size of the interpolated A cations upon the magnetoresistance properties of these two series of oxides is summarised on Figure 34 using ionic radii of Sr^{2+}, Pr^{3+} and Ca^{2+} according to Shannon [59].

In each series, *i.e.* for a constant Mn(III)/Mn(IV) ratio, one observes that the temperature T_{max} increases dramatically as the mean size of the interpolated cations increases, almost in a linear manner (Figure 34).

The influence of the hole carrier density upon the CMR properties is also spectacular. Assuming that δ does not vary significantly from one series to the other, and is close to zero in agreement with microthermogravimetric measurements performed on several samples, one can propose a carrier density of 0.30 hole per Mn mole for the $Pr_{0.7}Ca_{0.3-x}Sr_xMnO_{3-\delta}$ series and 0.34 hole per Mn mole for the $Pr_{0.66}Ca_{0.34-x}Sr_xMnO_{3-\delta}$ series. The consideration of the corresponding straight line $T_{max} = f(r)$ clearly shows that for a same mean radius r of the interpolated A cation, each oxide of the series $Pr_{0.66}Ca_{0.34-x}Sr_xMnO_{3-\delta}$ (Figure 34b) exhibits a temperature T_{max} 50K lower than the corresponding oxide of the series $Pr_{0.7}Ca_{0.3-x}Sr_xMnO_{3-\delta}$ (Figure 34a).

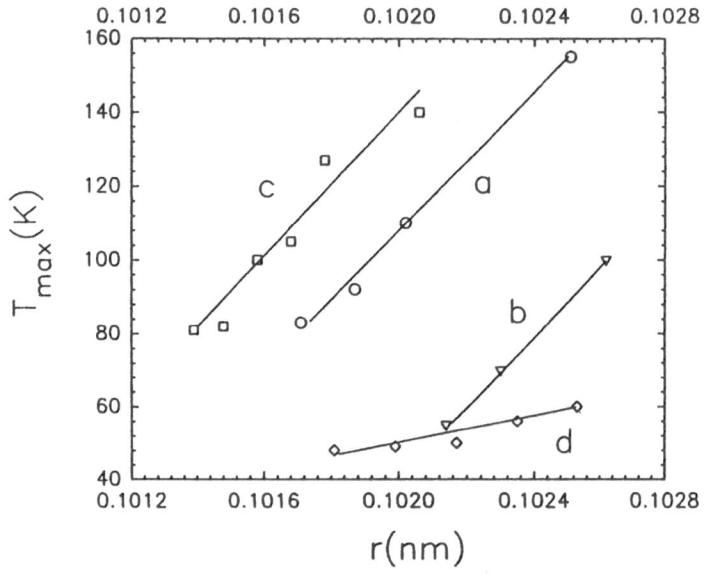

Figure 34: Temperature T_{max} corresponding to the R(T) maximum value *versus* the ratios of the interpolated cation for the series $Pr_{0.7}Ca_{0.3-x}Sr_xMnO_{3-\delta}$ (a), $Pr_{0.66}Ca_{0.34-x}Sr_xMnO_{3-\delta}$ (b), $Pr_{0.7}La_xCa_{0.3}MnO_{3-\delta}$ (c) and $Pr_{0.7}Ca_{0.3-x}Ba_xMnO_{3-\delta}$ (d).

5.2 THE PEROVSKITE $Pr_{0.7-x}La_xCa_{0.3}MnO_{3-\delta}$

If one takes into consideration the above results and their interpretation concerning the influence of the size of the interpolated cations, the introduction of strontium in $Pr_{0.7}Ca_{0.3-x}MnO_{3-\delta}$, should not be absolutely necessary to obtain a spectacular CMR effect. It should indeed be possible to use lanthanum whose size is close to that of strontium; in that case the substitution should be made on praseodymium in order to keep the Mn(III)/Mn(IV) ratio unchanged. For this reason the series $Pr_{0.7-x}La_xCa_{0.3}MnO_{3-\delta}$ was investigated.

The saturated magnetisation curves of these manganites (Figure 35) as well as the resistivity curves *versus* temperature (Figure 36) show that these compounds exhibit ferromagnetic properties, coupled with metallic conductivity at low temperature followed by a transition as the temperature increases.

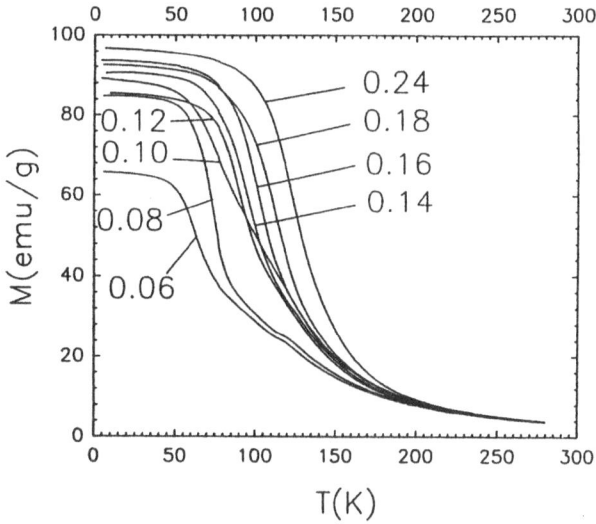

Figure 35: M(T) curves for samples of the series $Pr_{0.7-x}La_xCa_{0.3-x}MnO_{3-\delta}$ with $0.06 \leq x \leq 0.24$.

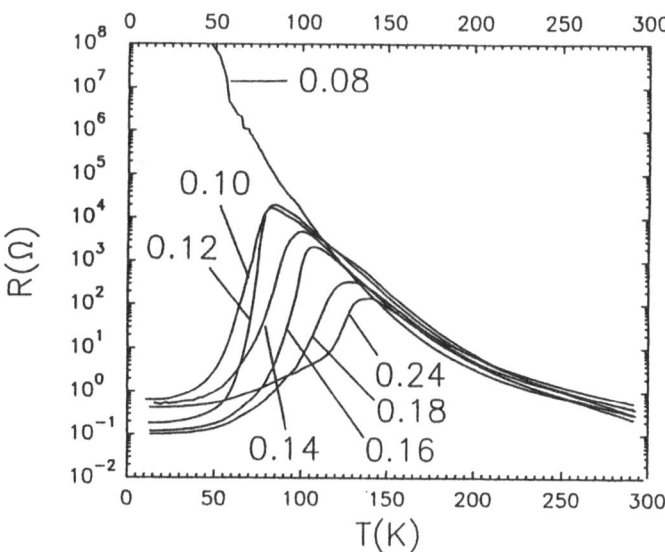

Figure 36 R(T) curves for samples of the series $Pr_{0.7-x}La_xCa_{0.3-x}MnO_{3-\delta}$ with $0.08 \leq x \leq 0.24$.

As for the phases $Pr_{0.7}Ca_{0.3-x}Sr_xMnO_{3-\delta}$, the transition is characterised by a peak at a temperature T_{max} which corresponds approximately to the temperature T_1. One again observes (Figure 36) that T_{max} increases as x increases. In the same way, the resistance ratio, RR, which characterises the CMR effect decreases as x increases, as for the oxides $Pr_{0.7}Ca_{0.3-x}Sr_xMnO_{3-\delta}$.

The exceptional CMR properties of these oxides are illustrated by the magnetoresistance curve of the compound $Pr_{0.58}La_{0.12}Ca_{0.3}MnO_{3-\delta}$, (x = 0.12) at T = 80K, corresponding to a variation of the resistance of $6.10^6\%$ (Figure 37).

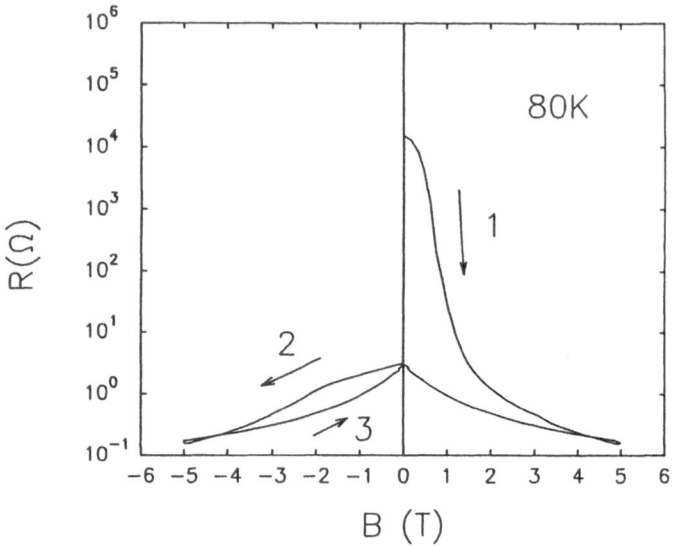

Figure 37: B dependence of the resistance of $Pr_{0.58}La_{0.12}Ca_{0.3}MnO_{3-\delta}$ at 80K after zero field cooling process.

These results clearly confirm the possibility of creating materials and optimising their CMR properties by adjusting the size of the interpolated cation. Nevertheless, plotting T_{max} *versus* the mean ionic radius of the A cation (Figure 34) shows that for the same size of the interpolated A cation the phases $Pr_{0.7-x}La_xCa_{0.3}MnO_{3-\delta}$ (curve c) exhibit a mean temperature T_{max} 30K higher than the phases $Pr_{0.7}Ca_{0.3-x}Sr_xMnO_{3-\delta}$ (curve a) despite the fact that both series are characterised by an identical mixed valence Mn(III)/Mn(IV). The small differences in the ionic radii of La^{3+} and Sr^{2+} may contribute partly to this difference, but are not sufficient to explain it. This suggests that there exists another parameter that may contribute to this different behaviour which relates to the fact that the ionic character of lanthanum is lower than that of strontium so that the competitive Mn-O-Mn bonds are changed when strontium is replaced by lanthanum.

5.3 THE ISSUE OF THE REVERSIBILITY OF THE GMR EFFECT

A most important point concerns the question of the reversibility of this phenomenon. The R(B) curve of $Pr_{0.7}Sr_{0.05}Ca_{0.25}MnO_{3-\delta}$ (Figure 33a) shows that the GMR effect of this phase is reversible at 80K, *i.e.* it recovers its initial resistivity in a zero magnetic field at this temperature. This is not the case for $Pr_{0.66}Sr_{0.08}Ca_{0.26}MnO_{3-\delta}$ (Figure 33b) at 50K, which after being submitted to a magnetic field of 5T, keeps its metallic conductivity when the field is suppressed at this temperature; a similar phenomenon is observed for $Pr_{0.58}La_{0.12}Ca_{0.3}MnO_{3-\delta}$ (Figure 37) and for $Pr_{0.7}Ba_{0.025}Ca_{0.275}MnO_{3-\delta}$. In the latter oxides, an annealing at a temperature superior to T_{max} is necessary to recover the initial resistivity. This hysteretic phenomenon is illustrated as an example for the phase $Pr_{0.66}Sr_{0.09}Ca_{0.25}MnO_{3-\delta}$ (Figure 38). The initial $R_0(T)$ curve in a zero magnetic field is characterised by the classical peak corresponding to a variation of the resistance close to five orders of magnitude; the application of a magnetic field of 5T, leads to a spectacular decrease of the resistivity, as shown from the R_H (T) curve, with a R_0/R_H ratio of 10^6 at 60K.

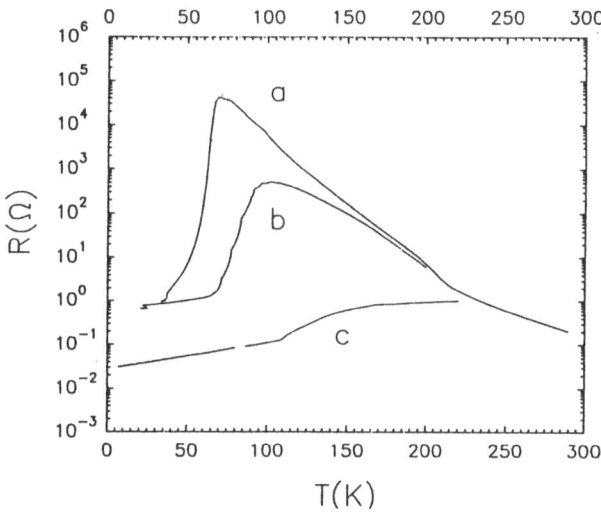

Figure 38: Temperature dependence of the resistance R for the $Pr_{0.66}Sr_{0.09}Ca_{0.25}MnO_{3-\delta}$ sample (registered in zero magnetic field B=0 (a) and in B=5T (b)).

The suppression of the magnetic field at the temperature <50K does not allow the $R_0(T)$ curve to be recovered directly; one obtains the $R'_0(T)$ curve, that also exhibits a peak but at a higher temperature of 100K and the $R_0(T)$ curve is only recovered beyond 150K. In fact the phase $Pr_{0.7}Sr_{0.05}Ca_{0.25}MnO_{3-\delta}$ also exhibits this hysteretic phenomenon but in a very narrow range of temperature below 80K. This result suggests that the CMR properties correspond to a structural transition similar to that previously described for $La_{1-x}Sr_xMnO_{3-\delta}$ [60]. This transition is induced here by the application of a magnetic field, and may be connected to Jahn Teller properties of Mn(III).

5.4   THE PARTICULAR BEHAVIOUR OF $Pr_{0.7}Sr_{0.04}Ca_{0.26}MnO_{3-\delta}$

A second issue concerns the limit of the R_0/R_H ratio that can be reached in these materials. In other words, the application of a magnetic field to a compound of this series does not exhibit a resistivity peak but only a semi–conductor behaviour at low temperature should be able to induce a still higher variation of the resistivity. For this reason, the manganite $Pr_{0.7}Sr_{0.04}Ca_{0.26}MnO_{3-\delta}$ which exhibits a $R_0(T)$ curve in a zero magnetic field characteristic of a semi–conductor was investigated (Figure 30), but ferromagnetic properties similar to those obtained for $Pr_{0.7}Sr_{0.05}Ca_{0.25}MnO_{3-\delta}$ (Figure 28). The application of a magnetic field of 5T is spectacular: as shown from the comparison of the $R_0(T)$ and $R_H(T)$ curves (Figure 39) at 30K the R_0/R_H ratio is 10^{12}.

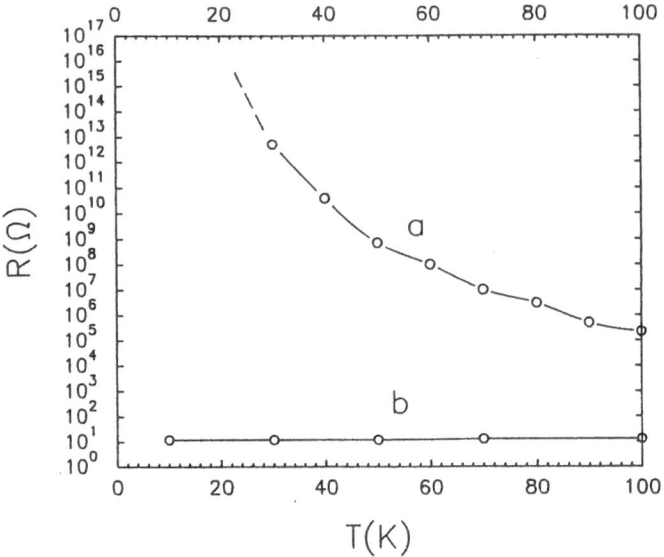

Figure 39: T dependence of the resistance for the $Pr_{0.7}Ca_{0.26}Sr_{0.04}MnO_{3-\delta}$ sample registered in zero magnetic field B=0 (a) and in B=5T (b).

The magnetoresistance measurements at 30K (Figure 40) confirm the colossal magnetoresistance effects, and show the existence of the two possible states in a zero field. An annealing above 80K is necessary to recover the initial zero field resistivity curve. In conclusion, CMR effects have been observed in praseodymium based manganites with the perovskite structure. The origin of this CMR effect can be related to the existence of a transition from a ferromagnetic metallic state to a non ferromagnetic semi–conducting state.

The magnetic phase diagram developed by de Gennes [61] to explain the particular properties of the mixed valent perovskite $La_{1-x}Ca_xMnO_3$ shows that two parameters are of capital importance in determining the temperature transitions T_C and T_1, the hole carrier density and the overlapping of the atomic orbitals. In the present study, the variation of the overlapping of the atomic orbitals is ensured by the control of the mean ionic radii and of the interpolated cations. In this phase diagram, it is predicted that a small range in carrier density or in band width where the system presents the ferromagnetic state, is necessary for the CMR properties; this is effectively observed here.

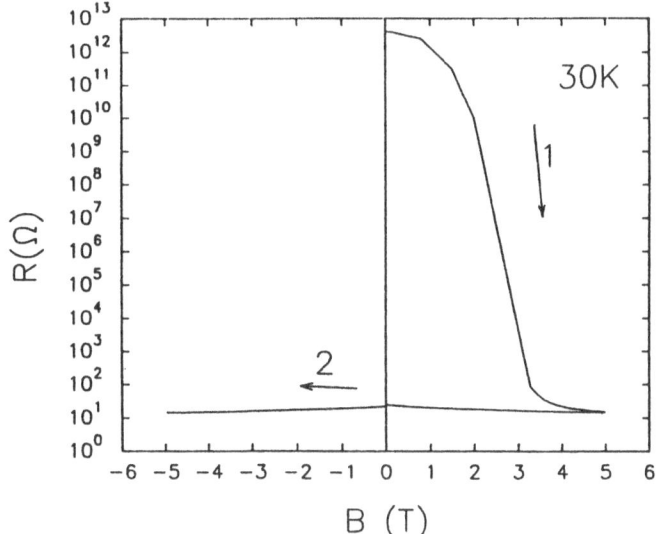

Figure 40: Magnetic field dependence of the resistance for the $Pr_{0.7}Ca_{0.26}Sr_{0.04}MnO_{3-\delta}$ sample registered at T=30K.

Although further investigations are needed to understand the rôle of the different parameters in the CMR properties of the manganites, it appears clearly that the size of the interpolated cation, and the mixed valence of manganese are the two main factors that govern these properties. Taking into consideration these two parameters, it will now be possible to predict the optimisation of various manganese perovskites with CMR properties. There is a narrow region in hole concentration close to 0.3 Mn(IV) per Mn and in cation mean size (depending on the hole concentration) which leads to the CMR properties.

The exploration of these materials is at its very beginning. Many other systems remain to be explored in order to understand the influence of the electronic configuration of the lanthanides upon the CMR properties of these phases. Systematic structural investigations, especially by neutron diffraction, will be essential to understand the evolution of the nuclear structures during this smooth transition.

References

1. S.N. Putilin, E.V. Antipov, Ochmaissen, M. Marezio, Nature, **362**, 226 (1993).

2. S.N. Putilin, E.V. Antipov, M. Marezio, Physica C, **212**, 266 (1993).

3. E.V. Antipov, S.M. Loureiro, C. Chaillout, J.J. Capponi, P.Bordet, J.L. Tholence, S.N. Putilin, M. Marezio, Physica C, **215**, 1 (1993).

4. M. Hirabayashi, K. Tokiwa, H. Ozawa, Y. Noguchi, M. Tokumoto, H. Ihara, Physica C, **219**,6 (1994).

282

5.   L. Bryntse, S.N. Putilin, Physica C, **212**, 223 (1993).

6.   F. Goutenoire, P. Daniel, M. Hervieu, G. van Tendeloo, C. Michel, A. Maignan, B. Raveau, Physica C, **216**, 243 (1993).

7.   D. Pelloquin, C. Michel, G. van Tendeloo, A. Maignan, M. Hervieu, B. Raveau, Physica C, **214**, 87 (1993)

8.   M. Hervieu, G. van Tendeloo, A. Maignan, C. Michel, F. Goutenoire, B. Raveau, Physica C, **216**, 264 (1993).

9.   S. Hahakura, J. Shimoyama, O. Shiinio, T. Hasegawa, K. Kitazawa, K. Kishio,

10. K.K. Singh, V. Kirtikar, A.P.B. Sinha, D.E. Morris, Physica C, **231**, 9 (1994).

11. A, Maignan, D. Pelloquin, S. Malo, C. Michel, M. Hervieu, B. Raveau, Physica C, **243**, 214 (1995).

12. D. Pelloquin, A. Maignan, S. Malo, C. Michel, M. Hervieu, B. Raveau, J. Mater. Chem., **5**, 701 (1995).

13. A. Maignan, D. Pelloquin, S. Malo, C. Michel, M. Hervieu, B. Raveau, Physica C, **243**, 233 (1995).

14. C. Michel, M. Hervieu, A. Maignan, D. Pleeoquin, V. Badri, B. Raveau, Physica C, **241**, 1 (1995).

15. D. Pelloquin, M. Hervieu, C. Michel, G. van Tendeloo, A. Maignan, B. Raveau, Physica C, **216**, 257 (1993).

16. C. Martin, M. Hervieu, M. Huvé, C. Michel, A. Maignan, G. van Tendeloo, B. Raveau, Physica C, **222**, 19 (1994).

17. S.F. Hu, D.A. Jefferson, R.S. Liu, P.P. Edwards, J. Solid State Chem., **103**, 280 (1993).

18. Z. Iqbal, T. Datta, D. Kirven, A. Lungu, J.C. Barry, F.J. Owens, A.G. Rinzler, D. Yang, F. Reindinger, Phys. Rev. B, **49**(17), 12322 (1994).

19. A. Maignan, G. van Tendeloo, M. Hervieu, C. Michel, B. Raveau, Physica C, **212**, 239 (1993).

20. A Miagnan, C. Michel, G. van Tendeloo, M. Hervieu, B. Raveau, Physica C, **216**, 1 (1993).

21. F. Goutenoire, A. Maignan, G. van Tendeloo, C. Martin, C. Michel, M. Hervieu, B. Raveau, Solid State Com., **90**, 1.47 (1994).

22. C. Martin, M. Huvé, G. van Tendeloo, A. Maignan, C. Michel, M. Hervieu, B. Raveau, Physica C, **212**, 274 (1993).

23. F. Letouze, S. Peluau, C. Michel, A. Maignan, C. Martin, M. Hervieu, B. Raveau, J. Mat. Chem.m **4**, 1353 (1994).

24. A. Maignan, C. Martin, C. Michel, M. Hervieu, B. Raveau, Chemistry of Materials, **7**, 1207 (1995).

25. M. Huve, C. Michel, A. Maignan, M. Hervieu, C. Martin, B. Raveau, Physica C, **205**, 219 (1993).

26. A. Maignan, M. Huvé, C. Michel, M. Hervieu, C. Martin, B. Raveay, Physica C, **208**, 149 (1993).

27. F. Letouze, C. Martin, A. Maignan, C. Michel, M. Hervieu, B. Raveau, Physica C — submitted.

28. D. Pelloquin, M. Hervieu, C. Michel, A. Maignan, B. Raveau, Physica C, **227**, 215 (1994).

29. A. Maignan, D. Pelloquin, S. Malo, C. Michel, M. Hervieu, B. Raveau, Phisica C, **249**, 220 (1995).

30. D. Pelloquin, M. Hervieu, S. Malo, C. Michel, A. Maignan, B. Raveau, Physica C, **246**, 1 (1995).

31. F. Goutenoire, M. Hervieu, A. Maignan, C. Michel, C. Martin, B. Raveau, Physica C, **210**[3], 359 (1993).

32. T. Noda, M. Ogawa, J. Akimitsu, M. Kikuchi, E. Oshshima, Y. Syono, Physica C, **242**, 12 (1995).

33. Y. Matsui, M. Ogawa, M. Vehara, H. Nakata, J. Akimitsu, Physica C, **217**, 287 (1993).

34. M. Uhera, S. Sahoda, H. Nakata, J. Akimitsu, Y. Matsui, Physica C, **222**, 27 (1994).

35. M. Huve, G.van Tendeloo, M. Hervieu, A. Maignan, B. Raveau, Physica C, **231**, 15 (1994).

36. M.H. Pan, M. Greenblat, Physica C, **184**, 235 (1991).

37. D. Pelloquin, M. Caldes, A. Maignan, C. Michel, M. Hervieu, B. Raveau, Physica C, **208**, 121 (1993).

38. D. Pelloquin, A. Maignan, M.T. Caldes, M. Hervieu, C. Michel, B. Raveau, Physica C, **212**, 199 (1993).

39. D. Pelloquin, M. Hervieu, A. Maignan, C. Michel, M.T. Caldes., B. Raveau, Physica C, **232**, 75 (1994).

40. J.L. Allen, B. Mercey, W. Prellier, J.F. Hamet, M. Hervieu, B. Raveau, Physica C, **241**,158 (1995).

41. T. Kawashima, Y. Matsui, E. Takayama-Muromachi, Physica C,**224**, 69 (1994).

42. M.A. Alario-Franco, P. Bordet, J.J. Capponi, C. Chaillout, J. Chevanas, T. Fournier, M. Marezio, B. Souletie, A. Sulpice, J-L. Tholence, C. Colliex, R. Argoud, J.L. Baldonedo, m.F. Gorius, M. Perroux, Physica C, **231**, 103 (1994).

43. K. Chahara, T.Ohno, M. Kasai, Y. Kozono, Appl. Phys. Lett., **63**, 1990 (1993).

44. R. von Helmont, J. Wecker, B. Holzapfel, L. Schultz, K. Samwer, Phys. Rev. lett., **71**, 2331 (1993).

45. S. Jin, T.H. Tiefel, M. McCormack, R.A. Fastnacht, R. Hamesh, L.H. Chen, Science, **264**, 413 (1994).

46. M. McCormack, S. Jin, T.H. Tiefel, R.M. Fleming, J.M. Phillips, R. Ramesh, Appl. Phys. Lett., **64**, 3045 (1994).

47. H.L. Ju, C. Kwon, Q. Li, R.L. Greene, T. Venkatesen, Appl. Phys. Lett., **65**, 2108 (1994).

48. S.S. Manoharan, N.Y. Vasanthacharya, M.S. Hedge, K.M. Satyalaksmi, V. Prasad, S.V. SubramanYam, J. Appl. Phys., **76**, 3923 (1994).

49. R. Mahesh, R. Mahendiran, A.K. Raychaudhuri, C.N.R. Rao, J. Solid State Chem., **114**,297 (1995).

50. R.M. Kusters, J. Singleton, D.A. Keen, R.M. Greevy, W. Hayes, Physica B, **155**, 362 (1989).

51. G.C. Xiong, Q. Li, H.L. Ju, S.N. Mao, L. Senapati, X.X. Xi, R.L. Greene, T. Venkatesan, Appl. Phys. Lett., **58**, 1427 (1995).

52. S. Jin, H.M. O'Bryan, T.H. Tiefel, M. McCormack, W.W. Rhodes, Appl. Phys. Lett., **66**, 382 (1995).

53. V. Caignaert, A. Maignan, B. Raveau, Solid State Commun., **95**, 357 (1995).

54. A. Maignan, V. Caignaert, C.H. Simon, M. Hervieu, B. Raveau, Journal of Materials Chem., **5**, 1089 (1995).

55. B. Raveau, A. maignan, V. Caignaert, Journal of Solid State Chem., **117**, 424 (1995).

56. A. Maignan, Ch. Simon, V. Caignaert, B. Raveau, Solid State Comm., – in press.

57. A. Maignan, Ch. Simon, V. Caignaert, B. Raveau, C. Rend. Acad. Sci. Fr., – in press.

58. A. Maignan, Ch. Simon, V. Caignaert, B. Raveau, Z. Phys. B, – in press.

59. R.D. Shannon, Acta Cryst., A, **32**, 751 (1976).

60. A. Asamitsu, Y. Moritomo, Y. Tomioka, T. Arima, Y. Tokura, nature, **373**, 407 (1995).

61. P.G. De Gennes, Phys. Rev.,**118**, 141 (1960).

DEFECTS AND MATTER TRANSPORT IN SOLID MATERIALS

A.V.CHADWICK* AND J CORISH⁺

*Chemical Laboratory, University of Kent, Canterbury, Kent CT2 7NH, U.K.
+Department of Chemistry, Trinity College, Dublin 2, IRELAND

Abstract

The rôle of defects in matter transport through solid materials is discussed and the nature of the principal types of point and extended defects, as well as the evidence for their existence, is examined. The equations governing the participation of defects in ionic conduction and diffusion are developed from first principles as is the relationship between these processes. The experimental methods available for the measurement of matter transport in solid materials and the techniques used for effective data analyses are described. Finally the current level of knowledge of diffusion processes in a range of different types of solids is reviewed.

1.   Introduction

The transport of matter *via* the normal translational and other motions of species in the liquid and gaseous states is easy to visualize. In contrast, atoms ions or molecules in the solid state, both crystalline and amorphous, are traditionally thought to occupy fixed positions within these structures. However, there are many important examples of matter transport through solids. These include self and impurity diffusion in crystals and in polymeric materials, the diffusion processes that may control high-temperature corrosion reactions when a growing compact product layer separates the reactants, and the very rapid ionic transport processes that occur in the fast-ion-conducting components of batteries and other devices. In general, in crystalline materials the movement of lattice species requires the presence of defects. Point defects, such as vacancy or interstitial sites with which the migrating species can interchange, as well as dislocations and grain boundaries, which provide areas of dilation through which facile migration can take place, may contribute to the overall transport process. Because both the nature and concentration of defects in a solid are crucial in the determination of the ability of its various components to move, this chapter begins with a description of the principal defect structures and processes that are important in this rôle. Since ionic crystals, with their charged defects, provide a very clear model that has been extensively studied, Section 3 will describe the measurement of ionic conductivity in these materials and the very detailed information that such data provide on

C.R.A. Catlow and A. Cheetham (eds.), New Trends in Materials Chemistry, 285–318.

the thermodynamic parameters that govern defect formation, interaction and migration. Section 4 will discuss the techniques currently in use to measure the rates of diffusion in solids and show how the results, when compared *via* the Nernst-Einstein equation with those from ionic conductivity data, can provide details on an atomistic level of the matter transport processes through the lattice. Computational studies using atomistic simulation techniques, as described in the chapter by Catlow, have also contributed to our detailed understanding of defect processes and some examples of these will also be discussed where relevant. The chapter concludes with a review of our current level of knowledge on defects and matter transport in a range of materials.

2. Defect Structures and Processes

2.1 POINT DEFECTS

2.1.1 Nature and Occurrence

Point defects are atomistic in nature and are confined to single or small aggregates of sites. As the temperature of a solid is increased, relatively small numbers of the lattice species receive sufficient energy to move off site and take up positions either on the surface or other sink or on an interstitial site. The nature of such Schottky and Frenkel defects in a simple ionic crystal is shown in Figure 1. The creation of such defects is

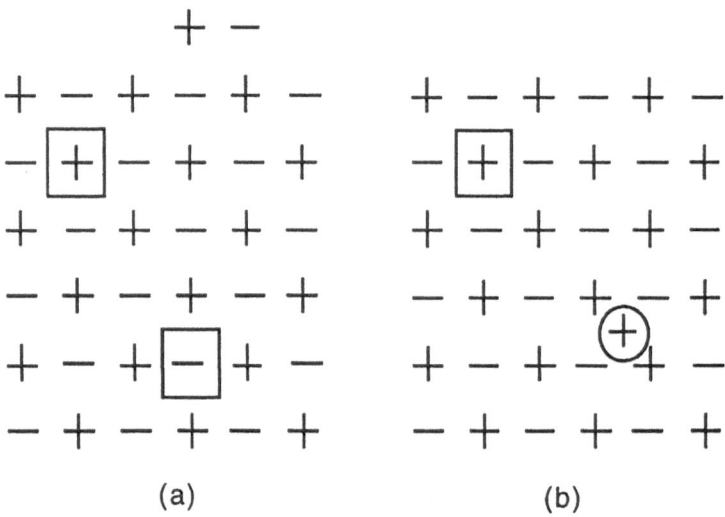

(a)   (b)

Figure 1. Schematic two-dimensional representation of Schottky and cationic Frenkel defects in a simple 1:1 ionic crystal: the squares represent vacant sites and the circle an interstitial cation.

thermodynamically feasible because the enthalpy required is compensated for by the resultant increase in the configurational entropy of the crystal. The Gibbs free energy of a real crystal containing n_v non-interacting mobile vacancies may be written as:

$$G(T,P) = G_o(T,P) + n_v g_v - kT \ln \Omega \tag{1}$$

where $G_o(T,P)$ is the Gibbs free energy of the perfect defect-free crystal, G_v ($\equiv h_v - Ts_v$) is the Gibbs free energy of formation of a single vacancy, k is the Boltzmann constant and Ω is the number of ways in which the vacancies can be arranged among the available sites. The equilibrium concentrations of these *intrinsic* defects can be calculated using statistical thermodynamical methods [1 - 4]. At thermal equilibrium $G(T,P)$ will be at a minimum with respect to the number of defects and when this condition, together with the need to preserve charge and mass balance, is applied to NaCl the defect equilibrum equation is given by

$$c_{V_{Na}'} \, c_{V_{Cl}^\bullet} = 2 \exp(-g_S / kT) \tag{2}$$

where $c_{V_{Na}}$ and $c_{V_{Cl}}$ are the site fractions of the cation and anion vacancies, respectively, and g_s is the Gibbs free energy change for the formation of each Schottky defect pair. For the Frenkel equilibrium in AgBr the corresponding equation is

$$c_{V_{Ag}'} \, c_{Ag_I^\bullet} = 2 \exp(-g_F / kT) \tag{3}$$

where g_F is the Frenkel defect formation energy. In addition to the equilibrium concentrations of *intrinsic* point defects which always occur in such crystals, the presence of aliovalent ions, either as impurities that occur adventitiously or as dopants added to alter the properties of the material, will give rise to *extrinsic* point defects. Such impurities can occupy interstitial sites or they may take up substitutional positions as, for examples, does the Cd^{2+} ion which as $Cd_{Ag}^\bullet$ occupies a silver lattice site in AgCl. In either case the aliovalent ion must be charge compensated and this is achieved by the creation of appropriate additional defects. As shown in Figure 2 the $Cd_{Ag}^\bullet$ is compensated for by a silver vacancy V_{Ag}' whereas trivalent substitutional ions in the fluorite structures, such as $Er_{Ca}^\bullet$ are balanced by fluoride interstitials F_i' which can occupy the centres of the empty cells that alternate with the cells containing the divalent host lattice ions.

The point defects present in an ionic crystal, both extrinsic and intrinsic, are typically charged mobile entities existing in a dielectric medium. As pseudo chemical species, they engage in simultaneous equilibria governed by the magnitudes of the free energies of the various reactions that are feasible. In general they tend to form higher aggregates and more complex defect structures which can be the forerunners in the formation of new phases [5,6]. The simplest of these aggregates is the vacancy pair which occurs, for example, in

288

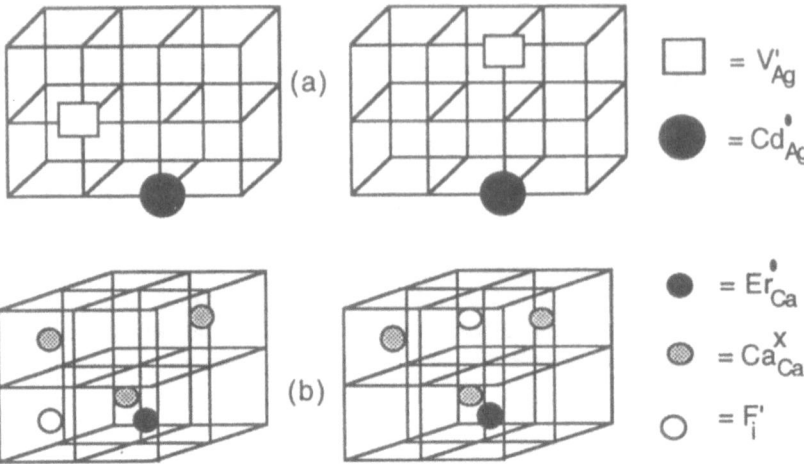

$$\square = V'_{Ag}$$

$$\bullet = Cd^{\bullet}_{Ag}$$

$$\bullet = Er^{\bullet}_{Ca}$$

$$\circledcirc = Ca^{x}_{Ca}$$

$$\bigcirc = F'_{i}$$

Figure 2. Nearest-neighbour and next-nearest-neighbour pairs in (a) AgCl:Cd^{2+} and (b) CaF$_2$:Er^{3+}crystals.

the alkali halides when the vacancies that constitute the Schottky pair occupy neighbouring sites to form a dipole. The presence of these pairs has been clearly demonstrated through its influence on self-diffusion measurements [6]. Impurity-vacancy dipoles also form in alkali- and silver halide crystals. The reaction in AgBr:Cd^{2+} can be represented by:

$$Cd^{\bullet}_{Ag} + V'_{Ag} \Leftrightarrow (Cd^{\bullet}_{Ag} V'_{Ag}) \tag{4}$$

for which the corresponding equilibrium constant, K_a, may be written as:

$$K_a = [(Cd^{\bullet}_{Ag} V'_{Ag})]/([Cd_{Ag}][V'_{Ag}]) = Z_a \exp(-g_a/kT) \tag{5}$$

Here Z_a is the number of distinct orientations of the complex and g_a is the non-configurational part of the Gibbs free energy associated with the process. The formation of such complexes removes free vacancy carriers and the resulting decreases in the measured ionic conductivities allow the magnitudes of the binding energies of these dipoles to be determined. Both nearest-neighbour dipoles, in which the constituent species occupy sites at diagonally opposite corners of the cube-face, and next-nearest neighbour dipoles, in which they are separated by twice the lattice distance along an axis (cf Figure 2), can form: their relative stabilities depend principally on the size of the aliovalent substitutional ion [5].

In systems which can accommodate much larger defect concentrations the defect clusters that form are correspondingly more varied and complex. Fluorite crystals can incorporate significant percentage concentrations of rare earth dopants and the resulting dopant clusters, containing up to six rare-earth ions, have been studied intensively experimentally using a variety of techniques [7,8]. Our understanding of the exact nature of the preferred complexes as well as detailed information on the relaxed positions of their constituent species has been greatly assisted by computational modelling techniques [6].

Non-stoichiometric compounds may also contain very high concentrations of defects. The change in valency exhibited by many transition metals in their oxides is analogous to doping by an aliovalent ion and likewise requires charge-compensating defects. Sørenson [9] has classified these oxides using the nature of their deviation from stoichiometry and the type of defects that they contain. Anion excess UO_{2+x}, in which the oxidation of the cation to $U_U^{\cdot}$ is compensated for by O_i'', contains clusters similar to these found in the rare earth-doped fluorites. Perhaps the most thoroughly understood systems are the rock-salt structured binary transition metal oxides $M_{1-x}O$ with M = Mn, Co Ni and Fe [10]. In these a combination of static lattice simulations, to calculate free energies for possible cluster formation reactions, and a mass action procedure was used to analyse the variation of defect concentrations with oxide composition. The very detailed models, each of which represented a different set of complex simultaneous defect equilibria, were tested by calculating the oxygen partial pressure dependence on composition and comparing the results with experimental data [11].

Intrinsic point defects also occur in molecular solids and the statistical thermodynamical treatment of their concentrations, which is analogous to that of ionic crystals, has also been discussed in detail [3]. Because of their close-packed structures, with only very small volumes being available to accommodate interstitial species, vacancy defects predominate in these materials and the vacancies can also associate to form divacancies. Recent years have seen much research effort into electroactive polymers which materials can exhibit electrical properties ranging from insulating through semi-conducting to metallic. A number of polymeric materials become electronically conducting when doped with ionic species. These dopants can act either as acceptors or donors with their interaction with the host represented as:

$$P + A \rightarrow P^+ + A^- \quad \text{or} \tag{6}$$

$$P + D \rightarrow P^- + D^+ \tag{7}$$

They can be introduced using either diffusion or electrochemical techniques [12,13]. Ionically conducting polymers in which the current is carried by the motion of ions typically consist of polymer-salt complexes with a coordinating polymer, usually a polyether, in which a salt e.g. $LiClO_4$ is dissolved. Both anions and cations may be mobile and these materials will be discussed in more detail in the following chapter. The driving force for the development of these newer materials has been their potential applications in advanced battery and other energy systems.

290

The vacancy is also the predominant point defect found in metals: again isolated defects may associate to form di- and tri-vacancies and, with time, even larger aggregates. Impurity atoms such as C, N, O and H can also take up interstitial positions with carbon in iron being the most important of these metal-impurity systems. When subjected to radiation metals can be damaged with the formation of self Frenkel pairs.

2.1.2 Evidence for Point Defects

As discussed in the introduction the most universal evidence for the existence of point defects is provided by the matter transport processes that they make possible. Detailed discussions of the nature of ionic conductivity and of diffusion processes form major parts of this chapter but there are several more direct techniques that reveal the presence of point defects. The first of these is measurement of the additional contribution to the specific heat of the crystal that is required to provide the enthalpy to create the point defects. If the enthalpy of the real crystal containing n_v vacancies each with an enthalpy of formation of h_v is written as:

$$H = H_o + n_v h_v \qquad (8)$$

where H_o is the enthalpy of the perfect defect-free crystal then c_p is given by

$$c_p = \left(\frac{\partial H}{\partial T}\right)_p = \left(\frac{\partial H_o}{\partial T}\right)_p + \left(\frac{\partial n_v h_v}{\partial T}\right)_p = c_{po} + \left(\frac{n_v h_v^2}{kT^2}\right) \qquad (9)$$

The additional contribution to the heat capacity, Δc_p, beyond that for the ideal crystal, c_{po}, is evident in the c_p vs T plots for many crystalline solids as an upward curvature which typically occurs as the melting temperature, T_m, is approached [14]. This is illustrated schematically in Figure 3. Although it is possible in principle to derive values for h_v, s_v and the site fraction of vacancies, c_v, from values of Δc_p as a function of temperature estimated by substracting extrapolated values of c_{po} from the experimental curve for the real crystal, it is now realized that such determinations of defect parameters are difficult. Anharmonic effects become more important as T_m is approached and this, in addition to effects caused in the c_p curve by even small concentration of impurities, can make attempted extrapolations of c_{po} unreliable.

A second direct manifestation of the presence of vacancy point defects in crystals in their contribution to the volume of the material. This can be detected when the lattice expansion determined using X-ray diffraction methods is compared with the macroscopic dilation of the material measured using a microscope. The x-ray measurements, which determine the average contents of a unit cell, determine only the microscopic thermal expansion of the

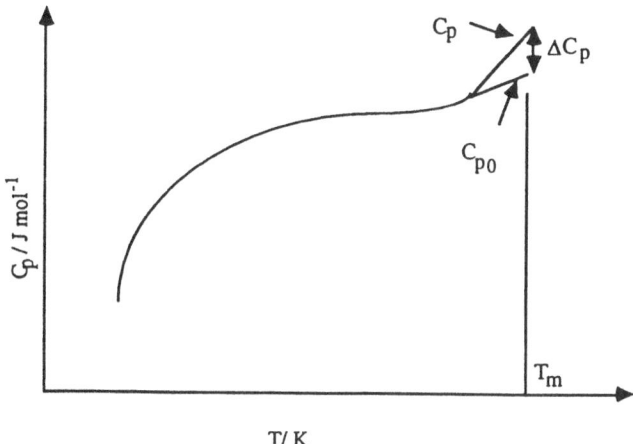

Figure 3. Schematic represenatation of the additional contribution to the specific heat as a result of the enthalpy required for defect formation: the terms are defined in the text.

lattice. When the bulk expansion $\Delta\ell$ is measured simultaneously with this measurement of the lattice expansion Δa it can be shown [15] that for a crystal containing vacancies

$$(\Delta\ell/\ell_o) - (\Delta a/a_o) = {}^1/_3 c_v \qquad (10)$$

where ℓ_o and a_o are the values of ℓ and a at a sufficiently low temperature at which the site fraction of the vacancies, c_v, is negligible and $\Delta\ell$ and Δa are the measured changes in these with temperatures. Simultaneous measurements of the relative changes in a and ℓ over a range of temperatures reduces the errors and provides the best chance of determining the variation in c_v with temperature as well as the values of h_v and s_v.

Two other experimental techniques, namely resistivity measurements in quenched samples and positron annihilation, have been used to study vacancy formation in metals: the latter technique has also seen rather limited use in plastic crystals [16]. When a metal is heated to a high temperature, T_q, and then quenched rapidly the equilibrium defect concentration at T_q will be frozen in. As a result the quenched sample will have a resistivity different from that of an untreated sample at the same temperature. The contribution made by the additional vacancies to the resistivity, $\Delta\rho$, is directly proportional to their concentration and a plot of $\ln\Delta\rho$ measured in samples quenched from a range of temperatures. vs. $1/T_q$ has a slope of $-\Delta h_v/k$.

The positron annihilation technique involves the measurement of the lifetimes of the particles from their emission to their annihilation by an electron. Both of these events produce gamma rays that can be detected and timed. Because vacancies in solids are regions in which the electron density is low they act as traps for positrons: the larger the concentration of vacancies in the solid the longer will be the positron lifetime [18]. The experiment can be carried out quickly and the technique has provided precise answers for the parameters governing the vacancy formation process in a number of metals.

2.2 EXTENDED DEFECTS

2.2.1 *Dislocations*

In addition to the intrinsic defects present at thermodynamic equilibrium and possible extrinsic impurity defects, solid materials may also contain extended defects such as dislocations and grain boundaries. Even though these defect structures increase the free energy, they cannot be easily eliminated even by annealing at high temperatures. Such extended defects, together with surfaces, provide essential sinks for both vacancies and insterstitials as well as regions to which point defects can aggregate. More importantly, in the context our present considerations, they can act as fast pathways or short circuits that facilitate considerably more rapid transport for host and impurity ions than do typical bulk lattice mechanisms. Dislocations are created and can multiply when a crystal is subjected to plastic deformation [18]. Typical densities are 10^8 to 10^{10} cm/cm^3 and because shear can occur readily along their glide planes they cause materials to be significantly less resistant to applied stress than theoretical estimates would predict. They are important to the performance of certain devices and also at interphases where dislocation arrays can accommodate the misfit between the lattices [19].

Dislocations are linear defects whose nature and occurrence are extensively discussed in a number of texts [18, 20, 21]. It is possible to visualize the formation of a dislocation by making a cut along any closed curve within a perfect crystal or along an open curve terminating on the surface at both ends. If the material on one side of this cut is translated by a vector **b** relative to the other side then it will be necessary either to insert or remove some material to restore the solid which will retain a discontinuity in the neighbourhood of the line. If **b** is a translation of the lattice then only a dislocation of Burgers vector **b** will result. When **b** is perdendicular to the dislocation line then an edge dislocation has been formed: a screw dislocation results when the directions of **b** and the dislocation line are parallel. Edge and screw dislocations are illustrated in Figure 4 : in between these ideal configurations the dislocation is termed as mixed. If **b** in the above consideration is not a lattice vector then a stacking fault is created. Stacking faults also result when perfect dislocations dissociate into two or more partials. Dislocations can also form arrays which can be regular. One of the simplest of these, which is comprised of a group of parallel edge dislocations, can give rise to a tilt grain boundary in which the angle of rotation is small (see § 2.2.2 below).

Dislocations emerging at the surfaces of crystals can have profound influences on both the rate at which the crystal can dissolve or react and on the rate of crystal growth. Emerging screw dislocations in general provide a step at which mobile adatoms can be easily accommodated. However this step can never grow to produce a completed atomic layer since the upper surface takes the form of a spiral ramp. In this way the presence of the dislocation removes the necessity to renucleate each monolayer as would be required if the growth was to occur on a perfect crystal face. This latter mechanism would typically require super- or undersaturations of the order of 25 to 40% for growth and dissolution, respectively, whereas these processes are found experimentally to take place at very small

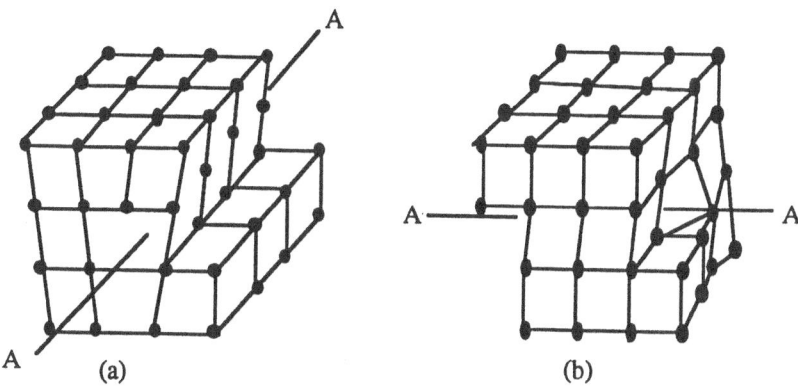

Figure 4. Simple Models of (a) edge and (b) screw dislocations: AA are the dislocation lines.

departures from equilibrium. This rationalization of the unexpectedly high observed rates of crystal growth and dissolution represented the most direct and clearcut early verification of dislocation theory and of the existence of dislocations in real crystals.

2.2.2 Grain Boundaries

Almost all of crystalline material exists in polycrystalline form with the small crystallites being separated from each other by grain boundaries. These interfaces represent extended defects in relation to a perfect crystal. In metals, alloys and semiconductors they tend to be 'clean' whereas in other ionic solids, and particularly in ceramics, they may contain specific intergranular phases. In general they are regions into which impurities tend to segregate and precipitate and they can considerably influence the overall properties of the material. A total of five degrees of freedom are used to describe a general phase boundary [22] though naturally occurring grain boundaries have not been found to take up all the possible configurations. The first three parameters (direction of rotation axis and the rotation angle) relate one of the crystallities to the other. The grain boundary is then fully specified by the two indices which determine its plane. There are two particularly simple grain boundaries. The first is the *tilt boundary* in which the axis of rotation is in the boundary and the two crystals are related by symmetrical rotation about the boundary plane. A small angle tilt boundary is illustrated in Figure 5 in which it is evident that it contains a row of parallel edge dislocations with the spacing between them, D, being given by b/θ where θ is the angle of tilt. The second simple boundary, the *twist boundary*, occurs when the rotation axis is perpendicular to the boundary plane: in certain special cases twist boundaries contain only sets of screw dislocations. Real grain boundaries will of course contain elements of both tilt and twist and the most general grain boundaries are described in terms of their coincidence lattice [23] which is defined by sites, occupied by atoms, which are common to both lattices.

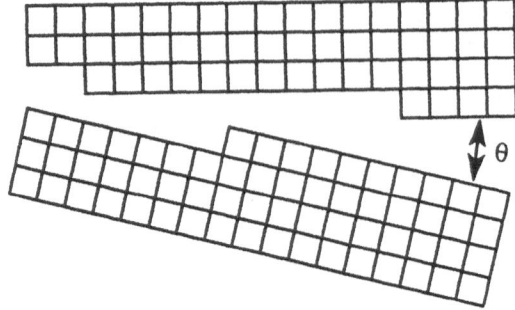

Figure 5. A small angle tilt grain boundary.

3. Ionic Conductivity

The principal objective of detailed measurements of ionic conductivity and of diffusion in ionic crystals is the determination of the complete set of thermodynamic parameters that govern the formation, interaction and migration of point defects that make such migration processes possible. Ionic conductivity measurements must be made on a range of pure and aliovalently doped crystals and over as wide a temperature range as is possible. The parameters are then determined by non-linear least squares fitting of the experimental data to a theoretical model which calculates the conductivity from the defect parameters. These parameters are adjusted during the data analysis to give the best fitting to all of the conductivity curves.

3.1 THEORETICAL

The overall conductivity , σ, of the crystal will be given by summing the contributions of all the mobile species as

$$\sigma = \sum_{r=1}^{i} c_r \mu_r e_r \tag{11}$$

where c_r is the concentration or site fraction of species r, μ_r its mobility and e_r is the charge which it carries. We have already seen that the concentration of intrinsic point defects can be evaluated from the principal defect equilibrium e.g. in pure AgBr at high temperature

$$C_{Ag_i^{\cdot}} = C_{V_{Ag}'} = \sqrt{2} \; \exp(-g_F / 2kT) \tag{12}$$

where g_F (=h_F - TS_F) is the Frenkel defect formation free energy. When the AgBr crystal contains cadmium $Cd_{Ag}^{\cdot}$ as a dopant and V_{Ag}' the possibility of the formation of $(Cd_{Ag}^{\cdot}V_{Ag}')$

complexes as described in equation (5) must be included as simultaneous equilibria. The site fraction of silver valencies, $c_{V_{Ag}'}$, is then given by solving the cubic equation

$$c_{V_{Ag}'}^3 + c_{V_{Ag}'}^2 / K_a - c_{V_{Ag}'}(K_F + c/K_a) - K_F/K_a = 0 \qquad (13)$$

where c is the concentration of the dopant added and K_F and K_a and the equilibrium constants for the formation of Frenkel defects and impurity-vacancy complexes, respectively [24]. The concentration of silver interstitials can then be found from the Frenkel equilibrium constant.

The mobility of each defect species arises from jumps the frequency of which can be calculated from the formalism developed by Wert [25] and Vineyard [26]. The same result

$$w_r = \upsilon \exp(-\Delta g_r/kT) \qquad (14)$$

where υ is an attempt frequency and Δg_r the change in free energy on passing from the ground state on a normal lattice site to the saddle point, is given by the transition-state theory [27] and the dynamical theory of Rice [28]. In the measurement of ionic conductivity an electric field, E, is imposed and this causes the jumps of particles to be biased so that the jump frequency in the favoured direction is given by

$$\overrightarrow{w_r} = w_r s_r (e_r Ea/kT) \qquad (15)$$

where s_r is a symmetry number of the jump and a the distance traversed in the field direction. This result which requires that $e_r Ea \ll kT$, gives the mobility of the species as

$$\mu_r = w_r s_r e_r^2 a^2 / kT \qquad (16)$$

From equation (11) we may express the contribution of a species r, σ, to the overall ionic conductivity. For V_{Ag}' in AgBr this is

$$\sigma_{V_{Ag}'} T = (\upsilon s_{V_{Ag}'} e^2 a^2 / k) \sqrt{2} \exp(-g_F/2kT) \exp(-\Delta g_{V_{Ag}'}/kT) \qquad (17)$$

where $\Delta g_{V_{Ag}'}$ is the activation energy for vacancy migration. Analogous expressions can be developed for contribution from other mobile species and all the free energy terms can be expressed in terms of the corresponding enthalpy and entropy terms. The summation of terms such as that in equation (17) provide a calculated value for the conductivity, σT_{calc}, in terms of the defect formation and migration parameters. Where necessary the doping level in the crystal can also be included as an adjustable parameter, equation (13), and can be adjusted with the thermodynamic parameters in the least squares minimisation of

$$\varphi = \Sigma_l \{\log \sigma T_{calc} - \log \sigma T_{expt}\}^2 \qquad (18)$$

3.2   EXPERIMENTAL AND RESULTS

The experimental measurements are made with an ac bridge with variable frequency to avoid polarization resistance and the data should be corrected for lead resistances.   In careful work several hundred data points at equal intervals of $1/T(K^{-1})$ are measured for each of a set of crystals containing a range of doping levels, preferably with both cation and anion dopants. The samples must be annealed reasonably close to their melting temperature until they yield consistent plots as the temperature is lowered from close to the melting point.   A set of curves for five single crystals of AgBr [29] is shown in Figure 6 .   Crystals A and B were supplied as nominally pure but exhibited well-developed extrinsic regions.   Crystal C was an off-cut of A doped with Cd^{2+}.   Crystal D contained in excess of Cd^{2+} and Crystal E was supplied as being very heavily doped also with Cd^{2+}.

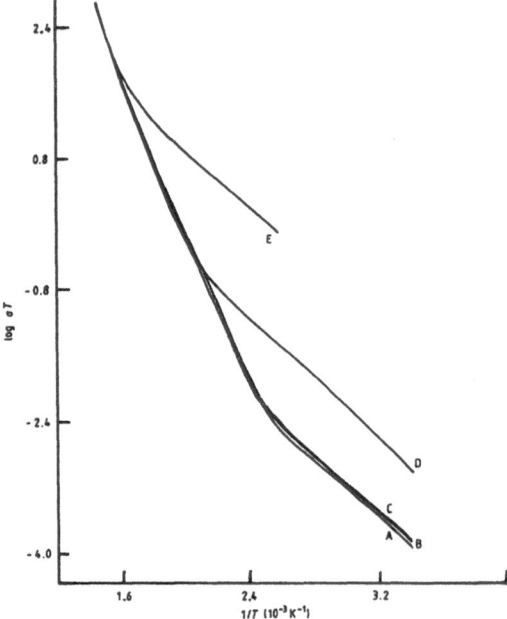

Figure 6. Experimental conductivity curves for five AgBr crystals $\log \sigma T (S\ cm^{-1}\ K)$ vs $1/T$ (K^{-1}). The crytals are described in the text. (From Reference 29 with permission).

The defect model for $AgBr:Cd^{2+}$ contains eleven parameters when the concentration of the impurity ion is included.   As has been found to be the case in a number of recent detailed studies of this type [30,31] it proved necessary to assist the fitting procedure for these $AgBr:Cd^{2+}$ crystals by fixing some of the parameters at values calculated using computational simulation techniques. In this case the upward curvature evident in the $\log\sigma T$ curve at the higher temperature is due to the variation with temperature of the defect formation energy. Since it proved to be impossible to determine this variation from the fitting and because failure to include it in the model made consistent fitting of the data for the five crystals impossible, theoretically calculated values for the energy of Frenkel defect formation, calculated in the quasi-harmonic approximation [32,33] were used in the fitting. The values then determined for the remaining ten parameters for the five crystals are listed in Table 1 together with the numbers of data points in each curve and the standard deviation achieved in the fitting procedure. The model, which is constructed on the basis of measurements of correlation coefficients for both vacancy and interstitial migration, contains activation enthalpies and entropies for vacancy motion and for the motion of insterstitials by both interstitialcy collinear and interstitialcy non-collinear mechanisms. The designations of the parameters are given in the table heading.

Table 1   Parameters obtained in non-linear least squares fitting of conductance data for the five AgBr crystals described in the text. h represents an enthalpy and s an entropy and the subscripts are as follows: F, Frenkel defect formation; cv, cation vacancy motion; ic, interstitialcy collinear; inc, interstitialcy non-collinear; and a, impurity-vacancy association. c is the total concentration of dopant, n, the number of data points and σ_{SD} the standard deviation. A calculated value which included temperature variation was used for the energy of Frenkel defect formation.

Parameter	Crystal A	Crystal B	Crystal C	Crystal D	Crystal E
S_F/k	0.045	0.006	0.036	0.060	0.974
$\Delta h_{cv}/eV$	0.296	0.290	0.318	0.284	0.288
$\Delta s_{cv}/k$	0.975	0.643	1.210	1.328	1.793
$\Delta h_{ic}/eV$	0.104	0.077	0.011	0.000	0.003
$\Delta s_{ic}/k$	0.430	1.051	2.718	3.702	2.407
$\Delta h_{inc}/eV$	0.565	0.525	0.449	0.383	-
$\Delta s_{inc}/K$	6.936	6.425	5.255	4.086	-
$10^6 c$	11.1	16.2	21.3	104.7	2200
$-h_a/eV$	0.287	0.283	0.279	0.268	0.298
$-s_a/k$	2.092	1.740	2.920	0.682	2.823
n	268	285	170	172	127
$10^3\sigma_{SD}$	4.167	5.493	4.077	2.878	4.181

In spite of the rather simple approach taken in the model of putting all of the temperature dependence into the defect formation energy and the fact that the two-body potentials used in the calculations cannot adequately represent AgBr, the results for the parameters in Table 1 are remarkably consistent and in line with other determinations. The treatment is, however, somewhat less successful than the analogous interpretations of conductivity data for AgCl [34]. But the most striking aspect is the very detailed information which the measurements and extensive analyses of ionic conductivity data in classical ionic conductors provide on the formation interaction and migration of point defects in these materials. In the case of the alkali halides [30,31] conductivity studies, again coupled with complementary computational simulations, have shown that complete defect models require the inclusion of minority carriers in certain cases.

4.   Diffusion in solids.

As mentioned at the outset of this chapter diffusion in solids is of immense technological importance as it controls a wide variety of processes, including corrosion, semiconductor processing, ceramics manufacture, and heterogenous catalysis. Therefore measurements of diffusion parameters have practical value to industry as they are required to optimise processing, product design, etc. From the academic viewpoint the prediction of diffusion parameters provides a severe test of the theoretical models of solids, particularly the ability to reliably represent interatomic potentials. In some crystalline solids diffusion data provide the only experimental information on the nature of the defects. This section will cover the theory of diffusion, the measurement of diffusion parameters and the identification of the mechanisms of diffusion. The focus will be on crystalline solids where there is, in general, a good understanding of the diffusion processes.

4.1   DIFFUSION THEORY FOR SOLIDS

4.1.1   Phenomenological theory

Translational diffusion in classical physics is governed by Fick's laws. In a one-dimensional system the flux of a given type of particle, i.e. the number passing through unit cross-sectional area in unit time, can be expressed as:-

$$J = -D\left(\frac{\partial c}{\partial x}\right) \tag{19}$$

where c is the number concentration of particles, $\partial c/\partial x$ is the gradient of the concentration of particles along the x-direction and D is the diffusion coefficient. A dimensional analysis shows that D has units of $L^2 t^{-1}$ and is usually reported as $cm^2 s^{-1}$ or $m^2 s^{-1}$. Equation (19) is Fick's first law. In three dimensions this can be written as:-

$$J = -D\nabla c \tag{20}$$

where D is now a second-rank tensor.

In the form of equations (19) and (20), Fick's first law is not particularly useful to the experimentalist as the evaluation of D would involve the measurement of a flux and a concentration gradient, both of which could be time dependent. If the particles are conserved, then the equation of material balance will apply and therefore a further condition must be obeyed, i.e.

$$\frac{\partial J}{\partial x} = -\frac{\partial c}{\partial t} \tag{21}$$

This continuity equation is simply derived by considering the net flow in and out of a small volume along the diffusion axis. Combining equations (19) and (21) yields:-

$$\frac{\partial c}{\partial t} = \frac{\partial}{\partial x}\left(D\frac{\partial c}{\partial x}\right) \tag{22}$$

If D is independent of concentration then equation (22) reduces to:-

$$\frac{\partial c}{\partial t} = D\frac{\partial^2 c}{\partial x^2} \tag{23}$$

which is often referred to as Fick's second law.

Equation (23) is a second-order differential equation to which a variety of solutions can be derived to correspond with the different experimental situations, i.e. different boundary conditions. These can be found in the standard texts (e.g. [35]) and the only solution quoted here is that for the initial condition of an *infinitely* thin source on the surface of an *infinite* sample. This geometry is commonly used in tracer diffusion experiments and the appropriate solution to equation (23) is:-

$$c(x,t) = \frac{A}{\sqrt{pDt}} \exp\left(-\frac{x^2}{4Dt}\right) \tag{24}$$

Here $c(x,t)$ is the concentration of particles at a penetration depth x into the surface after time t, and A is the initial concentration of particles on unit area of the surface. Thus the penetration profile, the plot of $c(x,t)$ versus x, will have a Gaussian shape and the diffusion coefficient can be determined from the plot of $ln\ c(x,t)$ versus x^2, i.e. the slope is $-1/4Dt$.

External forces acting on the system will perturb the diffusion and intuitively equation (19) may be modified by including an additional flux. This term is referred to as a *drift* and can be simply equated to the concentration of particles multiplied by an average drift velocity, $\bar{v}$, imparted by the force. Thus the resulting total flux can be expressed as:-

$$J = -D\left(\frac{\partial c}{\partial x}\right) + \bar{v}c \tag{25}$$

Examples of external driving forces are gradients of electric potential, temperature and stress. For particles of charge q in a gradient of electric potential $E\ (= -d\Phi/dx)$ then the drift velocity is given by:-

$$\bar{v} = \mu E \tag{26}$$

where μ is the mobility of the charged particle, i.e. the drift velocity in unit potential gradient. If the system is in a steady state with no net flux then from equation (25):-

$$D\left(\frac{\partial c}{\partial x}\right) = \bar{v}\,c = \mu E c \tag{27}$$

Assuming thermodynamic equilibrium in the system then the concentration of particles will obey Boltzmann statistics, hence:-

$$c(x) = a \,.\, exp\left[\frac{-q\,\Phi(x)}{kT}\right] \tag{28}$$

where a is a constant, k is the Boltzmann constant and T the temperature. Differentiating equation (28) with respect to x yields:-

$$\frac{\delta c}{\delta x} = \frac{-cq}{kT}\frac{d\Phi}{dx} = \frac{cqE}{kT} \tag{29}$$

Combining equations (27) and (29) gives the result:-

$$\frac{\mu}{D} = \frac{q}{kT} \tag{30}$$

and since the conductivity, σ, is defined by equation (11), i.e.:-

$$\sigma = cq\mu \tag{31}$$

the final result relating D and σ is

$$\sigma = \frac{cq^2 D}{kT} \tag{32}$$

Equation (32) is the Nernst-Einstein equation and is important in the study of atomic transport in ionic solids as it provides a quantitative link between diffusion and conductivity experiments. Diffusion resulting from other driving forces, e.g. a temperature gradient, can follow similar treatment starting from equation (25).

4.2   ATOMISTIC THEORY OF SOLID STATE DIFFUSION

The mechanisms of diffusion in crystalline solids which move an atom from one lattice site to another, termed *lattice* or *bulk diffusion*, involve the point defects. Diffusion will also occur along the other types of defect, such as dislocations and grain boundaries. In fact, it should occur more readily due to the open nature of their structures; however these processes will not be covered here. It is more difficult to treat dislocation and surface diffusion in a detailed analysis as the atomic pathway is less well-defined, although there has been some progress in recent years [36,37]. In addition, these defects are not thermodynamically inherent in crystals and their concentrations, and hence the contribution to the total diffusion, will depend on the history of the sample. At high temperatures lattice diffusion will usually dominate.

The diffusion mechanisms involving point defects are shown schematically in Figure 7. The simplest of these is the *vacancy mechanism*, shown in Figure 7a, in which an atom 'jumps' into a vacancy on a neighbouring site. This process is energetically favourable and is the dominant diffusion process in most close-packed crystal structures. For example, this is the mechanism that operates for self-diffusion in fcc metals. In systems where the dominant defects are interstitial atoms then the diffusion can occur via these defects by a number of possible processes. In the *direct interstitial mechanism*, shown in Figure 7b, the interstitial atom jumps directly from interstitial site to interstitial site. This is the mechanism by which small impurity atoms which occupy the interstitial sites in metals diffuse, for example the diffusion of carbon in iron. In pure solids interstitials can diffuse by the *interstitialcy mechanism*, shown in Figures 7c and 7d, where an interstitial moves on to a neighbouring normal lattice site with the displacement of the original occupant atom off on to a new interstitial site. There are two variants of this mechanism, the *collinear interstitialcy mechanism* (Figure 7c), where the motion of the two atoms is along the same direction, and the *non-collinear interstitialcy mechanism* (Figure 7d), where the

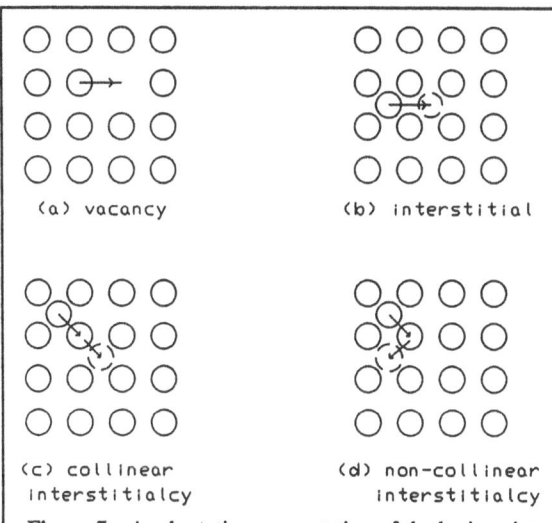

Figure 7: A schematic representation of the basic point defect diffusion mechanisms in solids

two atoms move along different directions. Proven examples of the interstitialcy mechanism can be found in the motion of silver ions in silver halides and fluoride ions in the alkaline earth fluorides.

Variations on the basic point defect mechanisms described above are also possible, such as the diffusion of atoms via divacancies and the *dumbbell interstitial* (two atoms located around the same lattice site). It is important to note the use of the term jump to describe the process of the atom's motion. This implies that the movement is rapid, occurring over the time-scale of the period of a lattice vibration, and that the time the atom is in 'flight' between its initial and final site is very short compared to its average 'residence' time on a lattice site.

The diffusion of atoms in a solid with no applied driving force is analogous to Brownian motion as each atom can be pictured as making jumps from lattice site to lattice site, the direction of each jump being in a random direction to preceding jumps. Thus the process can be treated by the random walk theory developed originally by Einstein and Smoluchowski. If an atom makes n consecutive jumps, each described by the translational vector r_i, then the total displacement vector between the initial and final sites of the atom is given by:-

$$R_n = r_1 + r_2 + r_3 \ldots\ldots = \sum_{i=1}^{n} r_i \qquad (33)$$

It follows that the displacement squared is simply:-

$$R_n^2 = \sum_{i=1}^{n} r_i^2 + 2\sum_{i=1}^{n-1} \sum_{j=i+1}^{n} r_i r_j \qquad (34)$$

The average displacement squared of many atoms after n jumps is given by:-

$$<R_n^2> = \sum_{i=1}^{n} <r_i^2> + 2\sum_{i=1}^{n-1} \sum_{j=i+1}^{n} <r_i r_j> \qquad (35)$$

If there is only one mechanism of diffusion operative then in a cubic crystalline solid the length of each jump step will be the same, r. In addition, if the jumps are truly random then the last term in equation (35) will be averaged out to zero. For a cubic solid $<r_i r_j>$ will be replaced by $<r^2 cos\theta_{ij}>$, where θ_{ij} is the angle between the ith and j^{th} jumps. Since all angles between jumps i and j will be found, the equation can be written simply as:-

$$<R_n^2> = nr^2 \qquad (36)$$

Thus the average displacement, the root mean square displacement, is given by:-

$$\sqrt{<R_n^2>} = \sqrt{n}\, r \qquad (37)$$

If the n jumps take place in a time t then the jump frequency, Γ, is given by:-

$$\Gamma = \frac{n}{t} \tag{38}$$

and:-

$$<R_n^2> = t\Gamma r^2 \tag{39}$$

The Einstein equation relates the diffusion equation to the mean square displacement, and for a cubic solid this is:-

$$D = \frac{<R_n^2>}{6t} \tag{40}$$

Thus combining this equation with equation (39) yields:-

$$D = \frac{\Gamma r^2}{6} \tag{41}$$

which is the *Einstein-Smoluchowski* equation.

It is the Einstein-Smoluchowski equation that allows the connection to be made between the measured diffusion coefficient in an experiment and the parameters involved in the atomistic mechanism of diffusion. Here the case of self-diffusion of atoms in a monatomic solid by a vacancy mechanism will be taken as the example.

The jump frequency of an atom, Γ, will depend on the probability that a vacant site is adjacent to the atom (the product of the site fraction of vacancies, c_v, and the coordination number, z) and the jump frequency, w_o, of the vacancy:-

$$\Gamma = zc_v w_o \tag{42}$$

The jump frequency was shown earlier (equation 14) to take the form:-

$$w_o = v \exp\left(\frac{-\Delta g_v}{kT}\right) \tag{43}$$

Combining equations (42) and (43), substituting into equation (41) with the expansion of c_v yields the final expression:-

$$D = \frac{vz\,r^2}{6} exp\left(\frac{(s_v + \Delta s_v)}{k}\right) exp\left(\frac{-(h_v + \Delta h_v)}{kT}\right) \qquad (44)$$

In experiments the temperature dependence of the diffusion coefficient in a crystalline solid usually obeys the *Arrhenius* equation:-

$$D = D_o exp\left(\frac{-Q}{kT}\right) \qquad (45)$$

where D_o is the pre-exponential factor and Q is the activation energy. Hence by comparing equations (44) and (45) the following identities between the experimental and microscopic parameters can be made:-

$$D_o = \frac{vz\,r^2}{6} exp\left(\frac{(s_v + \Delta s_v)}{k}\right) \qquad (46)$$

and

$$Q = (h_v + \Delta h_v) \qquad (47)$$

It is possible to make a rough estimates of h_v and Δh_v. The simplest approximation for a monatomic solid is that the interatomic forces are two-body central forces, then h_v is simply the latent heat of sublimation, L_S [3]. As an atom moves from its site the number of bonds it needs to 'break' with its neighbours is similar to those broken in vacancy formation. Thus Δh_v will also be approximately equal to L_S, i.e.:-

$$h_v = \Delta h_v = L_S \qquad (48)$$

and therefore:-

$$Q \cong 2L_S \qquad (49)$$

This prediction should be reasonable where there is no relaxation, either electronic or lattice; for example in molecular solids. If is assumed that $\Delta s_v \cong s_v \cong 2k$ then it is possible to estimate the value of D at T_m as 10^{-8} $cm^2\ s^{-1}$, which is the value typically found experimentally for most solids.

In order to measure the diffusion coefficient it is necessary to 'tag' atoms so that their motion can be followed, for example they may be isotopically labelled as in a tracer experiment. If the motion of an individual atom that is tagged is followed then it will be found that there can be correlations between successive diffusive jumps, depending on the mechanism. The simplest example that demonstrates correlation is that of diffusion in a pure solid via a vacancy mechanism. If the tagged atom makes a jump into a vacancy, then its *most probable next jump* is back to its original site as there is a vacancy there ready to accept it. Thus the tracer jumps will be correlated although the jumps of the vacancy are random. For the tracer it is necessary to modify the diffusion coefficient to:-

$$D_t = f \, D_{random} \tag{50}$$

Here D_t is the measured diffusion coefficient of the tracer, D_{random} is the diffusion coefficient given by random walk expressions and f is the correlation factor. Sometimes f is referred to as the *Bardeen-Herring correlation factor*, after the first workers [38] to note this correlation in the motion of tagged atoms. Since accurate values of f can be theoretically calculated for all the possible diffusion mechanisms in the different lattice structures, the determination of f provides an experimental route to the identification of diffusion mechanisms [39]. Theoretical values of f are given in Table 2.

Table 2   Values of the correlation factor f for self-diffusion in simple lattices

Mechanism of diffusion	Lattice	f
vacancy	fcc	0.7815
	bcc	0.727
	sc	0.653
	honeycomb	0.3333
divacancy	fcc	0.458
direct interstitial	independent of structure	1
collinear interstitialcy	fcc	0.9643
	sc	0.8000

4.3   EXPERIMENTAL METHODS OF STUDYING DIFFUSION

There is a wide variety of experimental techniques that can be used to obtain diffusion parameters for solids. Pre-eminent are those that will reliably yield the diffusion coefficient as a function of temperature and these will be the main focus of attention here as it is this information that is required to verify atomistic models of the diffusion process. In some systems it is not always possible to measure the diffusion coefficient and in such cases an alternative, although clearly less satisfactory procedure, is to employ experiments that will measure the activation energy for diffusion.

4.3.1   Tracer methods

The most reliable method of measuring the diffusion coefficient is to determine the profile of labelled atoms that have been diffused from the surface of a sample in a known time at a constant temperature. In order to ensure that the true bulk diffusion coefficient is measured, the sample should be a single crystal with the minimum number of dislocations. The majority of the accurate diffusion coefficients that are available have been measured by the *radiotracer serial sectioning technique*, or its variants, and it is applicable to a wide range of solid types. The full details can be found in a number of texts (see, for example [40,41]. The basic procedure of this technique are shown in Figure 8. Firstly a very thin layer of the radiotracer is applied to a flat surface of the sample by vacuum sublimation or by deposition from a solution. The tracer may be an isotope of the host crystal (e.g. ^{22}Na to measure self-diffusion in Na, ^{36}Cl to measure chloride ion self-diffusion in NaCl, 14C to measure carbon diffusion in steel, etc.). The

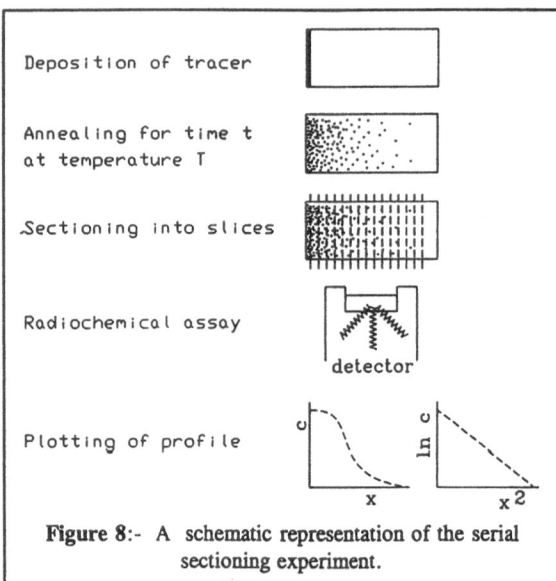

Figure 8:- A schematic representation of the serial sectioning experiment.

sample is then annealed at constant temperature, T, for a known time, t. Following the anneal the sample is sectioned parallel to the face of the tracer deposit and the sections subjected to a radioactive assay. This allows the determination of the penetration profile and the evaluation of the tracer diffusion coefficient from equation (24), i.e. using a plot of ln (*tracer concentration*) versus (*penetration*)^2. Two advantages of this technique are its simplicity and the relative cheapness of the equipment.

In addition, the determination of D_t directly from the penetration profile provides a good test that bulk diffusion is being observed; the diffusion plot should be linear and D_t should be independent of annealing time. An obvious disadvantage is the destructive nature of the sectioning while the technique is limited to systems where convenient radiotracers, in terms of half-life and radiological safety, are available.

The controlling parameters in this experiment are basically the anneal time and the thickness of the sections. Reasonable times for the anneal lie in the range 10^3 s (any shorter and heating-cooling corrections become large) to 10^6 s (longer times would generally be inconvenient). A variety of techniques can be used to section the sample. For bulk samples, the order of 10 mm dimensions the sections are generally taken off with a microtome (for reasonably hard materials, e.g. alkali halides), a lathe (for metals) or by grinding (very hard materials) and the individual sections are usually in the 1 to 100 micron range. Using the criterion [40] that the diffusion zone $(\sim 4(Dt)^{1/2})$ should be divided into at least 13 sections over which the activity falls by a factor of 100 gives the useful range of D_t that can be studied with these types of sectioning as roughly 10^{-14} to 10^{-6} cm^2 s^{-1}. For smaller values of D_t then thinner sections need to be taken and this is possible by sputtering (sections $\sim$ 5 to 100 nm) and electrochemical etching (sections $\sim$50 nm), putting the lower limit to D_t to around 10^{-17} cm^2 s^{-1}.

A very important variation of the technique is to use a stable isotope and secondary ion mass spectroscopy (SIMS) to study the penetration profile [42,43]. This has proved particularly useful in the study of oxygen diffusion as there is no radioactive isotope of this element and the diffusion is usually very slow. The sections taken in SIMS are in the range 1 to 100 nm allowing values of D_t as low as 10^{-19} cm^2 s^{-1} to be studied. In addition, very small samples, around 1 mm dimensions, can be used. However, the equipment required for a SIMS experiment is extremely expensive and requires a highly-trained operator. Other means of determining the penetration profile of a tracer that are particularly useful in studying slow diffusion are the use of Rutherford back-scattering and nuclear reaction analysis.

4.3.2   Nuclear magnetic resonance techniques

Nuclear magnetic resonance, NMR, diffusion experiments are best suited to systems where the NMR active isotope is in high abundance and has nuclear spin, s, equal to 1/2 (e.g. ^1H, ^{19}F) or a small quadrupolar moment (e.g. ^7Li, ^{23}Na). A reasonably large number of interesting systems fall into this category, particularly organic solids. There are a variety of NMR experimental procedures but those in current common usage involve the measurement of a spin relaxation time in a pulsed spectrometer [44,45]. The basis of the experiments involves disturbing the equilibrium spin population between a certain set of energy levels in the system by the application of a short burst (pulse) of radiofrequency power and then measuring the characteristic time required for the spins to return to equilibrium by monitoring the magnetization of the sample [46-49]. For example, the spin-spin relaxation time, T_I, is the time for the spins to equilibrate between the Zeeman levels whose separation is determined by the magnitude of the applied external magnetic field, B_o. A spin in an excited state can only relax down to the lower level by losing a quantum of energy equal to the separation between the states, $\Delta E = \hbar \omega$, where ω is the Larmor precession frequency. This involves the spin interacting with oscillating magnetic fields in the system which fluctuate at a frequency equal to w, i.e. it is a stimulated emission. In the system a spin will experience a *local field* due to dipole-dipole interactions with the neighbouring spins. One source of oscillating magnetic field is the modulation of the local field caused by relative motion of the spins. Diffusive motion provides a fluctuating magnetic field whose frequency components are similar to those of a 'noise spectrum'. Analysis of this spectrum into its Fourier components yields the spectral density function which depends on the atomic (spin) jump frequency, Γ. The precise relationship between the measured relaxation time and Γ involves an assumption about the correlation function (the probability that an atom at position $r=0$ at time $t=0$ will be at position $r=r$ after a time t) and various models have been developed. The diffusion coefficient, D_{NMR}, is calulated by using the evaluated Γ in the Einstein-Smoluchowski equation (41). In many cases an exact treatment of the correlation function is not possible and D_{NMR} may differ from the true D. However, it is clear that the relaxation will be most efficient (the relaxation time will be a minimum) when Γ is equal to ω. In addition, NMR will usually provide a precise value of the activation energy for diffusion, Q. By measuring different relaxation times, i.e. probing different sets of energy levels for the spins, a wide range of Γ, and hence D, can be determined. Typically T_I will monitor D in the range 10^{-11} to 10^{-6} cm^2 s^{-1}, the spin-spin relaxation time, T_2, the range 10^{-10} to 10^{-6} cm^2 s^{-1}, the spin-lattice relaxation time in the rotating frame, $T_{I\rho}$, the range 10^{-14} to 10^{-8} cm^2 s^{-1} and the spin-lattice relaxation time in the local dipolar field, T_{ID}, the range 10^{-20} to 10^{-12} cm^2 s^{-1}. Thus it is possible with one technique, using one spectrometer, to probe a very wide range of D.

Two further points concerning NMR methods are worthy of note. Firstly, for cubic systems a precise theory relating relaxation times to Γ has been developed, tested, shown to be reliable and can be used to obtain accurate values of D. This is the 'encounter model' as exploited by Wolf [47] (see, also [48]). Secondly, there is a further NMR technique, *pulsed-field gradient* NMR, that does not involve measurement of the relaxation time. In this experiment the spin is used to 'label' atoms, like a tracer, and the time-scale of the experiment is such that the atoms move macroscopic distances. Thus it is not subject to the criticisms outlined above and is particularly useful in the study of diffusion in liquids or solids where diffusion is rapid, i.e. $D > 10^{-9}$ cm^2 s^{-1} [48].

4.3.3   Other diffusion techniques

For ionic solids measurement of the ionic conductivity provides a reliable means of studying diffusion and this was treated in section 3. There is a number of techniques which can provide a measurement of the diffusion coefficient, but have limited applicability. Amongst these are Mössbauer spectroscopy [50] (limited to a few elements

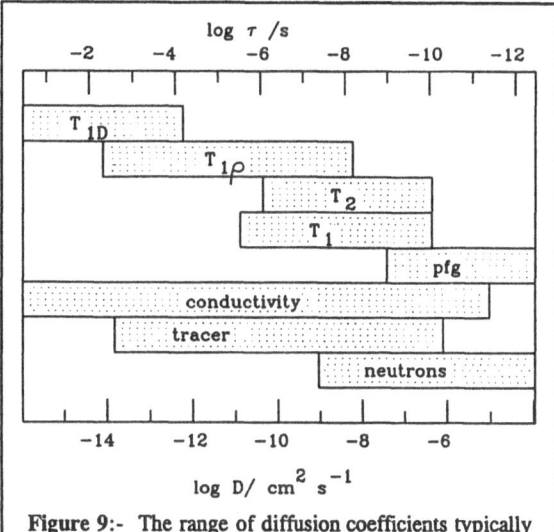

Figure 9:- The range of diffusion coefficients typically accessible by different techniques

with suitable isotopes, e.g. ^{57}Fe an ^{119}Sn) and neutron scattering [51,52] (also limited to certain elements and relatively fast diffusion) which, like NMR methods, determine the jump frequency of the atoms. In addition, there are techniques that monitor processes that rely on diffusion and hence the temperature dependence of these processes provides an estimate of the activation energy, Q, for diffusion. High temperature creep is controlled by dislocation climb, which occurs *via* atomic diffusion, and measurement of the strain rate provides a widely applicable method of determining Q. In the *secondary* or *steady-state* creep regime the strain rate, ε', at high temperatures is given by [53]:-

$$\varepsilon' = \frac{A}{T}\left(\frac{\sigma}{\mu}\right)^{n} \exp\left(-\frac{Q}{RT}\right) \qquad (51)$$

Here A is a constant, σ is the applied stress (extensive or compressive), μ is the shear modulus, and n is the creep exponent (typically 4 to 6). This technique requires single crystal samples and has been extensively used in the study of metals and molecular solids. In the case of multicomponent systems, for example an ionic crystal, the creep rate is controlled by the slowest diffusing species.

Computer simulation techniques now provide the means of studying diffusion processes and these are covered elsewhere in this volume. These have been particularly successful in the study of ionic solids [54,55], where static lattice methods provide reliable calculations of the activation energies of diffusion and for systems where diffusion is relatively rapid (e.g. as in fast-ion conductors) the diffusion coefficient can be directly evaluated in molecular dynamics simulations.

4.3.4 Mechanistic techniques

The determination of the atomistic mechanism of diffusion, i.e. vacancy, interstitial, interstitialcy, etc., is the peak of any diffusion study, however this is not always a simple procedure. Information on the nature of the dominant point defects in the solid will provide a starting point. For example, in close-packed metals the dominant defects are known to be vacancies so that mechanisms based on interstitials can be ruled out. In more complex situations one approach is to compare the experimentally determined diffusion parameters (predominantly activation energies but also activation volumes) with those obtained from theoretical calculations for the different mechanisms. The advent of the modern computer simulation procedures this is becoming a reliable procedure. For ionic solids the predictions from these calculations have been quite enlightening on some complex systems [54,55].

Experimental identification of the diffusion mechanism has usually been based on the determination of the correlation factor, f. In principle, this can be determined from the isotope mass effect for diffusion [56,57] which involves the measurement of the differences in diffusion coefficient of two radiotracers, A and B, with different masses.

The isotope mass parameter, $E_{A,B}$ is given by the relation:-

$$E_{A,B} = \frac{\dfrac{D_A}{D_B} - 1}{\left(\dfrac{m_A}{m_B}\right)^{1/2} - 1} \tag{52}$$

where m_A and m_B are the masses of the tracers. $E_{A,B}$ is equal to $f\Delta K$, where f is the correlation factor and ΔK is the kinetic energy factor (the fraction of the kinetic energy at the saddle point associated with the diffusing molecule. Therefore if f is evaluated it can be compared with the theoretical values (as listed in Table 2) and the mechanism identified. However, there are only a few elements with convenient isotopes (e.g. ^{22}Na, ^{24}Na) to use this approach and ΔK is not simply unity; it is typically 0.8 to 0.9. In some well-understood systems where the diffusion mechanism, and hence f, is known it is possible to work backwards to determine ΔK, however, in an unknown system it can be hard to separate f and ΔK and identification of the mechanism is not simple.

In ionic crystals the ionic conductivity arises from the movement of the effectively charged defects in the electric field. Since the diffusion of the defects is random the diffusion coefficient calculated from the ionic conductivity, D_σ, via the Nernst-Einstein equation (equation (32)) will be the diffusion coefficient of the ions if they were moving randomly. Thus using equation (50) yields:-

$$f = \frac{D_t}{D_\sigma} \tag{53}$$

and therefore in principle the mechanism can be identified. However, real systems are often more complex; the ions may diffuse by more than one mechanism, as seen earlier for the case Ag^+ ions in silver halides, or more than one ion may be involved in the conductivity. Thus the ratio D_t/D_σ is not simply f. This ratio is still useful and is now referred to as the *Haven ratio*.

5.   Review of diffusion in the different solid types

5.1   MOLECULAR SOLIDS

The largest group of all materials are the molecular solids, solids in which the intermolecular attraction is due to van der Waals forces. They include all the organic and biological materials, which are virtually infinite in number given the facility of carbon to covalently bond with itself and a range of other elements (e.g. O, N, H, S, etc.). Information on the point defects in these materials is sparse, a result of both the experimental difficulties in studying these crystals (i.e. high vapour pressures, low purity) and the lack of technological interest in the systems as solids. The interest has been primarily of an academic nature, particularly in the case of the rare gas solids where relative simplicity of the interatomic forces makes them ideal model systems.

It is usual to divide molecular solids into three types; *(i) rare gas crystals, (ii) plastic crystals -systems* where the molecules are roughly spherical, undergo a rapid endospherical reorientation and adopt highly symmetric crystal structures, usually f.c.c., b.c.c. or h.c.p. [58] and *(iii) non-plastic crystals* - the majority of organic crystals, where the molecular symmetry is low, the molecules are not reorienting and the crystal structure is determined by packing factors. In all three types the crystal structures are close-packed and the expected dominant point defects are vacancies.

5.1.1   *Rare gas crystals*

There has been considerable interest in the defects in rare gas crystals as they are model materials and the vacancy formation enthalpy, h_v, provides information on the magnitude on many-body forces [3,59]. Theoretical estimates of the vacancy formation and migration parameters in rare gas crystals using Lennard-Jones or more sophisticated empirical potentials predict that $h_v \cong 1L_s$ and $\Delta h_v \cong 1L_s$ consistent with the simple predictions in equations (48) and (49). There have been several experimental studies of vacancy concentrations in rare gas solids. The experiments are extremely difficult and there is still some confusion over the experimental vacancy formation parameters. For example, the earlier experiments for solid krypton yielded $h_v \cong 0.7L_s$, whereas the value for neon was $1L_s$. The experiments for krypton were over a limited temperature range, the data analysis has been questioned and there is a need for new studies on this system.

The diffusion data for rare-gas crystals from a number of sources, including radiotracers and NMR, suggest these systems are well-behaved [3,59]. The parameters are consistent with the simple theoretical predictions, i.e. Q is close to $2L_S$.

5.1.2 Plastic crystals

The rapid molecular reorientation averages out the interactions between atoms on separate molecules and these can be replaced by a single intermolecular potential acting between molecular centres of gravity, hence these materials are analogous to rare gas solids. The simplest predictions for defect parameters are the same as in the last section. Reliable information on vacancy formation parameters in plastic crystals is limited to a positron annihilation study of two materials, adamantane and bicyclooctane, which yielded $h_v = 1L_S$ in both cases [60].

There have been a large number of self-diffusion studies in plastic solids [58] as the motion is relatively rapid at moderate temperatures (D at T_m is typically 10^{-9} - 10^{-8} cm^2 s^{-1} and T_m is less than 600K) and a wide range of diffusion techniques can be employed. NMR measurements for these materials are particularly appropriate as D is in the required range and the organic plastic crystals usually contain a suitable nucleus (e.g. ^1H). In many cases the simple prediction of $Q \cong 2L_S$ seems to be observed in tracer and creep experiments, but there are a number of systems where NMR methods yield lower values of Q (e.g. $Q \cong 1L_S$ in cyclohexane and white phosphorus). This apparent anomaly is still unresolved.

5.1.3 Non-plastic crystals

Diffusion in non-plastic solids is at least 4 to 5 orders of magnitude *slower* than in plastic crystals. The diffusion experiments are therefore difficult and only a few systems, mainly simple aromatic ring systems like anthracene and naphthalene, have been studied by tracer techniques. The general results are that $Q \cong 2L_S$ and that the dominant point defects are single vacancies [59].

5.2 METALS AND ALLOYS

Because of their technological importance there have been extensive studies of point defects and diffusion in metals [61,62]. There is a recent comprehensive and up-to-date collection of data edited by Mehrer [63]. As expected for close-packed structures, the dominant disorder is single vacancies. Inevitably there will be vacancy aggregates (divacancies, trivacancies, etc.) present but the concentrations are significantly smaller and it is difficult to obtain reliable thermodynamic parameters for these defects. A variety of different techniques has been used to determine values of h_v in pure metals and in the well-studied systems the values obtained from different experiments are consistent (see, for example, the comparison between the values from the positron annihilation technique and quenching experiments [17]). Typically $h_v \cong 0.3$ - 0.4 L_S and smaller than the simple prediction of equation (48), the origin of the discrepancy being the atomic relaxation and electron charge redistribution that occurs on vacancy formation in a metal. Explicit calculations of defect parameters in metals are complicated by the complex electronic structure and the difficulty in describing the interatomic potential.

Table 3: Experimental values for the enthalpy of vacancy formation, h_V, enthalpy of vacancy migration, Δh_V, and self-diffusion in f.c.c. metals.

Metal	Experimental enthalpies/eV		
	h_V	Δh_V	Q
Cu	1.30	0.80	2.05
Ag	1.10	0.83	1.91
Au	0.98	0.82	1.81
Al	0.77	0.65	1.46

Thorough studies of the f.c.c. metals, like Al, Ag, Au, Cu and Pt, show the self-diffusion occurs predominantly *via* a vacancy mechanism. In cases where h_v and Δh_v can be determined separately the sum of the two parameters agrees well with experimental values of Q, as illustrated in Table 3. The table also demonstrates the empirical observation that $\Delta h_v \cong 0.8 \, h_v$ for f.c.c. metals. The Arrhenius plots of self-diffusion measurements for some of these metals exhibit slight curvature, the apparent activation energy increasing as the T_m is approached. The most common interpretation of this effect is that there is a contribution to the diffusion from the migration via divacancies [61,62]. Although the divacancy concentration is generally much lower than c_v it is argued that they are more mobile than the single vacancies and hence make a significant effect on D as the temperature approaches T_m. However, an alternative interpretation is that single vacancies alone are responsible for self-diffusion but the defect enthalpies and entropies are slightly temperature dependent due to anharmonicity of the lattice vibrations [64].

Self-diffusion studies in a number of b.c.c. metals, notably β-Ti, β-Zr, β-Hf and the rare-earth elements, reveal a strong curvature of the Arrhenius plots and these have long been regarded as 'anomalous' systems. A number of explanations have been proposed to explain the effect [61,62,64], including the presence of oxygen impurity, the presence of bulk defects, a contribution from divacancy diffusion and temperature dependent single vacancy parameters. The contribution from divacancies has been a favoured explanation but recent work [65] related the self-diffusion to soft phonon modes and provides a uniform explanation of the data in terms of single vacancies.

The literature on metallic alloys is also very extensive and a wide range of materials have been investigated [61-63]. It is appropriate here to consider only the dilute alloys. The sites occupied in the host lattice depends mainly on the size of the impurity; atoms of a similar size will dissolve substitutionally and small atoms can occupy interstitial sites (e.g.

C, N and O in b.c.c metals like α-Fe). Interstitial impurity diffusion in dilute alloys is consistent with the impurity migrating *via* the direct interstitial mechanism. In substitutional alloys the diffusion data are usually interpreted in terms of the '5 frequency model' [61,62]. In this model the diffusion coefficients of the host impurity atoms are derived as functions of the defect concentrations and the 5 distinct types of jump of the vacancy; (i) vacancy adjacent to impurity exchanging with the impurity, (ii) vacancy adjacent to impurity exchanging with host atom and remaining adjacent to vacancy, (iii) vacancy adjacent to impurity exchanging with host atom and moving away from vacancy, (iv) the reverse of (iii), and (v) vacancy exchanging with a host atom on sites not adjacent to the impurity. There are some well-documented systems where the impurity exhibits anomalously fast diffusion in the alloy, for example Cu, Pd, Au, Ni and Ag in Pb, Au in the alkali metals. In these cases the vacancy mechanism seems inappropriate and various forms of interstitial process have been suggested to explain the anomalies.

5.3   COVALENT SOLIDS

The nature of the point defects in both elemental (e.g. C, Si, Ge) and compound covalent crystals remains a matter of debate. In essence the problem arises in assigning the electronic structure around the defect e.g. what happens to the 'dangling bonds' when an atom is removed to form a vacancy. Each broken bond contains one electron which may occupy lower energy levels by forming new bonds across the vacancy with the dangling bonds of neighbouring atoms. Both self and impurity diffusion have been very actively studied in these materials, particularly for systems where there are semiconductor applications as the information is useful in device fabrication. In the elemental crystals the self-diffusion is extremely slow (D at T_m is typically 10^{-13} to $10^{-12} \, cm^2 \, s^{-1}$) and the mechanisms that are operative have not been fully resolved [66,67]. It is now believed that both self-interstitials and vacancies are involved in the self-diffusion and most impurity diffusion processes in Si and GaAs, a suggestion [68] made nearly 30 years ago which is now being substantiated by experiments.

5.4   IONIC SOLIDS

The point defects and diffusion are better understood in ionic crystals than in any other solid type. An experimental advantage is the availability of good model systems, like the alkali and silver halides, that are readily available in the form of large single crystals. In addition, as was seen earlier, it is possible to precisely control the defect concentrations by means of doping with aliovalent impurities. However, it is the applicability of the relatively simple, but reliable, technique of ionic conductivity to the determination of defect formation and migration parameters that is the major advantage. The technique was fully described in section 3. The combination of ionic conductivity and diffusion measurements also provides a means of identifying mechanisms via the Haven ratio method. From the theoretical viewpoint the interionic potential can be well represented by fairly simple mathematical functions and therefore computer simulation methods have proved highly successful.

Ionic solids are conveniently divided into two types. *Normal ionic conductors* are materials like the alkali and silver halides where the diffusion is well described by jump mechanisms of simple point defects and the ionic conductivity is around 10^{-3} S cm^{-1} at the melting point. These solids are the focus of atttention here. *Fast-ion conductors*, sometimes termed *superionic conductors*, are materials with unusually high conductivities, approaching the magnitude found in molten salts and aqueous solutions of strong electrolytes, and in which the classical jump diffusion model has been questioned. These materials are treated in a separate chapter of this Volume.

Most of the detailed conductivity and diffusion work has focussed on the simple systems like the metal halides, particularly those with the rock salt and fluorite structures, because of the ease of experiment, although very similar behaviour is expected for the higher melting oxides. For a large number of systems there is now a wealth of experimental information, including the nature of the dominant point defects, the thermodynamic parameters for point defect formation and migration, and the nature of the self-diffusion mechanisms [1,4,69,70]. In many of these systems static lattice computer simulations have extremely successful in calculating the defect parameters [55].

The most thorough experimental studies have been made for the alkali halides, the silver halides and the alkaline earth fluorides. The level of precision of the data and their analysis was demonstrated in section 3. Some of other data are shown as an example in Table 4. Only the principal point defect enthalpies are shown, however in many case the entropies are available and the parameters for defect association. The dominant defects in the alkali halides are Schottky pairs and both ions diffuse by a vacancy mechanism. The diffusion coefficients of both ions are similar. In systems like NaCl and KCl there is a wealth of data and the level of interpretation is now extremely sophisticated, including contributions to the ion transport from vacancy pairs and Frenkel defects (which are present in very low concentrations but are mobile enough to have a significant effect on the transport at high temperatures).

In silver chloride and bromide the dominant point defects are cation-Frenkel pairs and not surprisingly the cation diffusion coefficient is about 1000 times higher than that for the anion. In the intrinsic region the silver ions move predominantly by interstitialcy mechanisms, both collinear and non-collinear, whilst in the extrinsic region of crystals doped with divalent cations (i.e. containing excess cation vacancies) the silver ions diffuse by a vacancy mechanism. The alkaline earth fluorides at moderate temperature, up to about 0.8 T_m, behave as normal ionic crystals with anion-Frenkel pairs as the dominant defects and the fluoride ion diffuses predominantly by the vacancy mechanism, although there is a significant contribution from the non-collinear interstitialcy process [70]. The table also includes calculated enthalpies obtained from static lattice computer simulations [54,55]. The good agreement with the experimental values demonstrates the reliability of these calculations and gives confidence in the applicability of this method when dealing with more complex systems.

Table 4. Point defect enthalpies (in eV) for simple ionic crystals; h represents an enthalpy of defect pair formation and Dh represents an enthalpy of migration and the subscripts are as follows; S, Schottky pair formation; cF, cation Frenkel pair formation; aF, anion Frenkel pair formation enthalpy; c, cation; a, anion; v, vacancy; inc, interstitalcy non-collinear; ic, interstitialcy collinear

MATERIAL	DEFECT ENTHALPY	EXPERIMENTAL VALUE DETERMINED FROM IONIC CONDUCTIVITY	CALCULATED VALUE FROM COMPUTER SIMULATION
NaCl	h_S	2.41	2.32
	Δh_{cv}	0.67	0.65
	Δh_{av}	0.77	0.71
KCl	h_S	2.50	2.50
	Δh_{cv}	0.67	0.71
	h_{av}	0.85	0.69
AgCl	h_{cF}	1.49	1.4
	Δh_{cinc}	0.10-0.12	
	Δh_{cic}	0.002-0.004	
	Δh_{cv}	0.27-0.28	
CaF$_2$	h_{aF}	1.85	1.98
	Δh_{av}	0.52	0.40
	Δh_{ainc}	0.79	0.72
BaF$_2$	h_{aF}	1.91	1.98
	Δh_{av}	0.57	0.46
	Δh_{ainc}	0.76	0.72
MgO	h_S	5-7	7.5
	Δh_{cv}	2.2-2.6	2.18

References

1. Lidiard, A.B. (1957) in *Handbuch der Physik*, Vol 20, Ed. S.Flugge, Springer-Verlage, Berlin p. 246
2. Barr, L.W. and Lidiard, A.B. (1971) in *Physical Chemistry - An Advanced Treatise*, Vol 10, Eds, H. Eyring, D. Henderson and W. Jost, Academic Press, New York p. 151.
3. Chadwick, A.V. and Glyde, H.R. (1975) in *Rare Gas Solids*, Vol 2, Eds. M.L.Klein and J.A.Venables, Academic Press, New York, Ch. 19.
4. Corish, J. and Jacobs, P.W.M. (1973) in *Surface and Defect Properties of Solids*, Vol 2, Eds M.W.Roberts and J.M.Thomas, The Chemical Society, London, Chapter 7.
5. Bannon, N.M., Corish, J. and Jacobs, P.W.M. (1995) *Phil. Mag. A* **52**, 61.
6. Bendall, P.J., Catlow, C.R.A., Corish, J. and Jacobs, P.W.M. (1984) *J. Solid State Chem.*, 51, 159.
7. Catlow, C.R.A., Chadwick, A.V., and Corish, J. (1983), *J. Solid State Chem.*, **48**, 65.
8. Catlow, C.R.A., Chadwick, A.V., Corish, J., Moroney, L.M. and O'Reilly, A.N. (1989), *Phys. Rev. B.*, **39**, 1987.
9. Sørensen, O.T. (1983) in *Mass Transport in Solids*, Eds, F. Bénière and C.R.A.Catlow, Plenum Press, New York, Ch. 15.
10. Tomlinson, S.M., Catlow, C.R.A., and Harding J.H. (1990), *J. Phys Chem Solids* 51, 477.
11. Keller M. and Dieckmann, R. (1985) *Ber. Bunsenges, Phys. Chem.* **89**, 883.
12. Bénière, F. (1989) in *Adv. in Solid State Chem.*, Vol 1, Ed C.R.A.Catlow, JAI - Press Inc., London, p. 65.
13. Corish, J. *Phil Mag. A* **64**, 1073.
14. MacDonald, D.K.K. (1955) *Defects in Crystalline Solids*, The Physical Society, London.
15. Simmons, R.O. and Balluffi, R.W. (1960) *Phys. Rev.* 117, 52
16. Lightbody, D., Eldrup M. and Sherwood, J.N. (1980) *Chem. Phys. Lett.* 70, 487.
17. Eldrup, M. (1986) in *'Defects in solids - Modern Techniques'* Eds. A.V.Chadwick and M. Terenzi, Proc. NATO-ASI, Cetraro, 1985: Plenum Press, New York, 145.
18. Hull, D. (1975) *Introduction to Dislocations* Second Edition, Pergammon Press
19. Castaing, J. (1994) in *'Defects and Disorder in Crystalline and Amorphous Solids* Ed. C.R.A.Catlow Proc. NATO-ASI, Madrid 1991: Kluwer Academic Publishers, The Netherlands, Chapter 3.
20. Cottrell, A.H. (1953) *'Dislocations and Plastic Flow in Crystals'*, Oxford University Press.
21. Nabarro, F.R.N. (1989) Editor *Dislocations in Solids* Vols 1 - 8, North-Holland, Amsterdam.
22. Read, W.T. Jnr., (1953) *'Dislocations in Crystals'*, McGraw-Hill, New York.
23. Bollmann, W., (1970) *'Crystal Defects and Crystalline Interfaces'*, Springer, Berlin.
24. Corish, J. and Jacobs, P.W.M. (1972) *J. Phys. Chem. Solids*, 33, 1799-1818.
25. West, C.A. (1950) *Phys. Rev.*, 79, 601.
26. Vineyard, G.H. (1957) *J. Phys. Chem. Solids*, 3, 121.
27. Zener, C. (1952) in *Imperfections in Nearly Perfect Crystals* eds. W. Shockley, J.H.Hollomon, R. Maurer and F. Seitz, John Wiley, New York p 289.
28. Rice S.A. (1958), *Phys. Rev.*, **112**, 804.
29. Devlin, B.A. and Corish, J. (1987) *J. Phys. C: Solid State Phys.* 20, 705.
30. Acuña, L.A., and Jacobs, P.W.M. (1987) *J. Phys C: Solid State Phys.*, 41, 595.
31. Hooten, I.E. and Jacobs, P.W.M. (1990) *J. Phys. Chem. Solids*, 51, 1207.
32. Catlow, C.R.A., Corish, J., Jacobs, P.W.M., and Lidiard A.B., (1981) *J. Phys. C: Solid State Phys.*, 14, L121.
33. Jacobs, P.W.M., Corish, J. and Devlin, B.A. (1982) *Photographic Sci. Eng.*, 26, 50.
34. Corish, J. and Mulcahy D.C.A. (1980) *J. Phys C: Solid State Phys.* 13 6459
35. Crank, J.(1975), *The Mathematics of Diffusion*, Clarendon Press, Oxford.
36. Kaur, I. and Gust, W.(1988), *Fundamentals of Boundary and Interface Boundary Diffusion*, Zeigler Press, Stuttgart.
37. King, A.H.(1987), *Int. Materials Rev.*, **32**, 173.
38. Bardeen, J. and Herring, C.(1952), in: W. Shockley, (Ed.), *Imperfections in Nearly Perfect Crystals*, Wiley, New York, p. 261.
39. LeClaire, A.D.(1970), in: H. Eyring, D. Henderson and W. Jost, (Eds.), *Physical Chemistry - an advanced treatise*, X, Academic Press, New York, p.261.
40. Bénière, F.(1983), in: F. Bénière, and C.R.A. Catlow, (Eds.), *Matter Transport in Solids*, NATO-ASI, Plenum Press, New York, p. 21.
41. Mundy, J.N. and Rothman, S.J.(1983), in: J.N. Mundy, S.J. Rothman, M.J. Fluss and L.C. Smedskjaer, (Eds.), *Methods in Experimental Physics*, Volume 21, New York, Academic Press, p. 1.
42. Macht, M.P. and Naundorf, V.(1982), J. Appl. Phys., **53**, 7551.

318

43. Lodding, A.(1987), in: F. Adams, R. Gijbels and R. van Grieken (Eds.), *Inorganic Mass Spectrometry*, Wiley, New York, p. 125.

44. Farrar, T.C. and Becker, E.D.(1971), *Pulse and Fourier Transform NMR*, Academic Press, New York.

45. Fukushima, E. and Roeder, S.B.W.(1981), *Experimental Pulse NMR: A Nuts and Bolts Approach*, Addison-Wesley, Reading.

46. Kanert, O.(1982), *Phys. Reports*, **91**, 183.

47. Wolf, D.(1979), *Spin Temperature and Nuclear Spin Relaxation in Matter*, Clarendon, Oxford.

48. Strange, J.H.(1986), in: A.V. Chadwick, and M. Terenzi, (Eds.), *Defects in Solids - Modern Techniques*, NATO-ASI, Plenum Press, New York, p. 243.

49. Chadwick, A.V.(1988), Int. Rev. Phys. Chem., **7**, 251

50. Janot, C.(1976), *J. Physique*, **37**, 253.

51. Lechner, R.(1983), in: F. Bénière and C.R.A. Catlow, (Eds.), *Matter Transport in Solids*, NATO-ASI, Plenum Press, New York, p. 169.

52. Springer, T. and Richter, D.(1987), in: D.L Price and K. Sköld, (Eds.), *Methods of Experimental Physics*, Vol. 23B, Academic Press, Orlando, p. 131.

53. Dorn, J.E.(1957), *Creep and Recovery*, A.S.M., Cleveland, Ohio.

54. Harding, J.H.(1990), *Rep. Prog. Phys.*, **53**, 1403.

55. Catlow, C.R.A.(1987), in: A.K. Cheetham and P. Day, (Eds.), *Solid State Chemistry: Techniques*, Clarendon Press, Oxford, p. 231.

56. LeClaire, A.D.(1970), in: H. Eyring, D. Henderson and W. Jost, (Eds.), *Physical Chemistry - an advanced treatise*, volume X, Academic Press, New York, p.261.

57. Peterson, N.L.(1975), in: A.S. Nowick and J.J. Burton, (Eds.), *Diffusion in Solids, Recent Developments*, Academic Press, New York, p. 115.

58. Sherwood, J.N. (Ed.),(1979), *The Plastically Crystalline State*, Wiley, London.

59. Chadwick, A.V.(1983), in: F. Bénière and C.R.A. Catlow, (Eds.), *Matter Transport in Solids*, NATO-ASI, Plenum Press, New York, p.285.

60. Lightbody, D., Eldrup, M. and Sherwood, J.N.(1980), *Chem. Phys. Lett.*, **70**, 487.

61. Brebec, G.(1983), in: F. Bénière and C.R.A. Catlow, (Eds.), *Matter Transport in Solids*, NATO-ASI, Plenum Press, New York, p. 251.

62. Philibert, J.(1986), in: A.V. Chadwick, A.V. and M. Terenzi, (Eds.), *Defects in Solids - Modern Techniques*, NATO-ASI, Plenum Press, New York, p. 349.

63. Mehrer, H.(Ed.), (1990), *Diffusion in Solid Metals and Alloys*, Landolt-Bornstein, Berlin.

64. Mundy, J.N.(1992), in: G.E. Murch, (Ed.), *Unsolved Problems in Diffusion*, Trans Tech, Zurich, p.1.

65. Herzig, C. and Kohler, U.(1987), *Mater. Sci. Forum*, 15-18, 301.

66. Gosele, U., and Tan, T.Y.(1992), in: G.E. Murch, (Ed.), *Unsolved Problems in Diffusion Trans Tech*, Zurich, p. 189.

67. Leroy, B.(1990), in: A.L. Laskar, G. Brebec, J.L. Boucquet and C. Monty, (Eds.), *Diffusion in Materials*, NATO-ASI, Kluwer, Dordrecht, p. 525.

68. Seeger, A. and Chik, C.P.(1968), *Phys. stat. sol.*, **29**, 455.

69. Chadwick, A.V.(1991), *Phil. Mag. A*, **64**, 983.

70. Jacobs, P.W.M.(1983), in: F. Bénière and C.R.A. Catlow, (Eds.), *Matter Transport in Solids*, NATO-ASI, Plenum Press, New York, p. 81.

SOLID STATE IONICS

A.V.CHADWICK* AND J.CORISH+

*Chemical Laboratory, University of Kent, Canterbury, Kent CT2 7NH, U.K.
+Department of Chemistry, Trinity College, Dublin 2, IRELAND

Abstract

The nature and occurrence of fast-ion conduction (FIC) in solids is described and explained and the materials in which it is operative, ranging from crystals through amorphous glassy electrolytes to ionically conducting polymeric complexes, are classified. FIC processes in a number of materials which are of particularly technological interest e.g. oxygen ion conductors and glassy and polymeric electrolytes, are discussed in detail. A number of key applications of these materials in solid state ionics and, in particular recently reported progress is also described. The emphasis is on the overall physical properties that are required in the design of a successful optimized material.

1.   Introduction

Solid state ionics is the study of the transport of ions through solid materials and of the areas of technology in which the application of this phenomenon is important. It now constitutes a very significant part of the physics and chemistry of solids and, in particular, of solid state electrochemistry. There are direct analogies with solution electrochemistry and our principal interest here will be in solid state electrolytes. The past three decades have seen a worldwide and vigorous effort to produce a range of materials in which very facile and rapid ionic transport can take place. The initial driving force for these developments was the search for alternate electrical energy sources and storage systems and was a direct consequence of the oil shortages created in the early seventies. Environmental and conservation issues have sustained the search for new clean energy technologies: they have also championed the development of selective and portable sensors many of which rely on solid components in which specific ions exhibit very large conductivities. Frequently the rates of diffusion of the mobile ions are of the same order of magnitude as are observed in liquids. They are therefore many times greater than occur in the classical ionic conductors described in the previous chapter.

319

C.R.A. Catlow and A. Cheetham (eds.), New Trends in Materials Chemistry, 319–344.
© 1997 Kluwer Academic Publishers.

Solid materials of this type are now termed fast-ion conductors (FIC) and this chapter begins with a discussion of the nature of the fast-ion conduction process and a classification of the range of materials in which it has been observed: this includes inorganic solids, glasses and ionically conducting polymers. The emphasis will be on the factors that allow fast-ion conduction to occur and on the operative atomistic mechanisms where these have been determined. Examples of the different types of FIC will be discussed in detail, particularly where the design effort has been directed at the production of an electrolyte to carry a particular ion, e.g., oxygen. and with properties such as thermodynamic stability, ease of fabrication and cost that are compatible with commercial usage. There then follows descriptions of a number of applications in which the unique properties of some FIC materials are utilized in modern technology. The chapter concludes with a brief review of the current state of progress in the field and the prospects for the future.

2   Fast-Ion Conductors - Nature and Classification

The exceptionally high ionic conductivities that are typical of solid fast-ion conductors lie in the range ca. 0.01 - 10 S cm^{-1}. Several reviews and conference proceedings [1-3] of the field are available. The commercial pressure evident in the earlier development work meant that in some cases less than adequate attention was paid to understanding the fundamental transport mechanisms [4,5] but the situation in this regard has now significantly improved [6]. As with all conductors the overall conductivity of an FIC will result from a summation of the contribution of each of the migrating species as

$$\sigma = \sum_{r=1}^{n} c_r \, \mu_r \, e_r \tag{1}$$

where c_r is the concentration, μ_r the mobility and e_r the charge, respectively, on the species r. If σ is to be dominated by a contribution from a specific ion then either its concentration or the ease with which it can move, as determined by its mobility, should be exceptionally large. In an ideal case both of those factors can be favourable. In many instances the crystal structures involved contain a large excess of sites for the conducting ion over those actually required to accommodate its stoichiometric quantity and, as a consequence, the ions may move very rapidly between these sites. This situation is in contrast to that in a classical ionic crystal (cf. previous chapter) where a defect must first be created, with the expenditure of the appropriate defect formation energy, before ionic migration can take place. Figure 1, illustrates the conductivities of a selected range of FIC materials with some classical ionic conductors shown for comparison. The very shallow slopes of many of the Arrhenius curves are indicative of the small activation energies required for ionic migration in the materials.

The diversity of the materials in which FIC has been found to occur makes it difficult to arrive at a completely satisfactory classification. Nonetheless the categories listed

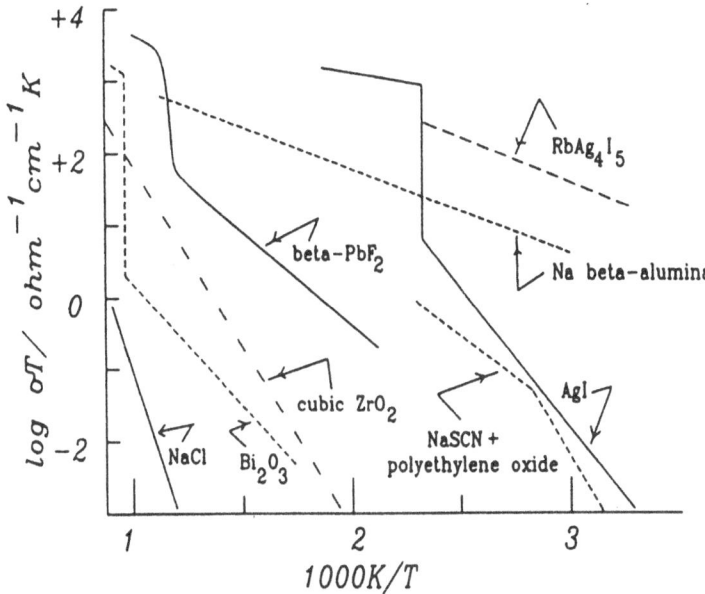

Figure 1. Some examples of the conductivities exhibited by Fast-Ion-Conductors

below and suggested by Catlow [6] provide a very useful division on the basis of the structural characteristics of the materials.

(i) *Solids with Phase Transitions.* This class of material exhibits FIC only at temperatures above a characteristic phase transition which may be first order, as in AgI (cf. Figure 1), or diffuse as in fluorite structured compounds. α-AgI, which is the form stable above 422 K, has a disordered cation state with only one-third occupancy and is the prototype for a number of Ag^+ and Cu^+ fast-ion conductors. Attempts to stabilize such high-temperature forms at lower temperatures by substitution of either anions or cations have been successful: the best known examples are provided by the MAg_4I_5 class of compounds [7].

(ii) *Tunnel- and Layer-structured Compounds.* FIC occurs in a range of compounds which have incompletely filled tunnels or layers in their structures. The joining of simple octahedra by corner-, edge- or face-sharing can provide the types of structures required, e.g., in the tungsten and vanadium bronzes [8,9], and the transport may be restricted to one or two dimensions. The very well known β and β''-alumina FIC ceramics have layered structures with the conduction planes being separated by spinel-structured alumina blocks. The non-stoichiometry of the conducting ion is accommodated in the conduction planes which also contain 'bridging' oxygen ions and the mechanism of conduction has been the subject of a number of studies using computational techniques [10-12].

(iii) *Heavily Doped and Massively Disordered Solids.* The fluorite structure in which each alternate cell is unoccupied can accommodate rather large concentrations of dopant ions. CeO_2 and ZrO_2 when doped substitutionally with trivalent or divalent cations such a Y^{3+} or Ca^{2+} respectively, contain very many charge-compensating oxygen vacancies. These vacancies are quite mobile and the materials provide fast ion conduction for oxygen ions. δ-Bi_2O_3, which is stable at temperatures $> 1023K$ and which has a fluorite-like structure, has 25% vacant anion sites and is an even better conductor for oxygen ions [13,14]. In $RbBiF_4$ an EXAFS study has shown the preferential generation of disorder around the Rb^+ ions, and the fluoride interstitials that are induced as a result of this have been shown by molecular dynamics modelling to move via complex correlated interstitialcy mechanisms [15,16].

(iv)   Ionically Conducting Polymers. These materials consist of a polymer-salt complex with a coordinating polymer, usually a polyether, in which a salt is dissolved. Both the anions and cations may be mobile and this contrasts with polymer electrolytes in which one of the charged groups is covalently attached to the backbone so that only the counterion is mobile. The physical properties of these materials, especially their flexibility, makes them particularly suitable for incorporation into devices and they have been the subject of an extensive research effort over the past fifteen years [17,18]. Polymers doped with ions may also conduct electronically (cf. previous chapter) and can be used as electrodes in advanced battery systems [19].

(v)   Amorphous Ionic Conductors. Although ionic conductivity has been known in glasses for well over one hundred years new halide, molybdate and sulphide glasses, among other compositions, with much higher ionic conductivities and which operate at ambient temperatures have been developed during the past two decades. Considerable progress has also been made in understanding the essential nature of the structure of these glassy electrolytes and of the conduction process [20,21].

(vi)   Proton Conductors. These materials are included as a separate class because of their distinctive conduction mechanisms. These can be conveniently divided into two types. The first consists of hydrated materials such as hydrogen uranyl phosphate (HUP) in which the presence of distinct $(H_5O_2^+)$ and H_4O_2 dimers which can exchange protons and also coupled rotations of pairs of water molecules, as revealed by neutron diffraction [22], have be proposed as being responsible for the proton conduction. In the second type, perhaps best exemplified by $SrCeO_3$ doped with Yb [23], incoming water molecules generate hydroxy ions in the distorted perovskite structure and the hydroxy protons conduct by tunnelling between neighbouring oxygen sites.

In addition to their classification on structural grounds Catlow [6] has pointed out that consideration of conduction mechanisms can lead to a useful mechanistic classification. Whereas this categorisation is very informative it is important to realize that there is no implied correspondence between mechanism and structure: examples of some of the mechanisms occur in more than one of the structural types. Catlow has suggested the following mechanisms as being important: (i) conventional hopping but rapid migration; (ii) highly correlated; (iii) liquid-like diffusion and (iv) intermediate mechanisms which appear to blend hopping and liquid-like characteristics. Mechanism (i) is not fundamentally different from that operative in classical ionic conductors but there may be very high concentrations of dopants or they may move with very small energies of activation. In correlated motion, (ii), several ions move together as, for example, has been revealed in Li_3N [24,25] and $RbBiF_4$ [16] by simulation studies. Examples of real liquid-like diffusion, in which the concept of a distinct lattice no longer holds, are very rare: the high-temperature α-Ag_2S is probably an exception [26]. To these can be added the assisted mechanism which occurs in polymer-salt complexes where the polymer segmental motion is important for ion transport. The temperature dependence of the ionic conductivity in these systems and in glassy electrolytes above the glass transition temperature is interpreted on a free volume basis and will be discussed later.

3. Some Examples

3.1 FIC OF OXYGEN IONS

The fast-ion transport of oxygen ions has important technological applications and there have been widespread efforts to find an ideal oxygen ion electrolyte. Here it will be possible to briefly consider only a few specific illustrative examples. The use of Ca^{2+} and Y^{3+} stabilized zirconia, which when doped changes from a monoclinic to the cubic fluorite structure, in oxygen gauges and pumps is well known. The aliovalent dopants substitute for the host cation, e.g., Ca_{Zr}'', and charge compensation is effected by oxygen vacancies, $V_o^{\cdot\cdot}$, which are mobile. ThO_2, UO_2, HfO_2 and the most widely studied CeO_2 all show high ionic conductivities when similarly doped in this way. The mechanism is a conventional hopping migration and the interest lies in the manner in which the ionic conductivity varies with the doping level. For the CeO_2:Y^{3+} system at low dopant concentrations (>1 mol %) the conductivity has been shown to be controlled by the formation of dopant-vacancy pairs in which an oxygen vacancy occupies the corner of the cube in the fluorite lattice that has the dopant ion at its centre. [27]. However such a simple model fails completely to predict the behaviour at higher dopant concentrations. This is shown in figure 2: the conductivity which peaks at about 8 mol % Y^{3+} then decreases abruptly as more dopant is added; the activation energy for the migration shows corresponding variations. An explanation has been provided for this behaviour by an extensive simulation study [28] involving static lattice calculations of all possible defect jumps followed by a Monte-Carlo simulation of the overall conduction process. The rapid decrease in the conductivity was shown to result from the fact that a very large fraction of the jumps undertaken by the vacancies in the more highly-doped materials occurred about the dopant ions or in a backwards and forwards mode between neighbouring sites: neither of these made any contributions to the conductivity. The computations reproduced the essential features shown in Figure 2 and were also in agreement with experimentally reported variations in the binding energies of dopant-vacancy pairs with the radii of the dopant [29].

Reference has already been made to δ-Bi_2O_3 as an oxygen ion conductor. It has a lower activation energy than has stabilized zirconia (0.3 as against $\sim 1eV$) and the $6s^2$ lone pair results is a very large polarizability of the anion which also favours its facile migration through the material. However δ-Bi_2O_3 can exist only in the very narrow thermal stability range of 1003 - 1103 K, is produced only after a drastic monoclinic to cubic phase transition and these properties make it quite unsuitable for use in real

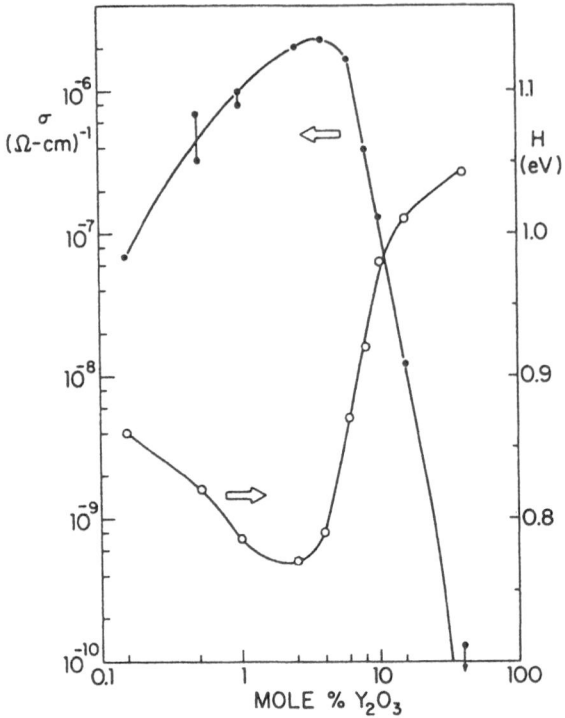

Figure 2. Behaviour of the conductivity and Arrhenius enthalpy for migration for the oxygen ion conductor $CeO_2{:}Y_2O_3$ as a function of the mole % age of the dopant. The data are from reference 52.

devices. Extensive attempts have been made to overcome these problems either by stabilizing the δ- form using dopants [30] or by producing novel structural types of bismuth-based materials of which the BIMEVOX family, derived from $Bi_4V_2O_{11}$, is perhaps the most promising [31]. Figure 3 shows the conductivity of ceramic samples of the parent compound and reveals three domains α, β and γ of which the γ -form existing above 833 K exhibits FIC of oxygen ions with a transport number close to unity. The activation energies measured were 0.24eV on heating and 0.17eV on cooling and the lower conductivities in the β and α phases were attributed to some ordering phenomena. Measurements on single crystals showed, as had been predicted from X-ray diffraction data, the material to be an anisotropic ionic conductor consistent with a bi-dimensional conduction mechanism in anion deficient $V{-}O_{3.5}$ sheets. The γ-phase was shown to consist of alternating $Bi_2O_2{2}$ layers and MO_4 perovskite-like sheets containing numerous oxygen vacancies [32]. Attempts were made to stabilize the γ-phase in a classical way by substituting a number of cations for the vanadium to prevent ordering of the structure. With Me signifying a divalent cation the reaction is written as:-

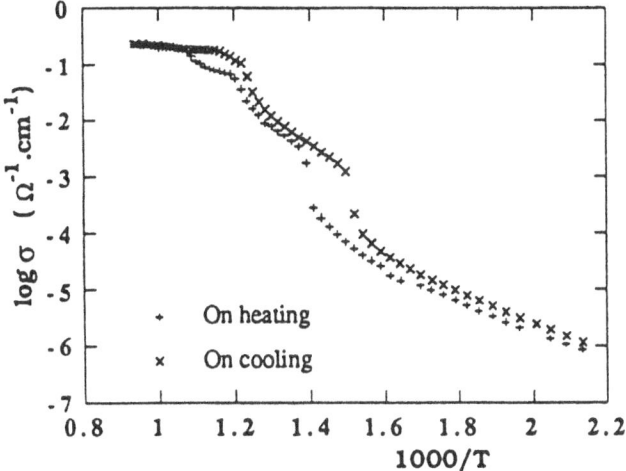

Figure 3. The ionic conductivity measured for ceramic samples of $Bi_4V_2O_{11}$ showing α, β, and γ domains (Reference 31).

$$Bi_2O_3 + 1/2(1-x)V_2O_5 + xMeO \rightarrow Bi_2V_{1-x}Me_xO_{(5.5\ -1.5x)} \qquad (2)$$

The new family of compounds has been given the acronym BIMEVOX with the level of doping being indicated as its amount relative to one vanadium in the parent compound $Bi_2VO_{5.5}$, e.g., $Bi_2V_{0.9}Co_{0.2}O_{5.2}$ is written as BICOVOX.20. Many elements were found to be capable of forming γ-type solid solutions and to give continuous FIC for oxygen over the entire temperature range without the emphatic changes in slope typical of phase transitions. Slight changes in slope are evident, as shown in Figure 4, but in all cases both the high and low temperature domains retain low activation energies. The best conductivities were always found to occur for a value of x close to 0.10 irrespective of the nature of the dopant and the range of suitable dopants is now quite extensive. The compounds have also been shown to be active catalytically with a correlation between their ionic conductivities and catalytic efficiency [33]. This family of compounds, produced specifically to produce FIC for oxygen ions over a wide range of temperatures, provide an excellent example of design in modern materials chemistry and of the procedures required to produce a useful fast-ion conductor.

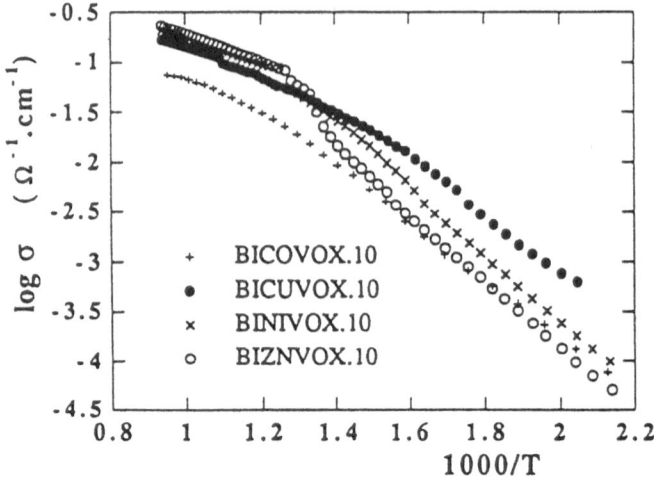

Figure 4. Ionic conductivities measured for four different BIMEVOX.10 compounds (Reference 31).

3.2. GLASSY ELECTROLYTES

Much of the recent development in solid state ionics has revolved around amorphous FIC materials such as glassy electrolytes, which will be briefly described here, and the polymer-salt complexes which are the subject of the following section. Ionically conducting glasses consist of a network former, such as P_2O_5, SiO_2, GeS_2 etc., that can form strongly cross-linked macromolecular chains. To this is added a network modifier chosen for its ability to interact strongly with the network former so that bridging bonds, such as those involving O and S, are broken and ionic bonds are introduced into the structure. Finally a doping salt, usually halides, sulphates or phosphates that contain the same cation as the network modifier may be added. This also causes changes in the conformation of the network and, more importantly, is found to influence the conductivity on relatively much larger scales than do the modifiers [20,21,34]. The mixed alkali-effect [35] is also observed in glasses e.g., in $K/CsSi_2O_5$ and in borate glasses. The exact nature of the environments surrounding the ions from the modifiers and any doping salts remain the subject of different views in the literature. Greaves [20] has discussed the Continuous Random Network (CRN) and Modified Random Network (MRN) models in detail with emphasis on the types of experimental information that are available and on the computational modelling techniques that have been used to attempt to reproduce the experimental results. Figure 5 shows the behaviour of the conductivity of an $(AgI)_{0.7}$ - $(AgMoO_{0.4})_{0.3}$ mixture [36] in an Arrhenius plot which spans the glass transition temperate, T_g, of the material. Below T_g the conductance is an activated process which is interpreted on the basis of a two-step mechanism. First a charged defect is created in a process that is analogous to the formation of a Frenkel defect in an ionic crystal (cf. previous chapter). A cation is envisaged as leaving its normal site and moving to a nearby site which is already occupied. The concentration of charge carriers, c_+, is then equal to that of the vacated

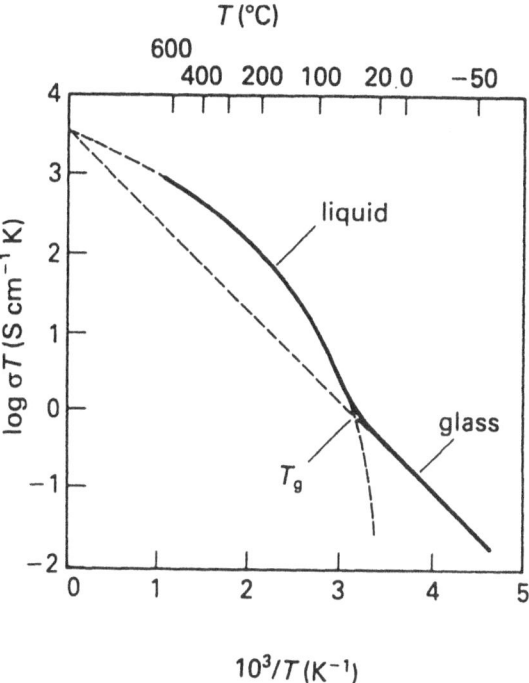

Figure 5. Arrhenius plot of σT for the $(AgI)_{0.7}$-$(AgMoO_4)_{0.3}$ mixture showing the different behaviour observed above and below Tg (Reference 36)

sites and also to that of the interstitial cationic pairs and, again by analogy, is given by

$$c_+ = c \exp\left[\frac{-g_f}{2kT}\right] \qquad (3)$$

where c is the mobile ion concentration and g_f the Gibbs free energy of formation of the defect pair. The migration, driven by the applied electric field, occurs via an interstitialcy non-collinear mechanism which is essentially the movement of one of the cations from a cationic pair to another cationic site. Measured Haven ratios [36] are consistent with this type of migration and from random walk theory the mobility is

$$u_+ = \left[\frac{el^2}{6kT}\right] \nu_o \exp\left[\frac{-\Delta g_{mig}}{kT}\right] \qquad (4)$$

where e is the charge on the ion, l the jump distance, ν_o the vibrational frequency of the interstitial ion and Δg_{mig} the activation free energy for the motion. Insertion into the standard conduction equation (eg. (1)) gives

$$\sigma T = \left[\frac{e^2 l^2 \nu_o c}{6k} \right] \exp \left[\frac{\frac{-g_f}{2} - \Delta g_{mig}}{kT} \right] \tag{5}$$

so that the Arrhenius slope would give the value of the activation enthalpy for the conduction process, h_a, as

$$h_a = \left[\frac{h_f}{2} + \Delta h_{mig} \right] \tag{6}$$

As has been pointed out previously [21], both h_f and Δh_{mig} may be expected to vary with the composition of the glass and, in the absence of any further information, there is no way in which their values can be individually determined from the value of h_a. It should also be emphasized here that the above treatment has, for simplicity, been limited to a single carrier and to simple defect-like mechanisms for its formation and migration. In real glasses these processes can be expected to be rather more complex though it is often the case that the dissociation of one salt will predominate. A number of other models, based for example on the dissociative behaviour of electrolytes, and proposed to account for the observed behaviour of h_a in certain cases have been reviewed by Ingram [38] and are also treated in detail by Souquet [34] and by Elliott [39].

Returning to Figure 5 it is evident that above T_g the conductivity process no longer exhibits Arrhenius behaviour but initially increases rather more rapidly with temperature. This type of behaviour can be represented by the empirical Vogel-Tamman-Fulcher (VTF) law [40-42] which is interpreted on the basis of the structural relaxation time, τ. This is the mean life time for the movement of a structural unit over a distance equivalent to its own size and is strongly temperature dependent. The VTF behaviour of all transport properties, and as will be evident later a similar behaviour is observed in polymer-salt complexes, may be interpreted using the free volume concept [43,44] in which the diffusing species is considered as being contained within an element of volume that is temperature dependent. The free volume is the excess volume that is redistributable about its mean value $<V_f>$ without any contribution from enthalpy. It is defined as $V_f = V - V_o$ where V_o is the volume of the element at a critical value of the temperature T_o. The temperature dependence of the mean free volume can be expressed by $<V_f> = \Delta \alpha V_o (T - T_o)$ where $\Delta \alpha$ is the difference in the volumetric dilation coefficient of the liquid and crystalline phases. Because the structural relaxation time for a diffusing species in a supercooled liquid is proportional to the probability that the species can attain access to a free volume greater than V_f^*, which is the minimum value required for the diffusion step to take place, we have

$$\tau \propto \exp\left[\frac{-V_f^*}{\langle V\rangle}\right] = \exp\left[\frac{-V_f^*}{\Delta\alpha V_o[T - T_o]}\right] \tag{7}$$

The value of V_f^* will depend on the diffusing ion and also on the local deformations of the macromolecular chain or chain segment motions that create the local free volume element and assist the migration. If the same defect creation mechanism considered previously for the ionic conductivity at temperatures $< T_g$ is operative then, by analogy with eq. (5), the full expression for cationic conductivity above the glass transition temperature is

$$\sigma_+ T = \left[\frac{e^2 l^2 \nu_o c}{6k}\right] \exp\left[\frac{-g_f}{2k}\right] \exp\left[\frac{-V_f^*}{\Delta\alpha V_o[T - T_o]}\right] \tag{8}$$

where the last exponential term predicts the type of non-Arrhenius behaviour observed experimentally. The temperature dependencies of the conduction processes in FIC glasses, both above and below the glass transition temperature, are therefore reasonably well understood. The materials offer conductivity values that are comparable to those in crystalline materials especially when the dopants added, such as AgI, themselves exhibit exceptionally high ionic conductivities in their crystalline states. Because of the marked dependence of the conductivities of the glassy electrolytes on their dopant content they can be designed to attain particular conductivity values often over considerable ranges. In addition, being amorphous they exhibit isotropic conductivity which is in contrast to some of the crystalline FIC materials.

3.3   IONICALLY CONDUCTING POLYMERS

Research and development work on ionically-conducting polymer-salt complexes has now been actively pursued for some twenty years [45] and there are number of recent reviews available [17, 18, 46-48]. The most widely studied host polymer is polyethylene oxide (PEO) which consist of $-O-CH_2-CH_2-$ repeat units and is found in both crystalline and amorphous forms depending on its composition and on the method of preparation. X-ray data from oriented fibres have shown the crystalline form to consist of an extended helix with seven OCH_2CH_2 units repeating in two turns.

Many other polymers and salts have been successfully complexed and in appropriate circumstances very high concentrations of salts can be incorporated. Standard thermodynamic arguments are involved to determine the feasibility and likely extent of complex formation. The relevant energy terms include the lattice energy of the salt, the energy required to create suitable sites on the polymer, the energy of cation solvation and finally any interaction energies between the ions that are dissolved. Predominant among these are the lattice energy of the salt and the energy of solvation of the cations by the polymer chains. Although polymer-salt complexes can be prepared by direct diffusion from the salt into the polymer their preparation is more often achieved through evaporation of the non-aqueous solvent from solutions containing their carefully dried components. Preferred polymers are those that may be expected to readily coordinate cations, e.g., polyethers and polyesters with the most suitable salts being those that contain singly-charged anions. Shriver and Bruce [17] and Bruce and Gray [18] have reviewed the general principles of polymer-salt complex formation. Recent diffraction data on a range of PEO-based systems have provided sufficient information on which to build some basic guidelines governing the structures of polymer electrolytes in general. For example they show that all cations sized from Li^+ to Rb^+ are accommodated within the PEO helix with varying coordination numbers including, in some instances, coordinations with atoms from the anionic groups. These reviews also describe a range of specific examples of the types of interactions that govern the formation of these materials and the transport mechanisms that facilitate their fast-ion conduction processes.

It is not possible to discuss all of these here but the basis for the transport of the ions is that their migration is intimately coupled with the microscopic viscosity of short segments of the polymer. This is in contrast to ion transport in analogous liquid solvents where solvent sheaths move with the ions. It has superseded the earlier concept of a hopping motion of the ions through a rigid polymer matrix or through channels. Such a mechanism, in which the polymer segmental motion assists the ion transport, requires that ions dissociate from their coordination spheres. This implies that for an effective conductor a suitable balance exists between the strength of the polymer-cation bonding, which is essential for the incorporation of the salt, and the ease with which these cations can be released for migration in an electric field. The *log* σT vs *1/T* plots commonly observed for polymer-salt complexes are found to be curved but the data can be adequately represented by an equation of the form:

$$\sigma = \sigma_o \exp\left[\frac{-B}{[T-T_o]}\right] \qquad (9)$$

This is analogous to that used for the ionic migration process above T_g in glassy electrolytes, eq. (8) above, and can again be interpreted using the free volume concept. The segmental polymer motion is thought to assist the ionic transport by facilitating the breaking of the coordination sphere of the solvated ions and by making available the free volume necessary for the diffusion to take place. This approach is perhaps the simplest way to understand the conduction process in these materials but it is not wholly satisfactory and several extensions to it have been proposed [50,51] which attempt to take account of activated steps in the ion transfer process. In general the model does not relate to the atomistic parameters of the motion so that factors such as ion size, polarizability, solvation and the possibility of the formation of pairs or larger aggregates by the ions cannot be included. Other theories, based on a configurational entropy model, have also been used as a basis for the interpretation of the VTF equation but again refer only to the properties of the polymer. The formation of ion pairs and triple ions in polymer-salt electrolytes as well as the effects of long-range ion interactions have been studied electrochemically [18].

4. Applications of FIC

The possible application of ionically conducting solids in devices has been considered since the start of this century, although the majority of the early proposals failed to reach fruition. Exceptions were the *Nernst glower* [53], where bright, white light is emitted on passage of an electric current through hot Y doped ZrO_2, and the use in the measurement of the thermodynamic properties of solids [54,55]. In many cases the lack of success was due to the low magnitude of the ionic conductivity of the materials available at the time. Research initiated in the mid-sixties and driven in the seventies by the need for better energy management has vastly increased the number and range of solids which exhibit FIC. Many of the materials discussed in section 2 have only been the subject of serious study in the last two decades. It is now evident that a high conductivity is a necessary but not sufficient requirement of a FIC for usage in a commercial device. Depending on the application other important criteria can include chemical and mechanical properties, processability and cost. These factors will be considered in the following discussion of the major areas of potential application of FIC, namely *fuel cells, batteries, sensors and electrochromic displays*.

4.1 FUEL CELLS

Fuel cells have the advantage over other power sources of a highly efficient electrochemical generation of energy from fossil fuels. The use of oxygen ion conductors in high temperature *solid oxide fuel cells* (SOFC) has a long history reaching back to the use of a zirconia electrolyte in a cell by Baur and Preis in 1937 [56]. Real interest in these cells began in the sixties when they were considered as power sources for space vehicles and then as large-scale (i.e. MW range) electrical power generators. Research in this area has been extremely active in several large, international companies, such as Westinghouse, Siemens and Brown Boveri, and the developments are summarised in several reviews [57-61]. Prototype cells have performed well and it is hoped that commercial systems will soon be in manufacture.

The production of energy from fossil fuels currently involves the use of some form of heat engine to yield electrical energy from the thermal energy produced in a combustion reaction. The efficiency, η, of this process is governed by the Second Law of Thermodynamics and is given by:-

$$\eta = \frac{T_{source} - T_{sink}}{T_{source}} \times 100 \%$$ (10)

In a modern power station the efficiency is around 50% and increasing this value, by increasing T_{source}, presents considerable problems in materials requirements. If a fuel cell is used for the oxidation of hydrogen by the reaction:-

$$H_2 + 1/2 \; O_2 \rightarrow H_2O$$ (11)

the efficiency is given by:-

$$\eta = \frac{\text{Change in Gibbs free energy of reaction, } \Delta G}{\text{Change in enthalpy of the reaction, } \Delta H} \times 100 \%$$ (12)

The theoretical efficiency can therefore be around 95% at room temperature. A real cell will be operating irreversibly and other factors need to be considered, including ohmic losses due to the internal resistance, R_i, and polarisation losses at the electrodes. If a fuel cell is operating at a current I and the voltage is E (which will be less than the open-circuit, no load voltage E_o) then:-

$$E = E_o - IR_i - V_p$$ (13)

A voltage efficiency, η_E, can then be defined as:-

$$\eta_E = \frac{E}{E_o} = 1 - \frac{IR_i + V_p}{E_o}$$ (14)

which can be as high as 75%. In addition to SOFC, fuel cells that are currently being explored include those with a phosphoric acid electrolyte, with a polymer electrolyte membrane (e.g. Nafion), with a molten carbonate electrolyte and with an alkaline (e.g. KOH) electrolyte. There are various problems with each of these systems, particularly the requirement of an expensive electrocatalytic electrode (e.g. platinum) and associated difficulties with poisons, and corrosion of materials.

The general arrangement of a fuel cell is shown schematically in Figure 6. The open-cell voltage, E_o, can be calculated from the Gibbs free energy change of the oxidation reaction:-

$$E_o = -\frac{\Delta G}{nF}$$ (15)

where n is the number of electrons involved in the reaction and F is the Faraday. This can be re-expressed in terms of the oxygen partial pressures, p_{O2}, at the electrodes as:-

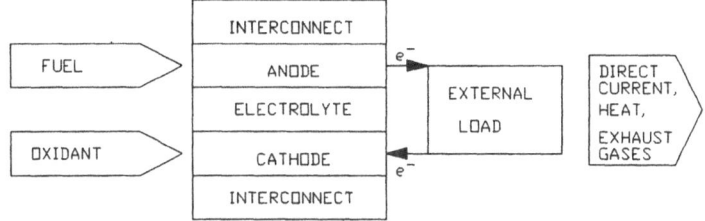

Figure 6. The schematic representation of a fuel cell. In a stack individual cells will be connected together with interconnects.

$$E_o = \frac{RT}{2F} \ln \left[\frac{(p_{O_2})^{1/2}_{cathode}}{(p_{O_2})^{1/2}_{anode}} \right] \qquad (16)$$

where R is the gas constant. This is the equation for the open-circuit voltage of a reversible oxygen concentration cell with a transport number for O^{2-} ions of unity.

In a SOFC the electrolyte that has been favoured has been cubic stabilised ZrO_2 with 8 mole per cent yttria as the dopant. This has the necessary very low electronic conductivity but the ionic conductivity is only sufficiently high enough to be useful at temperatures between 800 to 1000°C, around 0.1 S cm^{-1} at the latter temperature. If air is used as the oxidant then the open-circuit voltage is 1.14V.

A variety of designs of the fuel cell stack have been investigated [58-61]. The electrolyte, electrodes and interconnects can be formed on a supporting ceramic in very thin films if the cells are to operate at relative low temperatures. Since cubic stabilised zirconia is a very tough ceramic, it is possible to produce strong sheets that are self-supporting in the stacks that have a low enough resistance at relatively high temperatures to give an acceptable internal resistance. Different configurations of the stacks have also been exploited, including single tube multicells, multichannel monoliths and planar arrangements. The requirement to run a fuel stack with zirconia at around 1000°C places serious materials demands on the other components of the overall system. A primary concern in choosing the stack components is the matching of thermal expansivities. If this is not achieved then leaks will develop between the anode and cathode compartments on heating and cooling the stack. Another concern is the chemical stability of the components. This is a particular concern for the cathode and interconnect as these operate at high oxygen partial pressures and therefore must be stable to oxidation at the high temperatures involved. In addition, the electrodes must have good electronic conductivities and have low polarisation losses. A list of materials that are currently favoured for the high temperature SOFC operating at 1000°C are given in Table 1.

Table 1. Typical components for a SOFC operating at 1000°C with CH_4 fuel

Component	Material	Comments
Electrolyte	$Zr_{0.92}Y_{0.08}O_{1.96}$	Satisfactory as 200 micron sheet
Cathode	$La_{1-x}Sr_xMnO_3$	Satisfactory as 20-40 micron film
Anode	$Ni-ZrO_2$ cermet	Used as 10-30 micron film
Interconnect	$LaCr_{0.8}Mg_{0.2}O_3$	Used as 1-2 mm sheet

The materials in the Table 1 have a good thermal expansivity match ($\sim 1 \times 10^{-5}$ K^{-1}) and prototype cells have run for thousands of hours. However, there are still materials problems to be solved, particularly the difficulty in producing a sintered, dense interconnect of the chromate.

Some materials problems are reduced by operating at lower temperatures, for example metal alloy interconnects could be employed. A cell using methanol fuel and operating at 400 to 500°C is a current target of research. In this case the ionic conductivity of zirconia is not sufficiently high. In this case gadolium doped ceria is a suitable electrolyte, as seen from the data shown in Figure 7. However, at the lower temperatures the kinetics of the electrode reactions are slower, increasing the polarisation losses and markedly lowering the efficiency. Investigation of new electrode materials is an intensive area of current SOFC research [61].

4.2   BATTERIES

Commercially viable rechargeable solid-state batteries have been a prime target of research in solid state ionics since the sixties. The applications can broadly be divided into two categories; those requiring high-energy and low-energy density cells. In the former category are load-levelling at electricity power plants and for vehicular traction. Power stations currently are constructed with capacities capable to meet peak demands. Considerable efficiencies are possible if a power station is run at a constant output with the 'spare' electricity generated when demand is low (e.g. during the night) put into storage that can be accessed during periods of high demand. Efficient batteries are obviously one method of storing electricity (another approach is to use the spare power to pump water to higher altitude reservoirs and then, when power is required, release

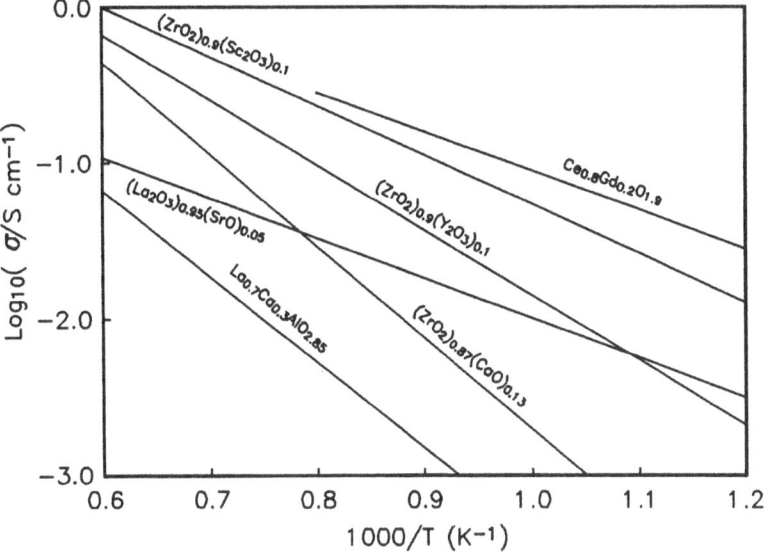

Figure 7. The ionic conductivity of some selected oxides.

the water through turbines). Originally it was the efficiency gains in fuel usage that motivated research on electric vehicles, however in recent years an additional impetus has been the growing awareness of environmental impact of the internal combustion engine. A battery-powered car would be emission-free. There are a wide variety of devices for which low-energy density cells would be the power source, including portable equipment (e.g. mobile phones) and heart pacemakers.

4.2.1 High-energy density cells

During the seventies the Na-S cell, first patented by Ford Motor Company [62], was the system that was the focus of attention as the most promising cell. The anode material is molten sodium, the cathode molten sulphur and the electrolyte is Na β-alumina ($Na_2O.11Al_2O_3$), or in later designs Na β''-alumina ($Na_2O.5.33Al_2O_3$ stabilised with additives of MgO or Li_2O). A variety of cell designs have been published [58,63] and a schematic arrangement is shown in Figure 8. The cell reaction is:-

$$2Na + xS = Na_2S_x$$

and the electrode reactions are reversible. The first stages of discharge yield Na_2S_5 and the open circuit voltage is 2.08V at 350°C. The solid sodium sulphides, Na_2S_2, Na_2S_4, Na_2S_2, Na_2S are poor conductors and if they were to build up on the electrolyte wall would rapidly yield an unacceptably high internal cell resistance. Hence the cell is operated at high temperatures where the reaction products will be polysulphides that are liquids [64]. The theoretical energy density is 360 W h kg^{-1}, and although this is not achievable it is possible to produce cells with values of ~130 W h kg^{-1}, which is three times that of the lead-acid cell.

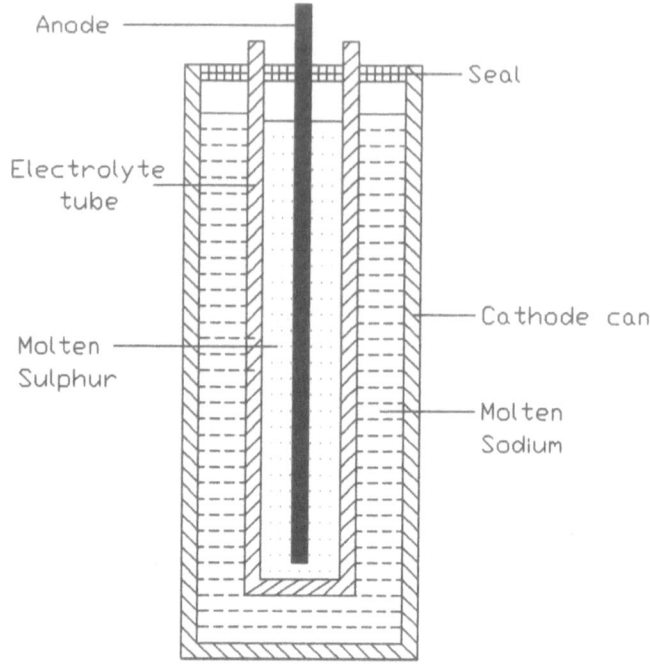

Figure 8. A schematic reepresentation of the Na/S cell.

Several technical problems were encountered in the production of reliable Na-S cells, many of which arose from the requirement of high temperature operation. These included the choice of material for the seals, with appropriate matching expansivities, that have to bond the components and prevent leakage of molten sodium and sulphur [63]. Molten sulphur is very corrosive and stainless steel has to be coated with a protective layer. If high current densities are drawn from the cell solid Na_2S_2 can from locally and crack the electrolyte tube. Molten sulphur is an insulator and it was necessary to incorporate current collectors (carbon felt) in the anode compartment [63]. Most of these difficulties have been resolved but the main problem, construction of reliable electrolyte tubes with long lifetimes, still remains. The tubes have to thin, around 600 micron wall thickness, in order to keep the internal resistance low and relatively large, the order of 10 cm diameter and 50 cm long, to have a high capacity. This puts severe demands on the sintering procedures to produce a high quality flaw-free ceramic [58,63]. In addition, the inspection of tubes for flaws is extremely expensive. During usage there is a degradation of the electrolyte tube and failures of the cells is usually due to the rupture of the tube. Currently the number of charge-discharge cycles that can be achieved with these cells is around several hundred [58]. This is not sufficiently high for load-levelling applications, which is the order of thousands of cycles. It is sufficient for electric vehicles, where the required number of cycles is expected to be 500 to 1000. However the current cost of production cells is much higher than original projections and further development is necessary to make them economical in comparison to alternative power sources.

The proposal by Armand in 1978 [65] that polymer electrolytes could be used as the membranes in solid state batteries caused immediate interest. An obvious advantage of these materials compared to the brittle ceramics was their mechanical flexibility and polymer electrolyte batteries became the focus of interest for high energy density cells for electric vehicles. The first batteries were based on the lithium cell with the arrangement

$$Li \ / \ LiX.P(EO)_x \ / \ intercalation \ electrode$$

The schematic arrangement of the cell is shown in Figure 9. The electrolyte was a film of lithium perchlorate or triflate, x (the number of monomer units per molecule of salt)

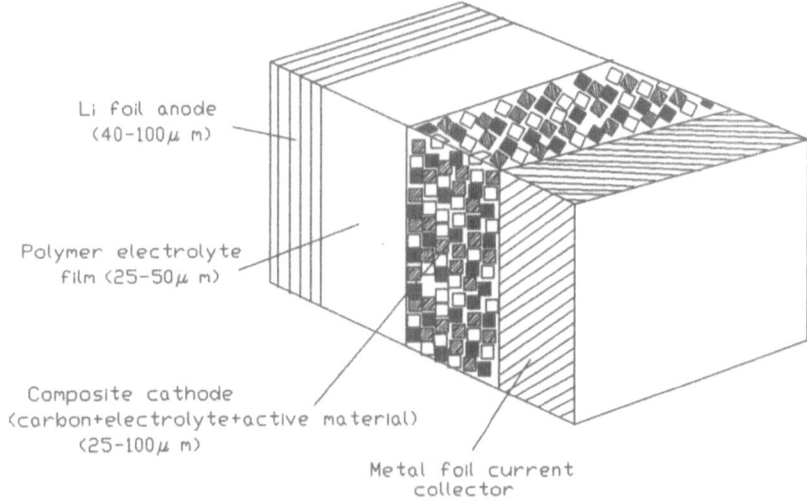

Li foil anode
(40–100μ m)

Polymer electrolyte
film (25–50μ m)

Composite cathode
(carbon+electrolyte+active material)
(25–100μ m)

Metal foil current
collector

Figure 9. A schematic arrangement of the polymer electrolyte cell (after [66]).

was typically in the range 8 to 20 and the intercalation electrode was TiS_2, later to be replaced by V_6O_{13}. The theoretical energy density of these cells is around 800 W h kg^{-1} the anticipated practical values is 425 W h kg^{-1} [Tofield]. In order to minimize the volume of the batteries several configurations were envisaged, including 'concertina' and 'swiss roll' arrangements [67,68].

A major problem with the PEO based electrolytes is their low conductivity at ambient temperature ($\sim 10^{-6}$ S cm^{-1}) and it is only above the melting point of PEO ($\sim 60°C$) that the conductivity is sufficient for battery applications ($\sim 10^{-3}$ S cm^{-1} at 100°C). The operating temperature can be lowered by using plasticizers or other polymer electrolytes [68]. There were also problems with build up of passivating layers at the electrodes on cycling these cells. The most recently developed polymer electrolyte batteries are showing considerable promise on the laboratory scale and their commercial exploitation mainly awaits improvements in maintaining the energy capacity on long-term cycling.

4.2.2   *Low-energy density cells*
A variety of solid-state batteries have been proposed for low energy applications based on a range of mobile ions, including Ag^+, Cu^+, Li^+, Na^+ and F^- [58,69], and types of separating membrane, including crystalline, glassy and polymer electrolytes. Silver-based batteries are usually quite successful in the laboratory and there are several highly conducting electrolytes that have been employed. Some examples of the cells are $Ag/RbAg_4I_5/RbI_3/C$ [70] and $Ag_xV_2O_5/Ag_2WO_4.4AgI/Ag_xV_2O_5$ [71]. The high cost of silver mitigates against the commercial exploitation of the cells.

In recent years considerable attention has focused on polymer electrolyte batteries using arrangements similar to those for high-energy density cells. The advantage is that the cell can be made both light and thin. A target has been a lithium-based battery. An interesting development has been the construction of batteries which employ electronically conducting polymers [47] as one or both the electrodes. A cell designed by [72] used doped polyacetylene , $(CH)_x$, as the electrodes with the formal arrangement:-

$$(CHNa_y)_x \mid NaI.P(EO)_5 \mid (CHI_y)_x$$

A cell using a lithium-aluminum anode, a polymer electrolyte separator and a polyaniline cathode has been constructed as a 'button' cell, 20 mm diameter and total thickness 1.6mm [73]. All plastic batteries are under development in the US (Allied-Signal), Japan (Hitachi and Bridgestone) and Germany (Varta). The applications would be those requiring a long life and low power consumption, e.g. mobile phones, timers, etc.

Presently the only commercially available all solid-state battery utilises lithium iodide as the electrolyte. The cell arrangement is:-

$$Li \mid LiI \mid I_2 - poly\ (\ 2vinyl\ pyridine)(PVP)$$

LiI is does not have a high specific conductivity however in the cell it is formed *in situ* by reaction of the lithium and the iodine-PVP complex. The open-circuit voltage is 2.8 V at 25°C. The cell is marketed by Wilson Greatback Ltd and is the major power source for heart pacemaker.

4.3   SENSORS

A very successful application of FIC has been in chemical sensing. The most common approach has been to use the FIC as the membrane of a potentiometric gauge with the sample of unknown concentration of chemical, X, on one side and a reference concentration on the other side. The system is then essentially a concentration cell and the EMF, E, across the membrane is given by the Nernst equation:-

$$E = \frac{RT}{nF} \ln \left[\frac{X_{unknown}}{X_{reference}} \right]$$

A high ionic conductivity of the membrane is necessary so that the steady electrical potential is rapidly established. This principle is used in the ion-selective electrode used to analyze for fluoride ions in aqueous solution where the membrane is a europium doped lanthanum fluoride crystal.

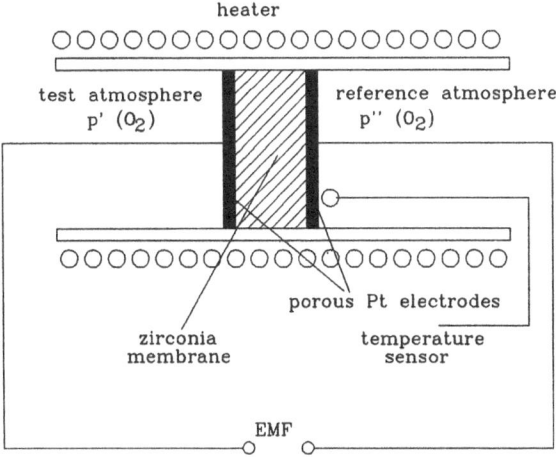

Figure 10. Basic structure of a zirconia-based oxygen monitor.

A variety of oxygen sensors used widely in industry are based on a solid electrolyte membrane of zirconium oxide, ZrO_2 stabilized in the cubic fluorite structure by the addition of yttria, Y_2O_3, calcia, CaO, or magnesia, MgO. The applications include the measurement of oxygen in gaseous atmospheres [74,75], the determination of oxygen levels in molten metals [76,77] and in automotive engine management - the so-called *'lambda'* sensor [78]. A sketch of the essential features of an oxygen monitor are shown in Figure 10 and these can be summarised as follows:-
(i) A heater to keep the zirconia membrane at temperatures above 600°C so that it is a

good oxide-ion conductor.

(ii) Two porous Pt electrodes at which the oxygen gas/oxide ion equilibria can be established and which provide the electrical contacts to the electronics of the sensor.

(iii) A reference oxygen concentration on one side of the membrane, against which the oxygen concentration in the test atmosphere is measured.

The cell shown in Figure 10 can be used in three modes for gas detection; namely as a concentration cell, as a fuel cell and as an oxygen-ion pumping cell [74]. The simplest mode is the concentration cell method in which the EMF of the cell is given by the appropriate form of Nernst equation, i.e.:

$$E = \left[\frac{RT}{4F} \right] \ln \left[\frac{p''_{O_2}}{p'_{O_2}} \right] \qquad (22)$$

where p'_{O_2}, p''_{O_2} are the oxygen partial pressures in the test and reference atmospheres, respectively. This mode of operation is often referred to as the *'potentiometric'* mode.

A more sophisticated means of operating the sensor is to use an *'amperometric'* mode. This involves applying a voltage across the membrane and electrochemically pumping oxygen from the cathode (the test gas chamber) to the anode and measuring the current flow. If a diffusion barrier is placed in front of the cathode (a porous frit or pin-hole) then above a threshold voltage the current measured will be independent of applied voltage, i.e. there will be a limiting current which will be directly proportional to the oxygen concentration in the test atmosphere [75]. The advantage of this mode is clearly the direct proportionality of the signal to the oxygen concentration rather than the logarithmic dependence in the potentiometric mode. Other methods of employing zirconia devices can be found in the review by Maskell [[75].

A whole variety of designs of the sensing element in zirconia sensors have been employed (see, for example, [74]). These range from ceramic tubes of zirconia to thin film devices using semiconductor technology.

Although the exploitation of FICs in gas sensors has mainly focused on the measurement of oxygen concentrations the same principle can be applied to the detection of a wide range of gases, e.g. H_2, SO_2, and CO_2.

4.4   ELECTROCHROMIC DISPLAYS

In an electrochromic display a colour change is created in a material by applying a current. The most commonly used material with FICs has been WO_3. MO_3 has also been employed and the colouring mechanism is similar in both cases. The insertion of protons or alkali ions into WO_3 creates a 'tungsten bronze' by the reaction:-

$$xA + WO_3 = A_xWO_3$$

The tungsten bronzes are highly coloured (Li_xWO_3 is dark blue) in comparison to WO_3 which is pale yellow. A schematic diagram of the arrangement of a display is shown in Figure 11. A voltage applied across the two indium tin oxide (ITO) electrodes can drive ions in or out of the WO_3 layer from the ion source (a thin film of alkali metal) through the FIC. FICs that have been demonstrated as useful in electrochromic displays are $LiAlF_4$ [79], $LiNbO_3$ [80], $LiTaO_3$ [80], LiN_3 [81], β-alumina [82], $RbAg_4I_5$ [83] and NASICON [84]. For large area displays these materials would have to be sputtered as films, which is difficult processing technology, and recently there has been considerable interest in using polymer electrolytes, e.g. $LiClO_4.P(EO)_8$ [85] and $H_3PO_4.PEO$ [86], which are easily solvent-cast as high area films. The use of electronically conducting polymers as the electrodes in the displays offering the possibility of introducing a wide range of colours [73,87]. For example, undoped polyaniline and polythiophene are light yellow and red, respectively, and change to green and blue when doped with perchlorate ions.

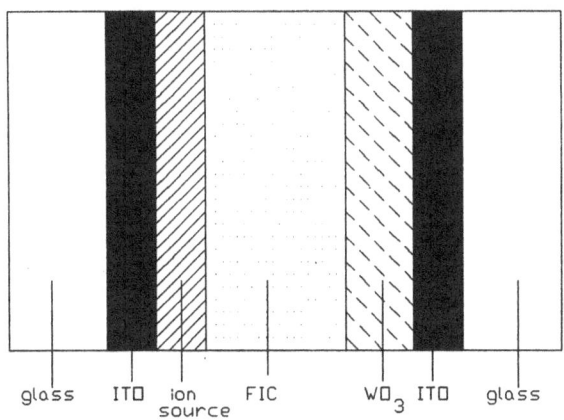

Figure 11. A schematic diagram of an electrochromic display.

Electrochromic devices are relatively slow in the switching of the colour, typically the order of seconds, and they would not be used for fast displays where liquid crystal devices are pre-eminent. However, in contrast to liquid crystal displays, the contrast of an electrochemical display is the same at all viewing angles and is a passive device, i.e. the colour remains when the driving current is switched off and only changes on reversing the current direction. The major potential applications for an electrochromic device are as 'smart windows' or large area displays. A smart window in a building or automobile will form part of the energy management system, being switchable from transparent to opaque in response to external conditions. There is a large market for these devices and a commercial device is expected in the near future.

5 Summary

The research since the sixties has revealed FIC in a wide range of physical and chemical structures. As a result the early attempts to develop general theories FIC, such as the 'molten sub-lattice' model were not particularly fruitful for the diversity of systems that exhibited high ionic conductivities [1,2]. The most useful approach has been to try to understand the mechanism of conduction in a specific class of fast-ion conductors, such as the fluorite structured material, amorphous conductors, etc. In this respect the classification of FIC proposed by Catlow [1] is particularly useful and was followed in this article. For the inorganic, crystalline fast-ion conductors the understanding of the mechanisms of conduction is now quite good, to the extent that it possible identify the structures and compositions that will produce the best results. The BIMEVOX class of materials is a good example of the design fast-ion conductors. The understanding of FIC in glassy materials is less good, although when the effect is produced by a dopant it is possible to control the conduction by the nature and concentration of the dopant. Currently, the more general phenomenon of ionic motion in glasses is a very active research area[20] and the developments will have considerable influence on FIC in glassy electrolytes. Polymer electrolytes are the newest class of materials and are the least well-understood. There are clearly analogies with the glassy materials, however there is the added complexity of the flexibility of the framework. Attempts to develop analytical theories will therefore be difficult and the preliminary results [88] from molecular dynamics simulations suggest this will provide a means of modelling the conduction process.

It is clear that there are a wide range of important technological applications for fast-ion conductors. Most of these have still to reach the commercial market and are still in the early stages of development. Amongst the materials the polymer electrolytes are particularly promising as they are easily processed as large area, flexible films. For many applications the currently available materials have a sufficiently high conductivity for a useful device and it is other factors that are limiting their exploitation, e.g. the development of suitable electrode and interconnect materials for the SOFC. The general opinion in this field is that in the next decade the technological problems will be resolved and the promise of these relatively new materials will be fully realised.

References

1.   Catlow, C.R.A. (1983) *Solid State Ionics*, **8**, 89.
2.   Chadwick, A.V. (1983) *Radiation Effects* ,**74**, 17.
3.   Nicholson, P.S., Whittingham, M.S., Farrington G., Smeltzer, W.W., and Thomas, J. (1992) Eds *'Solid State Ionics -91'*, North Holland, Amsterdam.
4.   McGeehin, P. and Hooper, A. (1975) *AERE Report R8074*
5.   Whittingham, M.S. (1975) *Electrochim. Acta*, **20**, 575.
6.   Catlow, C.R.A. (1990) *J. Chem. Soc. Faraday Trans.*, **86**, 1167.
7.   Bradley, J.N. and Greene, P.O. (1967) *Trans. Faraday Soc.* **63**, 2516.
8.   Dickens, P.G. and Whittingham, M.S. (1968) *Quart. Rev.* **22**, 30.
9.   Hagenmuller, P. (1971) Progr. *Solid State Chem.*, **5**, 71.
10.   Wang, J.C., Gaffari, M. and Choi S.I. (1975) *J. Chem. Phys.* **63**, 772.

11. Walker, J.R. and Catlow, C.R.A. (1982) *J. Phys. C: Solid State Phys.*, **15**, 6151.

12. Skotniczny, Z., Moscinski, J. and Rycerz, Z. (1986) *J. Phys. C: Solid State Phys.* 19, 4781.

13. Takahashi, H. and Iwahara, H. (1978) *Mat. Res. Bull.* **13**, 1447.

14. Battle, P.D., Catlow, C.R.A., Drennan, J. and Murray A.D. (1983) *J. Phys. C: Solid State Phys*, **16**, L561.

15. Cox, P.A., (1989) Ph.D. Thesis, University of Keele.

16. Cox, P.A., Catlow, C.R.A. and Chadwick, A.V. (1994) *J. Mater. Sci.* , **29**, 2725.

17. Shriver, D.F. and Bruce, P.G. (1995) in *'Solid State Electrochemistry'*, Ed P.G.Bruce, Cambridge University Press, Chapter 5.

18. Bruce, P.G. and Gray, F.M. (1995) in *'Solid State Electrochemistry'*, Ed. P.G.Bruce, Cambridge University Press, Chapter 6.

19. Scrosati, B. (1995) in *'Solid State Electrochemistry'*, Ed. P.G. Bruce, Cambridge University Press, Chapter 9.

20. Greaves, G.N. (1994) in *'Defects and Disorder in Crystalline and Amorphous Solids'* Ed. C.R.Catlow, Proc. NATO-ASI, Madrid 1991: Kluwer Academic Publishers, The Netherlands, Chapter 5.

21. Souquet J.L. (1994) in *'Defects and Disorder in Crystalline and Amorphous Solids'* Ed. C.R.A.Catlow, Proc. NATO-ASI, Madrid 1991: Kluwer Academic Publishers, The Netherlands, Chapter 10.

22. Fitch, A.N. (1986) *Mat. Sci. Forum*, **9**, 113.

23. Iwahara, H., Uchida, H. and Tamaka, S. (1983) in *'Solid State Ionics -83'* Eds M. Kleitz, B. Sapoval and D. Ravaine, North Holland, Amsterdam, p. 1021.

24. Wolf, M.L., Walker J.R. and Catlow, C.R.A. (1984*) J. Phys. C: Solid State Phys.*, **17**, 6623.

25. Wolf, M.L. and Catlow, C.R.A. (1984) *J. Phys. C: Solid State Phys.*, **17**, 6635.

26. Rickert, H. (1982) *'Electrochemistry of Solids'*, Springer-Verlag, Berlin.

27. Gerhart-Anderson, R. and Nowick. A.S. (1981) *Solid State Ionics*, **5**, 547.

28. Murch, G.E., Catlow, C.R.A. and Murray, D.A. (1986) *Solid State Ionics* **18-19**, 196

29. Butler, V., Catlow, C.R.A., Fender, B.E.F. and Harding, J.H. (1983) *Solid State Ionics* , **8**, 109.

30. Verkerk, M.J., Keizer, K. and Burggraaf, A.J. (1980) *J. Appl. Electrochem.* **10**, 81.

31. Mairesse, G. (1993) in *'Fast Ion Transport in Solids '*, Eds. B.Scrosati, A.Magistris, C.M.Mavi and G.Mavitto, NATO ASI Series, Vol. 250, Kluwer Academic Publishers, The Netherlands, p. 271.

32. Abraham, F., Boivin, J.C., Mairesse, G. and Nowogrocki, G. (1990) *Solid State Ionics*, **40-41**, 934.

33. Cherrak, A., Hubaut, R., Barbaux,Y. and Mairesse, G. (1992), *Catalysis Letts.*, **15**, 377.

34. Souquet, J.L. (1995) in *'Solid State Electrochemistry '*, Ed. P.G.Bruce, Cambridge University Press, Chapter 4.

35. Day, D.E. (1976) *J. Non-Cryst. Solids* , **21**, 343.

36. Kawamura, J.and Shimoji, M. (1986) *J. Non-Cryst. Solids* **88**, 281.

37. Kant, H., Kaps, C. and Offermann, J. (1988) *Solid State Ionics*, **31**, 215.

38. Ingram, M.D. (1987) *Phys. Chem. Glasses*, **28**, 215.

39. Elliott, S.R. (1990) *'Physics of Amorphous Materials '* 2nd Edition, Longman Group U.K., p. 246.

40. Vogel, H. (1921) *Phys Z.* **22**, 645.

41. Tammann, G. and Hesse, W. (1926) *Z. Anorg. Allgem. Chem.* , **156**, 245.

42. Fulcher, G.S. (1925) *J.Amer. Ceram. Soc.*, **80**, 5059.

43. Doolittle, A.K. (1951) *J.Appl. Phys.*, **22**, 1471.

44. Cohen, M.H. and Turnbull, D. (1959) *J. Chem. Phys.*, **31**, 1164.

45. Fenton, D.E., Parker, J.M. and Wright, P.V. (1973) *Polymer*, **14**, 589.

46. Linford, R.G. (1987, 1990) Ed. *'Electrochemical Science and Technology of Polymers '*, Vols. 1 and 2.

47. Skotheim, T.A.(1986) Ed. *'Handbook of Conducting Polymers'* Vols. 1 and 2, Marcel Dekker Inc. New York.

48. Bruce, P.G. and Vincent, C.A. (1993) *J. Chem. Soc. Faraday Transactions*, **89**, 3187.

49. Yoshihara, T., Tadokoro, H. and Murahashi, S. (1964) *J. Chem. Phys.* , **41**, 2902.

344

50. Mujamoto, T. and Shibayama, K. (1973) *J. Appl. Phys.* **44**, 5372.
51. Cheradame, H. and Le Nest, J.F. (1987) *in 'Polymer Electrolyte Review -1'* Eds J. R. McCallum and C.A. Vincent, Elsevier, London.
52. Nowick, A.S., Wang, D.Y., Park, D.S. and Griffith, J. (1979) in *'Fast Ion Transport in Solids'* Eds P. Vashista, J.N.Mundy and G.K.Shenoy, Elsevier, North Holland, New York, p. 673
53. Nernst, W. (1900) *Z. Elektrochem.*, **6**, 41.
54. Katayama, M. (1908) *Z. Phys. Chem.*, **61**, 566.
55. Kiukkola, K. and Wagner, C. (1957) *J. Electrochem. Soc.*, **104**, 379.
56. Baur, E. and Preis, H. (1937) *Z. Elektrochem.*, **43**, 727.
57. Rohr, F.J. (1978) in *'Solid Electrolytes'*, Eds. P. Hagenmuller and W. van Gool, Academic Press, New York, Chapter 25.
58. Yamamoto, O.(1995) in *'Solid State Electrochemistry'*, Ed. P.G. Bruce, Cambridge University Press, Chapter 11.
59. Singhal, S.C. (1989) Ed. *'Solid Oxide Fuel Cells'*, The Electrochemical Society, Pennington, N.J.
60. Yamamoto, O., Kaneko, S. and Takahashi, H. (1988) Eds. *'Solid Oxide Fuel Cells'*, Science House, Tokyo.
61. Steele, B.C.H. (1992) *Mat. Sci. and Eng.*, B13, 79.
62. Weber, N. and Kummer, J.T. (1967) in *'Advanced Energy Conversion'*, ASME Conference, Florida, p.913.
63. Scholtens, B.B. and van Gool, W. (1978) in *'Solid Electrolytes'*, Eds. P. Hagenmuller and W. van Gool, Academic Press, New York, Chapter 27.
64. Gupta, N.K. and Tischer, R.P. (1976) *J. Electrochem. Soc.*, **119**, 1033.
65. Armand, M.B., Chabagno, J.M. and Duclot, M.J. (1978) *paper presented at the Second International Conference on Solid Electrolytes*, St. Andrews, Scotland; Armand, M.B., Chabagno, J.M. and Duclot, M.J. (1979) in *'Fast Ion Transport in Solids'* Eds P. Vashista, J.N.Mundy and G.K.Shenoy, Elsevier, North Holland, New York, p. 131.
66. Linford, R.G. (1991) in *'Solid State Materials'* Eds R. Radakrishna and G. Daud, Narosa Publishing, New Delhi.
67. Tofield, B.C., Dell, R.M. and Jensen, J. (1984) *AERE Harwell Report* 11261.
68. Gray, F.M. (1991) *'Solid Polymer Electrolytes'*, VCH Publishers, New York.
69. Scholtens, B.B. and van Gool, W. (1978) in *'Solid Electrolytes'*, Eds. P. Hagenmuller and W. van Gool, Academic Press, New York, Chapter 26.
70. Argue, G.R., Owens, B.B. and Groce, J.J. (1968) *Proc. Ann. Power Sources Conf.*, **22**, 103; Owens, B.B. (1971) *Adv. Electrochem. Eng.*, **8**, 1.
71. Takada, K., Kanbara, T., Yamamura, Y. and Kondo, S. (1990) *Solid State ionics*, **40/41**, 982.
72. Chiang, C.K. (1981) *Polymer Comm.*, **22**, 1454.
73. Kanatzidis, M.G. (1990) *C & E News*, December 3, p. 36.
74. Kocache,R.M.A. (1987) in *'Solid State Gas Sensors'*, Eds P.T. Moseley and B.C. Tofield (Adam Hilger, Bristol), p. 1.
75. Maskell, W.C. (1987) *J. Phys. E: Sci. Instrum.*, **20**, 1156.
76. Pluschkell, W. and Engell, H.J. (1965) *Z. Metalkunde*, **56**, 450.
77. Fischer, W.A. and Ackermann, W. (1965) *Arch. für das Eisenhätten messen*, **36**, 346.
78. Velasco, G.V. and Schnell, J.P. (1983) *J. Phys. E: Sci. Instrum.*, **16**, 973.
79. Oi, T. and Miyauchi, K. (1981) *Mater. Res. Bull.*, **16**, 1281.
80. Glass, A.M., Nassau, K. and Negran, T.J. (1978) *J. App. Phys.*, **49**, 4808.
81. Miyamura, M., Tomura, S., Imai, A. and Inomata, S. (1981) *Solid State Ionics*, **3/4**, 149.
82. Green, M. and Kang, K.S. (1977) *Thin Solid Films*, **40**, L19.
83. Green, M. and Richman, (1974) *Thin Solid Films*, **24**, 45.
84. Barna, G.G. (1979) *J. Electron. Mater.* **8**, 153.
85. Bohnke, O., Bohnke, C. and Amal, S. (1989) *Mater. Sci. Eng.*, **B3**, 197.
86. Pedone, D., Armand, M.B. and Deroo, D. (1988) *Solid State Ionics*, **28/30**, 1729.
87. Patil, A.O., Heeger, A.J. and Wudl, F. (1988) *Chem. Rev.*, **88**, 183.
88. Mills G.E. and Catlow, C.R.A. (1994) *J.C.S. Chem. Comm.*, **18**, 4745.

AN INTRODUCTION TO MOLECULAR HETEROGENEOUS CATALYSIS

R.A. VAN SANTEN, R.J. GELTEN
Schuit Institute of Catalysis,
Eindhoven, University of Technology
The Netherlands

Abstract:

Fundamental concepts of heterogeneous catalysis are presented. Sabatier's principle relates *activity* with an *optimum* interaction energy between reactant and catalyst. Selectivity is determined by the *mechanism* of reaction and *surface state* of the catalyst. Surface reconstruction may lead to facetting of surfaces and non-ideal mixing behaviour of adsorbed surface layers. This may have large consequences for catalyst behaviour. Principles of chemisorption and dissociation will be highlighted.

1. A Summary of our Understanding of Surface Chemical Reactions

Catalysis has only recently started to outgrow the level of a purely empirical discipline. The reason for this slow change from "alchemy" to science is the complexity of practical catalysis as well as the complexity of the catalytic reactions. Several developments in the past two decades have led to a major advance in the knowledge of solid surfaces [1]. These developments are:

- Application of modern spectroscopic methods to the characterisation of solid state surfaces such as catalysts [2]. Catalysts are usually highly porous and high surface area materials. On the surface of the catalyst support, catalytically active compounds are distributed that are often ill defined and complex in composition.

- The study of single crystal surfaces under well defined conditions, made possible by the development of Ultra High Vacuum Technology.

- Application of molecular beam technology [3], in which the reactivity of small particles can be studied as a function of particle size.

- Computational chemical methods that enable predictions on the interaction energies and structure of intermediates in heterogeneous and homogeneous catalytic reactions [4].

- Advances in synthesis (manipulation of molecules) have led to techniques to fabricate well defined surfaces and model systems, so that theoretical insights can be implemented on a molecular level.

C.R.A. Catlow and A. Cheetham (eds.), New Trends in Materials Chemistry, 345–362.
© *1997 Kluwer Academic Publishers.*

346

In this chapter we will focus our attention on the molecular basis of catalysis as a phenomenon in kinetics [5].

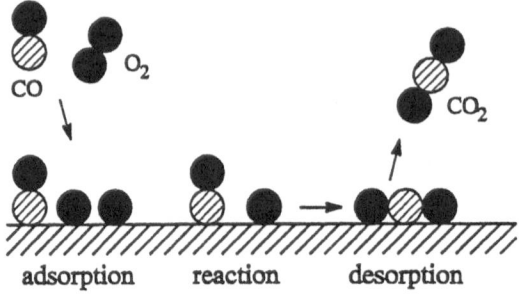

Figure 1. Schematic representation of the oxidation of CO.

The essence of a catalytic reaction is that it is a self regenerating cycle of successive elementary reaction steps. The catalytically active site, free of reactants before reaction, is reproduced after the reaction. This is illustrated in Figure 1 with the example of the catalytic oxidation of CO. In this reaction scheme we discern four different elementary reaction steps. First the reactant molecules CO and O_2 have to adsorb on the surface (step 1: Adsorption). Then O_2 has to dissociate (step 2: Dissociation). In the third step an oxygen atom recombines with CO to produce CO_2 (step 3: Recombination). In a final step the CO_2 molecule desorbs and the free surface vacancy is regenerated (step 4: Desorption).

Most practical reactions are more complicated, since also surface diffusion and often surface reconstruction occurs. When there is no synergy between reactions on different sites, there is no correlation between the phases of the reaction cycles and the overall reaction is stationary, not showing oscillations on a macroscopic level. However, examples of cyclic surface reactions are known in which the cyclic nature of the total surface composition could be experimentally followed [6].

The example mentioned above can also be used to introduce the principle of Sabatier, which is the basis of catalytic kinetics. According to this principle the rate of a reaction is maximum for an optimum value of the interaction between adsorbate and catalysis surface. This stems from the competing nature of some of the different reaction steps that form the catalytic reaction cycle. In the case of the CO oxidation reaction, the elementary reactions that compete are dissociation and desorption.

When a weak interaction between reactant and surface is increased, the rate of dissociation will increase. This may occur by changing to a more reactive surface such as metal. However, when the interaction of reactants or products with the surface becomes too strong, no molecules will desorb and the surface becomes blocked by adsorbates. For the example of the CO oxidation reaction, chemisorbed CO will block the surface at low temperatures. Also the interaction with oxygen may be too strong so that surface or bulk oxide formation may occur.

In the case of surface blocking by CO, a weaker interaction between CO and the surface will shift the equilibrium of absorbed CO, the surface coverage of CO will decrease, and the overall reaction rate will increase. The principle of Sabatier is illustrated schematically in Figure 2. The reaction rate has a volcano type dependence on the interaction strength of adsorbate and catalyst, a phenomenon that has been often observed experimentally. The reaction rate has a positive order in the reactant to the left of the maximum where the interaction strength is small. To the right of the maximum the reaction order becomes negative in reactant or product, because the surface sites become blocked by adsorbates. For the oxidation reaction of CO, platinum is one of the most effective catalysts.

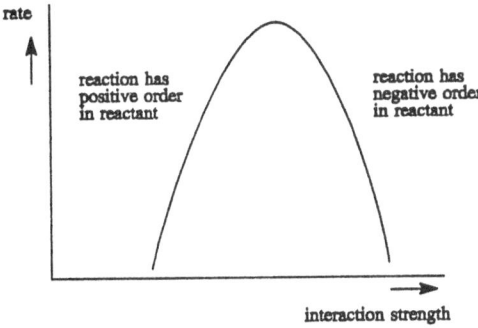

Figure 2. Principle of Sabatier

When more than one product is formed, the selectivity of the reactions is to be considered. We will discuss this qualitatively for the reaction of CO with NO in the presence of H_2, a reaction of interest to automotive exhaust gas catalysis. The selectivity of a reaction is defined as the ratio of conversion to desired products *versus* total conversion. The important overall reactions are given in equations (1) and (2).

$$NO + CO \rightarrow CO_2 + \frac{1}{2}N_2 \qquad (1)$$

$$NO + 2\frac{1}{2}H_2 \rightarrow NH_3 + H_2O \qquad (2)$$

Reaction (1) is desirable and reaction (2) has to be suppressed. For both reactions NO has to dissociate. It appears that platinum and rhodium [7] show a completely different selectivity with respect to these two reactions. On platinum a significant amount of NH_3 is formed, whereas on rhodium the selectivity to N_2 is larger.

On a rhodium surface NO dissociates much easier than on platinum. As a result at steady rate Rh has a high coverage of adsorbed nitrogen atoms, whereas on the much less reactive platinum, the steady state coverage of N_{ads} is low. Hydrogen, though dissociating easily on every transition metal surface, has little chance to recombine with nitrogen, because of the high nitrogen coverage on the Rh surface. On Pt, however, due to its low N coverage, many sites are available, resulting in a high hydrogen coverage on this metal. Together these effects result in an enhancement of reaction (1) and suppression of reaction (2) on the surface of rhodium.

In summary, two phenomena are of importance to the selectivity of a reaction:

- *Mechanism*, which controls which molecules dissociate and also which surface intermediates recombine and which do not;
- *Surface composition*, which influences the recombination probabilities.

In our example, the rate of NH_3 production is high when the nitrogen surface coverage is low. A high nitrogen coverage corresponds to a high N_2 *versus* NH_3 production ratio.

The classical picture of catalysis by a surface has been provided by Langmuir [8]. He proposed a surface with a finite number of equal vacant sites. Adsorbates were supposed not to interact. Surface scientific investigations and especially STM studies have convincingly demonstrated that the reality of the reactive surface is significantly more complex [9]. Generally the interaction between adsorbates cannot be ignored. Lateral interactions may lead to non-ideal mixing behaviour of adsorbate layers. Surface adsorbate layers appear often to form ordered structures that develop into overlayer islands. Also atoms or molecules adsorbed near to each other may develop strong repulsive interactions that have to be overcome when a reaction occurs.

As we will discuss more extensively later, the interaction energy of an adatom with a surface is a strong function of the surface structure. Two geometric parameters are important:

- The number of surface substrate atoms that coordinate to the atom: the *coordination number* of the *adatom;*
- The degree of coordinative unsaturation of the surface substrate atoms: the coordination number of the *surface atoms* with *substrate neighbour* atoms.

Generally the reactivity of a surface atom decreases with decreasing coordinative unsaturation. For example, for the dense surfaces of a face centred cubic crystal, a surface atom with 9 surface atom neighbours ((111) surface) will be less reactive than one with 8 neighbours ((100) surface). It is in essence due to the law of Bond Order Conservation [4, 10] which states that bonds become weaker the more the valence electrons have to be shared. The interaction energy of an adsorption complex is determined by the embedding of the surface complex. The lower reactivity of the surface atoms compared to those in an organometallic complex tends to lead to weaker interactions in the surface complex [13].

The increased interaction of an adsorbate atom when the coordination of surface atoms with neighbouring surface atoms decreases can provide the driving force for surface rearrangements with important consequences for catalysis. An example is the oxidation of CO on the (110) surface of platinum [11]. As is shown in Figure 3 in ultra high vacuum the Pt (110) surface reconstructs to a less reactive structure, indicated with surface–crystallographic notations Pt(110)–(1x2), in which a row of atoms is missing. Coordinatively unsaturated atoms contract their surface bonds, thus generating surface stress forces. The driving force for the reconstruction is surface stress relief.

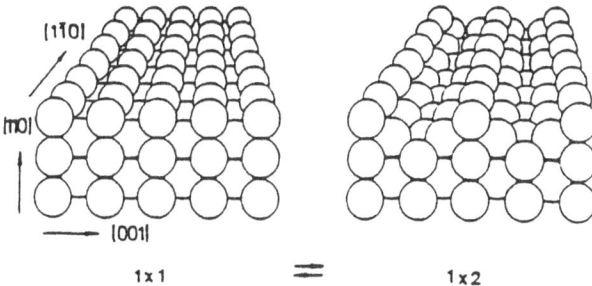

1x1 $\rightleftharpoons$ 1x2

Figure 3. The (110) surface of platinum (left) and its (1x2) reconstruction (right)

As a consequence, a less reactive surface is formed. During reaction the surface alternates between the nonreconstructed and reconstructed phase. Surface conversions to the non–reconstructed phase is driven by the lower degree of coordination of the surface atoms and hence a higher reactivity. Surface reconstruction may also lead to more favourable coordination of the adatom itself.

At low surface coverage an adsorbed atom will usually not disrupt the surface lattice, but pull surface atoms outwards from the surface to form a surface complex embedded in the metal surface [12]. Adsorbates tend to from chemisorbed complexes that are structurally close to the analogous coordination complex. The local surface structural change is counteracted by surface stress forces that respond to the local surface strain of the adsorption site. When more atoms adsorb, surface stress can be released by rearrangement of surface atoms [14], as we discussed earlier for the reconstructed Pt (110) surface. The reconstructed surface will create sites that have stronger interaction with the adsorbate. This may lead to the formation of ordered overlayers with transport of surface atoms. Facetting of surfaces to give surface phases stable during reaction may occur after an initial period.

The chemical reactivity of adsorbed atoms may also change as a function of surface concentration and composition. For adsorbed oxygen Roberts [15] has proposed a difference between the reactivity of isolated oxygen atoms and oxygen atoms that are part of an oxygen overlayer structure. Isolated oxygen atoms adsorbed to a Cu (111) surface have been found to be highly reactive and to promote dissociative adsorption of NH3. However, oxygen atoms that are part of an oxidic overlayer are found to be non–reactive. This may be due to the extra stabilisation of oxygen atoms in reconstructed overlayers. It is illustrated by an interesting reactivity study of Bowker [16] of the reactivity of adsorbed methoxy species with an overlayer of adsorbed oxygen on a reconstructed Cu (110) surface. Using STM it was found that only atoms on edges into the direction of an oxygen overlayer would be activated by interaction with the methoxy CH bonds (Figure 4).

The presence of lateral interactions between atoms adsorbed in the surface layer can also have an important consequence for the selectivity of a reaction. For instance, the selectivity of the epoxidation reaction of ethylene by oxygen appears to be strongly dependent on the composition of the surface oxygen layer present on the silver catalyst [7]

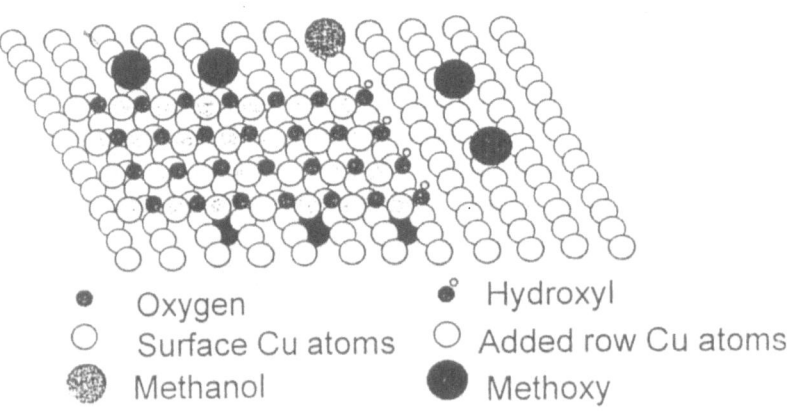

● Oxygen		◉ Hydroxyl	
○ Surface Cu atoms		○ Added row Cu atoms	
⬤ Methanol		● Methoxy	

Figure 4. Reactivity of methoxy species on Cu (119) (reference 16)

At high oxygen surface coverage selective conversion of ethylene to epoxide occurs. The surface layer then has a stoichiometric composition of AgO, with part of the oxygen atoms located in a subsurface layer and other oxygen atoms adsorbed in the external surface layer. The nature of adsorbed oxygen atoms has become electrophyllic and favourable for insertion into the π bond of ethylene.

The microscopic picture of catalysis that evolves is a dynamic one. The timescale of a catalytic event is typically a millisecond or larger. However, surface atoms will locally rearrange on a picosecond timescale, the same timescale at which adsorbate molecules will equilibrate to the surface. On a microsecond timescale, surface diffusion and surface reaction events can take place as illustrated in Figure 5.

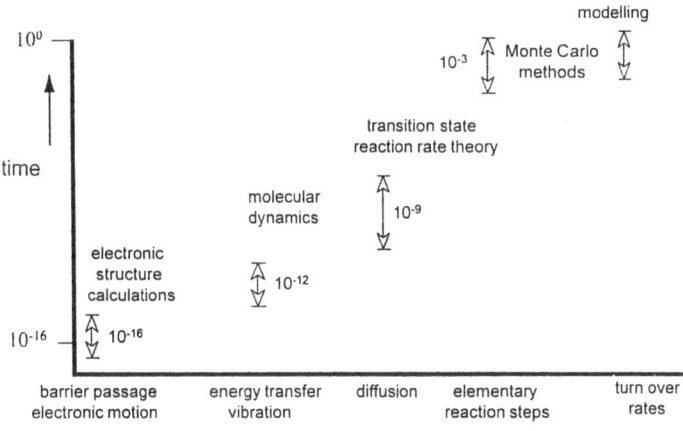

Figure 5. Time scales in catalysis

This leads to a heirachy of events. Parameters deduced from short time scale processes can be used in the equations corresponding to the slower processes. This provides a means to bridge the molecular and macroscopic sciences gap (see Figure 6). The kinetic equations contain rate parameters (equilibrium constants, activation energies, pre–exponents) that can, in principle, be deduced from molecular theory [5]. In the next paragraph we will describe essential features of the four main elementary steps that contribute to the catalytic reaction cycle.

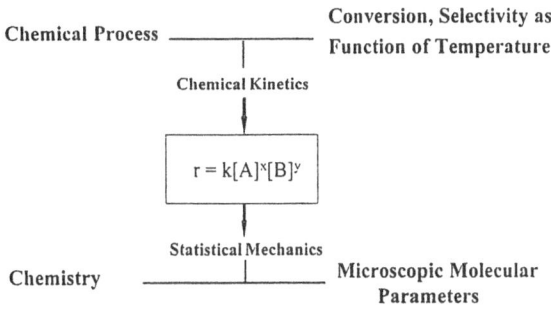

Figure 6. Heirachy of events

2. Chemisorption

Whereas the bond strength of atoms such as O or C to a transition metal surface typically are of the order of 500 kJ/mol, the interaction energy between an adsorbed molecule and a transition metal surface is much weaker. It may typically vary between 50 kJ/mol as for NH_3 to Cu and 160 kJ/mol as for CO on cobalt. The much stronger bonding of adatoms can be readily understood. In a molecule, the valence electrons are saturated by molecular bonds, whereas the valence electrons of an atom are freely available. The asymmetric $2p_x$ and $2p_y$ atomic orbitals of C and O (the z axis is chosen perpendicular to the surface) require asymmetric surface orbitals for bonding. On a surface asymmetric orbitals can be d–orbitals centred on the atoms or asymmetric linear combinations of s–atomic orbitals (surface group orbitals [4]) between atoms.

As is also illustrated in Figure 7, symmetric p_z orbitals can only overlap with symmetric surface group orbitals. In contrast to the $2p_x$ and $2p_y$ atomic orbitals, the p_z orbital has also a finite overlap with an s–atomic orbital when adsorbed on top of an atom.

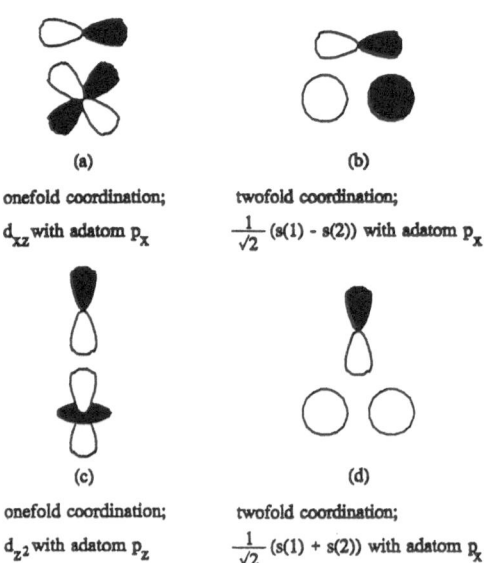

(a)

onefold coordination;

d_{xz} with adatom p_x

(b)

twofold coordination;

$\frac{1}{\sqrt{2}}(s(1) - s(2))$ with adatom p_x

(c)

onefold coordination;

d_{z^2} with adatom p_z

(d)

twofold coordination;

$\frac{1}{\sqrt{2}}(s(1) + s(2))$ with adatom p_z

Figure 7. Symmetry combinations of surface orbitals

On most of the transition metal surfaces the overlap between adatom valance electrons and metal s–valence electrons (occupied by approximately one electron per atom, the contribution of p–atomic orbitals is usually small) dominates the bond energy. This favours high coordination (twofold or higher) for adatoms, otherwise the asymmetric atomic p_x and p_y orbitals could not contribute to the chemical surface bond.

The interaction with the metal d–valence electrons (see Figure 8) gives a small but significant contribution to the interaction energy. It determines largely the differences in bonding to different metals because of the variation in the total number of d–valence electrons and d–atomic orbital spatial extension. This will be illustrated for the interaction of the CO valence electrons with the d–valence electrons of a transition metal. The HOMO (Highest Occupied Molecular Orbital) of CO is a 5σ orbital. When CO is chemisorbed to a surface, the 5σ orbital is directed towards the surface and can only overlap with symmetric surface orbitals. CO has two LUMOs (Lowest Unoccupied Molecular Orbital), The $2\pi^*$ orbitals that in the adsorbed state are asymmetric with respect to the surface normal. The resulting orbital interaction scheme is sketched in Figure 9. Bonding as well as anti–bonding orbital combinations are formed.

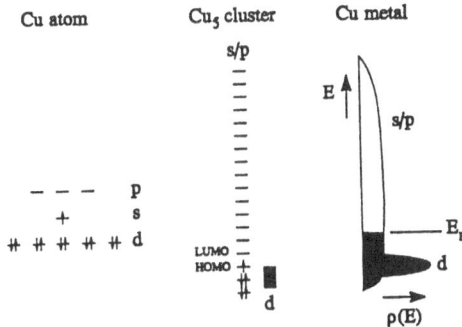

Figure 8. Schematic representation of the s–, p–, and d–orbitals on a copper atom, cluster and metal

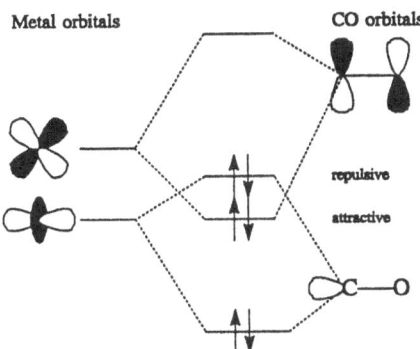

Figure 9. Interaction of CO HOMO (5σ) and LUMO ($2\pi^*$) with copper orbitals $3d_{z^2}$ and $3d_{xz/yz}$

Whereas the metal electrons will mainly occupy bonding orbital fragments between the adsorbate LUMO ($2\pi^*$) and surface orbitals (the energy position of the anti–bonding fragments is too high), in the case of interaction with the adsorbate LUMO orbital, antibonding orbital fragments will become occupied to a significant extent [4].

Whereas electron occupation of bonding orbitals strengthens a bond and because of the larger orbital overlap, will favour bonding to high coordination sites, electron occupation of anti–bonding orbital fragments will weaken the chemical bond. It leads to repulsive interactions that are minimised when bond overlap is least, *i.e.* in low coordination sites. The larger the number of d–valence electrons, the more anti–bonding surface orbital fragments become occupied and the weaker the surface bond becomes. Whereas for adatoms, dominated by the interaction with the adatom 2p orbitals this results in significant changes in the bond energy, these changes may be much less for adsorbed molecules. With an increase in the number of surface d–valence electrons the increase in the occupation of the bonding surface–LUMO ($2\pi^*$) orbitals, which are attractive, will partially off–set the repulsive increase in the occupation of the anti–bonding surface–HOMO (5σ) orbitals.

Also the differences in energy for molecules adsorbed one-fold, two-fold or higher are small and often molecular adsorption to one-fold coordination is found to be preferred.

A similar feature is found when one compares the bond energies of adatoms between different surfaces of the same metal. Because of the decrease in delocalisation of the metal valence electrons when their coordination number decreases, the overlap with adsorbate orbitals increases. This again enhances bonding interactions. However, it destabilises the anti-bonding orbitals. The overall effect is an increase in the interaction energy with adsorbates, but less for molecules than for atoms.

Of importance for our next subject, dissociation, is also the weakening of intramolecular bonds by adsorption. The occupation of valence electrons of orbital fragments that are formed between the metal valence band and unoccupied adsorbed $2\pi^*$ orbitals, implies that in the chemisorbed state the CO $2\pi^*$ orbitals become occupied. Electron population of these anti-bonding orbitals give a weakening of these bonds, experimentally observed by a decrease in the CO bond frequencies.

The balance between the interaction of HOMO and LUMO adsorbate orbitals with surface orbitals may vary largely between adsorbates. For adsorbed ammonia molecules the interaction with the HOMO dominates. In molecules as NO and O_2 the interaction with LUMOs is strong and more enhanced than in CO, because of their lower energies.

The rate of molecular adsorption is usually large. The sticking coefficient, the ratio between the number of collisions that lead to adsorption and the total number of collisions varies between 0.1 and 1. Molecular adsorption is usually unactivated and determined by the rate of energy exchange between adsorbing molecule and the surface.

The rate of desorption is slow compared to adsorption because the heat of adsorption has to be overcome. The activation energy for desorption is often close to the heat of adsorption. The pre–exponent for desorption is often high $\sim 10^{16}$ sec^{-1}, because of the gain in entropy of the desorbing molecule.

3. Dissociation

3.1 THERMODYNAMICS

Figure 10 illustrates schematically the differences in the heats of adsorption of CO on surfaces of Cu, Ni and Co and the energies of the dissociation products. As expected from the previous analysis, the differences in energies between the adatoms are much larger than the differences in the molecular adsorption energies. The energies are found to decrease with d–electron valence occupation.

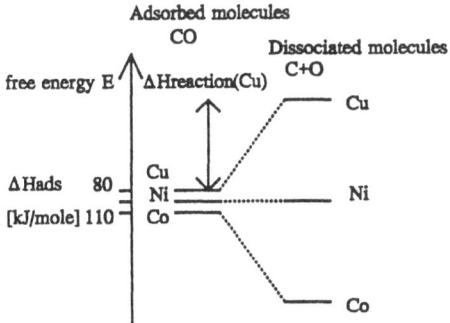

Figure 10: Reaction energy scheme for the dissociation of CO

As a consequence, CO will not dissociate on Cu; dissociation is thermodynamically neutral on Ni and exothermic on Co. For the conversion of synthesis gas (a mixture of CO and H_2), it explains the difference in selectivity between these metals. Cu is an active component of methanol synthesis catalysts. Ni is a methylation catalyst and Co is a Fischer–Tropsch catalyst that produces higher hydrocarbons. In order to produce higher hydrocarbons, C_1 intermediates are needed that are strongly bonded, so that chain growth will compete with methylation [4]. Thermodynamics also relates to large differences in the overall reaction rates, when dissociation is rate–limiting. An example is provided by the data of Somorjai [18] (Figure

11). He studied the reaction of N_2 with H_2 to produce NH_3 and found that surfaces with reactive surface atoms (low surface coordination) are most active. On such a surface the rate of dissociation is highest. The relation between the thermodynamics of a reaction and the rate of a reaction is given by the Brønsted–Polanyi relation. According to this relation, as long as the reaction path remains similar, the change in activation energy of the reaction rate constant is proportional to the change in reaction energy:

$$\Delta E_{act} = {}^1/_2 \Delta E_{reaction} \qquad (3)$$

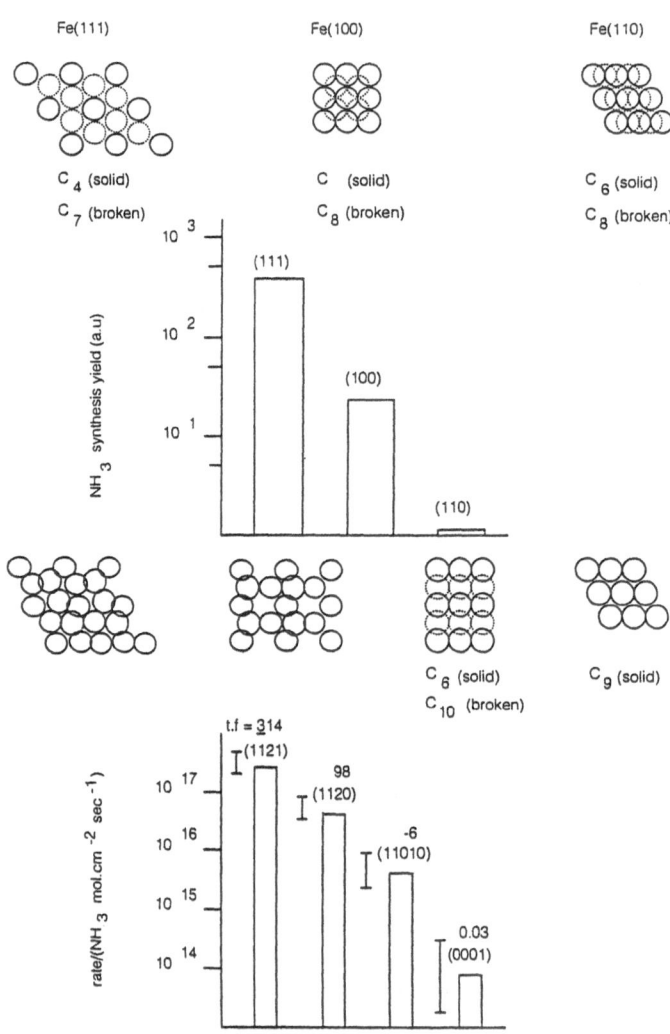

Figure 11. Reactivity of different surfaces of Fe for the formation of NH_3 from N_2 and H_2 [8]

3.2 KINETICS

3.2.1 Transition Metal Catalysis:

For CO, NO and CH_4 calculations at different levels of accuracy have provided [4] a detailed picture of paths for dissociation. The dissociation path is controlled by two factors:

a.   Minimisation of the Transition State Barrier

b.   Stabilisation of dissociation products

Figure 12 [4] illustrates this for the transition state of NO on a cluster representative of the (111) surface of Cu. In the transition state the LUMO of NO has to have optimum overlap with surface orbitals. Dissociation over the top of a surface atom provides maximum overlap with asymmetric d–atomic orbitals. The N and O atoms that are generated upon dissociation adsorb in three fold coordination sites. As a consequence for NO dissociation a surface ensemble of five surface atoms is needed. This is a very general result and is called the ensemble effect. The ensemble effect in catalysis is very general for dissociation reactions. It has been discovered by its extreme sensitivity for site blocking, and has been widely discussed for alloy catalysis.

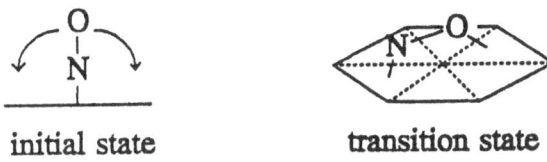

initial state   transition state

Figure 12. Transition state for dissociation of NO

One distinguishes the primary and secondary ensemble effects [19]: the primary ensemble effect is illustrated in Figure 13. It concerns the suppression of heptane cracking when a reactive metal surface such as Pt is alloyed with a non–reactive metal such as Au.

Chemisorption of heptane to a transition metal occurs by C–H bond cleavage and coordination of several of the molecular carbon atoms to the metal surface. A large ensemble of surface atoms is needed to accommodate such an adsorbed molecular fragment. Carbon–carbon bond cleavage to produce smaller hydrocarbon fragments occurs from such a multiple bonded state. Alloying with inert atoms (*e.g.* Au) but also coadsorption of S, will decrease the surface atom ensemble size and hence decrease the probability of multi atom interaction of the molecule.

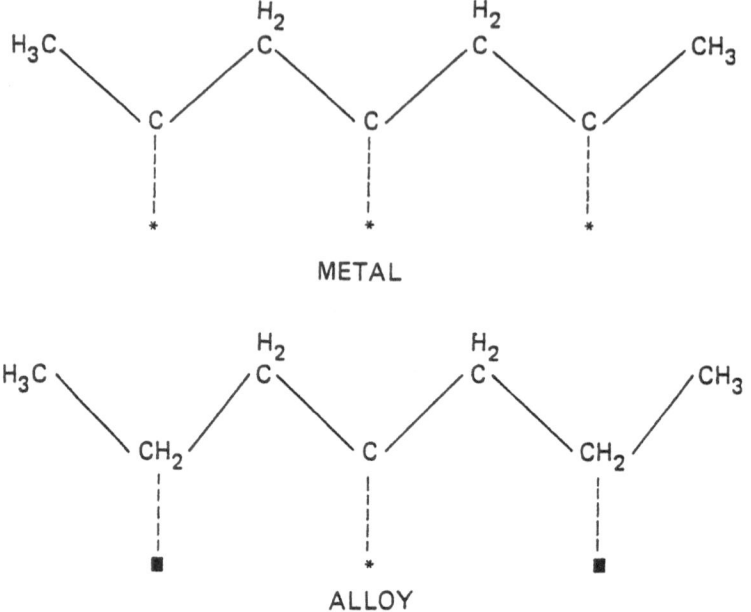

Figure 13. Primary ensemble effect

The secondary ensemble effect considers the number of surface atoms that participate in a bond with an adsorbate atom. It is illustrated in Figure 14. On a pure Pd surface, CO prefers two–fold coordination to surface Pd atoms. The bond energy of CO with the Pd surface is weakened when Pd is alloyed with Au. Alloying of Au with Pd decreases the ensemble size of the Pd surface sites. It forces the CO molecule to bind only to single Pd atoms. As a consequence, the bond energy of adsorbed CO weakens.

Figure 14. Secondary ensemble effect. Change in coordination of CO upon alloying of Pd with Au (schematic)

The rate constant for dissociation of adsorbed molecules sometimes has an activation energy that is close to or higher than that of desorption. This is the reason that sometimes molecules cannot be activated when in their chemisorbed state. The rate of desorption then is fast compared to that of dissociation. In contrast to the pre–exponent for the rate of desorption, the–exponent for the rate of surface dissociation is usually low $\sim 10^{13}$ sec^{-1}, because of the tight nature of the transition state [4].

The overall rate for dissociation is extremely coverage dependent because of the need for a surface ensemble. A typical form for the overall rate equation of dissociation, r_{diss} is:

$$r_{diss} = k_{diss}\theta(1-\theta)^x , \qquad (4a)$$

where k_{diss} is the rate constant for dissociation of the adsorbed molecule, θ the surface coverage, x depends on the ensemble size requirement. Substituting the Langmuir expression for the coverage θ gives:

$$r_{diss} = k_{diss} \frac{K_{eq} \cdot p}{(1+K_{eq}\cdot\beta)^{x+1}} \qquad (4b)$$

where K_{eq} is the molecular adsorption equilibrium constant, p is its pressure. One notes that as predicted by Sabatier, expression (4a) has a maximum for a particular coverage (e.g. $x = 1$, $\theta_{m\,ax} = 1/2$).

From expression (4a) it follows that at low coverage the apparent activation energy of the reaction is:

$$E_{app} = E_{act}(diss) - E_{ads} \qquad (\theta<<1) \qquad (5a)$$

At high coverage, one finds:

$$E_{app} = E_{act}(diss) + xE_{ads} \qquad (\theta\approx1) \qquad (5b)$$

To obtain the apparent activation energy at low coverage, one has to subtract the heat of adsorption from the surface activation energy for dissociation; at high coverage it has to be added. It reflects the need to remove adsorbed molecules in order to create surface vacancies for the dissociating molecule. Desorption will cause a decrease in the apparent activation energy with temperature.

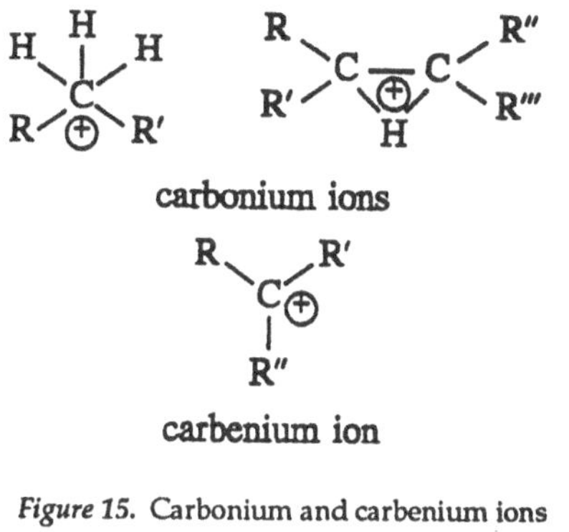

Figure 15. Carbonium and carbenium ions

A consequence of the coverage dependent apparent activation energy is that it will depend on temperature. The apparent activation energy will decrease with increasing temperature. Interestingly, according to (2), an increase in catalytic reactivity can have several reasons. An increase in heat of adsorption enhances the rate at low coverages, or a decrease in the heat of adsorption will increase the overall rate at high coverage. We recognise the principle of Sabatier. Alternatively, the intrinsic rate of the elementary surface reaction step can increase. It is becoming apparent in metal as well as zeolite catalysis that often the adsorption effect dominates.

3.2.2 Transition States in Proton Transfer Reactions

Whereas on a transition metal surface a low activation barrier for reaction is due to the large overlap between adsorbate and surface orbitals, activation of acid catalysed reactions is quite different. Reactions that are activated by proton addition proceed *via* intermediate carbonium ion or carbenium ion formation. For a carbonium ion a proton has been added to a saturated hydrocarbon. A carbenium ion is planar three coordinated positively charged ion (Figure 15), that can be generated by proton addition to an olefin. Such ions have been observed in mass spectrometers. Computational chemistry has recently significantly elucidated the nature of these intermediates in solid acid catalysis [20]. We will illustrate this for the protonation reaction of an olefin [21] and cracking and dehydrogenation [22] of ethane by zeolitic protons. Protonation of an olefin initiates many acid catalysed reactions as oligomerisation or isomerisation reactions. The initial transition and final state of the protonation reaction of ethylene are sketched in Figure 16. Initially ethylene adsorbs π bonded to the proton. Proton transfer to the molecule implies separation of charge. Due to the strong zeolitic OH bond, an energy barrier has to be overcome for dissociation of the proton. The negative charge around the alumino–ion generated by proton–abstraction, becomes redistributed on the surrounding oxygen atoms. The energy cost for proton transfer becomes minimised when the positive charge that develops on the protonated ethylene can become compensated by the negative charge of the basic oxygen atoms around aluminium. The resulting carbenium–ion like intermediate appears to be a transition state on the potential energy maximum of a reaction path that leads to an adsorbed σ–bonded ethoxy intermediate. Consecutive reactions may occur from the ethoxy intermediates.

The transformation of carbonium ions into carbenium ions is seen in the two reactions of ethane (Figure 17). Reaction of the proton with ethane can result in different reaction paths. In one reaction path the C–C bond breaks, methane desorbs and a surface methoxy species is formed (Figure 17a). In another reaction path ethane is dehydrogenated, a CH bond breaks and a surface ethoxy species is formed (Figure 17b).

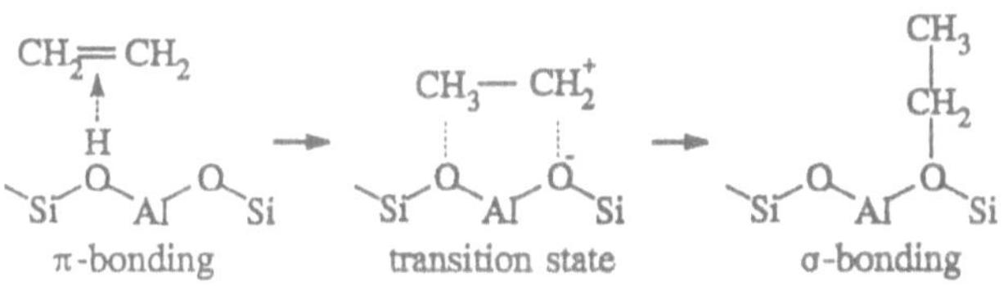

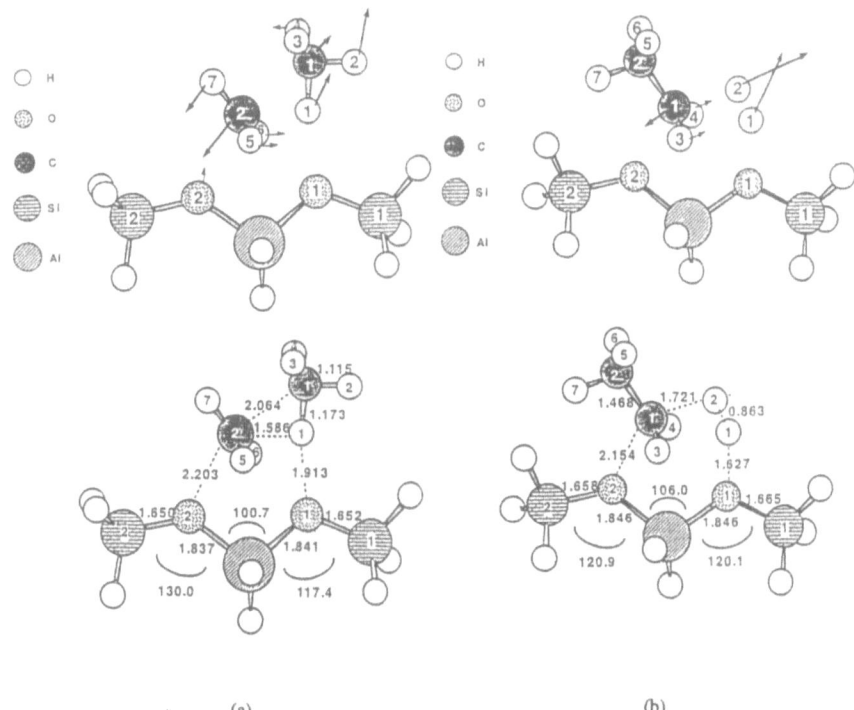

Figure 16. Protonation of ethylene; π– and s–bonded ethylene [21]

Again the protonated intermediate species are transition states. In the transition state of the ethane cracking reaction, a carbenium type CH_3^+ is formed, that adsorbs to zeolitic oxygen to form the adsorbed methoxy species. The carbonium ion transition state for ethane dehydrogenation (barrier high $\approx$ 290 kJ/mole) is very different from that of cracking. It resembles a carbenium ion with weakly bonded H_2. The carbenium intermediate adsorbs to a basic oxygen atom in the product state to form ethoxylate.

Figure 17. Transition states for ethane cracking (a) and ethane dehydrogenation (b) [22]

The Zwitter–ionic reaction paths are seen to involve several oxygen atoms around an Al site. Electrostatic stabilisation is kept optimum by the closeness of positive charge in the transition state molecule and negative charge of the lattice. The protonation reaction can be considered a Lewis base–Brønsted acid catalytic reaction.

REFERENCES

1. Somorjai, G.A. (1994) *Introduction to Surface Chemistry and Catalysis*, John Wiley and Sons, New York.

2. Niemantsverdriet, J.W. (1993) *Spectroscopy in Catalysis, an Introduction*, VCH, Weinheim; Thomas, J.M. and Lambert, R.M. (eds.) (1980) *Characterisation of Catalysts*, Wiley, Chichester.

3. Kaldor, A., Cox, D.M. and Zakin, M.R. (1988) Molecular surface chemistry: reactions of gas-phase metal clusters, in I. Prigogine and S.A. Rice (eds.), *Advances in Chemical Physics, Volume LXX, Evolution of size effects in chemical dynamics, Part 2*, John Wiley & Sons, New York, pp. 211-261.

4. Van Santen, R.A. and Neurock, M., (1995), Concepts in theoretical heterogeneous catalytic reactivity, *Catal. Rev. – Sci. Eng.*, **37**, 557.

5. Van Santen, R.A. and Niemantsverdriet, H. (1995), *Chemical Kinetics and Catalysis*, Plenum Press, New York.

6. Cox, M.P., Ertl, G., Imbihl, R. (1985) Spatial self-organisation of surface structure during an oscillating catalytic reaction, *Phys. Rev. Lett.*, **54**, 1725.

7. Nieuwenhuys, B.E., Siera, J., Tanaka, K.-I, Hirano and H. (1994) Differences in behaviour of Pt, Rh and Pt-Rh alloy surfaces toward NO reduction, in J.N. Armor (ed.) *ACS Symposium Series 552: Environmental Catalysis*, American Chemical Society, Washington, p. 114-140.

8. Langmuir, I. (1916) Constitution and fundamental properties of solids and liquids: I.–Solids, *J.Am. Chem. Soc.*, **38**, 2221.

9. Besenbacher, F., Springer, P.T., Ruan, L., Olesen, L., Stensgaard, I. and Loegsgaard, E. (1994) Direct observations of changes in surface structures by scanning tunnelling microscopy, *Topics in Catal.*, **1**, 325.

10. Shustorovich, E. (1990) The bond–order conservation approach to chemisorption and heterogeneous catalysis: applications and implications, *Adv. Catal.*, **37**, 101.

11. Engel, T. and Ertl, G. (1970) Elementary steps in the catalytic oxidation of carbon monoxide on platinum metals, *Adv. Catal.*, **28**, 1.

12. Somorjai, G.A. (1992) Correlations and differences between homogeneous and heterogeneous catalysis – a surface review, in J.M. Thomas and K.I. Zamaraev (eds.) *Perspectives in Catalysis*, Blackwell Scientific Publications, London p. 147–167.

13. Van Santen, R.A., Zonnevylle, M.C. and Jansen, A.P.J. (1992) The quantum–chemical basis of the catalytic reactivity of transition metals, *Phil. Trans. R. Soc. Lond. A* **341**, 269.

14. Norskov, J. (1993) Chemisorption on metal surfaces, *Rep. Prog. Phys.*, **53**, 1253.

15. Roberts, M.W. (1994) Chemisorption and reactions at metal surfaces, *Surf. Sci.*, **299/300**, 769.

16. Francis, S.M., Leibsle, F.M., Haq, S., Xaing, N. and Bowker, M. (1994) Methanol oxidation on Cu(110), *Surf. Sci.*, **315**, 284.

17. Van Santen, R.A. and Kuipers, H.P.C.E. (9187) The mechanism of ethylene epoxidation, *Adv. Catal.*, **35**, 265.

18. Somorjai, G.A. (1986) Surface science and catalysis, *Phil. Trans. R. Soc. Lond. A*, **318**, 81.

19. Sachtler, W.M.H. and Van Santen, R.A. (1977) Surface composition and selectivity of alloy catalysts, *Adv. Catal.*, **26**, 69.

20. Van Santen, R.A. and Kramer, G.J. (1995) Reactivity theory of zeolitic Brønsted acidic sites, *Chem. Rev.*, **95**, 637.

21. Kazansky, V.B. and Senchenya, I.N. (1989) Quantum chemical study of the electronic structure and geometry of surface alkoxy groups as probable active intermediates of heterogeneous acidic catalysts: what are the adsorbed carbenum ions?, *J. Catal.* **119**, 108.

22. Blaszkowsky, S.R., Nascimento, M.A.C. and Van Santen, R.A. (to appear) Activation of C–H and C–C bonds by an acidic zeolite: a density functional study, *J. Phys. Chem.*

THE CHANGING FACE OF MODERN CATALYSIS

JOHN M. THOMAS and THOMAS MASCHMEYER*

Davy Faraday Research Laboratory
The Royal Institution of Great Britain,
21, Albemarle Street, London, W1X 4BS, U.K.

After enumerating the many practical challenges facing the materials chemist interested in catalysis, and outlining ways in which those challenges could be addressed, this review then focuses on the design of new catalysts. It is shown that considerable dexterity already exists in the design of uniform heterogeneous catalysts. The prospects for the more distant future are also critically analyzed.

1.   Introduction

Two main factors are responsible for the central role of catalysis in pure and applied science. First, there is the fascination of the phenomenon itself and the ever-present desire in the mind of the academic to understand its mode of operation. Second, applied catalysis has been a reality almost from time immemorial; the ancients unconsciously harnessed biological catalysts (enzymes) in processes of fermentation; Arab alchemists used mineral acid for the catalytic dehydration of alcohol; and even before Berzelius in 1835 coined the word catalysis, Döbereiner had already commercialized his discovery of the catalyzed synthesis of water from hydrogen and air by platinum (to produce a flame) and the production of his famous lighter, the so-called "feuerzeug" or tinder box.

It would be difficult to overemphasize the importance of catalysts as twentieth century materials. So comprehensively are they woven into the fabric of our lives that it is hard to picture civilized life without them. Fertilizers, fuels, foodstuffs, pharmaceuticals and fabrics all require catalysts for their manufacture. Moreover, the growth of so-called clean technology, the need to utlilize "sustainable" materials, and the environmental demands for more efficient chemical processes that not only greatly diminish solid and other wastes – eliminating them completely, if possible – but also permit new ways of generating energy without generating atmospheric pollutants, have all added greatly to the changing face of modern catalysis (See Table 1).

---

* Presenting Author.

C.R.A. Catlow and A. Cheetham (eds.), New Trends in Materials Chemistry, 363–376.
© 1997 *Kluwer Academic Publishers.*

TABLE 1. A selection of environmental challenges

- Development of 'zero waste' processes
- Minimization of hazardous products and 'greenhouse' gases
- Replacement of corrosive liquid acid by benign solid acid catalysts
- Evolution of sustainable systems
- Development of processes requiring less 'consumption' of catalysts
- Reduction or elimination of voluminous by-products

(From *"Principles and Practice of Heterogeneous Catalysis"* by Thomas, J. M. and Thomas, W. J., (1996), VCH, Weinheim)

2. Responding to the Challenges

Table 2 contains a selection of the technological challenges to which a deeper knowledge of the materials chemistry of catalysis can contribute greatly. It is to be noted that although inorganic materials loom large on the catalytic horizon, biological (including genetically engineered) materials also have a part to play.

TABLE 2. A selection of topics in which new catalysts have a vital role to play [a]

- Cheaper and safer methods of generating important "chemicals" such as H_2 and H_2O_2
- Better electrocatalysts for the fuel cell consumption of plentiful hydrocarbons such as CH_4
- New catalytic membranes
- Fischer-Tropsch catalysts for sharply defined reaction products (alkanes, alkenes, and alkanols)
- Better methods of isomerizing linear alkanes into branched-chain ones
- Development of processes using CO_2 as a reactant
- Reformulated transport fuels (containing lower amounts of aromatics and volatile components and larger amounts of completely combustible additives)
- Development of catalysts for automobiles operating on methanol dissociation
- Efficient routes to cheaper feedstocks for the chemical and pharmaceutical industries
- Single-step syntheses of important products (e.g. phenol from benzene, formaldehyde or acetic acid from methanol)
- Functionalization of alkanes (especially of methane and cyclohexane)
- Hydrodesulphurization; hydrodenitrefication of light oils, heavy oils, tars and coal
- New shape-selective catalysts (to circumvent thermodynamic restrictions in product distribution)
- Families of solid (acid) catalysts for tuneable conversion of methanol to either ethene or propene
- Uniform, molecular sieve (of redox or Brønsted acid type) possessing well-defined larger pores (30 to 100 Å diameter)
- Development of robust, re-usable chiral catalysts
- New selective oxidation (especially epoxidation) catalysts operating at low temperatures
- Development of modified enzymes, organisms or transgenic plants for 'natural' production of polymers [b]
- 'Targetable' antibody catalysts as therapeutic products [c]

[a] Based on Table 1.5 of Ref 1.
[b] Zeneca PLC produced 100 tons of the natural polymer Biopol in 1993. (Biopol is a bacterial storage polymer, produced by fermentation of an alkaligenes strain. It consists of a copolymer with polyhydroxyvalerate. It is both biocompatible and biodegradable.
[c] Zeneca PLC has recently illustrated the potential power of biotechnology in general and biocatalysis in particular in a medical context: it developed a toxin-antibody conjugate,known as DO490, aimed at coloectal cancer. The idea in using antibody conjugates in this way is to link a 'warhead' to a sophisticated guidance system in a modern version of the so-called 'magic bullet'. Zeneca chose ricin as the warhead, a natural plant toxin with two protein strands. Ricin depends on its carrier B-chain to convey the toxic A-chain to its site of action. Recombinant DNA technology enables the A-chain of the toxin to be produced independently of fermentation. The toxin acts in a very effective catalytic fashion and kills off the tumour.

Apart from the examples quoted in the footnotes cited in Table 2, there have been some effective responses proposed for several of the challenges outlined in the Table itself. Thus recent Japanese work – see Yamanaka *et al.* [2] – has shown that methane may be catalyzed to methanol with O_2 at room temperature using a catalytic system that consists of $EuCl_3$ - Zn - CF_3CO_2H, a catalyst that is also reported [3] selectively to oxidise higher alkanes (excluding cyclohexane) and to convert alkenes (but not propene) to their corresponding epoxides. Fe^{3+} cations, substituted into framework sites of FeZSM-5 (a shape-selective catalyst) also functions as a very good catalyst for converting CH_4 to methanol using N_2O as oxidant.[4] Many examples of "molecularly designed" catalysts useful for meeting some of the challenges enumerated in Table 2 – for further details see the contributions of Cusumano, Zamaraev, Likholobov, Basset, Moiseev and Vangaftik, Shilov and Ono [5] – are now known.

Knowledge of the magnitude of the electronic band gap and precise position of the top of the valence band and the bottom of the conduction band edges, together with estimates of the bending that inevitably occurs owing to space-charges at the subsurface of the semiconductor, permits one to assemble an appropriate photoelectrolytic cell of the type schematized in Figure 1. The desired oxidation ensues as a result of the electronic hole created by photo-absorption at the semiconductor-electrolyte interface and reduction by the mobile electron at the counter electrolyte-electrolyte interface. Microcapsules and other appropriate miniature devices housing a tailored photocatalyst like the one shown sandwiched between colloidal Pt and RuO_2 microelectrodes in Figure 2 can be so structured, as outlined by the Swiss worker M. Gratzel, for the continuous photocleavage of water, or indeed for the destruction of a wide variety of undesirable pollutants in fresh waters.

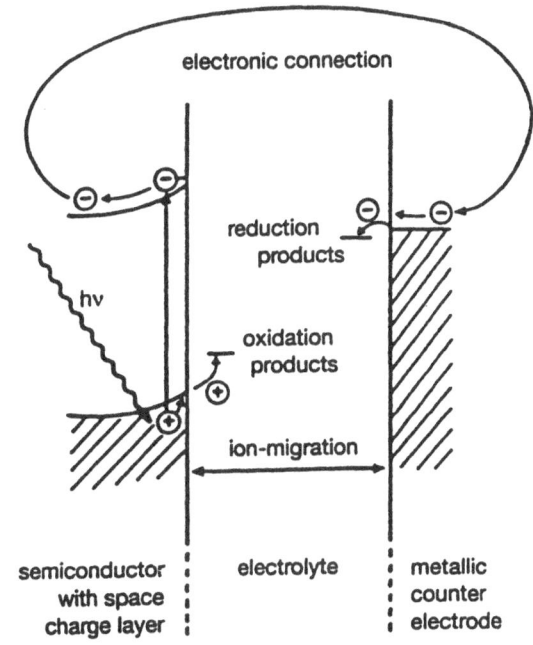

Figure 1. In photoelectrolysis light is used to create electrons and holes which then serve to effect reduction and oxidation respectively.

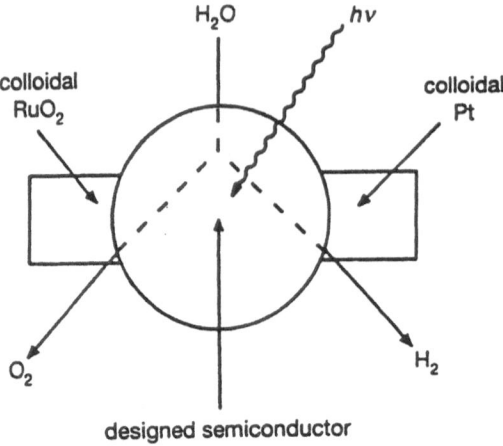

Figure 2 It should be feasible to design a semiconductor catalyst microcapsule for the photocleavage of water and other abundant materials [5].

Heterogeneous photocatalysis is used not only for degradation of pollutants and other unwanted organic materials, it can be harnessed for synthetic purposes. Bard *et al.* [6] have, for example, photocatalytically synthesized methane from acetic acid. (Whilst this demonstrates skilled technical virtuosity, in a practical sense the need, at present, as mentioned in Table 2, is to go the other way i.e. to convert, by appropriate catalytic means, methane (or methanol) to acetic acid.) Of late, German workers have utilized ZnS and CdS semiconductor catalysts in the photoaddition of cyclic enol ethers to 1,2-di-azenes to afford hitherto unknown hydrazine derivatives [7]. Great scope exists in "band-gap engineering" of solid inorganic catalysts through the agency of the quantum size effect. It is a quantum mechanical fact [8,9,10,11] that the width of the band gap for a given semiconductor widens as the particle size is decreased. Micro and ultramicrocrystalline specimens of semiconductors may therefore be designed – by the recently developed techniques of nanotechnology – so as to yield the required band gap for solar-driven (or any other suitable light-induced) chemical process be it synthesis (of an organic material) or degradation (either of an undesirable organic contaminant or of water to generate H_2).

Elegant work by Parmon and his colleagues [12] in Russia has led to an effective means of harnessing solar energy so as to photosplit naturally occurring impurity H_2S in the Black Sea. By ingeniously fashioning a heterojunction (see Figure 3) between minute particles of the sulphide of Cu and the sulphide of solid solutions of Zn and Cd, light from the entire solar spectrum may be fruitfully harnessed for the separation of H_2S into H_2 and water.

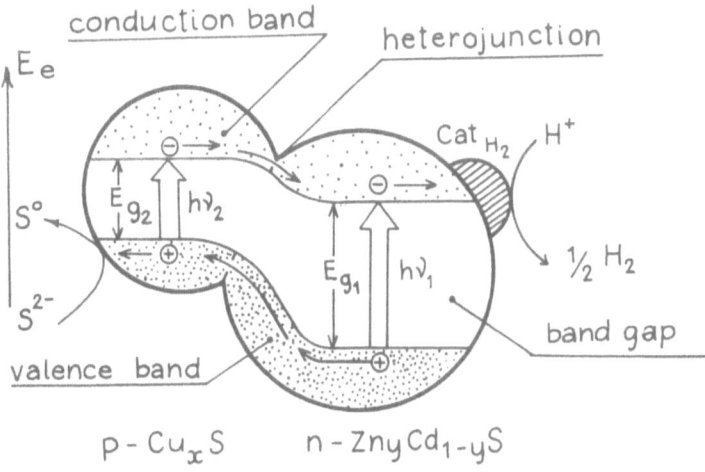

Figure 3. Energy diagram and scheme (after Parmon *et al.* [12,13,14]) for the photoseparation of charges and catalytic processes in a suspended particle of a semiconductor with microheterojunction $CuS_x/Zn_yCd_{1-y}S$ which is one of the most efficient photocatalysts for H_2S cleavage in aqueous solutions. E_e is the electrochemical potential of the electron, E_{g1} and E_{g2} are the widths of the band gaps of the respective n- and p-type semiconductors.

3. How Well Can One Design Inorganic Catalysts?

Inorganic catalysts fall into two large categories: those that are (a) multicomponent and multiphase, and (b) monophasic and spatially uniform.

Typical of those in the first category are the Haber (ammonia synthesis) catalyst, in which Fe and traces of other solid phases are present, Fischer-Tropsch catalysts (consisting of Co or Ru [or alloys] supported on a high-area non-metallic oxide), the Cu on ZnO (ICI) catalyst for the synthesis of methanol from CO + H_2 (syngas) mixture, and, of growing importance, the auto-exhaust catalyst which contains minute particles of Pt, Rh and Pd on a variety of oxides, the most crucial being non-stoichiometric CeO_x ($x \leq 2.0$).

The most effective way of illustrating catalysts in the second category, i.e. those that are spatially uniform, is to describe a specific example, the so-called pentasil molecular sieve known as ZSM-5. This catalyst, which in its acid form has an idealized stoichiometry $H_n Si_m O_{2m} Al_n O_{2n}$ (m/n = 20 to 20,000 with m > 60) as well as compositional (but still monophasic) variants of it, is used commercially to effect a number of commercially important conversions – alkylation of aromatics, isomerization of xylenes, synthesis of petrol from methanol, conversion of toluene to benzene and xylene. It is microcrystalline in the sense that the individual particles are in the size range of 0.1 to 50 μm. A typical high-resolution electron micrograph is shown in Figure 4. Each of the white circular patches denotes a 5.5Å diameter aperture running in a direction perpendicular to the plane of the micrograph. Depending upon the precise size of the particles, some 95 percent or more of the accessible surface area, usually 400 to 500 m^2g^{-1} of the catalyst is internal. In effect such a catalyst possesses a three-dimensional (largely internal) surface; and there is a spatially uniform distribution of active sites inside the microcrystals and these sites are accessible to all the molecules capable of diffusing through the extensive three-dimensional network of micropores.

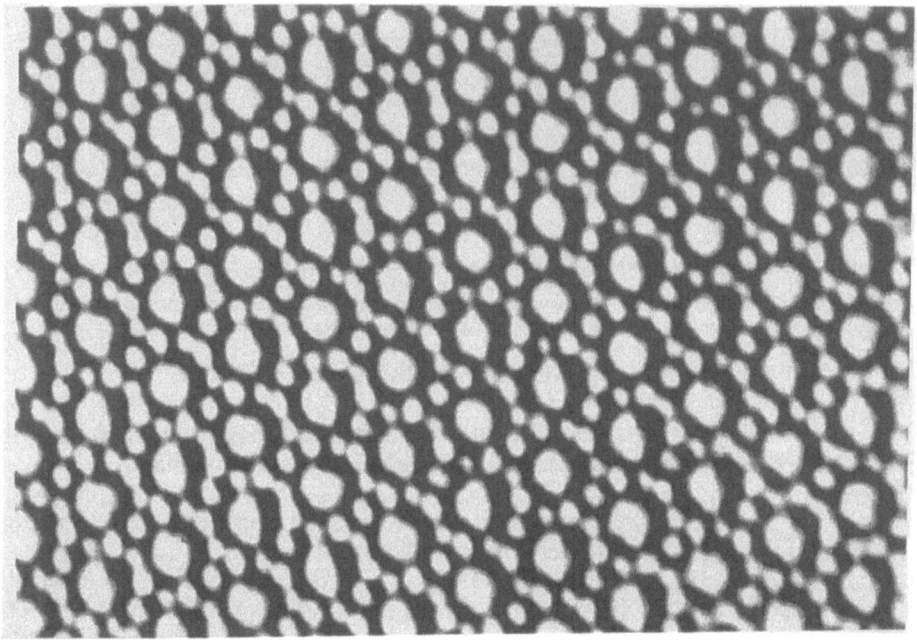

Figure 4. High-resolution electron micrograph of a zeolite catalyst showing projected structure of pores line with active sites. Large pores are 5.5A in diameter (from Thomas, J. M. and Gai-Boyes, P. L., (1993), *Nature*, **364**, 478.

Although it is often possible to assemble effective and selective multicomponent catalysts through accumulated knowledge of general catalytic principles – and the auto exhaust-catalyst is a prime illustration of that fact – in general it is not at all straight forward to design such catalysts with the degree of dexterity that, say, the organic chemist or enzymologist can invest into his or her creations. With uniform, monophasic catalysts, however, the situation is much more promising. Indeed, considerable progress has already been made in this direction; and in the remainder of this article we focus on several illustrative examples and also on new strategic principles.

By introducing certain key elemental substituents into the ZSM-5 framework, many other highly selective catalytic reactions of profound petrochemical significance may be effected [15]. Thus (see Figure 5) with Ga in the ZSM-5, propane and butane may be effectively dehydrogenated to yield benzene and toluene, this being the basis of the British Petroleum cyclar process. With Zn in the ZSM-5 framework, dehydrogenation of isobutane to isobutene (also known as 2-methyl propene) is effected. This alkene (first prepared in this Laboratory by Michael Faraday) is of central importance industrially as it is the precursor of MTBE (methyl tertiary butyl ether), a major constituent these days of motor car fuel since it replaces the deleterious lead-containing additives (MTBE also facilitates the complete combustion of the fuel to CO_2 rather than CO. As mentioned earlier, with Fe in the framework, the ZSM-5 functions as a catalyst for CH_4 oxidation to CH_3OH using N_2O as oxidant.

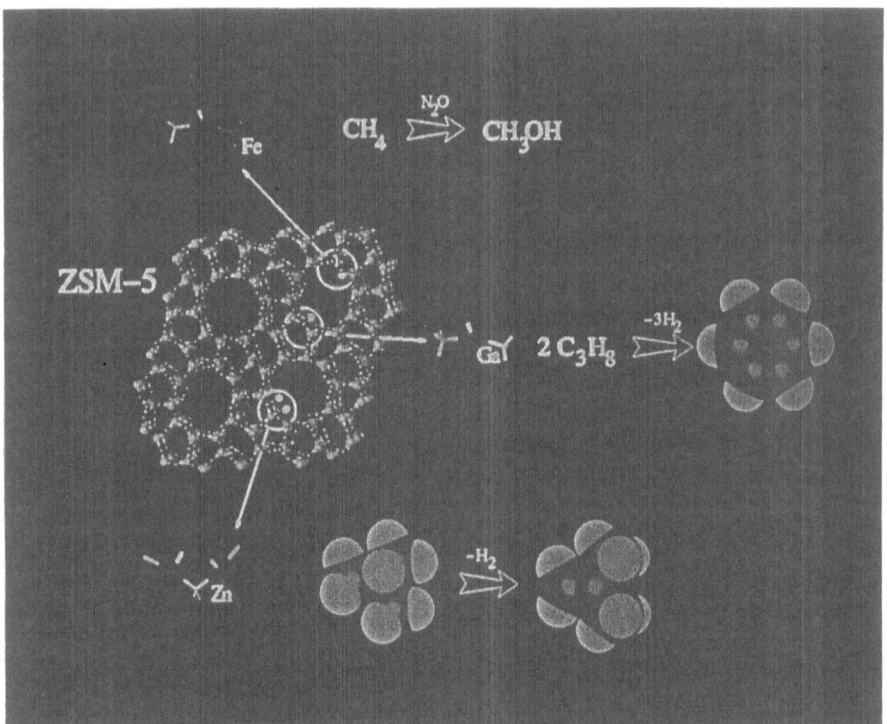

Figure 5. ZSM-5 and its use as a modified catalyst for methane oxidation to methanol, propane to benzene and isobutane to 2-methyl propene.

All these new catalytic variants of ZSM-5 as well as the titanosilicalite (known as TS-1) prepared by Enichem in Italy, in which the silicious extreme of ZSM-5, i.e. silicalite, contains occasional tetrahedrally co-ordinated Te^{4+} ions in place of Si^{4+}

in the framework), are made by the agency of templating, a process that requires adroit choice of an organic molecule to be added to the gel or mother liquid from which the microporous solid crystallizes [15,16].

4.1 De Novo Synthesis of new, shape-selective inorganic catalysts

As well as being a critical determinant of the catalytic performance of enzymes and abzymes (catalytic antibodies), shape-selectivity is also a key feature of molecular-sieve inorganic catalysts. They are examples of uniform heterogeneous catalysts, in the sense previously outlined, and they represent one of the most rapidly growing categories of inorganic catalyst. The dimensions of the pores of the particular molecular sieve under consideration govern the size and shape of the reactant molecules that may enter it, the product molecules that may leave it, as well as the nature of the reactive intermediates that can form within it. The acidic zeolite H+-ZSM-5 may be viewed as a three-dimensional network of corner-sharing SiO_4 tetrahedra, the spaces between these forming molecular-scale micropores of variable size and dimension dependent upon the precise nature of the crystal architecture (which, in turn, as outlined below, is influenced by the properties of the template molecule that pre-determines its formation). Whenever Si^{4+} ions are replaced by Al^{3+} at the centre of the oxygen tetrahedra protons are required locally for electroneutrality. These are localized (as Brønsted acid sites) at one of the four oxygen atoms of the AlO_4 tetrahedron forming bridging hydrogen groups.

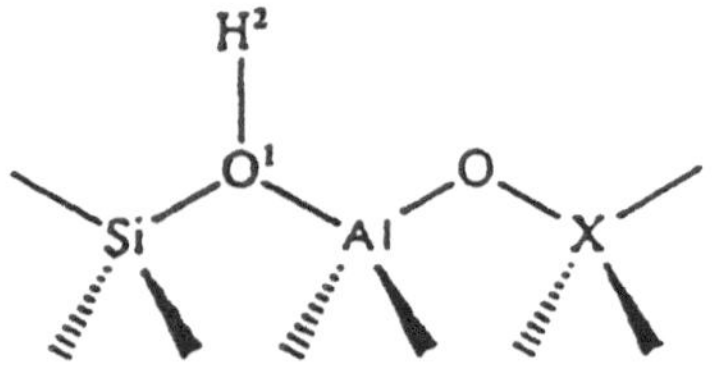

Scheme 1. X = Si, P.

Figure 6. A typical Brønsted acid site.

The Brønsted sites, shown in Figure 6, have been established to be catalytically active in a wide range of conversions of alkanes, alkenes and alkanols.

Figure 7 shows the protonated (acidic) form of mordenite, inside the pores of which are molecules of diisopropylnaphthalene (DIIP) a vitally important building block in the synthetic polymer industry [17]. (DIIPN is formed when propene and naphthalene are allowed to react in the presence of an acid catalyst). Over acidic silica-alumina gel (i.e. non-crystalline) catalysts numerous other, undesirable products are formed as well as DIIPN. Over the environmentally harmful $AlCl_3$, again there is a multiplicity of products. Acidic mordenite, however, is shape-selective in that it yields predominantly 2, 6 DIIPN. In Figure 7 the tightly fitting product (DIIPN) gives a clue as to the mode of action of organic templates in the synthesis of microporous catalysts.

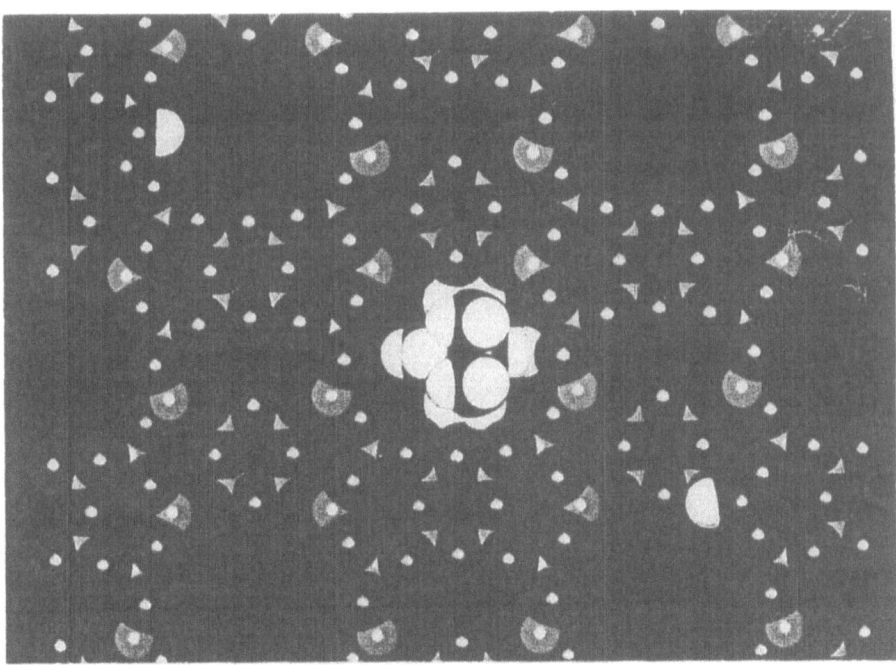

Figure 7. Inside one of the large pores of a protonated mordenite zeolite catalyst (of high Si/Ac ratio), there is a molecule of DIIP (see text). On environmental grounds this acid zeolite is the catalyst of choice in the conversion of napthalene and propene to DIIPN.

The degree to which templates are critical in the synthesis of a particular framework is not fully understood – as has been described by Catlow *et al.* [18]. Certain templates (of which ethylene diamine is an example) may be regarded simply as void-filling species that do not contribute significantly to the preferential formation of a desired structure. We also find that a particular framework is formed by several different template molecules, a process which might be more correctly termed structure-directing rather than genuine templating. In general, however, all the templates suitable for a particular framework will possess similar properties – shape, size, basicity, etc. which direct the gel medium (usually under hydrothermal conditions) from which the crystals nucleate towards the formation of particular structural motifs. A good deal of accumulated chemical experience, supplemented by computer graphics and computational endeavours [18,19] has been acquired to enable an inorganic preparative chemist to design a wide range of zeolitic material. In particular, microporous

372

aluminium phosphates (ALPOs) and metal-substituted (in the framework) ALPOs – otherwise designated MeALPOs or MALPOs – may now be routinely prepared. Some of them have structures similar to those that have been characterized in the aluminosilicate zeolites (natural and synthetic). Some, however, are quite new.

The framework structure of one novel MeALPO that is known as DAF-1 (Davy Faraday-One) [20,21] is schematized in Figure 8. It is a solid acid the precise formula of which can be $H^+_{0.22}$ $Mg_{0.22}$ $Al_{0.78}$ PO_4 or $H^+_{0.22}$ $Co_{0.22}$ $Al_{0.78}$ PO_4. Many MeALPOs (with Me = Co, Ni, Zn, Mo, Cu, Mg and other divalent cations partially replacing the trivalent Al^{3+} of the parent ALPO) are very good solid acid and redox catalysts [15,22].

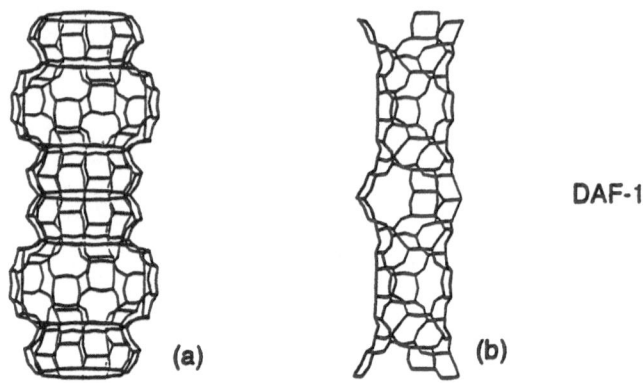

(a)   (b)   DAF-1

Figure 8. Schematic illustration of the novel double-barrelled microporous magnesium aluminophosphate known as DAF-1 [20].

About a hundred distinct microporous structures, with channel apertures falling in the range from 4 to 14Å have been prepared. Into such structures, about a third of the elements of the Periodic Table may be incorporated at framework sites. And into the sites occupied be extra-framework, usually exchangeable, cations a further third of all known elements may be placed. These architectonically exquisite inorganic repositories have a prodigality of chemical compositions that defy precise description – millions of distinct compositions are possible. Many thousands have already been prepared. The scope for delicate variation is enormous.

4.2   Uniform solid acid catalysts and clean technology

Clean technology [23] currently provides much stimulus for a multitude of new developments and initiatives in applied catalysis, not the least important of which is the growing environmental concern about the use of corrosive liquid acids (such as hydrofluoric acid and 96 percent sulphuric acid) as catalysts for the alkylations or isomerizations of hydrocarbons. Selective and active solid acids are therefore attractive as replacements; and the rapidly growing family of microporous (molecular sieve) catalysts with strong proton-donating sites distributed uniformly within the pores and cavities offer a promising range of viable alternatives [24].

Apart from the considerable family of aluminosilicate zeolitic acids (typified by H^+ mordenite, H^+-Y and H^+-ZSM-5), the large family of mircoporous MeALPOs and SAPOs loom larger with the passage of time. (In a SAPO, silicon atoms replace some of the phosphorous atoms – in tetrahedral co-ordination – thereby yielding a $H^+[AlP(Si)O_4]$ solid acid). SAPO-18, for example [25], with a framework composition consisting of typically one Si per twenty tetrahedral sites as well as its MeAPO analogue [22] (MeALPO-18, with Me=Mg, Zn or Co) is a very efficient catalyst in converting methanol to mixtures of ethene and propene. And DAF-1, as well as H^{+-} Theta-1 and, best of all, H^{+-}ferrierite, are all good catalysts [26] for the skeletal isomerization of but-1-ene at lowish temperatures (to produce the desirable 2 methyl propene that functions as a precursor to MTBE, mentioned earlier). The Shell Company is currently commercializing H^+-ferrierite, and Mobil and BP are jointly involved in the use of H^+-Theta-1 for the industrial production of 2-methyl propene from but-1-ene, a substantial by-product from oil refineries (Figure 9).

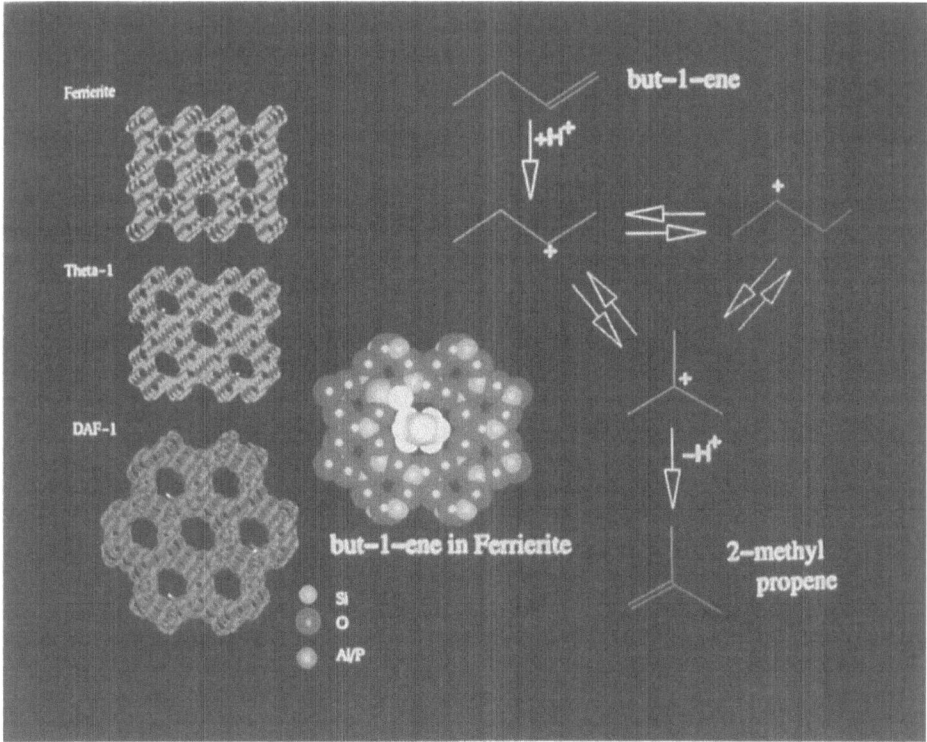

Figure 9. The four isomers of butene. Of these, but-1-ene is plentifully available from catalytic oil crackers, 2-methylpropene is the desirable product, as it may be converted to MTBE (see text) [26].

The production of DIIPN is, as mentioned earlier (see figure 7), also dependent on the agency of a microporous, solid acid catalyst, H+-mordenite.

5.   Whither Uniform Heterogeneous Catalysts?

All the solid-acid, microporous catalysts mentioned above – DAF-1, ZSM-5, mordenite, ferrierite, theta-1, SAPO-18, MeALPO-18 – are examples of underlined uniform heterogeneous catalysts [15,27,28] where the active sites, such as bridging hydroxyls of the type Si-O(H)-Al or Me-O(H)-P are distributed in a spatially uniform and accessible (to reactants) fashion throughout the bulk of the solid. They have other noteworthy attributes: exceptionally large surface areas (often in excess of 600 m^{2}g^{-1}) more than ninety percent of which is inside the solid. In effect, as mentioned earlier, they are solid catalysts with three-dimensional surfaces permeating the entire material and accessible to all reactant (and product) molecules small enough to enter the apertures at their outer surfaces.

This being so, these uniform heterogeneous catalysts are amenable to the entire panoply of spectroscopic diffraction and scattering techniques devised and refined by chemists and physicists in recent years. Such catalysts present extensive opportunities to explore the relationship between structure and catalytic performance, much in the manner that modern enzymologists and biochemists pursue their investigations.

In situ X-ray crystallography is readily adapted to such work, as was originally demonstrated for lipozyme by D. C. Phillips [29]. But when, as is often the case, the concentration of the active sites is low, some serious practical obstacles have to be overcome. X-ray diffraction alone, even when imaginative use [30] is made of anomalous scattering, is not sufficiently sensitive to probe the atomic environment of the active site. But, thanks to the availability of synchrotron radiation, X-ray absorption spectroscopy, embracing pre-edge, near-edge and extended-edge fine structures is a viable technique for elucidating the nature of the active site. (Depending upon the precise nature of the active site, other powerful tools may also be employed, notably solid-state NMR. [31]) Moreover, we have evolved methods of probing catalysts of this kind under operating conditions using combined X-ray diffraction and X-ray absorption spectroscopy (XRD and XAS respectively), recorded with the same sample under realistic conditions of catalytic use. [32-35] (It is even possible, with high flux synchrotrons equipped with appropriate wigglers to record parallel XRD and XAS patterns of the nucleation and growth of templated microporous and mesoporous catalysts [36]). There is every reason to suppose that this dual *in situ* approach to the study of uniform heterogeneous catalysts will provide deeper insights into their mode of operation.

Details have been given elsewhere of several sub-categories and specific examples of uniform heterogeneous catalysis – see refs. [27,28,37] – suffice it to say that zeolites are typical uniform catalysts, as are intercalated clays, heteropolyacids, and certain perovskite oxides [38,39]. But, as is described more fully in the chapter by

Maschmeyer, Thomas and Masters [40], mesoporous siliceous solids with channel apertures from 25 to 100Å have opened up new possibilities in heterogeneous catalysis. The large diameter channels of the so-called MCM-H mesoporous silicas that we and others have used for selective oxidation and oxidative dehydrogenation permit, in principle, the direct grafting of complete metal complexes and organometallic moieties onto the inner walls of these high-surface-area (typically >800 m^2g^{-1}) solids. The chapter [40] also gives examples of how these novel uniform heterogeneous catalysts may be fashioned to order so as to place large concentrations of active sites in accessible, well spaced and a structurally well-defined manner on the inner walls of the mesoporous support.

With MCM-H based catalysts its is possible not only to anchor a reactive (catalytic) centre designed to order according to the principles of organometallic chemistry, but also to tether them as schematized in Figure 10. Here the active site is situated at the extremity of the tether, and is free to flutter, so to speak, in the stream of reactant and product species in which it is immersed during the course of catalytic conversion. By deliberately restricting the spatial freedom in the vicinity of the active centre, it should be feasible to design highly stereoselective (enantiomeric) catalysts. We and our colleagues are currently investigating such possibilities.

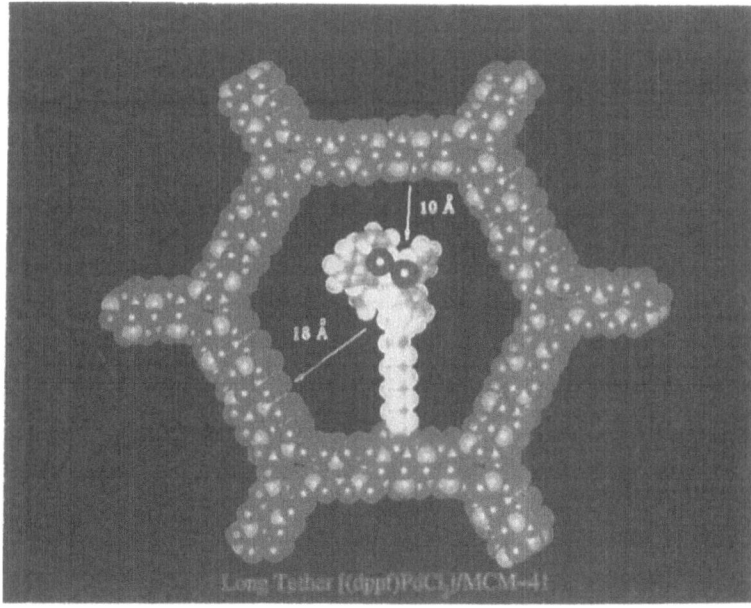

Figure 10. An example of a tethered organometallic catalyst where a chiral centre can be conveniently introduced at the extremity of the tether. (Maschmeyer *et al.* to be submitted).

376

We acknowledge with gratitude continued support from EPSRC. Thomas Maschmeyer thanks the AVCC and the Sir Robert Menzies Foundation for the Australian Bicentennial Research Fellowship.

REFERENCES

1. Thomas, J. M. and Thomas, W. J., (1996),'*Principles and Practice of Heterogenous Catalysis'*, VCH Weinheim, Chapter 1.
2. Yamanaka, I., Soma, M. and Otsuka, K., (1995), *J. Chem. Soc., Chem. Commun.*, 2235.
3. Yamanaka, I., Nakagaki, K., Akimoto, T. and Otsuka, K., (1994), *Chemistry Letters*, 1717.
4. Zamaraev, K. I., *Topics in Catalysis*, in press
5. Thomas, J. M. and Zamaraev, K. I. (Eds), (1992),"*Perspective in Catalysis*", Blackwells and I.U.P.A.C.
6. Kraeutler, B. and Bard, A. J., (1978), *J. Amer. Chem. Soc.*, **100**, 2239.
7. Kunneth, R., Feldmer, C. and Kisch, H., (1992), *Angew. Chem. Int. Edn. Eng.*, **31**, 1039.
8. Brus, L. E., (1984), *J. Chem. Phys.*, **80**, 4403.
9. Nedeljkovic, J. M., Nenadovic, M. T., Micic, O. I. and Nozik, A. J., (1986), *J. Phys.Chem..*, **90**, 12.
10. Weller, H., Koch, V., Gutierrez, M. and Henglein, A., (1984), *Ber. Bunsenges Phys. Chem.*, **88**, 649.
11. Linsebigler, A., Lu, G. and Yates, Jr., J. T., (1995), *Chemical Review*, **95**, 735.
12. Parmon, V. N., Ismagilov, Z. R., Kenzhentsev, M. H., ref 5. p337.
13. Parmon, V. N. and Zamaraev, K. I., (1989), *Photochemical Energy Conversion* (ed Norris, J. R. and Meisel, D.), Elsevier, New York, p316.
14. Parmon, V. N. & Zamaraev, K. I., (1989), *Photocatalysis Fundamentals and Applications*, (Eds. Serpone, N. and Pillozzetti, E.), Wiley, New York, 565.
15. Thomas, J. M., (1994), *Angew. Chem. Int. Edn. Eng.*, **33**, 913.
16. Bell, R. G., Lewis, D. W., Catlow, C. R. A. *et al.*, (1994), *Zeolites and Related Microporous Materials, State of the Art* , Eds. Weitkamp, J., Karge, H. G., Pfeiffer, H. and Hölderich, W., Elsevier, **84**, 2075.
17. Thomas, J. M. and Zamaraev, K. I., (1994), *Angew. Chem. Int. Ed. Eng.*, **106**, 316.
18. Lewis, D. W., Catlow, C. R. A. and Freeman, C. M., (1995), *J. Phys. Chem.*, **99**, 11194.
19. Lewis, D. W., Catlow, C. R. A. and Thomas, J. M., submitted
20. Wright, P. A., Jones, R. H., Natarajan, S., Poole, R. G., Chen, J., Hursthouse, M. B. and Thomas, J. M., (1993), *J. Chem. Soc. Chem. Commun.*, 633.
21. Wright, P. A., Sayag, C., Rey, F., Lewis, D. W., Gale, J. D., Natarajan, S. and Thomas, J. M., (1995), *J. Chem. Soc. Faraday Trans*, **91**, 3537.
22. Chen, J. and Thomas, J. M., (1994), *J. Chem. Soc. Chem. Commun.*, 603.
23. Thomas, J. M., (1992), *Scientific American*, **266**, 118.
24. van Bekkum, H., Flanigen, E. M. and Jansen, J. C.(Eds), (1991),"*Introduction to Zeolite Science and Practice*", Elsevier.
25. Chen, J., Wright, P. A., and Thomas, J. M., (207)
26. Natarajan, S., Wright, P. A., and Thomas, J. M., (1993), *J. Chem. Soc. Chem.Commun.*, 1861.
27. Thomas, J. M., (1988), *Angew.. Chem. Int. Edn. Eng.*, **27**, 1673.
28. Thomas, J. M., Chen, J. and George, A. R., (1992), *Chem. in Britain*, **28**, 991.
29. Phillips, D. C., (1967), *Proc. Natl. Acad. Sci. USA*, **57**, 484.
30. Cheetham, A. K. and Wilkinson, A. P., (1992), *Angew. Chem. Int. Edn. Eng.*, **31**, 1559.
31. Bell, A. T. and Pines, A. (Eds), *NMR Techniques in Catalysis* 1994, Dekker.
32. Couves, J. W., Thomas, J. M., Greaves, G. N., Jones, R. H., Waller, D., Dent, A. J. and Derbyshire, G. E., (1991), *Nature*, **354**, 465.
33. Sankar, G., Wright, P. A., Natarajan, S., Thomas, J. M., Greaves, G. N., Dent, A. J., Dobson, B. R., Ramsdale, C. A., Jones, R. H., (1993), *J. Phys. Chem.*, **97**, 9550.
34. Thomas, J. M., Sankar, G., Wright, P. A., Chen, J., Marchese, L., Greaves, G. N., (1994), *Angew. Chem. Int. Edn. Eng.*, **33**, 639.
35. Thomas, J. M. and Greaves, G. N., (1994), *Science*, **265**, 1675.
36. Sankar, G., Rey, F., Thomas, J. M. and Greaves, G. N., (1995), *J. Chem. Soc. Chem. Commun.*, 2449.
37. Thomas, J. M., (1990), *Phil. Trans. Roy. Soc. Lond.*, A**333**, 173.
38. Pickering, I. J. and Thomas, J. M., (1991), *J. Chem. Soc. Faraday Trans.*, **87**, 3067.
39. Ramesh, S. and Hegde, M. S., *J. Phys. Chem.*, in press
40. Maschmeyer, T., Thomas, J. M. and Masters, A.. F., this volume, Developments in Silica-supported Organometallic Catalysis.

THE APPLICATION OF MATERIALS SCIENCE IN CATALYTIC PROCESSES: RECENT DEVELOPMENTS

I.E. MAXWELL and P.W. LEDNOR

Shell Research and Technology Centre, Amsterdam,
Shell International Chemicals B.V. ,
P.O. Box 38000, 1030 BN Amsterdam, The Netherlands

1. Introduction

Materials science is a prime driver for innovation in the field of catalytic processes. Developments in materials science, particularly in the last decade have yielded a broad range of porous solids which have found application in the industrially important area of catalysis. The areas of application have been traditionally in oil refining and petrochemicals, but more recently there has also been rapid growth in the development of new and often quite novel catalyst systems for environmental processes.

Catalytic chemistry has traditionally been highly empirical but due to advances, in not only the development of new materials, but also the associated growth in understanding of the surface chemistry and structural features of these porous materials has led to a sounder scientific basis in this field of endeavour. By means of examples, from a wide variety of applications it will be shown how the functionality of materials can often be tailored to develop optimal catalyst systems for use in many different process technologies.

The dimensional space of catalytic chemistry is very broad and spans the following; atomic surface chemistry, bulk composition, microporous, macroporous and finally external shaping and structure. While recognizing that the interplay between these various dimensional levels is very important in this article the impact of materials science will be discussed from the perspective of these different dimensions. Clearly, the scope of this subject is very broad and within the limited framework of this review a selection of topics is required. Topics and examples have therefore been chosen which are either relatively recent innovations or of generic industrial relevance where the discipline of materials science has played an important role.

The primary processes involved in catalysis are as follows:

(i) the diffusion of reactant(s) from the external catalyst interface through the porous solid to the internal catalytic surface
(ii) the reaction(s) of the reactant(s) at the catalytic site(s)
(iii) the diffusion of the product(s) through the porous solid to the external catalyst interface

C.R.A. Catlow and A. Cheetham (eds.), New Trends in Materials Chemistry, 377–402.
© *1997 Kluwer Academic Publishers.*

378

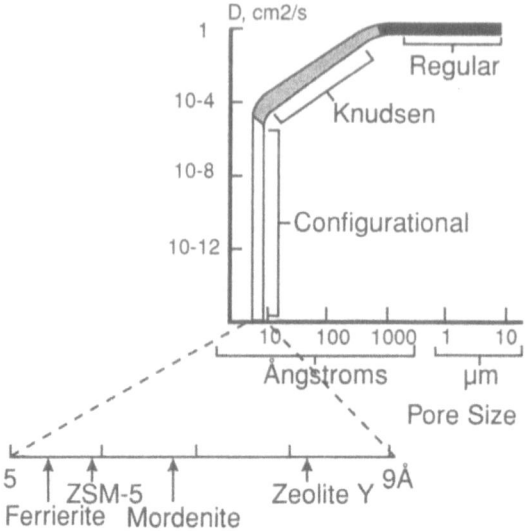

Figure 1. Zeolite shape selectivity.

Thus, the chemistry of the catalyst surface and the detailed geometry of the pore structure are important parameters in determining catalyst performance. Weisz [1] showed that catalyst systems typically span many orders of magnitude in molecular diffusion and average pore diameter space. As shown in Fig. 1 these can be described in terms of regular (bulk), Knudsen (wall effects) and configurational (molecular sized pore diameters) diffusion regimes. Behrens [2] recently demonstrated the range of materials available to span this wide diffusional regime (see Fig. 2).

Catalysis clearly imposes high demands on materials to enable both the surface chemistry and the pore structure to be accurately tailored for the particular reaction involved.

2. Surface Chemistry

Surface chemistry is very important in catalysis since the nature of the active site(s) plays a prime role in determining the intrinsic activity and selectivity for a particular reaction. The fundamental understanding of the relationship between surface chemistry and catalysis has advanced significantly in recent years aided by computational chemistry. For the major heterogeneous catalyst systems which are applied industrially the active surfaces are metals, metal oxides, metal sulphides and acidic hydroxyl groups or combinations thereof.

In view of the limited scope of this review only three aspects of surface chemistry which are of industrial importance will be highlighted which will include acidity, promotors and metal functions.

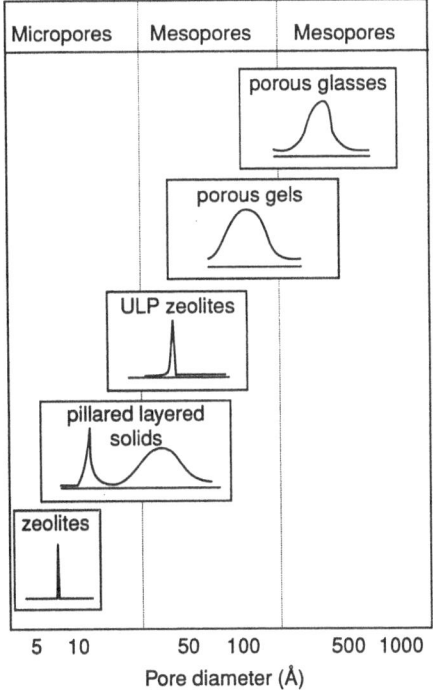

Figure 2. Catalytic porosity [2].

2.1. SURFACE ACIDITY

Many of the most important applications of heterogeneous catalysis, such as in refining and chemicals, involve the use of solid acid catalyst systems as has been demonstrated in a recent review article by Corma [3]. For example, in oil refining solid acid catalysts are applied in a number of key process technologies such as catalytic cracking, hydrocracking, hydrotreating and paraffin isomerization. In each of these applications the specific tailoring of the surface acidity is crucial to obtain optimal performance for that particular catalyst system.

Correlations of catalytic properties with acidity are often very complex with a multiple of factors contributing to the overall measured catalytic performance. However, in a few cases direct correlations have been found between material properties which relate directly to acidity and catalytic reactions. For example, Mobil researchers found [4] that the activity of the hydrogen form of ZSM-5 exhibited a linear relationship with the aluminium content of the zeolite (see Fig. 3) for hexane cracking (so-called alpha test). The aluminium content of the ZSM-5 zeolite is correlated with the number of acid sites hence in this particular case a direct relationship is observed between activity and acidity.

As previously emphasized in most cases where solid acid catalysts are applied the relationships with some measure of acidity are generally much more complex. In addition, for many heterogeneous acid catalyst systems dual functions are applied

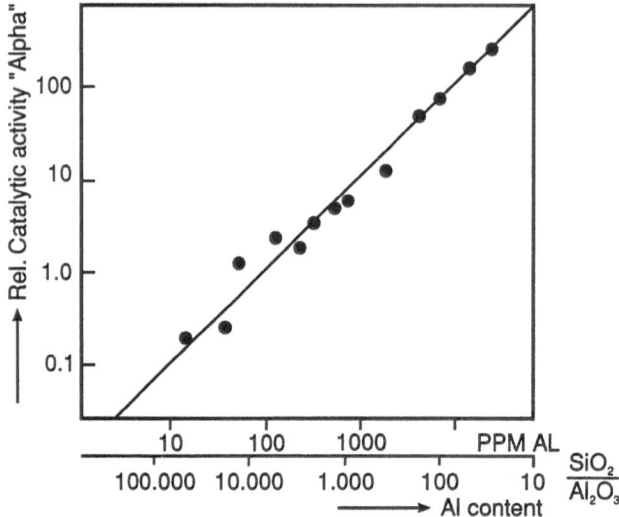

Figure 3. Cracking activity/Al content relationship for ZSM-5 [6].

whereby a metal hydrogenation function is also included. The role of the hydrogenation function can be multiple which includes establishing hydrogenation/dehydrogenation equilibria and enhancing catalyst stability by reducing coke formation. Some examples of both mono-functional and bi-functional solid acid catalysts will be discussed.

2.1.1. Catalytic Cracking

Fluid catalytic cracking (FCC) is a key process in the modern refinery for converting heavy feedstocks into lighter products and primarily gasoline. The world wide scale of this process is such that FCC is the single largest market for refining catalysts. Unlike many other applications of acidic catalysts the FCC process does not employ a bifunctional catalyst under a hydrogen partial pressure. This results in a relatively fast rate of coke build-up on the catalyst which necessitates a regeneration step and a moving bed reactor system as integral parts of the FCC process.

The prime solid acid component of an FCC catalyst is zeolite Y in the hydrogen form. In recent years it has been discovered that the performance of FCC catalysts can be influenced by tailoring the acid density of the zeolite Y component. The acid density of zeolite Y can, to a first approximation, be correlated with the crystallographic unit cell dimension. Thus, a number of critical FCC catalyst performance parameters, such as activity, coke make, gas make, and gasoline octane can be quiet well correlated [5] with the zeolite Y unit cell dimension as illustrated in Figs. 4 and 5. Another important parameter related to the acidity of the zeolite Y is the Na_2O content which is also found to correlate with the gasoline octane quality (see Fig. 6).

Although the microporous zeolite component is a critical component of a FCC catalyst the overall pore size distribution including the mesopore structure also plays an important role [5] particularly for metals uptake and pre-cracking of large feed molecules as is illustrated in Fig. 7. Modern FCC catalyst systems are becoming

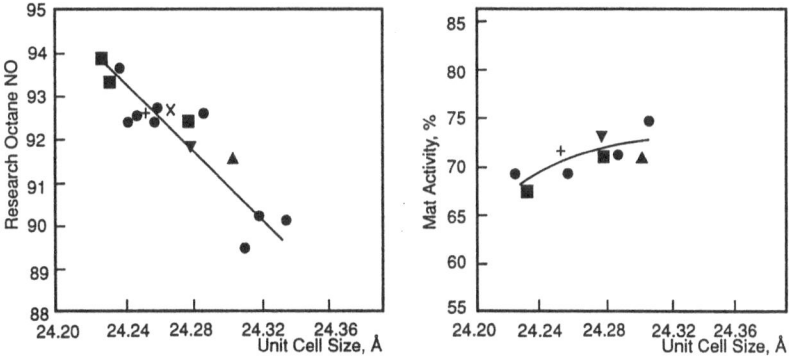

Figure 4. Catalytic cracking/unit cell relationship for zeolite Y [5].

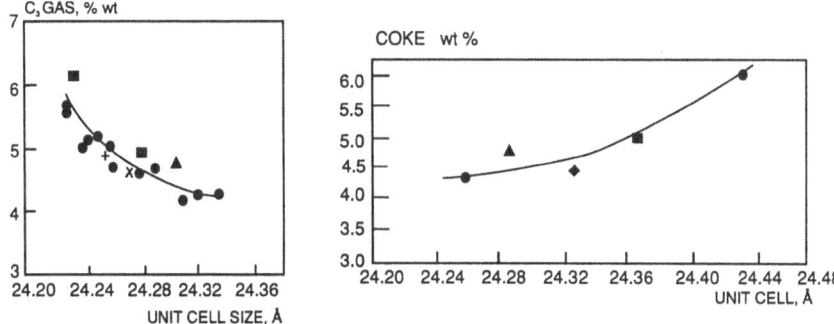

Figure 5. Catalytic cracking/unit cell relationship for zeolite Y [5].

increasingly complex where, in addition to the generic zeolite Y catalyst, a range of co-catalyst are added separately with a variety of functions such as metal traps, gasoline octane enhancers, light olefin boosters and SO_x traps. All these and future developments in this important field of catalysis will be enhanced by innovation in materials science.

2.1.2. Xylene Isomerization

Xylene isomerization is gaining in importance as a chemicals process due to the increasing demand for p-xylene is an intermediate for the production of terephthalic acid which is in turn the base material for polyester fibres and the fast growing polymer, PET. Thus, in recent years there has been some attention focussed on optimizing catalyst systems for the selective isomerization of mixed xylenes to p-xylene. Two acidic zeolitic catalyst systems are suited for xylene isomerization and these are based on ZSM-5 and mordenite.

The mordenite based catalyst has slightly larger pores than the ZSM-5 catalyst and is therefore more suited to mixed xylene isomerization when there are also significant quantities of ethylbenzene in the feedstock. Although the xylene isomerization reaction is purely acid catalysed as shown in Fig. 8 for the conversion of ethylbenzene a dual

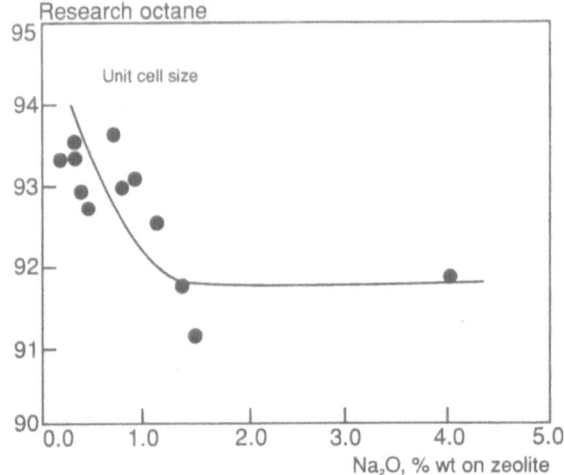

Figure 6. Catalytic cracking/Na$_2$O relationship for zeolite Y [5]

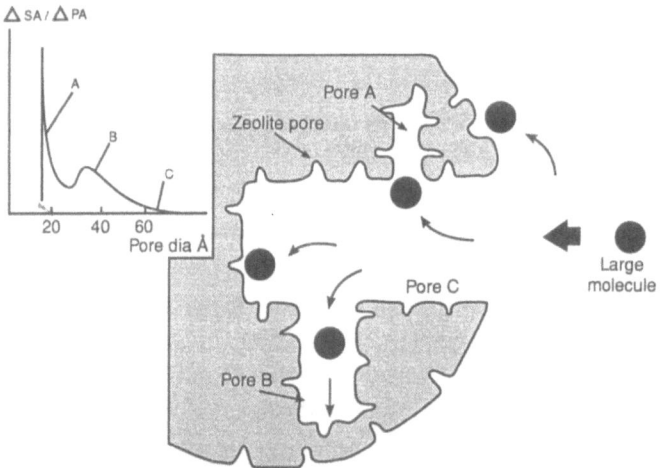

Figure 7. Pore size distribution in a catalytic cracking catalyst [5].

functional catalyst is required (see Fig. 9). To obtain an optimal mordenire based catalyst for the combined isomerization of xylenes and conversion of ethylbenzene a very delicate balance is required between acidity and hydrogenation power. Excessive acidity results in hydrocracking of intermediate napthenic products to low value by-products. Insufficient acidity will not give complete conversion to establish the desired equilibrium between the various xylene isomers.

The development of such a balanced C$_8$ aromatics isomerization catalyst requires careful control of the acidity of the mordenite which has been recently accomplished for a commercial catalyst system [6].

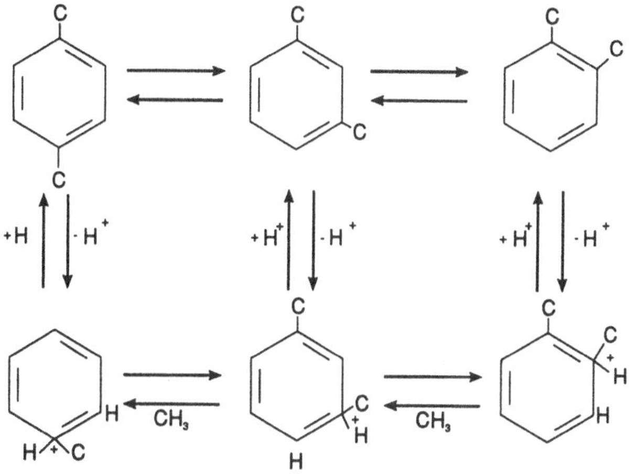

Figure 8. Mechanism for xylene isomerization [6].

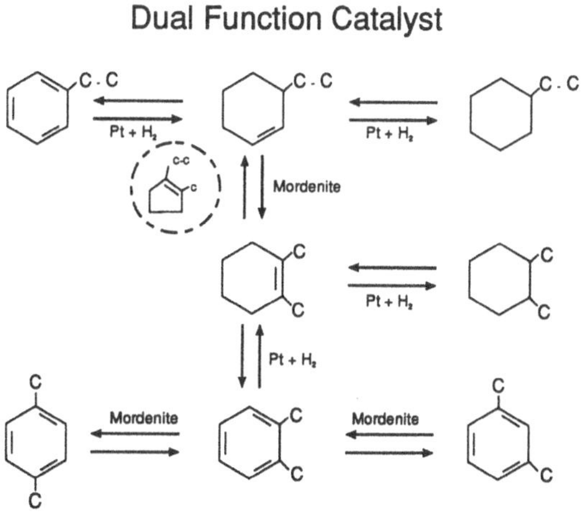

Dual Function Catalyst

Figure 9. Mechanism for ethyl benzene isomerization over bi-functional mordenite catalyst [6].

2.2. PROMOTORS

Another important aspect of surface chemistry as relates to catalysis is the so-called "promotor effect". For particular catalyst systems the addition of a secondary element in addition to the primary catalytic metal function results in significantly enhanced performance such as activity or selectivity improvement.

For example, the chemical process for the production of ethylene oxide (EO), from ethylene and oxygen, employs a heterogeneous catalyst based on silver metal dispersed on an inert support material, such as Al_2O_3. As shown in Fig. 10, the discovery of promotor elements has had a major impact on improving the selectivity and activity of EO catalyst systems [7] The early catalysts, which were based purely on supported

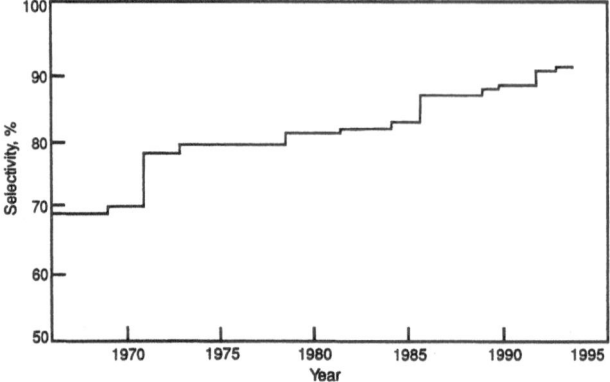

Figure 10. Improvements in ethylene oxide silver-based catalysts due to promotor effects [7].

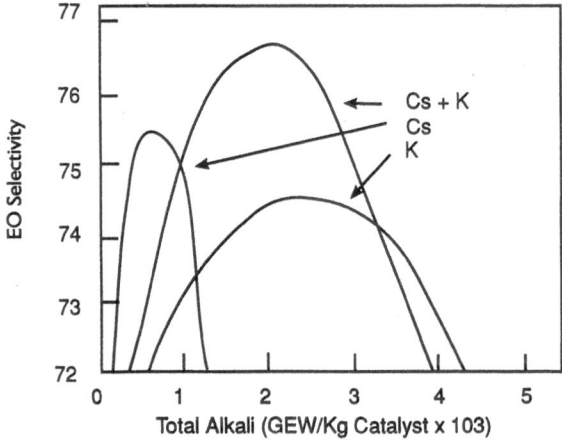

Figure 11. Alkali metal ion promotion for ethylene oxide silver-based catalyst [9].

silver, exhibited EO selectivities in the range of 50% whereas the modern promoted silver catalysts have selectivities as high 90%. Such dramatic improvements have resulted in significant savings in ethylene and therefore a major reduction in process costs.

Promotors include alkaline earth metal ions [8] such as barium which enhances activity and alkali metal ions such as potassium and cesium (see Fig. 11) which dramatically improve selectivity [9]. Although the fundamental understanding of these systems is somewhat limited there is evidence that these promotors act by modifying the electronic properties of the silver surface chemistry.

2.3. METAL FUNCTIONS

For some catalyst systems there are a range of metals which are active for a particular reaction. The preferred choice of metal in such cases often depends on a number of

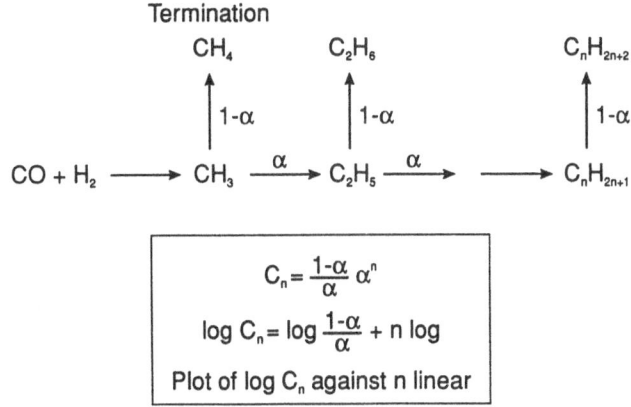

Figure 12. Synthesis gas Fischer–Tropsch catalysis. Anderson–Schultz–Flory kinetics.

factors such as the feedstock, process conditions and the desired product selectivity. Some areas of industrial importance where the choice of metal function is used to steer catalytic performance include; catalytic reforming (metal alloys), hydrotreating (mixed metal sulphides) and syngas conversion (Fischer–Tropsch). An example will be given of the latter process in view of the growing importance of catalytic technology for the conversion of natural gas into chemical and fuel products.

Quite a number of catalytic technologies are currently under development for natural gas conversion and most of these are based on Fischer–Tropsch (F–T) catalysis. This technology route involves two separate steps being the production of syngas (CO/H_2) from natural gas followed by the conversion of syngas to the desired products which almost invariably employs F–T catalysis [10].

The F–T reaction is based on a step-wise chain growth mechanism (so-called Anderson–Flory–Schultz kinetics) in which the relative probabilities of chain growth and chain termination (α and $1 - \alpha$, respectively) are independent of the chain length (see Fig. 12). The statistical growth model predicts a linear relationship between the logarithm of the molar amount of a paraffin and the carbon number of the paraffin as found in practice.

The type of metal employed in F–T catalysis has a significant influence on the value of the chain growth probability parameter, α, and thereby the product distribution. As shown in Fig. 13, for example, iron, ruthenium and cobalt based F–T catalysts, in general, exhibit different α values. In addition, promotors can also be used,in combination with different metals to steer both product distribution and, for example, product olefin to paraffin ratios. This is indicative of the dominant role of surface chemistry in determining the activity and selectivity behaviour of F–T catalyst sytems.

In terms of commercial developments Shell, for example, have recently built a first plant in Malaysia to convert natural gas into liquids based on F–T technology [10] which produces high quality middle distillate fuels together with other high added value products such as waxes, solvents and lube oils.

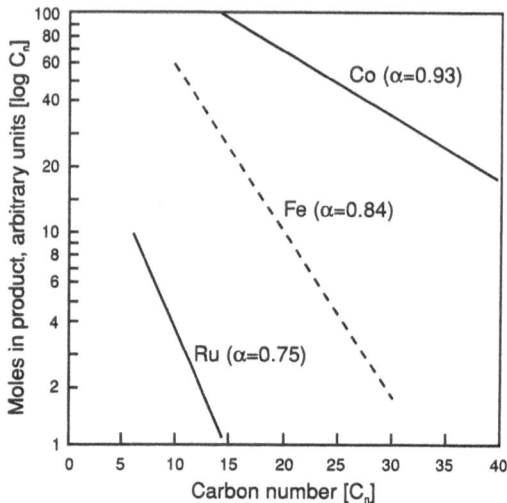

Figure 13. Fischer–Tropsch catalysis. Anderson–Schultz–Flory kinetics.

3. Micro-porosity

At the micro-porosity level (5–15 Å) the field of catalysis has seen major new developments with the availability of a broad range of synthetic zeolite and molecular sieves. Although there are several hundred of these synthetic crystalline microporous materials available to the catalytic chemist to date only a relatively small number have found industrial applications. Nevertheless, these materials have had a major impact on important refining processes (e.g. catalytic cracking, hydrocracking, paraffin isomerization and catalytic dewaxing). In addition, zeolites are also used in some petrochemicals processes, for example, for alkylation and isomerization of aromatics.

Some recent examples of the industrial application of microporous materials will be discussed. Both of these involve the use of so-called medium pore zeolites or molecular sieves which exhibit shape selective properties. Such catalyst systems with pores of near molecular dimensions can, in addition to the surface chemistry, also influence the overall catalytic reaction selectivity. At least three mechanisms have been defined and are termed reactant, transition state and product shape selectivity. The examples chosen include butene isomerization and hydroisomerization for which novel catalyst systems have been developed and recently applied in new commercial processes.

3.1. BUTENE ISOMERIZATION

The development of a process for the selective isomerization of butene was driven by environmental legislation, particularly in the United States, which has led to a rapid growth in the use of oxygenated components in gasoline. In particular, ethers such as me-thyl-tert-butyl ether (MTBE) have become an important part of the motor gasoline pool.

This growth in MTBE, which is manufactured from *i*-butene and methanol, has led to an increasing demand for the precursor components and, in particular, *i*-butene.

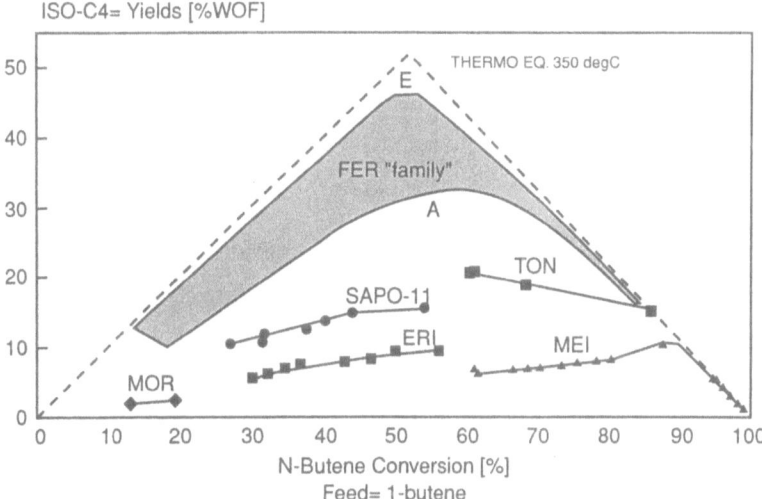

Figure 14. Skeletal isomerization of butenes: effect of zeolite/molecular sieve catalyst. Unoptimized and optimized FER vs. "typical" unmodified zeolites/molecular sieves [11].

Although the skeletal isomerization of an olefin such as *n*-butene would, at first sight, appear to relatively straightforward until recently catalysts for this reaction performed poorly [11]. The reason being that the selective isomerization is hampered by competitive reactions such as dimerization and oligomerization of butenes. It was therefore reasoned that a shape selective zeolite might be employed which would suppress oligomerization without inhibiting skeletal isomerization. Scouting experiments [11], in which a variety of medium pore zeolites and molecular sieves were tested, showed that the zeolite ferrierite (FER) indeed exhibited the required shape selective properties (see Fig. 14). A mechanism has been proposed in which the highly branched C_8 olefins are trapped within the zeolite pores due to geometric constraints and are selectively cracked to yield *i*-butene (see Fig. 15). Molecular modelling studies provided support for this mechanism in which it was demonstrated that the diffusion of such C_8 intermediate molecules was the most retarded in the FER structure compared to other medium pore zeolites and molecular sieves (see Fig. 16).

3.2. CATALYTIC DEWAXING

The presence of longer chain *n*-paraffin components in gasoils and lubeoils impairs their product properties and these were conventionally removed by means of a solvent dewaxing process. Some years ago a catalytic dewaxing process was developed and commercialized by Mobil researchers [12] which is based on the medium pore zeolite ZSM-5. This catalyst system primarily make use of reactant shape selectivity whereby the *n*-paraffins (wax molecules) are more readily hydrocracked [13] relative to branched isomers. However, this process has the disadvantage that valuable *n*-paraffin feed molecules are often cracked to low value lighter components.

388

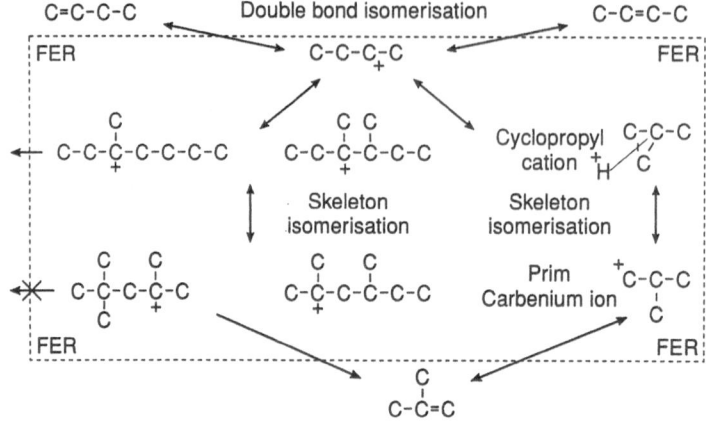

Figure 15. Skeletal isomerization of butenes: proposed mechanism [11].

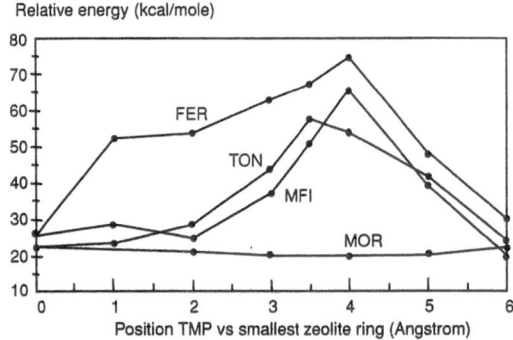

Figure 16. Molecular modelling: diffusional barriers for 2,4,4,trimethyl-3-pentane [11].

Recently researchers at Chevron [14] have developed a novel catalyst system, based on a medium pore molecular sieve, in which wax molecules are isomerization rather than cracked. The unique property of this silicoaluminophosphate based catalyst (Pt/SAPO-11) is that it exhibits a high selectivity for isomerization at high conversion levels but has a low selectivity for multiply branched isomers. These catalyst characteristics are precisely what is required to effectively remove wax molecules and produce molecular components, for example, which exhibit UHVI lube oil properties.

The enhanced performance of this novel catalyst is demonstrated in Fig. 17 in which the Pt/SAPO-11 is compared with ZSM-5 for lubeoil dewaxing. The much higher retention of viscosity index (VI) as a function of pour point for the Pt/SAPO-11 catalyst system is direct evidence of the enhanced wax isomerization performance. This enhanced selectivity for isomerization with minimal multi-branching is attributed to the constrained one-dimensional pore structure of SAPO-11 which induces the desired transition state selectivity properties with minimal secondary cracking. By contrast, within the two-dimensional pore system of ZSM-5, multiple branch isomerization and

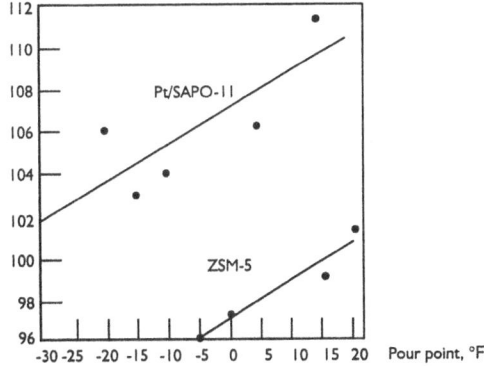

Figure 17. Comparison ISO-dewaxing of lube oil for ZSM-5 and Pt/SAPO-11 catalysts [14].

secondary cracking reactions can likely readily occur at the more spacious channel intersections.

Chevron has commercialized [15] a new process to selectively isomerize wax in lubeoils based on this mew molecular sieve catalyst technology. This nicely illustrates recent developments in the catalytic application of advanced shape selective materials.

4. Meso-porosity

The mesopore systems of a catalyst is often very important, particularly in refining catalysts where relatively high molecular weight and therefore bulky molecules are being processed. For such feedstocks a mesopore system is required to obtain acceptable rates of diffusion within the catalyst pore structure. Depending on the particular catalyst the mesopore system may also be used in combination with a micropore structure (e.g. hydrocracking and catalytic cracking).

Some examples will be given where mesopore structures are vital for optimal performance in industrial catalyst systems.

4.1. HYDROCRACKING

Hydrocracking, rather like fluid catalytic cracking, FCC is a process for converting heavy oil feedstocks into lighter transportation fuels products such as gasoline, diesel and jet fuel. However, in contrast to FCC, hydrocracking makes use of bifunctional (i.e. acidic and hydrogenation functions) catalysts which operate at high hydrogen partial pressures (i.e. 40–120 bar).

Many modern hydrocracking catalysts, similar to FCC catalysts, employ zeolite Y based systems as their prime acidic function [16]. Also in a similar manner to FCC catalysts, the acidity of the zeolite can be tailored, as measured by the unit cell parameter, to influence the catalyst performance. The kerosine selectivity can be controlled to a significant degree by adjusting the acid site density as indicated by the unit cell dimension [17].

390

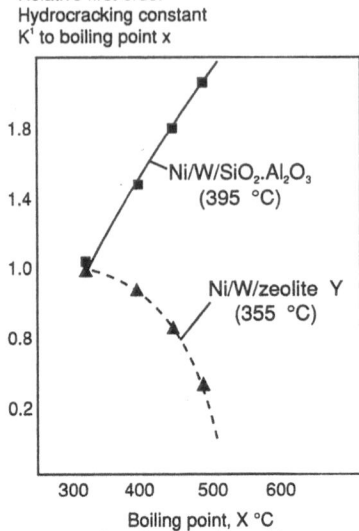

Relative first order
Hydrocracking constant
K^1 to boiling point x

1.8

1.4

Ni/W/SiO$_2$.Al$_2$O$_3$
(395 °C)

1.0

Ni/W/zeolite Y
(355 °C)

0.8

0.2

300   400   500   600

Boiling point, X °C
Two-Stage hydrocracking

Figure 18. Relative hydrocracking rates for SiO$_2$·Al$_2$O$_3$- and zeolite Y-based catalysts [16].

However, hydrocracking catalysts which are based only microporous zeolite Y components suffer from diffusional limitations when processing heavy feedstocks. This phenomenon is nicely demonstrated [16] by comparing the relative hydrocracking rates using a pure (microporous) zeolite and an amorphous (mesoporous) silica-alumina catalyst. As shown in Fig. 18, as the average boiling point (i.e. average molecular weight) of the feedstock increases the zeolite hydrocracking activity decreases markedly which is indicative of diffusional limitations.

This problem can be resolved, for example, by incorporating into the catalyst formulation, a mesoporous component such as a silica-alumina together with the zeolite Y (so-called composite catalysts). An alternative or even complementary approach is to subject the zeolite Y to a hydrothermal treatment [18] whereby a mesopore structure can be introduced into the zeolite itself with significant retention of crystallinity (see Fig. 19).

Although not discussed in detail in this article, the type, location and dispersion of the hydrogenation function (i.e mixed metal sulphide or noble metal) is also critical to obtaining optimal performance of a hydrocracking catalyst. The tailoring of both the acidity and the porous texture of zeolites is a field of catalyst materials development which has a major impact on important refinery processes such as catalytic cracking and hydrocracking and may find application in other areas in the future.

4.2. RESIDUE HYDROPROCESSING

Residue hydroprocessing catalysts are also, in principle, bi-functional but under the severe operating conditions feed impurities tend to rapidly poison the acidic function. For residue catalysts the meso-pore structure and hydrogenation function are of more importance.

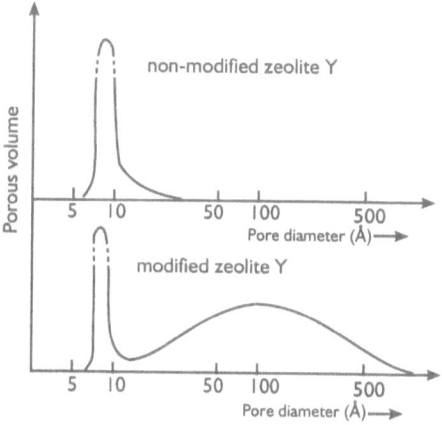

Figure 19. Zeolite Y catalyst with and without meso-pore structure [18].

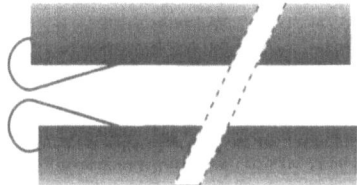

Small pores and high metals deposition
rates lead to pore mouth plugging

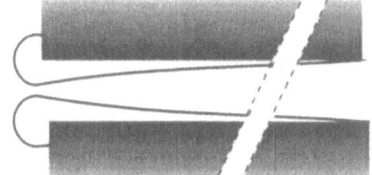

Larger pores and lower metals deposition
rates result in pore poisoning

Figure 20. Metal pore mouth plugging and core poisoning [19].

Residue feedstocks not only contain very bulky high molecular weight components (e.g. asphaltenes) but are also quite heavily contaminated with nitrogen and sulphur compounds as well as metals such as Ni, V and Fe. Furthermore, despite the relatively high hydrogen partial pressure of operation coke deposition on residue catalysts is quite severe. All these aspects in combination require careful design of the meso-pore structure to avoid rapid deactivation of such catalyst systems [19].

For example, the demetallization reactions that occur under hydroprocessing conditions result in the deposition of feed metal contaminants onto residue catalysts. These metals tend to deposit in the catalyst pore mouths and result in plugging. As illustrated in Fig. 20 under these conditions small pores will end to result in excessive pore mouth plugging whereas large pores lead to catalyst core poisoning. This effect is shown graphically as a function of average pore diameter in Fig. 21. Clearly, an optimal average meso-pore diameter is required which achieves a balance between these two modes of catalyst deactivation.

In fact, due to the often conflicting design demands on residue catalyst systems they are often deployed in a reactor as three separate catalysts. The first is designed primarily for demetallization, the second for desulphurization and denitrogenation and the third for hydrocracking.

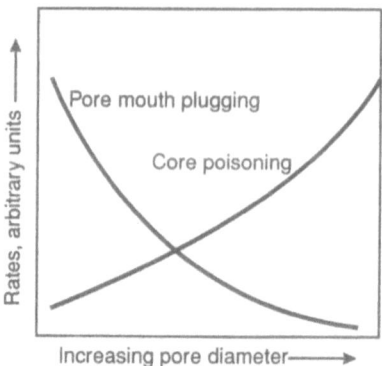

Figure 21. Relative rates of metal pore mouth plugging and core poisoning as a function of pore diameter [19].

5. Macro-structures

Commercial catalyst systems are normally manufactured in the form of macro-structures (shaped particles) which are then most often loaded into fixed bed reactors. The shapes vary considerably but cylinders, rings, pellets and trilobes are the most common and are achieved by means of extrusion technology. In addition, some major processes, such as catalytic cracking, make use of moving beds for which fine particles are required in order to enhance the fluidization properties. These are manufactured to narrow particle size range specifications using spray drying technology.

In recent years there has been a growing trend towards employing so-called mono-lithic structures, particularly in environmental applications of catalysis. Monolithic structures [20] generally have a honeycomb-like shape and are formed by an extrusion process. The catalytic component is either added onto the surface of the monolith by means of a washcoat (see Fig. 22) or is an integral part of the monolith itself. An important characteristic of the monolith structures is their low pressure drop which is very often an important consideration in catalytic gas treating processes.

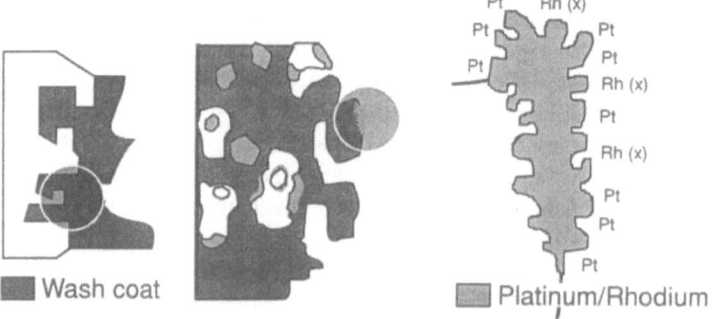

Figure 22. Pt/Rh washcoated monoliths [20].

Some examples will be given of catalytic applications of monoliths where the developments in materials science and, in particular, ceramics are of importance.

5.1. AUTO-EXHAUST EMISSIONS

The emissions caused during the combustion of transportation fuels in vehicles is significant source of air pollution particularly in densely populated areas. However, in the case of gasoline engines, the development of the so-called three-way exhaust catalyst has been highly effective in achieving major reductions in these emissions. For example, as shown in Fig. 23 a typical three-way catalyst when operated under stoichiometric conditions (i.e. $\lambda = 1$ or air/fuel ratio $14.7:1$) is capable of high conversions for the oxidation of CO, hydrocarbons and simultaneous reduction of NO_x [21]. An important component of an auto-exhaust catalyst is the control system which ensures that a stoichiometric air/fuel ratio is maintained at the catalyst inlet.

The catalyst systems are based on monoliths which are washcoated with Pt/Rh as catalytically active components (see Fig. 22). Cerium is often added to the wash coat as a promotor since it is capable of storing oxygen during lean (fuel-deficient) engine excursions and release oxygen to oxidize CO during rich excursions. This is an interesting demonstration of the use of promotors to improve catalyst performance under rapidly changing operational conditions. Another rather unique feature of the three-way catalyst is the ability to simultaneous catalyze oxidation and reduction reactions. Platinum is the active component for the oxidation reactions whereas rhodium is highly selective for NO_x reduction. A typical exhaust catalyst contains less than 0.1 troy ounce Pt, less than 0.001 troy ounce Rh and up to 15% wt Cerium. Catalytic converters account for a very significant percentage of the world demand for both the noble metals which are applied [21].

The monolith component is typically composed of corderite, $(2MgO \cdot 2Al_2O_3O \cdot 5SiO_2)$ which has a thin walled honeycomb design and incorporates the features of low pressure drop, mechanical strength and high temperature resistance. The trends in modern auto-

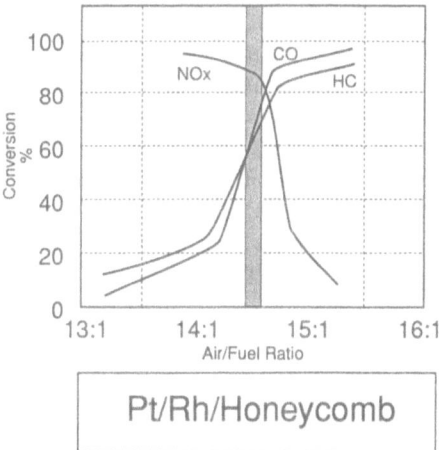

Figure 23. Three-way auto exhaust catalysis.

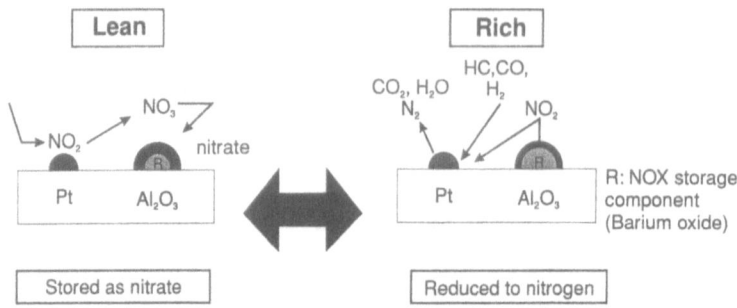

Figure 24. NO$_x$ storage-reduction (NSR) lean burn catalyst system [22].

exhaust catalysts are towards reduced emissions during cold start and systems which operate under lean burn (more fuel efficient) engine operating conditions. Both of these trends are making significant demands on new developments in materials science.

The next generation of auto-exhaust catalyst systems are aimed at further reducing emissions, particularly during cold start operation. In addition, there is a drive towards catalysts which will operate effectively under lean burn engine conditions. Lean burn engines offer significant potential fuel savings but higher NO$_x$ emissions since the three-way catalysts can no longer operate in the stoichiometric (so-called $\lambda = 1$) regime. There are some promising new developments in progress, where materials science plays an important role, aimed at solving the above problems.

One promising approach to reducing cold start emissions is to shift the catalyst closer to he exhaust manifold and thereby achieve faster catalyst light-off. However, this makes severe temperature resistance demands on the monolith systems. Current developments of new ceramic materials for application in, for example catalytic combustion, promises to provide the improvements in materials performance which are needed for the higher temperature operation of auto-exhaust catalysts.

For lean burn auto-exhaust catalysts an interesting approach is being developed by Japanese researchers at Toyota [22] called NO$_x$-storage-reduction (NSR). This NSR catalyst system includes a barium component (see Fig. 24) which is used to store NO$_x$ as a nitrate under lean operation which is then subsequently rapidly regenerated by a very short period of stoichiometric or fuel rich operation during which the NO$_x$ is released and converted across the catalyst component (Pt/Rh). This adsorption and reduction cycle is then repeated with cycle times of some about 1 minute for adsorption and 0.5 seconds for catalytic NO$_x$ reduction. Lean burn catalyst systems based on this NSR principle have been shown to exhibit high (90%) NO$_x$ conversion efficiencies when operating in the 300–450°C temperature range (see Fig. 25). However, during durability testing it was found that there was significant deterioration of catalyst performance due to the presence of SO$_2$ in the exhaust emissions caused by barium sulphate formation. The durability has been significantly improved by reducing the particle size of the storage component (barium oxide) whereby the barium sulphate formed is decomposed during the catalytic NO$_x$ reduction cycle (see Fig. 26). Clearly, the optimal development of such lean burn systems clearly multi-disciplinary whereby materials science, catalysis and engine control systems all play a role.

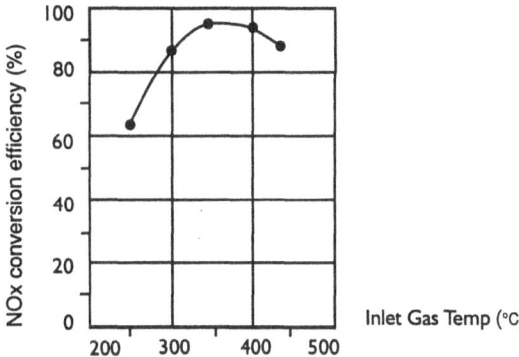

Figure 25. NO$_x$ conversion efficiency as a function of inlet gas temperature for lean burn (NSR) catalyst [22].

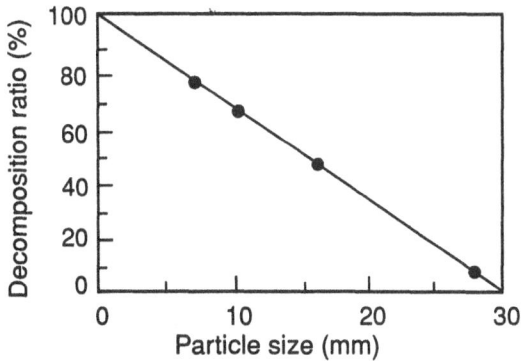

Figure 26. Relationship between decomposition ratio for barium sulphate and particle size (700°C, reduction) for lean burn (NSR) catalyst [22]

5.2. NO$_x$ REDUCTION

So-called end-of-pipe technologies are those which are added to off-gas streams to remove environmentally harmful components. An example of such a technology is selective catalytic reduction (SCR) of NO$_x$ whereby ammonia is used as the reducing agent. This technology was originally developed in Japan employing Ti/V and Ti/W based catalyst systems. These catalyst components are applied as wash coats onto a monolith support in order to minimize the pressure drop across the reactor system. More recent developments in materials science have enabled catalysts to be developed in which the active components are incorporated directly as part of the monolith structure.

Some further recent developments in this field include a low temperature (200–250°C) catalyst system based on Ti/V [23] which can be readily added on to off-gas streams. Interestingly this low temperature DeNO$_x$ system makes use of a lateral flow reactor (filled with normal catalyst pellets) rather than a monolith to achieve a low pressure drop.

In addition, a zeolite (Cu-ZSM-5) based catalysts have been developed [24] which can operate in the high temperature (350–450°C) range without excessive oxidation of ammonia to NO_x as occurs with conventional catalysts. Cobalt-zeolite based catalyst systems have been shown [25] to exhibit the interesting property of being able to reduce NO_x using methane as a reducing agent. Although such catalyst systems would be advantageous they do not yet appear to be sufficiently robust to be applied commercially.

An interesting development which integrates the use of zeolite catalysts in a low pressure drop system is the use of supported zeolite crystals, grown on the surface of stainless steel gauze [26]. ZSM-5 crystals are grown in situ on the surface of the gauze which is present as rolls in the zeolite synthesis mixture. The wire thickness was 35 μm and the zeolite loading achieved was some 3.5% wt. Following exchange of copper ions into the zeolite, the catalyst was tested for the reduction of NO by ammonia at 350°C and found to show improved activity and selectivity compared to a conventional catalyst consisting of Ti and V supported on amorphous silica. Although this approach has not yet been commercialized this does provide a good demonstration of the novel application of materials science to tackle problems in catalytic environmental technology.

5.3. CATALYTIC COMBUSTION

Natural gas is a clean and abundant fuel which can be used, for example, to drive turbines for power generation. There is, however currently an environmental drawback which is the formation of nitrogen oxides at the high temperatures (e.g. 1800°C) generated by the adiabatic thermal combustion of the gas/air mixture. NO_x concentrations at these temperatures are about 165 ppm in the turbine exhaust, whereas regulatory requirements in California, Japan and Europe are in the range of 5–80 ppm [27].

A potential solution to this problem is to carry out the combustion at somewhat lower temperature which could, in principle, be achieved by employing a catalyst [27,28]. Recognition of this attractive approach to reduce gas turbine NO_x emissions has led to a considerable effort in catalytic combustion in recent years [29]. One approach, under investigation by the Catalytica company in collaboration with Tanaka Kikinzoku Kogyo K.K. [30], is to have a low temperature catalytic combustion step at relatively low temperature followed by a second thermal combustion step in the gas phase downstream of the catalyst (Fig. 27). This two step combustion process results in significantly reduced overall temperatures. In this system a catalyst is used consisting of Pd supported on high surface area mixed oxide support, which in turn is supported on FeCrAl metal foil shaped into the form of a honeycomb.

An alternative approach is to develop a catalyst system which has sufficient thermal stability to operate at temperatures up to 1300°C. This is a major challenge in catalyst development, requiring not only high thermal stability (sintering or volatilization of both active sites and support components) but also adequate resistance to thermal shock under the fluctuating temperature conditions of gas turbine operations.

A rather striking example of progress which has been made in this field and which further reflects the importance of the interface between catalyst development and materials science is provided by the research of Osaka Gas Company in collaboration with Kobe Steel and Catalysts and Chemicals Inc. [31]. This consortium has developed

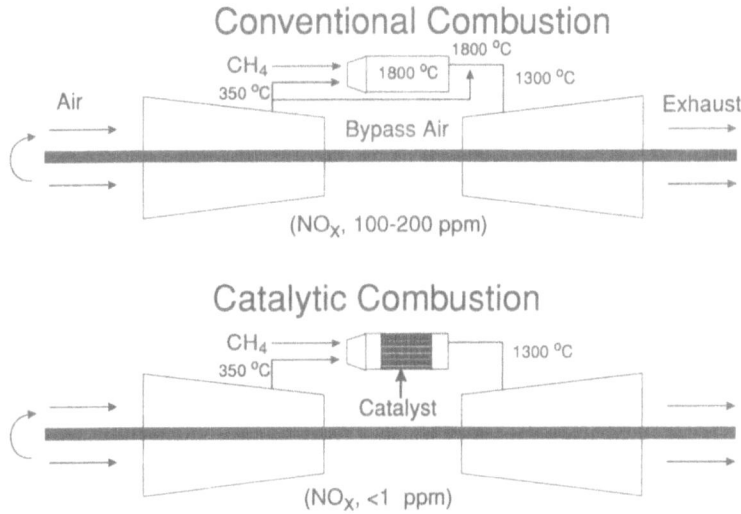

Figure 27. Catalytic combustion gas turbine [27].

an active catalytic monolith which has been successfully tested in a 160 kW gas turbine. The outlet temperature of the catalyst bed was 1100°C, and the system was tested continuously for some 215 hours during which the NO_x emissions were maintained below 60 ppm (Fig. 28).

The catalyst system developed for this high temperatures application has several features of interest being based on Mn-substituted hexa-aluminates, a class of compounds developed in recent years by Arai and co-workers [28]. These mixed oxides maintain surface area at very high temperatures, for example, about 10 m^2/g after calcination at 1600°C (Fig. 29). It is thought that this stability is derived from the anisotropic crystal structure which helps to prevent sintering. Interestingly, the high temperature stability depends on the method of synthesis, with hydrolysis of metal alkoxides proving to be a better route than solid state reactions between the precursors. This is related to the degree of homogeneity which can be obtained and is an example of the general trend to improve the properties of inorganic materials through the use of molecular precursors [32]. The hexaluminate powder was extruded into a honeycomb shape as is required to maintain a low pressure drop. Further, the monolithic pieces which were 220 mm cross-section were built up from segments in order to accommodate thermal shock.

These developments would appear to be sufficiently promising that catalytic gas turbine technology will likely be commercialized in the near future.

5.4. SYNGAS MANUFACTURING

The efficient and cost effective conversion of natural gas (primarily methane) into higher added value products such as fuels and chemicals remains a challenge especially at current relatively low oil prices. However, in the medium to longer term, as oil reserves become depleted the exploitation of natural gas, particularly at remote sites, will become increasingly important.

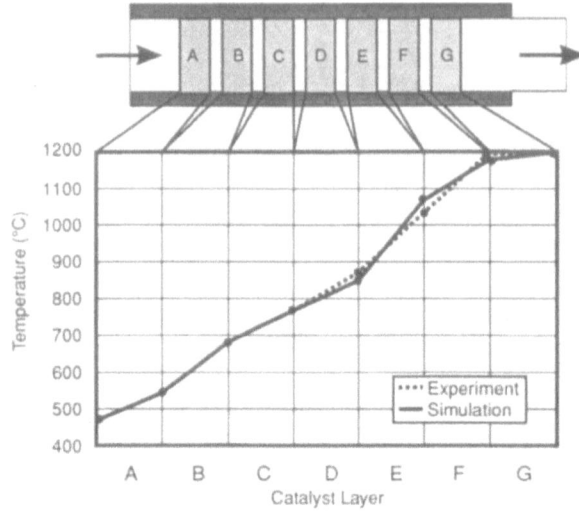

	Kind of Catalyst	Cell number	Height
A B C	Precious metals-carrying cordierite honeycomb	200 cpi	20 mm X 3
D E	Mn substituted hexaaluminate ceramic honeycomb sintered at 1200 °C	300 cpi	20 mm X 2
F G	Mn substituted hexaaluminate ceramic honeycomb sintered at 1300 °C	300 cpi	20 mm X 2

Figure 28. High pressure (10 bar) catalytic combustion of natural gas [31].

Some recent work at the University of Minnesota [33] by Professor Schmidt has shown that methane can be selectively oxidized to syngas (CO/H$_2$) using a rhodium on ceramic foam catalyst. As shown in Fig. 30 rhodium is significantly more selective for this reaction than platinum. Under these rather extreme conditions of temperature and space velocity the properties of the ceramic foam in terms of both porosity, mechanical strength and temperature resistance are vital for optimal catalyst performance. Although an industrial process has not yet been developed this work could possibly provide the basis for catalytic partial oxidation of methane to syngas sometime in the future.

6. Future Trends

In terms of future trends there are some more recent developments will be discussed which indicate that the interface between the disciplines of materials science and catalysis are fertile ground for innovation.

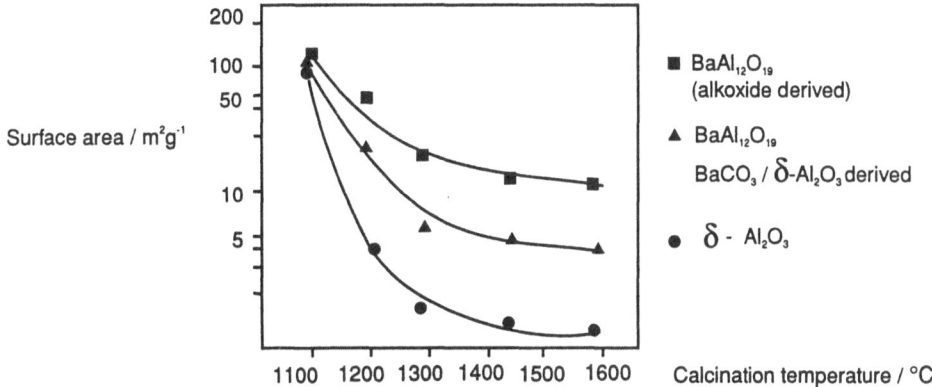

Figure 29. Thermal stabilities of δ-Al$_2$O$_3$ and hexa-aluminates [28].

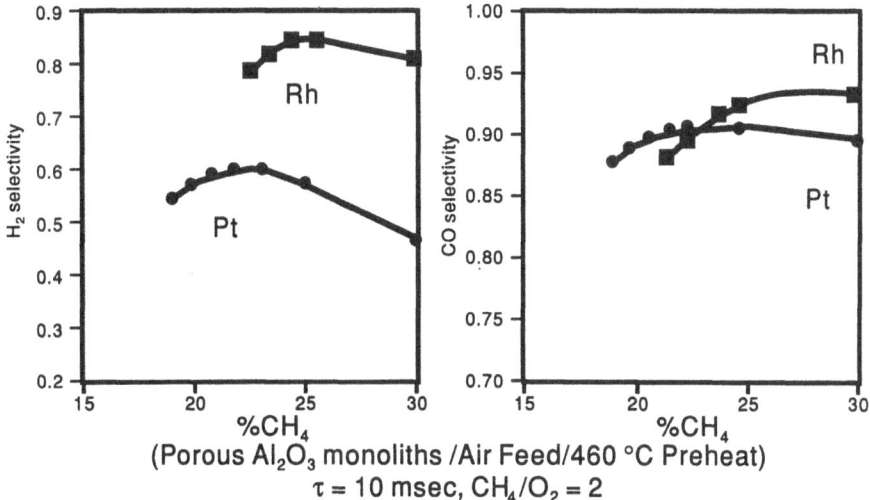

Figure 30. Catalytic partial oxidation (CPO) of methane [33].

For example, in the environmental catalysis area [34] new monolith materials are being developed which incorporate the catalytic components (e.g. zeolites and titanium dioxide) within the structure rather than applying these components as a washcoat. Monolith cell densities (i.e. number of cells per square inch) are being increased to levels as high as 200 cells/sq. in. which enable higher catalyst activity levels to be achieved without significant increases in pressure drop. In addition, the wall thicknesses are being reduced and the operational temperature ranges of these new monolith materials are being extended.

Catalytic membranes which will enable equilibrium limitations to be overcome are being developed. A recent example was reported by Japanese workers [35] who employed a palladium membrane together with a Ga/ZSM-5 catalyst for the conversion

Homogeneous / Heterogeneous

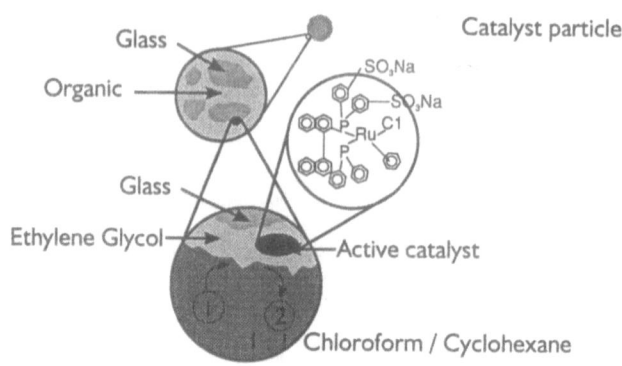

Figure 31. Schematic of heterogeneous enantio-selective catalyst Ru-BINAP [36].

of butane into aromatics. The membrane/catalyst system enabled the hydrogenation-dehydrogenation equilibrium to be shifted and thereby increase the yield of the desired aromatic products.

The interface between homogeneous and heterogeneous catalysis has been perceived for some time to be potentially fruitful whereby, in principle, the advantages of both systems could be exploited. However, practical solutions to problems, such as the leaching of the homogeneous component from the heterogenous support, have limited progress in this field in the past. Recent, however, some novel approaches to this problem appear to be emerging which could possibly revitalize this catalysis interface. For example, recent work by Wan and Davies [36] has shown that a heterogeneous system could be synthesized which incorporated a homogeneous hydrogenation catalyst component that was capable of chiral transformations with high activity levels and no apparent leaching problems (Fig. 31). This system was applied to achieve both high activity and chiral selectivity for a critical hydrogenation step in the synthesis of (S)-naproxen which is a commercially important anti-inflammatory agent.

Conventional materials used in heterogenous catalysis are strongly represented by compositions based on silica, alumina or combinations thereof (i.e. clays, zeolites and amorphous silica-aluminas). The application of non-oxide materials in heterogenous catalysis has been limited largely by the lack of high surface area forms. This situation is changing, due to the interest in the ceramics field in fine powders as precursors to highly dense materials and new synthetic routes to high surface area non-oxides [37].

For example, interesting catalytic results have recently been obtained over a high surface area molybdenum oxycarbide [38]. This material was obtained with a surface area of 140 m^2/g from the reaction of low surface area molybdenum oxide (4 m^2/g) with hydrogen/hydrocarbon mixtures. The oxycarbide is an effective catalyst for isomerization of n-hexane or n-heptane, a reaction of importance for octane enhancement of gasoline [39]. The oxycarbide shows a higher isomerization selectivity at high conversion than a conventional catalyst based on Pt/zeolite-beta (Fig. 32). Interestingly, it is proposed [40] that the enhanced selectivity of the oxycarbide is due

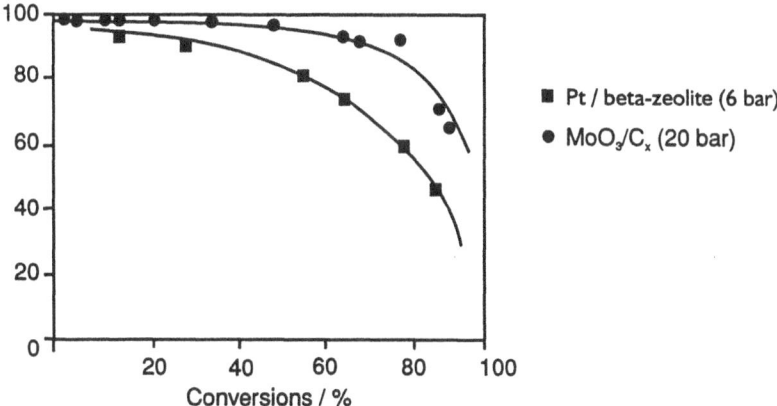

Figure 32. C$_7$ isomerization selectivity for MoO$_3$/O$_x$ and Pt/zeolite β [38].

to a mechanism involving a metallocyclobutane intermediate rather than a conventional bifunctional mechanism involving carbenium ions.

The above examples from the recent literature serve to reinforce the future importance of materials science for innovation in catalysis. In order to maximize this synergy there will need to be a high level of cross-fertilization between these two disciplines of endeavour.

References

[1] Weisz, P.B. (1973) *Chemtech.* **13**, 498.

[2] Behrens, P. (1993) *Adv. Mater.* **5** (2), 127.

[3] Corma, A. (1995) *Chem. Rev.* 559–614.

[4] Haag, W.O., Lago, R.M. and Weisz, P.B. (1984) *Nature* **309**, 589.

[5] Biswas, J. and Maxwell, I.E. (1990) *Appl. Catal.* **63**, 197.

[6] de Vries, A.F. and Eggers-Borkenstein, P. (1995) in *Proceedings Solid Acid/Base Process Conference '95*, Feb. 26–28th, 1995, Houston, USA.

[7] Rothwell, W.P., Shearer, C.J. and Taylor, G.L. (1995) *Chemtech.* June, 6.

[8] Yinsheng, P., Shi, Z., Liang, T. and Jingfa, D. (1992) *Catal. Lett.* **12**, 307.

[9] Bhasin, M.M. and Hendrix, C.D. (1993) in L. Ghuzi et al. (eds.), *Proceedings 10th Int. Congress on Catalysis*, Elsevier, p. 1431.

[10] van Wechem, V.M.H. and Senden, M.M.G. (1994) in H.E. Curry-Hyde and R.F. Howe (eds.), *Natural Gas Conversion II, Studies in Surface Science and Catalysis*, Elsevier, p. 43.

[11] Mooiweer, H.H., de Jong, K.P., Kraushaar-Czarnetzki, B., Stork, W.H.J. and Krutzen, B.C.H. (1994) in J. Weitkamp et al. (eds.), *Zeolites and Related Microporous Materials: State of the Art 1994*, Studies in Surface Science and Catalysis, Elsevier, p. 2327.

[12] Smith, K.W., Starr, W.C. and Chen, N.Y. (1980) *Oil Gas J.* **78**, May 26, 75.

[13] Chen, N.Y., Gorring, R.L., Ireland, H.R. and Stein, T.R. (1977) *Oil Gas J.* **75**, June 6, 165.

[14] US Patent 4,921,594 (1987) to Chevron Research Company.

[15] Miller, S.J. (1994) *Microporous Mater.* **2**, 439.

[16] Maxwell, I.E. (1987) *Catal. Today* **1**, 385.

[17] Sie, S.T. (1994) in J.C. Jansen et al. (eds.), *Advanced Zeolite Science and Applications*, Studies in Surface Science and Catalysis, Elsevier, p. 587.

402

[18] Hennico, A., Billon, A. and Bigeard, P-H. (1992) *Hydrocarbon Technology International*, p. 19.
[19] Adams, C.T., Del Paggio, A.A., Schaper, H., Stork, W.H.J. and Shiflett, W.K. (1989) *Hydrocarbon Processing*, Sept., p. 57.
[20] *Catalysts for Stationary Engines*, Brochure, Degussa AG.
[21] Taylor, K.C. (1990) *Chemtech*, Sept., 551.
[22] Matsumoto, S. (1995) Pre-prints 2nd Japan–EC Joint Workshop on the Frontiers of Catalytic Science and Technology **1**, 39.
[23] van der Grift, C.J.G., Woldhuis, A.F., Maaskant, O.L. (1995) G. Centi et al. (eds.), *Proceedings of the 1st World Congress Environmental Catalysis*, SCI, p. 19.
[24] Armor, J.N. (1992) *Applied Catalysis B: Environmental* **1**, 221.
[25] Armor, J.N. (1993) *13th North American Catalysis Meeting*, Presentation, May 2–6, Pittsburg, USA.
[26] Calis, H.P., Gerritsen, A.W., van den Bleek, C.M., Legein, C.H., Jansen, J.C. and van Bekkum, H. (1995) *Can. J. Chem. Eng.* **73**, 120.
[27] Cusumano, J.A. (1992) *Chemtech* **23**, 482.
[28] Arai, H. and Machida, M. (1991) *Catal. Today* **10**, 81.
[29] Arai H. (ed.) (1994) *Proceedings of the International Workshop on Catalytic Combustion*, April 18–20th, Catalysis Society of Japan.
[30] Dalla Betta, R.A., Schlatter, J.C., Chow, M., Yee, D.K. and Shoji, T., in ref. [29], p. 154.
[31] Sadamori, H., Tanioka, T. and Matsuhisa, T., in ref. [29], p. 158.
[32] Sanchez, C., Mecartney, M.L., Brinker, C.J. and Cheetham, A. (eds.) (1994) *Better Ceramics Through Chemistry* Vol. 1, Materials Research Society Symposium Proceedings Vol. 346, and earlier volumes in this series.
[33] Hickman, D.A. and Schmidt, L.D. (1993) *Science* **259**, 343.
[34] Armor, J.N. (1994) *Chem. Mater.* **6**, 730–738 (membranes, etc.).
[35] Dirass Report E16-8 (1994) *Recent Catalyst and Catalytic Technologies*, DIA Research Institute Inc., Japan.
[36] Wan, K. and Davis, M.E. (1994) *Nature* **370**, 449.
[37] Chorley, R.W. and Lednor, P.W. (1991) *Adv. Mater.* **3**, 474.
[38] Ledoux, M.J., Delporte, P. and Pham-Huu, C. (1995) in E. Iglesia, P.W. Lednor, D.A. Nagaki, and L.T. Thompson (eds.), *Synthesis and Properties of Advanced Catalytic Materials*, Materials Research Society Symposium Proceedings, Vol. 368.
[39] Maxwell, I.E. and Naber, J.E. (1992) *Catal. Lett.* **12**, 105.
[40] Pham-Huu, C., Ledoux, M.J. and Guille, J. (1993) *J. Catal.* **143**, 249.

MICRO- AND MESO-POROUS MATERIALS AS CATALYSTS

A. CORMA, D. KUMAR
Instituto de Tecnologia Quimica, UPV-CSIC,
Universidad PolitEecnica de Valencia, 46022 Valencia, Spain

1. Introduction

One of the greatest revolutions in catalysis took place when RE exchanged zeolite Y illustrated the superior cracking activity of a microporous crystalline Silica–Alumina structure over the formerly used Silica–Alumina catalysts. Since then, the application of zeolites and zeotypes in adsorption and catalytic processes has increased enormously, such that it is possible to say that they represent today the largest volume used of a given type of catalyst. The practical importance of these materials has impulsed fundamental studies on the synthesis, characterisation and applications of microporous materials up to the point that they are today an important part of material science, inorganic chemistry, and catalysis.

Zeolites are crystalline aluminosilicates in which Si and Al are tetrahedrally coordinated by oxygen atoms. These tetrahedral building blocks are linked through oxygen atoms producing a three dimensional network containing channels and cavities of molecular dimensions. Crystalline structures of the zeolite type but containing tetrahedrally coordinated Si, Al, P, as well as transition metals and many group elements with a valence ranging from I to V such as B, Ga, Fe, Cr, Ti, V, Mn, Co, Zn, Cu *etc.* have also been synthesised. They are referred to with the generic name of zeotypes including $AlPO_4$, SAPO, MeAPO and MeAPSO type molecular sieves[1-7].

From a catalytic point of view, it is evident that for molecules that can diffuse into the channels and cavities of the zeolites and zeotypes, the latter will become very special types of microreactors, if one is able to generate the catalytic active sites within the zeolite or to use the cavities as a support or cage to generate extra framework types of active sites. Following this idea, one can consider the following types of catalytic sites in zeolites:

a) Active sites in framework positions: acid; basic; redox.

b) Active sites in extra framework positions: Lewis acid sites; alkaline metal clusters; transition metal clusters; transition metal complexes occluded in cavities.

C.R.A. Catlow and A. Cheetham (eds.), New Trends in Materials Chemistry, 403–460.
© 1997 *Kluwer Academic Publishers.*

The different active sites named above will generate the corresponding type of catalysts, and in the following section the generation, characterisation and catalytic properties of these types of catalysts will be briefly discussed.

2. Acid catalysis on zeolites and zeotypes

Brønsted acid sites are generated on the surface of the zeolite when Si^{4+} is isomorphically substituted by a trivalent metal cation, for instance Al^{3+}, and a negative charge is created in the lattice, which is compensated by a proton. From a structural and chemical point of view, the Brønsted acid site in a zeolite can be seen as a resonance hybrid of structures I and II:

where structure I is a fully bridged oxygen with a weakly bonded proton, and structure II is a silanol group with a weak Lewis acid interaction of the hydroxyl oxygen with an Al atom[8]. This model is adequate to explain the increase in acidity of the bridged hydroxyl with respect to the silanol group.

On the basis of this, and taking into account Gutmanns first rule[9], Gajda and Rabo[10] concluded that an amorphous material has no long range stabilisation as for structure I, and so structure II can dominate. On the contrary, in the case of zeolites a significant stabilisation results from the electronic structure of the Al–O and Si–O bonds becoming more equivalent, and consequently a dominance of structure I in the resonance should occur.

The presence of Al–OH–Si entities in zeolites has been confirmed by neutron diffraction studies11, which proposed an averaged geometry for the protonic site, being in reasonable agreement with the geometry proposed on the basis of NMR results[12], and theoretical calculations[13] (Table 1).

TABLE 1. Geometries of bridged hydroxyls.

Technique	distances (pm)			angles (deg)	
	OH	OSi	OAl	SiOAl	SiOH
NMR (HZSM–5)	96.5	168.4	184.0	124.4	114.5
RHI	96.4±0.4	170.0±0.2	194.5±2	131.5±5	117.5±3
ND (HY)	105.1	164.8	141.3	108.5	8.417

It appears then that the acid properties of these materials should be correlated with the electronegativity of the solids as well as with the corresponding proton abstraction energies. The deprotonation energy is sensitive to lattice relaxation and depends on local geometric constraints due to the long range ordered structure of the zeolites, as well as the zeolite composition[14].

Taking this into account, and from a catalytic point of view, it is clear that the number and strength of the Brønsted acid sites will strongly depend on both framework chemical composition and geometrical structure.

If the Brønsted acid sites are associated with the presence of framework Al, it can be directly deduced that the number of potential active sites in an acid catalysed reaction will be the same as the number of framework Al atoms. Thus, if the purpose is to maximise the total number of Brønsted acid sites, one should prepare samples with the lowest Si/Al ratio possible, which turns out to be a ratio of one. However, considering the average Sanderson electronegativity of the resultant zeolite framework, the strength of the resultant acid sites will be the lowest possible[15], and therefore, regardless of possible stability limitations, the acid sites will only catalyse weakly demanding reactions. When more demanding reactions are to be catalysed, stronger acid sites will be required and zeolites with higher framework Si/Al ratios will have to be prepared, since a decreasing heterolytic proton bond dissociation energy correlated with an increasing Si/Al ratio[16]. For a given structure, the maximum strength of the Brønsted acid sites will occur when the Al tetrahedron is completely isolated. Therefore when the number of strongest acid sites has to be maximised, samples with a specific framework Si/Al ratio will have to be prepared, and this will change from one structure to another[17].

Besides framework compositional effects, the geometrical structure of the molecular sieve can also have an influence on acidity. In this way, the effect of the structure on acidity has been related to differences in the T–O–T angle[18], such that the deprotonisation energy decreases with increasing T–O–T angle, and consequently the proton becomes more acidic. However Sauer et al.[19] have noticed that when examining the results for mordenite and ZSM-5, one has to conclude that energy differences between acidic sites at the various lattice sites are small.

Mortier[20] has considered the influence of the zeolite structure on the deprotonation energy by means of the Sanderson's average effective electronegativity parameter. Thus he concludes that the highest values of effective electronegativity were calculated for zeolites with the most open frameworks (lower framework densities).

Another way of changing the strength of the acid sites in a given zeolite structure while keeping constant the number of acid sites, is to carry out the isomorphic substitution with either T^{IV} or T^{III} framework atoms (where T = transition metal) in such a way that the larger the electronegativity of the atoms the stronger will be the acidity of the resultant site. Thus, substituting

Al^{3+} by Ga^{3+} and B^{3+} atoms, or Si^{4+} by Ge^{4+} atoms, the strength of the resultant sites will be modified.

SAPO type microporous materials can be considered to be derived from the corresponding electrically neutral ALPO materials by introducing Si into the framework. In this case, the replacement of P^{5+} by Si^{4+} produces a negative charge in the framework which can be compensated by a proton in an Al–OH–Si centre similar to that of zeolites. Therefore, in the case of SAPOs, the number of acid sites will be related to the pressure of Si atoms, but in this case (owing to the different substitution mechanism[21]) it is not possible to establish a direct correlation between the number of Si and the number of H$^+$ species, as could be done in the case of zeolites for the number of framework Al^{3+} atoms.

In the case of SAPOs, the acid strength of the material will be lower for structures in which the Al supporting the bridging OH is surrounded by P and the Si by Al (Structure I). On the other hand, the strongest acid sites will be present in SAPOs containing Si rich zeolitic regions in which both the Al an Si are surrounded by Si atoms (Structure II).

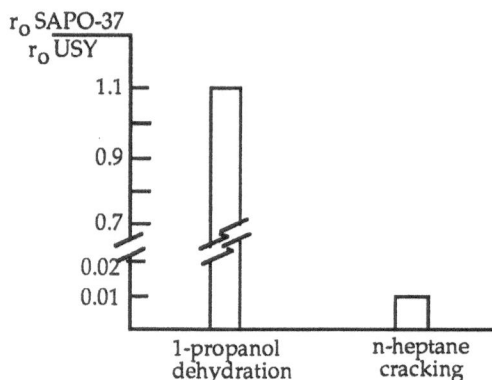

STRUCTURE I   STRUCTURE II

The effect of compositional changes and acid strength in SAPOs and their consequences on catalytic activity are illustrated when the activity of SAPO–37 for catalysing reactions requiring increasing acidity such as alcohol dehydration, aromatic alkylation, alkyl aromatic isomerisation and paraffin cracking are compared with that of a USY zeolite (Figure 1).

Figure 1. SAPO–37 to USY ratio of initial reaction rates for *i*–propanol dehydration and *n*–heptane cracking

It is possible to see that the activity of SAPO–37 is larger for a reaction such as isopropanol dehydration which requires weak acid sites. However, this difference becomes smaller and is even reversed when the reactions catalysed by stronger acid sites are considered. Thus, it appears that by considering the range within the group of zeolites and zeotypes one has the possibility of tailoring not only the dimensions of the pores, but also the acidity needed for carrying out selectively a given reaction.

In this way, zeolites and zeotypes can be used to catalyse a simple reaction such as the double bond isomerization of olefins. A reaction like this does not involve a "free" carbenium ion which could be detected as a stable species, but involves a carbenium-like intermediate in which the C-O bond stretching occurs simultaneously as the carbenium-like intermediate is transferred to one of the neighbouring basic oxygens.

Stronger acid sites are needed if instead of the double bond isomerisation one desires to isomerise butene to *iso*butene. Indeed, from the viewpoint of a classical isomerisation mechanism, as could be that occurring on superacids, a primary carbenium ion intermediate should be formed:

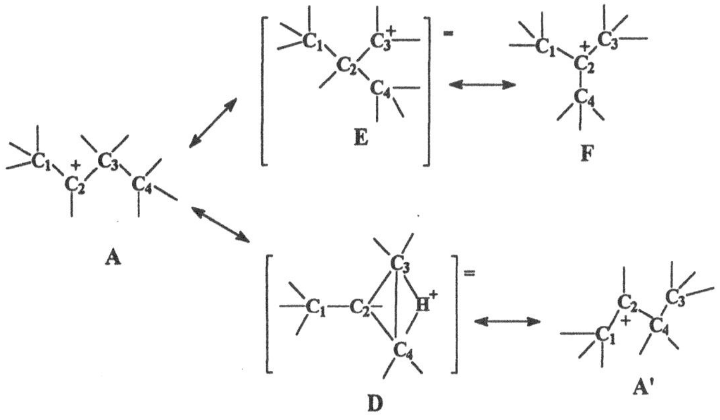

However, in the case of zeolites, one could envisage a more concerted reaction mechanism in which the more basic oxygens of the framework play a specific role and could avoid the formation of a formal primary carbenium ion.

However, with the above reaction, as with many others, there are competing reactions which are also catalysed by the same acid sites and render the selectivity below the desired values. An example is the branching isomerisation of butene which is an important conversion reaction directed to produce *iso*butene which is a starting material for the production of methyl *tert*-butyl ether (MTBE) and is used as an oxygenated octane booster additive in the new reformulated gasolines. Besides the branching isomerisation, oligomerisation also occurs[22]. This takes place especially at the desired low reaction temperatures. Indeed, at lower temperatures the thermodynamic equilibrium is shifted towards branched products[23] (Figure 2).

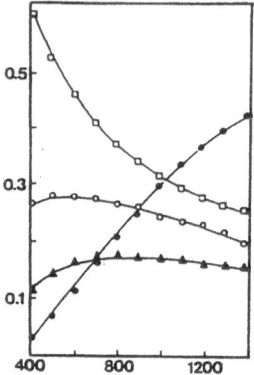

Figure 2. Temperature dependence of the equilibrium concentration of butene isomers: (•) 1-butene, (□) isobutene, (o) trans-2-butene, and (Δ) cis-2- butene.

On the other hand if the reaction temperature is increased in order to avoid olefin oligomerisation, other competing reactions such as cracking, hydrogen transfer, and coke formation can compete with a corresponding negative effect on selectivity. At this point, molecular sieves can be particularly useful catalysts since they can combine the control of acid sites, with the control of pore dimensions which can restrict some undesirable reaction pathways. In this way, medium pore size zeolites and SAPOs geometrically restrict the formation of large oligomers and coke. Moreover, it has been claimed that unidirectional ten member ring (10MR) pore zeolites such as ferrierite, ZSM-22, ZSM-23, Theta-1, SAPO-11 *etc.*[24-28], can catalyse much more selectively the branching isomerization of 1-butene than the larger pore zeolites. In Table 2 we have compared the behaviour of zeolites with different topologies and pore dimensions for this isomerisation reaction. It is seen that among the different structures only those zeolites with unidirectional 10MR pores are selective for the formation of *iso*butane.

The zeolite catalytic properties for acid catalysed reactions can be further modified by changing the framework Si/Al ratio. This is particularly important in those reactions where uni- and bi-molecular reactions are competing. Indeed, while unimolecular reactions can occur on a single acid site, bimolecular reactions seem to prefer the presence of pairs or, at least, proximal acid sites. To visualise this effect we can use the same reaction discussed above, *i.e.* the isomerisation of 1-butene to *iso*butene. The results presented in Figure 3 show that when increasing the framework Si/Al ratio, *i.e.* when decreasing the density of acid sites, the ratio of isomerisation to oligomerisation plus cracking strongly decreases. Again a compromise should be found since when decreasing the number of framework Al the catalyst activity decreases, while the selectivity to *iso*butene increases.

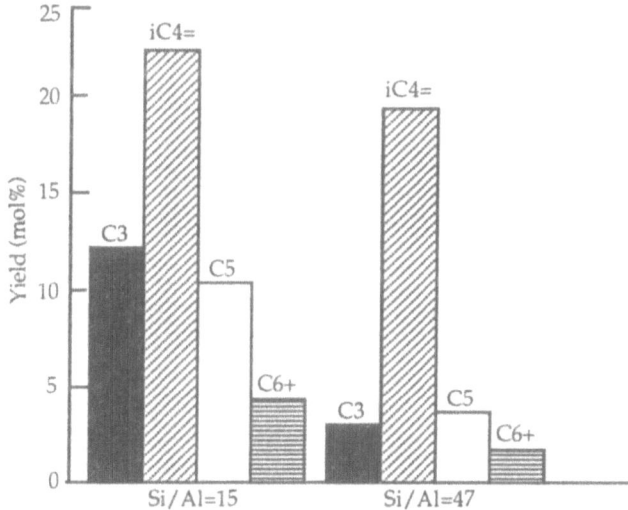

Figure 3. Product yields obtained during the isomerisation of 1-butene on MCM-22 zeolite catalysts with two different Si/Al ratio. Reaction conditions: T=350°C, N_2/1-butene = 9/1, 1-butene WHSV = 28 h^{-1}, TOS = 30 min.

Zeolites are used to catalyse many more commercially important reactions such as electrophilic alkylation of aromatics (benzene, naphtalene, biphenyl *etc.*) with olefins (ethylene, propylene, C_7 olefins)[29-41]. Alkylation of aromatics is a good example of reactions where the zeolite geometrical effects leading to diffusion and transition shape selectivity play a predominant role in controlling the selectivity of zeolite catalysts. In the references given one can see that depending on the size of the reactant and product molecules, shape selectivity does not only apply to medium pore size, but also to large-pore size zeolites, and selective alkylation processes have been developed with both types of zeolites.

One of the most demanding acid catalysis reactions in terms of acidity is the cracking of paraffins and more specifically the cracking of short chain paraffins. Zeolites with high framework Si/Al ratios are able to carry out this type of reaction at moderate temperatures (>350°C). Again the combination of strong acidity, high activity and controlled pore sizes has allowed zeolites to replace amorphous Silica-Alumina in most cracking units. The large pore Y Zeolite is able to deal with bulky molecules present in the gasoil feed and to crack them predominantly within the gasoline fraction[42]. ZSM-5 zeolite is added selectively to crack linear *versus* branched alkanes

TABLE 2. Influence of zeolite topology on isobutene selectivity during the skeletal isomerization of 1-butene at similar n-butene conversion level[a].

Zeolite	Structural characteristics		Si/Al ratio	n-Butene conv. (mol%)	Isobutene selectivity (mol%)
	Topology	Channel system			
Mordenite	MOR	[001] 12 6.5 x 7.0* ↔ [010] 8 2.6 x 5.7*	20	53.9	18.3
Beta	BEA	[001] 12 7.6 x 6.4* ↔ <100> 12 5.5 x 5.5**	13	48.8	34.1
ZSM-12	MTW	[010] 12 5.5 x 5.9*	49	44.1	18.6
ZSM-5	MFI	[[010] 10 5.3 x 5.6 ↔ [100] 10 5.1 x 5.5]***	50	51.5	18.4
MCM-22		{12 7.1 x 7.1 x18.2}*Ò 10**	47	49.9	53.4
ZSM-22	TON	[001] 10 4.4 x 5.5*	77	37.0	50.7
Theta-1	TON	[001] 10 4.4 x 5.5*	64	44.6	59.2
Ferrierite	FER	[001] 10 4.2 x 5.4* ↔ [010] 8 3.5 x 4.8*	5	53.7	66.3

a) Reaction conditions: 350°C, P_{1-C_4} = 0.1 atm, TOS= 30 min.

and alkenes present in gasoline, increasing, therefore, the octane number of the resultant naphtha, while increasing the yield of propylene in the gas fraction[43-46]. Again, by controlling the pore size of the zeolite cracking additive it is possible to change the cracked product distribution. Indeed, by using a large pore zeolite such as Beta, or a zeolite with unconnected 10 and 12 MR pores, such as MCM-22 one can retain a good gasoline selectivity while increasing the RON of the gasoline and shifting the C_4 and C_5 products to the more desirable isobutylene and isoamylene products[47,48].

It can thus be concluded that by combining the adequate zeolite acidity, acid site density and pore size, it is possible to direct the cracking of hydrocarbons towards the most desired products.

In some cases it is possible to prepare bifunctional catalyst by combining on a molecular sieve the acidity function with a hydrogenating-dehydrogenating (metal or transition metal) function. For instance, zeolites supporting metal platinum are used to isomerise n-paraffins or to hydrocrack gasoil. In such cases the reaction mechanism involves first the dehydrogenation of the paraffin on the metal site to form the n-olefin, followed by the isomerisation of the n-olefin on the acid site to form the isoolefin, which is then finally hydrogenated on the metal site:

$$\text{n-paraffin} \underset{\text{metal}}{\overset{-H_2}{\rightleftarrows}} \text{n-olefin} \overset{+H^+}{\rightleftarrows} \text{sec } C^+$$

$$\text{isoparaffin} \underset{\text{metal}}{\overset{+H_2}{\rightleftarrows}} \text{isoolefin} \overset{-H^+}{\rightleftarrows} \text{tert-C}$$

In the bifunctional mechanism described above, and if the adequate balance between the metal and the acid function is achieved (Pt atom/acid site $\geq$ 0.15)[49], the rearrangement of the carbenium like-ion becomes the controlling step. In this case large pore zeolites with very strong acid sites are required, and mordenite has given excellent results to isomerise short chain paraffins50, while large pore less acidic zeolites such as USY and Beta are more adequate to isomerise long chain paraffins and naphthenes, as well as to hydrocrack vacuum gasoil[51-53].

Bifunctional zeolite catalysts also involve Ga_2O_3 on ZSM-5. This catalyst is used to convert propane and butane into aromatic molecules. The n-alkane is first dehydrogenated to propylene on the gallium function and then the olefin oligomerizes and cyclizes giving the final aromatics. It is claimed[54], that propane forms a protonated pseudocyclopropane species:

$$CH_3H_8 + \underbrace{Ga^{3+}O_2^-}_{\substack{Ga_pO_r \\ Ga_{ionex}}} \quad \underset{Z}{\overset{H}{\underset{|}{O}}} \rightleftarrows \left[\begin{array}{c} \overset{CH_2}{\diagup} \searrow H^+ \\ CH_2-CH_2 \\ H^\delta \quad H^{\delta+} \\ Ga^+ \quad O^{2-} \quad O^- \\ \underset{Z}{|} \end{array} \right]$$

These species are formed by interaction with a $Ga^{3+} O_2^{2-}$ ion pair and on adjacent Brønsted acid site. The $Ga^{3+} O_2^-$ ion pair is probably situated on a highly dispersed gallium oxide species. The protonated pseudocyclopropane species can evolve in various ways, the most important being towards a $C_3 H_7^+$ carbenium ion and molecular H_2. A general reaction scheme for this process can be summarised in the following diagram, but other hydrogenating functions such as ZnS, Zn/Pt and Pt/Ga_2O_3 have also been used.

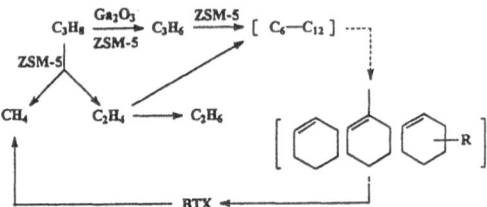

Up till now we have seen that it is possible to generate acid sites on the framework of the microporous molecular sieves within a large range of controlled acidities. They are able to catalyse a large number of hydrocarbon reactions. While, the possibility of controlling number and strength of acid sites is important for acid catalysis, what really makes zeolites and zeotypes unique, is the possibility of combining the adequate acidity with the required pore size requirements. When the reactions occur in the confined space of the pores and cavities, shape selective effects together with activation of the molecules due to the strong electric fields and to an electronic confinement[55,56] take place, converting the zeolites into highly active and selective acid catalysts.

3. Basic Sites in Zeolites: Base Catalysed Reactions

3.1 FRAMEWORK BASIC SITES

Brønsted basic sites are not observed in zeolites. Although there are hydroxyl groups in zeolites that are not acidic, as for instance those associated with exchanged di- and tri-valent cation and extra framework Al species, it is not possible to claim that they are Brønsted basic sites. In the case of zeolites and zeotypes one has to look for the potential basic sites on the framework oxygens wich are negatively charged due to the isomorphous substitution of a T^{IV} by T^{III} in zeolites, and a P^{+5} by Si^{+4} in ALPO-SAPO type structures. In a zeolite, the basicity corresponds to the conjugated base of a given acid centre. Therefore, the stronger the acid site the weaker will be the conjugated base. Following on from this, the criteria given above to rationalise the acidity in zeolites will also be adequate to rationalise their basicity. Thus, considering compositional factors, it was claimed above that the larger the average Sanderson electronegativity of a given zeolite, the more acidic it is. So it is possible to say that the lower the average Sanderson electronegativity of a given zeolite the more basic it will be. This is reflected, for instance in a series of faujasite zeolites in which the framework Si/Al ratio and the counter cation has been altered (Figure 4)[57,58]. It is shown that the lower the Si/Al ratio and the lower the charge to radius ratio of the countercation the more basic the zeolite will be. Moreover, in analogy with acidity in zeolites in which the T atoms are changed, as for instance Al by Ga or Si by Ge, the basicity changes with an inverse relationship to the average Sanderson electronegativity[59,60].

With respect to the zeolite structure, the electronic charge on oxygen *i.e.* the basicity, decreases as the TOT angles are narrower and the distances longer[61]. In the Electronegativity Equalisation Method (EEM) not only the composition but also the influence of the structure account for the oxygen charge[62].

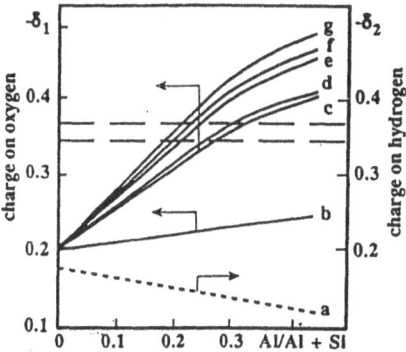

Figure 4: Changes in the calculated charges of proton (a), and oxygen (b-g), as a function of Al content. Zeolites exchanged with protons (a,b), Li (c), Na (d), K (e), Rb (f) and Cs (g).

3.2 ZEOLITE EXTRA FRAMEWORK BASIC SITES: METAL AND ALKALINE OXIDE CLUSTER

Sodium metal clusters can be generated in zeolites by treating the latter with alkali metal vapour, either as such or by decomposition of NaN_3. Other methods to produce alkali metal clusters involve adding organoalkali metals to NaY or using sodium solutions in liquid ammonia[63-65]. With this method Na_6^{5+} and Na_4^{3+} paramagnetic centers are formed in alkali X and Y Zeolites[66], and simultaneously small neutral metal particles may be generated inside and/or outside the zeolite framework[67-69]. These particles induce strongly basic sites in zeolites which are believed to be basic framework oxygen in the vicinity of a neutral Na^0 cluster entrapped in the zeolite cages, rendering the ionic clusters inactive[70,71].

Another way of producing strong basic sites in zeolites involves the formation of extra framework fine particles of alkali oxides inside the cavities of zeolites[72,73]. These particles are obtained by impregnation of faujasite zeolite with an aqueous solution of CS acetate and then decomposing the salt by calcination at 723K. The preparative conditions are very important to preserve the crystallinity of the zeolite. Such alkali ion zeolites are easily destroyed by exposure to water vapour at high temperatures, and furthermore zeolites of high Si/Al ratios are unstable to alkali treatment[74].

Small crystallites of alkaline-earth, and more specifically MgO have been prepared within the cavities of zeolites[75]. The basicity developed by these zeolites is stronger than that of the original zeolites but lower than bulk MgO.

There is no doubt that basic zeolites and zeolites containing basic catalysts can possess shape selectivity which is produced by the microporous molecular sieves. However, as with acid catalysts one needs to measure the basicity of the sites in order to adapt then to the needs of a given reaction. However, measurement of the basicity of basic zeolites is not an easy task. Many techniques have been used up till now, and new ones are being developed, suggesting that none of them gives results good enough to become the most adequate or the preferred technique.

3.3 CHARACTERISATION OF BASIC SITES IN MICROPOROUS SOLIDS.

The ultimate purpose of the direct methods used for the characterisation of basic sites is to determine the density of charges on the oxygens (basic sites) of the zeolites. One can do this by XRD by measurement of the TOT bond angles which are correlated to the oxygen charges[76]. However, it has to be taken into account that the measured angles represent only an average value for any oxygen type considered, and therefore it is very limited.

The binding energy (BE) of the oxygen in a zeolite structure (O_{1S}) is related with the charge on the atom, in such a way that the lower the value of O_{1S} BE, the electron pair donation (and therefore, the basic strength) becomes stronger. Several authors[77-81] have studied the basicity of zeolites by XPS, and they conclude (Figures 5 and 6)[82] that the O_{1S} BE increases when increasing the Al/Si ratio, and more generally when decreasing the average Sanderson electronegativity. However, two things that one must consider are: when using this technique to measure the basicity of oxygens in zeolites it may occur that since XPS examines only the surface of the zeolite crystal, and since compositional gradients may exist, the values calculated may not be representative of the zeolite bulk. Also the presence of coke has to be avoided in XPS studies since a splitting of the O_{1S} peak may result from the charge donation of coke to the framework oxygen[83].

There are other indirect methods of measuring basicity that use probe molecules whose interaction is studied by different techniques such as temperature programmed desorption, spectroscopic (I.R., U.V., X.P.S., N.M.R.) methods, and test reactions. Highly simplifying the matter, it can be said that the most commonly used characterisation methods for measuring the basicity of zeolites, use probe molecules such as chloroform, pyrrol, CO, CO_2, CH_4, acetonitrile, alcohols, and organic acids (benzoic, acetic, trichloroacetic), and boric acid trimethylesters which can adsorb on the basic sites and react with them forming species which can be detected by the different techniques named above[84-96]. Benzene, has also been used as a probe molecule, since it interacts with framework oxygens and shifts both the CH and the C-C band[97-99].

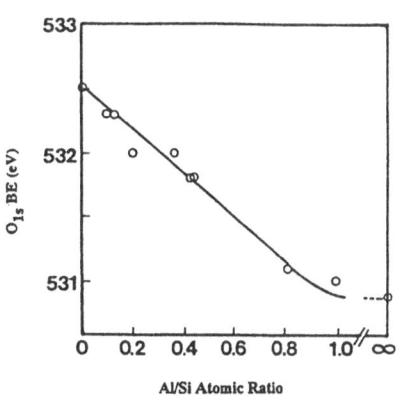

Figure 5: Correlation between the binding energy of the O_{1s} band and the Al/Si atomic ratio.

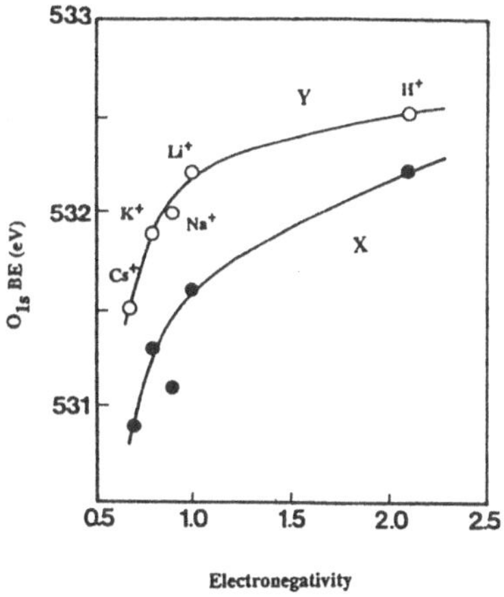

Figure 6: Binding energy of the O_{1s} band for cation exchanged zeolites as a function of the cation electronegativity (×): (l) Y-zeolite and (O) X-zeolites

From the point of view of reaction testing, double bond isomerisation, isopropanol dehydrogenation, toluene alkylation by methanol, cyclisation of diketones, and hydrogen transfer from cyclopentanol to cyclohexanone have been used[100-104]. Recently, a test reaction which can measure the base strength of the catalyst has been developed. This involves the use of esters containing methylenic groups in which the activated H to be abstracted by the basic site have different pKa (different acidities) values. These esters (ethylcyanoacetate, ethylacetoacetate, and diethylmalonate) are reacted with benzaldehyde by means of a Knoevenagel type condensation. The more basic a given catalyst will be a less acidic methylenic group it will be able to activate. Thus, by comparing the results obtained with the solid catalysts to those obtained with amines with well defined PK_b values, the basicity of the catalysts can be evaluated[105-107].

This test reaction has been successfully applied to study the basicity of zeolites with different framework and non framework compositions.

3.4 CATALYSIS ON BASIC MICROPOROUS MOLECULAR SIEVES

The intrinsic basicity of zeolites and zeotypes is rather weak. The strongest basicity is obtained for zeolites with a T^{III}/T^{IV} ratio close to one, in which the T^{IV} atom is Ge instead of Si, and Cs is used as the countercation[108]. Basic zeolites can be used for catalysing a large number of reactions such as alkylation of alkylbenzenes, aromatic amines, and phenol by methanol, hydrogenation of unsaturated hydrocarbons or CO, dehydrogenation of alcohols, and condensation reactions[109-124].

Larger activities due to a stronger basicity are obtained when using occluded alkali metals in zeolites[126,127]. Meanwhile the strong basicity combined with the limited space within the zeolite cavities can give high selectivities to the products desired.

Therefore, in the case of basic zeolites, one prompts a series of base catalysts in which the basicity can be controlled from a very low value such as in the case of NaY zeolites, to progressively higher basicities in CS-GeX zeolites, and culminating with the very strong basicities generated when occluded in the cavities of faujasite alkaline metals and alkaline oxides. This control of basicity and restricted reaction space, can be of particular interest in the case of producing fine chemicals where the most important feature is high selectivity.

In order to illustrate this, two types of reactions are presented, one catalysed by the intrinsic basicity of an alkaline exchanged faujasite zeolite, and the other in which alkaline oxide particles have been introduced into a ZSM-5 structure.

We have seen above that faujasite exchanged with alkali cations are able to catalyse selectively Knoevenagel type condensations. This ability can be used to prepare calcium antagonist intermediates. Indeed, for the preparation of 4-aryl-1, 4-dehydropyridine derivatives, there are two general routes (a, and b) which involve variants of the Hantzsch reaction:

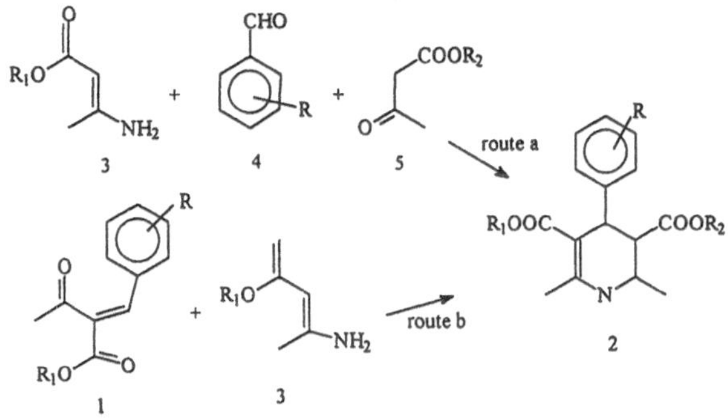

In route a, an alkyl aminocrotonate (3) is condensed with a substituted benzaldehyde (4) and an alkyl acetoacetate (5) under reflux in methanol or ethanol, to give the corresponding 1,4-dihydropyridine (2) in *c.a.* 35% yields. On the other hand, route b involves the reaction of an alkyl benzylideneacetoacetate (1) with an alkyl β-aminocrotonate (3), which involves consistently better yields (34-71%) and easier purification. By using zeolites as basic catalysts, it was possible to carry out the synthesis of intermediate (1) using a series of alkaline exchanged zeolites X, Y and GeX:

4a-e   1a-e

It was thus possible to produce the intermediates 1a-e with conversions close to 70% with 100% selectivity using a NaXGe zeolite:

	R_1	R_2	R_3	Conv. (%)
a	H	H	H	66
b	NO_2	H	H	50
c	H	NO_2	H	55
d	Cl	H	Cl	56
e	CF_3	H	H	25

In the case of zeolites containing small alkaline oxide particles, Gortsema *et al.*128, using Cs modified ZSM-5, synthesised 4-methylthiazol, which is an intermediate in the production of pharmaceuticals:

$$SO_2 + (CH_3)_2 C = NCH_3 \xrightarrow{470°C} \text{(4-methylthiazol)} + 2H_2$$

It appears, therefore, that zeolites either exchanged with alkaline cations, or containing alkaline oxide particles are promising base catalysts and their possibilities need to be studied for other reactions of interest, both from the fundamental and the applied point view.

4. Redox Microporous Molecular Sieves

One of todays main driving forces in chemistry is to find new processes and/or to modify existing ones to render them environmentally friendly. In the case of catalytic oxidations, there are many liquid phase processes which still use soluble transition metals which in some cases are used even in stiochiometric amounts[129]. The tendency, therefore, has been to support them in a way that leaching does not occur. In this sense Shell developed a Ti-Silica catalyst in which amorphous Silica was treated with $TiCl_4$ such that in the resultant material the Ti atoms were bonded to four Si through oxygen atoms:

$$Si-O \diagdown \diagup O-Si$$
$$Ti$$
$$Si-O \diagup \diagdown O-Si$$

These isolated Ti atoms were the active sites for carrying out epoxidation of olefins using organic hydroperoxides as oxidants. This catalyst, however, was inactive when H_2O_2 was used as the oxidant.

A logical step was to try to introduce Ti into the framework of molecular sieves either during the hydrothermal synthesis or in during post synthetic chemical treatment. The first attempt was as early as 1967[130], but the claimed crystalline titanosilicates were dismissed by different authors[132]. It was in 1981 that the efforts were successful and Tamarasso et al.[133] described the synthesis of a titanosilicate with MFI topology, which was called TS-1. Other Ti containing zeolites with topologies MEL[134], BEA[135], and ZSM-48[136] have also been synthesised, with the Ti-Beta zeolite being the only large pore zeolite showing catalytic activity that has been synthesised up till now.

4.1   NATURE OF ACTIVE SITES IN Ti-ZEOLITES

The active oxidation sites in Ti-zeolites are considered to be the Ti^{IV} atoms isolated in a silica matrix with every single Ti^{IV} atom surrounded by O-Si-O-Si-O in all directions. By taking this into account the formation of Ti-O-Ti bonds are avoided. More specifically, in the case of TS-1 two structures have been proposed for these isolated Ti^{IV} species: the titanyl form (A) and the coordinated form (B):

A   B

By means of UV-Vis spectroscopy it has been claimed that the titanyl form is unlikely to be present, as the tetrahedral coordination is the most probable. The tetrahedral geometry of Ti in dehydrated TS-1 has been confirmed by

EXAFS-XANES[137]. However, the Ti^{IV} atom can coordinate molecules such as H_2O and NH_3 generating Ti^{IV} species with a six coordination fold[138]. In the presence of H_2O, Ti-O-Si bonds can easily be hydrolysed to form Ti-OH and Si-OH species[139]:

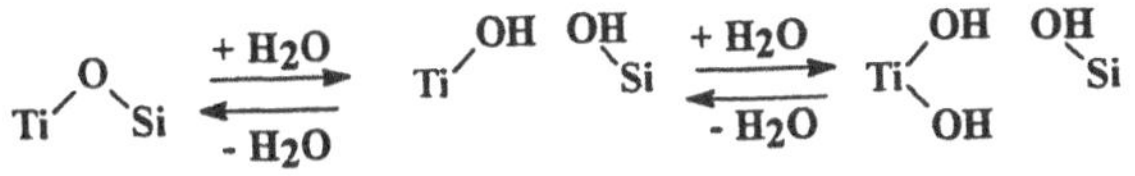

In the case of Ti-Beta zeolites the structure of the framework Ti has been studied by EXAFS-XANES[140], and it has been shown that Ti^{IV} can be tetrahedrally penta- and hexa-coordinated, as is presented in the following scheme:

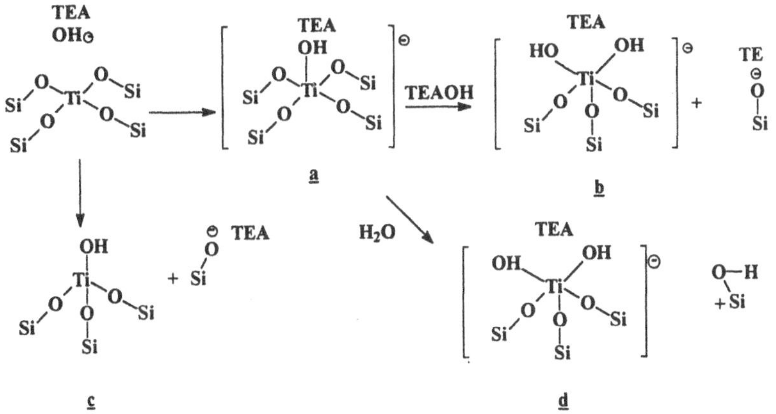

The isolated Ti^{IV} sites interact with H_2O_2 and peroxo and hydroperoxo species are formed[141]:

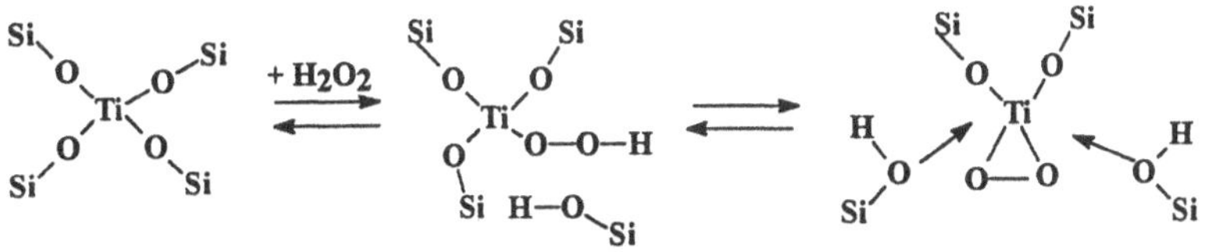

These two types of species have been proposed for group IV-VI transition metal compounds, and the hydroperoxo species have been proposed as intermediates or active species in oxidation reactions with hydroperoxides[142].

Ti has also been introduced in SAPO structures and the presence of framework Ti is claimed[143]. Besides Ti-zeolites, other molecular sieves containing elements with potential oxidation activity such as V, Cr, and Co have been prepared[144-148]. In Table 3 a series of molecular sieves prepared with different transition metals are presented:

Table 3. Redox Molecular Sieves with incorporated transition metals.

Structure Type	Framework Structure	Ring number	Pore size	Dimensionality	Metals incorporated
MFI	ZSM-5	10	5.6X5.4	3	Ti, Zr, V, Fe, Cr, Ti
MEL	ZSM-11	10	5.1X5.5	3	Ti, Zr, V, Fe, Cr
ZSM-48	ZSM-48	10	5.4X4.1	1	Ti
BEA	Beta	12	7.6X6.4	3	Ti, V, Fe
MCM-41	MCM-41	--	30-100	1	TI, V, Fe
MCM-48	MCM-48				
AEL	ALPO-11	10	6.3X3.9	1	Co, Mn, Cr, V, Ti
AFI	ALPO-5	12	7.3	1	CO, Mn, Cr, V, Ti

4.2 CATALYTIC ACTIVITY

4.2.1. Ti-zeolites

Redox molecular sieves have been used to catalyse the oxidation of many organic molecules, using H_2O_2, organic hydroperoxides and O_2 as oxidising agents. When using H_2O_2, TS-1 has been shown to be an active and selective catalyst for the hydroxylation of phenol to catechol and hydroquinone[149,150].

The results obtained with TS-1 are compared with those produced using the current technology involving HO_4Cl, and Fe^{2+}/Co^{2+}, in Table 4.

TABLE 4: Results obtained with the different commercial process.

Catalyst	TS-1	Fe^{2+}/Co^{2+}	HO_4Cl
H_2O_2 Conv. (%)	100	-	-
H_2O_2 Select. (%)	82	50	70
Phenol Conv. (%)	27	10	8
Phenol Select (%)	91	80	90
Para/ortho	1.1	0.50	0.70

The use of the heterogeneous TS-1 catalyst, besides having environmental advantages, gives a high selectivity to Catechol and hydroquinone with a larger amount of the most desired para substituted (catechol none) product. These results can only be attained when TS-1 samples are well prepared and no extraframework Ti is present. In the case of the larger pore Ti-Beta zeolite, the hydroxylation reaction also occurs[151], but the selectivity to H_2O_2 is somewhat lower and the process is less selective towards the formation of catechol. It is therefore apparent that the pore dimensions play an important role on the selectivity.

Another important process in which TS-1 has shown good catalytic properties up to the point that it has started to be commercialized is the preparation of cyclohexanone ammoximation. This is a precursor chemical of E-caprolactama which is used to manufacture nylon-6. The classical route forms the cyclohexanone oxime by reacting hydroxylamine and cyclohexanone producing high amounts of ammonium sulfate as the by product. However, when TS-1 was used as a catalyst a new process was developed[152] which used NH_3, H_2O and cyclohexanone as reactants:

The most probable mechanism for this reaction involves the formation of hydroxylamine on Ti sites, followed by the attack of this to form cyclohexanone:

Ti zeolites are also active and selective catalysts to carry out the oxidation of primary and secondary alcohols to the corresponding aldehydes and ketones, respectively. In the case of TS-1 Zeolite Catalysts it has been observed that the order of reactivity is:

$$CH_3OH << \text{primary} < \text{secondary alcohol}$$

Furthermore, it has been observed[153], that the closer to the end of the chain the hydroxyl group is situated in a secondary alcohol the more reactive it is. Other factors which imply a slower diffusion of the reactant, such as increasing the chain length of the alcohol or the presence of branched alkyl chains give slower rates than linear chains.

In the case of a large pore Ti-Beta zeolite it has been found that its activity for oxidizing alcohols[154] follows in general terms the same behavior as TS-1. However, and due to the larger pore diameter of Ti-Beta, it is more active when dealing with branched alcohols.

The full kinetic study shows that the reaction is first order with respect the alcohol and H_2O_2, and depends inversely on the concentration of water.

A possible mechanism for the oxidation of alcohols by H_2O_2 on Ti zeolites which would be consistent with the reaction kinetics found would involve the interaction of one molecule of alcohol adsorbed on the Ti, with the hydroperoxide group formed by the interaction of H_2O_2 with the Ti155,153.

Ti atoms in the Silicalite and Beta structures are active and selective catalysts for the epoxidation of olefins. When short chain lineal olefins are to be oxidized,TS-1 is more active than Ti-Beta exhibing a higher intrinsic activity, however, when cycloalkenes and branched alkenes are epoxidized Ti-Beta is more active than TS-1 due to the larger pore diameter of the former zeolite. This shows again the advantage of having large pore Ti-zeolites for oxidizing bulk molecules. Geometrical effects are also demonstrated by the fact that organic hydroperoxide (terbutyl hydroperoxide) can be used as an oxidating agent on Ti-Beta but is inactive when TS-1 is used as a catalyst156. Furthermore, the TS-1 gives a very high selectivity to epoxides, whereas the Ti-Beta, due to the presence of framework Al and therefore Brønsted acid sites opens the epoxide forming glycols and glycolethers156. There is one way to increase epoxide selectivity and this is through elimination of the Brønsted acidity in Ti-Beta. One way is to poison the acid site by Na^+; this has been done by treating the zeolite with sodium acetate157,158. In this way the selectivity to the epoxide was increased, however, the conversion was decreased. A high increase in epoxide selectivity has been obtained by using a polar and aprotic solvent such as acetonitrile instead of one protic solvent such as methanol (159). This indicates the important role of the solvent when the activity and selectivity of a hydrophobic zeolite such as TS-1 and a hydrophilic zeolite such as Ti-Beta have to be compared.

There is no doubt, however, that the most elegant way to supress the acidity in aluminum containing Ti-Beta would be to synthesize the sample without any framework Al or any other M^{III} element. This has been done recently160, and it has been shown that, indeed, the selectivity to epoxide increases when carrying out the epoxidation of olefins using H_2O_2. It has to be remarked that when tert-butyl hydroperoxide is used instead of H_2O_2 very high selectivities to epoxide are obtained161.

A reasonable mechanism for the epoxidation of olefins on Ti-zeolites has been proposed by Clerici et al.162:

It has been presented that CH_3OH has an accelerating effect in the epoxidation of olefins, and this has been explained considering the participation of a CH_3OH molecule instead of water as a ligand[163].

A much more difficult reaction is the oxidation of paraffins. With respect to this, secondary and tertiary carbon atoms of paraffins are efficiently oxidized by H_2O_2 on TS-1 to the corresponding alcohols and ketones[164,165]. Ti-Beta on the other hand gives very low activity for oxidising paraffins and some better results are obtained with cycloalkanes, for which TS-1 is much less active, indicating the limited reactivity of TS-1 for the oxidation of the more bulky hydrocarbons. With respect to the reactivity of different secondary carbon atoms, it has been presented166 that their reactivity is not equivalent but dependent on the solvent in which the reaction is carried out. More specifically, in protic solvents (alcohols) the positions nearest to the end are preferentially oxidized, while on aprotic polar solvents (acetronitrile) γ-CH_2 reacts at a higher rate than β-CH_2. In the case of alkyl benzenes, there is a preferential oxidation of n-and C_4-hexane with respect to benzene.

A proposed mechanism for paraffin oxidation is the following[167]:

$$Ti\overset{O}{\diagup}O \;+\; H-\overset{|}{\underset{|}{C}} \;\longrightarrow\; Ti\overset{O}{\diagup}OH \;+\; \overset{|}{\underset{|}{\underset{C}{C}}}- \;\longrightarrow\; Ti{=}O \;+\; HO-\overset{|}{\underset{|}{C}}-$$

While it is accepted that the reaction proceeds through an homolytic mechanism, it does not appear that a free radical chain mechanism occurs.

4.2.2. V-, Cr-zeolites

It has been claimed that V can occupy framework positions in zeolites such as ZSM-5, ZSM-11, and Beta168,169; however, one is not totally sure about the isomorphous substitution of vanadium in zeolites, but rather it seems that isolated vanadium species occupy defect sites forming the so-called framework satellites170 which can be represented as:

The V-Silicalites have been used for hydroxylation of phenol[171] with H_2O_2 and they are both active and selective. However, their selectivity with respect to H_2O_2 is clearly lower than in the case of Ti-silicalites, and they also form larger amounts of para benzoquinone. A similar phenomena occurs when n-hexane is oxidized. However, despite their lower selectivity, V-silicalites are more active in the oxidation of primary atoms in paraffins and toluene[172,173].

It has also been claimed that it is possible to carry out isomorphous substitution of chromium into the framework of Silicalite. This is effective for carrying out the oxidative cleavage of olefins with aq. - H_2O_2. However, there are still doubts that Cr retains its framework position after calcination.

4.2.3. Metal containing ALPOs

Several CoAPO molecular sieves have been shown to be active for oxidation of hydrocarbons using O_2 as the oxidating agent[174-178].

These catalysts offer new possibilities for the oxidation of cyclohexane to produce cyclohexanol, cyclohexanone and adipic acid. Indeed, the autooxidation of cyclohexane is an important bulk process which is normally carried out in the presence of small amounts of dissolved cobalt salts. The autooxidation proceeds through radical chain reactions. In the liquid phase and at low temperatures the kinetic chains are long and hydroperoxides are the major products. The rôle of the catalyst is not to activate oxygen, but to catalyse the decomposition of the hydroperoxides and consequently to generate more radicals and thus increasing the overall reaction rate.

CoAPO-5, CoAPO-11 and CoAPO-16 are active and selective catalysts for cyclohexane oxidation. This necessitates that the Co sites in the framework are completely isolated. When the samples are properly prepared, they could be reused several times after recalcination without any apparent loss in activity. When Co is not immobilised in framework positions and exists, for instance, in an exchangeable position of a given zeolite, this highly mobile cobalt ion is not only inactive but can even inhibit the reaction under conditions necessary for spontaneous autoxidation. The solvent used for carrying out the oxidation plays an important role. Table 5 shows that the amount and type of acid used as a promoter strongly affects catalyst stability[176].

TABLE 5: Conversion of cyclohexane in the presence of air (total p= 6.5×10^5 Pa) and acid (acetic = 99%, Sulfuric = 0.1 M., hydrochloric = 0.5 M) at volume ratio substrate/acid = x and with CO (III) APO-5 as a catalyst T=373K; TOF at 10% conversion.

Acid type	X	TOF [mol /(mol CO.h)]	Major Product	Co-Loss of Catalyst
acetic	1	44	Cyclohexanol cyclohexanone	50%
acetic	2	40	cyclohexanol, cyclohexanone, adipic acid	20%
acetic	4	36	cyclohexanol, cyclohexanon, adipic acid	no loss
H_2SO_4	4	not determ.	traces cyclohexanol, cyclohexanone	total damage
HCl	4	not determ.	traces cyclohexanol, cyclohexanone	total damage

When high concentrations of acetic acid are used as a promoter, acid leaching of cobalt from its isolated sites and reprecipitation at the surface can form clusters which can quench the reaction.

Another factor which is of crucial importance for the use of CoAPOs as oxidation catalysts is the Co content. Upon varying Co content, activity increases with cobalt content at low loading. Above a Co/(Co+Al) ratio of 0.01, higher leaching occurs but no further increase in activity is observed. This can be ascribed to a decrease in site isolation at higher cobalt contents.

In conclusion, we have shown in this chapter that the use of transition metal microporous materials as oxidation catalysts, is an increasingly important field which besides its fundamental interest also has remarkable practical repercussions in industrial processes. In all these materials there are two points which are the decisive factors. Firstly the metal has to form part of the framework (this avoids or retards leaching), and secondly site isolation is vital, in order to retain activity and selectivity.

TABLE 6: The typical larger pore zeolites/zeotypes

Material	Ring size	Year discovered	Synthesis media	Inorganic framework composition	Channels/Pores
Cacoxenite	20 - TO_4 ring	-	Naturally occurring	Al, Fe, P	14.2 Å pore diameter.
Zeolites X/Y FAU (178)	12 - TO_4 ring	1950s		Al, Si	7 Å diameter pore. 12 Å diameter cavity. 3-D channel system.
ALPO$_4$-8 AET (178-181)	14 - TO_4 ring	1982	n-dipropylamine template	Al, P	1-D channel system.
VPI-5 VFI (178,182,183)	18 - TO_4 ring	1988	Tetrabutyl ammonium/ n-dipropylamine templates.	Al, P	13 Å channel diameter. Hexagonal arrangement of 1-D channel system.
Cloverite CLO (178,184-186)	20 - TO_4 ring	1991	(a) Quiniclidinium template. (b) F^- rather than OH^- as mineraliser.	Ga, P	Largest aperture of window is 13 Å. 30 Å cavities. 3-D channel system.
JDF-20 (6,187)	20 - TO_4 ring		(a) Triethylamine template. (b) Glycol solvent.	Al, P	Hydroxyl groups protruding into channel system.

5. Mesoporous Molecular Sieves

We have seen that microporous molecular sieves are of great help as catalysts for reactions involving reactants and products with molecular diameters below 8-9 Å. However, there are many reactants with sizes above these dimensions whose diffusion, and therefore whose reactivity, is restricted when the microporous zeolites are used as catalysts. Owing to this researchers have increased their efforts to synthesise molecular sieves with larger pores. The strategy used was based on the fact that most of the organic template molecules affect the gel chemistry and act as void fillers in the growing porous solids. Thus, attempts were made at using larger organic templates which would hence result in larger voids. However, such attempts failed and no ultralarge pore zeolites were synthesized following this procedure, however, only ultra-large pore zeotypes were actually synthesised by substituting Al and Si atoms in the traditional aluminosilicate gels with Al and P, or Ga and P atoms (Table 6). However, it has very recently been presented[177] that a 14 MR unidirectional zeolite could be synthesized using a Co organometallic complex as the template. This obeyed to the following strategies as illustrated in Table 7.

TABLE 7: The major routes employed by zeolite synthetic chemists to increase the pore size of microporous zeolites and zeotypes.

Method employed to increase pore size	Examples	Structures
Use specific spacing units to build the inorganic framework	Addition of further four ring building units to six ring units in porous aluminophosphates	$ALPO_4$-5 framework further extended to VPI-5 (182,183,188)
Use different oxide systems	Use two sorts of tetrahedral atoms to yield different T-O bond lengths	VPI-5 (aluminium and phosphorous) cloverite (gallium and phosphorous)
Use specially designed templates	Exploit the specific structure directing effect of an organic template	Use of quinuclidine to form cloverite

There exist about 85 framework topologies[178] that have been synthesised and structurally determined and of these the largest free dimensions of windows are present in the porous phosphates[179-181]. The discovery of the first of these compounds in 1982 ($AlPO_4$-8) prompted the question as to whether crystals with mesoporosity can be synthesised.

To summarize so far, that although the unrivalled ability of zeolitic porosity and ion-exchange properties have been exploited in the field of cracking petroleum[189] and catalysing a variety of important reactions many efforts have been made to design regular pore systems in the 20-30 Å range for the cracking of large molecules. Up until the discovery of VPI-5 the largest cavity in a zeolite was in zeolites X and Y. The 12 Å diameter of VPI-5 pores and the 13.2 Å pore in cloverite pioneered a new field to synthesize large pore molecular sieves.

Cacoxenite[190] is a naturally occurring mineral with a 15 Å pore system that, however, is unstable and cannot be used as a catalyst owing to its thermal instability.

Although cloverite possesses potentially large pores the unusual shape of the pore openings (due to protruding hydroxyl groups) potentially restricts the size and shape of molecules that can pass into this mesoporous molecular sieve. Likewise in VPI-5, stacking disorders or deformation of some of the 18 membered rings during dehydration result in a decrease in pore size from 12 Å to about 8.3 Å. So it can be seen that irrespective of the astounding progress made in producing large pore molecular sieves, the materials are still not suited to the current catalytic processes due to a fundamental lack of acidity and stability.

5.1 PILLARED LAYERED SOLIDS

The previous section dwelt on the permanently porous frameworks. A second group of ultra-large pore materials consists of layer structures which can be made permanently porous. These are known as pillared layered solids (PLS)[191-197], simply because they are solids that have a layered structure and molecular pillars in the interlamellar region.

In order to synthesise zeolitic compounds one has to start from the raw materials at the very first step of the hydrothermal synthesis. The gel or solution consequently undergoes the processes of nucleation and crystallization to form the porous crystalline material. In the case of PLSs large pores are formed by introducing compounds in between the layers of inorganic solids that already exist. The pillars should theoretically be of a specific length and be distributed in an ordered manner as to ensure the formation of pores of a specific diameter.

It was Barrer and Macleod in 1955 who first prepared such materials by exchanging alkali and alkaline-earth cations in the smectite clay montmorillonite for quaternary ammonium compounds. However, the resulting material was thermally unstable and it was the introduction of oxyhydroxyaluminium compounds into the interlamellar regions of the solid that yielded PLSs with a better thermal and hydrothermal stability as well as active catalytic centres. Since then it has been discovered that the type of pillaring material to be used is one that:

(a) Cements the layers of the host compound together to provide mechanical and thermal stability;

(b) Provides active centres for a variety of uses including catalysis, sorption and sensor devices.

The literature to date indicates that the materials most investigated as substrates to form PLSs[195] are:

(a) Smectite clays[198] :

(b) Phosphates and phosphates of tetravalent metals (*e.g.* zirconium phosphate)[199-203] :

(c) Layered double hydroxides (*e.g.* hydrotalcite)[204-206].

The types of pillaring species used include:

(a) Organic cations :

(b) Oxyhydroxyaluminium cations[207] :

(c) Zirconium, chromium, silicon, titanium, iron and transition metal complexes.

The fundamental principle is the same irrespective of the pillar used or the substrate used. Most of the pores generated are micropores (10-20Å). This is due to the small size of the pillars. A broad size distribution of mesopores, however, is also introduced. The mesopores are generated due to two process's that are illustrated in Figure 7.

Regardless of the type of mesoporosity the processes occur in rather an uncontrollable manner so that ultimately it is not possible to design pore shapes and broad pore size distributions result. As the three types of lamellar host are chemically distinct it is necessary to discuss each one separately.

5.2 SMECTITE CLAYS

The preparation of molecular sieves based on pillared interlayered clays originally used aluminum Keggin ions $(Al_{13} O_4 (OH)_{24} (H_2O)12)^{7+}$ [195,207] zirconium tetramer $[Zr (OH)_2.4 H_2O]^{8+}$ species and chromium oligomers.

Clays are naturally occurring 3-layer sheet silicates, each sheet of which comprises of two layers with tetrahedrally coordinated atoms surrounding a layer that is formed of octahedrally coordinated atoms. It is due to the substitution of Al^{3+}, Fe^{3+} for Si^{4+} in the tetrahedral layer or Mg^{2+} , Fe^{2+} for Al^{3+} or Li^+ for Mg^{2+} in the octahedral layer that the net negative charge generated in the structure needs to be compensated by interlayer cations, most commonly Na^+ and Ca^{2+}.

Smectite clays195,198 possess a relatively small negative charge and swell spontaneously in water. After swelling in water the interlamellar cations can be exchanged with Keggin, zirconium or chromium cations. Calcination of the exchanged clay induces the formation of stable interlamellar and interpillar links. Since the number of ion-exchange cations necessary to charge compensate is relatively small a porous material with small sized pores is formed. Pore size distributions suggest that the majority of pores obtained are usually in the micropore region.

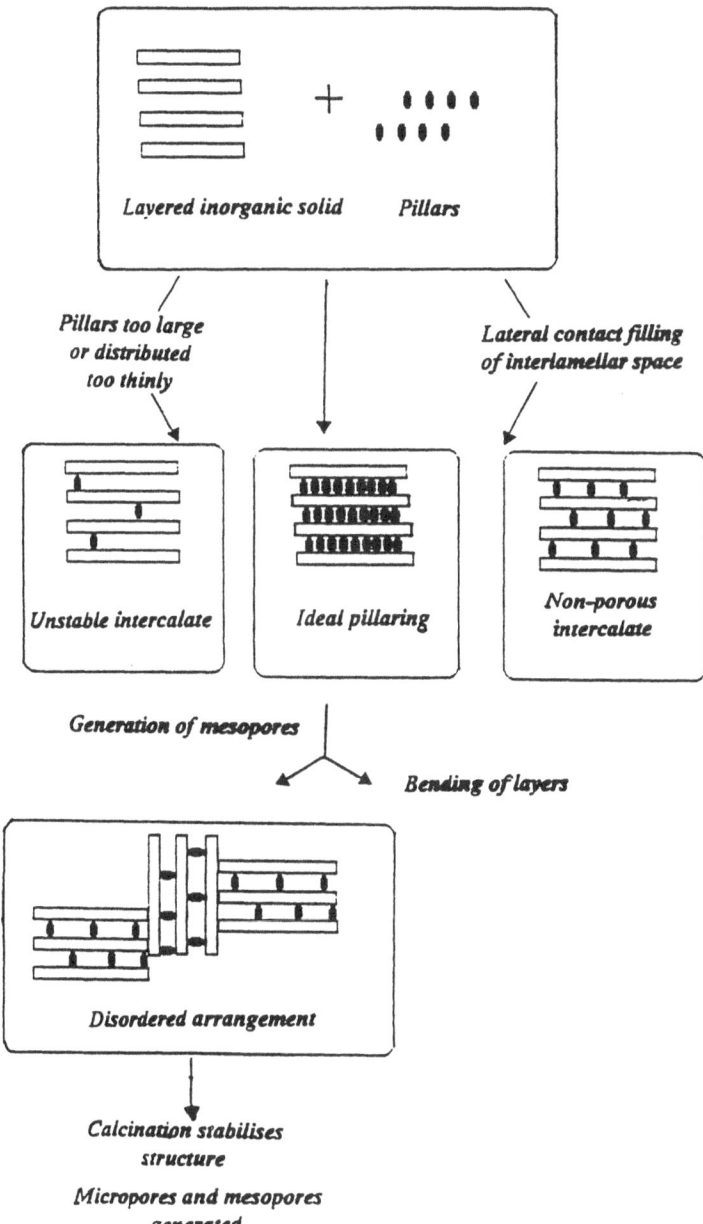

Figure 7: The generation of mesopores in PLSs.

By knowing the charge of the layers and the size and charge of the pillars it is possible to calculate the distance between the pillars. Sorption isotherms can be used to determine the diameter of the pores and also their shape. An advantage to using polyhydroxy species as pillars is that even though the charge on the layers is not homogeneously distributed (which depends on the degree of lower valence cation substitution) the pillars fill the interlamellar region to the same extent, irrespective of the layer charge. This is because the polynuclear hydroxo complex adjusts its own charge to that of the layer charge by hydrolysis so that relatively uniform pores are formed. Mesopores, however, result due to the distortion of layers or due to the gaps in between the sheets of layers that are not fully extended. However, such mesopores are formed as a direct result of the pillars bringing together the layers.

The rather uncontrollable manner of formation of mesopores suggests that novel synthetic routes have to be investigated to ensure uniform large pores with narrow size distributions. One method to attempt this has been to employ more extensively hydrolyzed pillars or metal oxide sols in which they are allowed to react with clays to achieve interlayer spacings even as high as 28Å. Larger pillars, for example, are used in the DIMOS - Direct Intercalation of Metal Oxide Sols - route[208-210]. With this method the exfoliated clays are allowed to react directly with silica or titania particles. The resulting compounds are characterised by the sorption of large molecules.

5.3 TETRAVALENT LAYERED COMPOUNDS

The group IV phosphates are layered compounds that also possess exchangeable cations. They can be represented by the formula $Me(HPO_4)_2 \cdot n\ H_2O$, where Me is a tetravalent metal such as Ti, Zr or Sn[195,199-203]. The oxygen atom of the phosphate ions are octahedrally coordinated to the metal ion, so that both the metal and phosphate ions form the layers. The phosphate groups bridge the metal atoms with three of their four hydrogens and the fourth hydrogen projects into the aqueous interlamellar space. Since these substrates do not swell in water it is necessary to induce swelling so that the pillars may enter the interlamellar regions. The pre-intercalation step involves the introduction of an alkylamine compound that accepts a proton from each phosphate group. This step ensures that the number of cations intercalated is not too high (ensuring porosity) and that the pillaring is more ordered. The alkylamine is then ion exchanged with a Keggin ion that can enter the artificially "swelled" phosphate layer.

The products possess a broad size range of mesopores and because the phosphate groups tend to hydrolyse, a disordered irregular structure is formed. Also in the case of α-zirconium phosphates, for example, the Keggin ions completely fill the interlamellar region leaving no possibility for the formation of pores. Therefore, many new approaches to obtain more ordered mesoporous materials have been investigated and include the following:

(a) Pillar phosphates with chromium acetate $Cr_3(OH)_2$ $(CH_3COO)_7$, as this polymerises and is a suitable size to leave more space within the pillars than Keggin ions that "stuff " the layers.

(b) Pillar zirconium phosphate with silica. For example, it is possible to intercalate zirconium phosphate with an alkylamine such as propylamine and then to reflux the resulting gel with a silylamine to yield a product with for example 33.2Å interlayer spacing that upon heating shrinks to 21.5Å. The isotherm of such a product has a hysteresis loop that indicates well defined mesopores with 40Å diameter.

(c) Pillar zirconium phosphates with organic fillers such as biphenyl groups that link the inorganic layers.

5.4   LAYERED DOUBLE HYDROXIDES

These layered compounds possess structures based on that of $Mg(OH)_2$, brucite. A sequence of neutral OH-Mg-OH-Mg-OH species are found perpendicular to the layers. Substitution of some of the Mg^{2+} ions with Al^{3+} induces a charged layer that is charge compensated with anions such as carbonate and hydroxide. It is due to the net negative charge of the substituted layers that LDH's are known as anionic clays. Since LDH's can be prepared with a variety of atoms (Fe^{2+}, Ni^{2+}, Zn^{2+}, Cr^{3+} and Fe^{3+}) they possess a chemical diversity which can be exploited as novel inorganic *anion* exchange compounds.

5.5   MESOPOROUS MOLECULAR SIEVES

Although there has been an ever increasing interest in the use of pillared clays as catalysts and molecular sieves in processes where larger pores are required the disadvantages are that the pore sizes and pore size distributions of pillared clays cannot be easily controlled.

Very recently, however, two independent research groups shook the zeolite community by reporting the syntheses and characterisation of a new and exciting class of mesoporous inorganic solids. Table 8 briefly summarizes the former information regarding new attempts to synthesize mesoporous materials in the early nineties. Figure 8 illustrates the X-ray diffractograms of these new mesoporous solids and their corresponding structures.

5.6   SYNTHESIS OF MESOPOROUS MATERIALS FROM KANEMITE

In 1990 Kurado[211] and his co-workers at Waseda University, Tokyo reported the synthesis of microporous materials by using the single layered polysilicate, kanemite and alkyltrimethylammonium cations as the starting materials. They were further able to alter the pore size of the material in a controllable manner by varying the chain length of the alkyl trimethyl ammonium cation.

Following on from this in 1992[212-214] they synthesised highly ordered mesoporous silicates in a similar manner utilizing a single layered polysilicate and alkyltrimethylammonium cations as the precursor species. These mesoporous materials possessed surface areas in the 1100 m^2/g range and nitrogen adsorption isotherms were type IV indicating the presence of mesopores. The structure of these materials was assigned to a unidimensional hexagonal array of uniform channels of diameter 30Å and furthermore the framework composition could be manipulated to introduce acidity into the material. The authors claimed that "The silicate layers of kanemite wind around the exchanged alkyltrimethylammonium cations. This causes the condensation of silanol groups on the adjacent silicate layers of kanemite, since the silicate layers of have the flexibility to wind due to its single layered structure." Since then, however, due to its poor phase and structure characterisation not much research has been conducted on the kanemite derived mesoporous material made by Kurado and his co-workers.

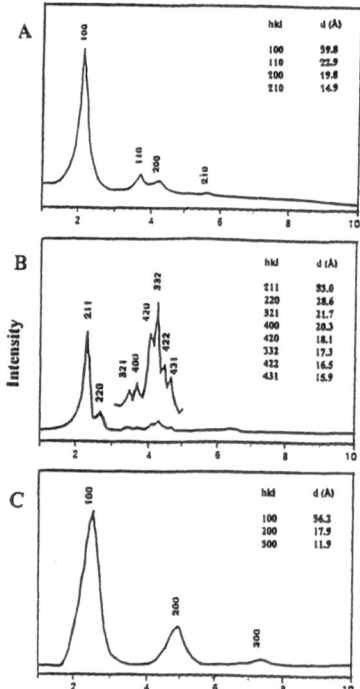

Figure 8 X-ray diffractograms of the new mesoporous molecular sieves[222] where A is the calcined MCM-41 (40 Angstrom pore diameter); B is the calcined cubic material (MCM-48) and C is the as-synthesised lamellar material.

TABLE 8: Mesoporous molecular sieves of the 1990s

Research group	Structure	BET surface area (m^2/g)	Synthetic strategy
Kurado and his co-workers (1992)	Silicate with an hexagonal array of uniform channels	~1100	Ion-exchange reaction of interlayer Na^+ ions of a layered silicate kanemite with alkyltrimethylammonium ions; and subsequent calcination.
Beck and his co workers (1992)	M41S molecular sieves: MCM-41 - one-dimensional hexagonal array of uniform mesopores whose diameter can be varied from 15-200Å MCM-48 three dimensional channel system with a periodic mesoporous structure which can be envisaged as a single silicate sheet separating two equal and disconnected volumes. Lamellar - Inorganic oxide sheets separated with an interlayer distance of 26 - 36Å	~1040	The synthesis mechanism employed cationic surfactants as structure directing agents in hydrothermal syntheses (similar to zeolite syntheses) from aluminosilicate/silicate gels. This family of materials offered an array of structures that were thermally stable analogues of organic, lyotropic liquid crystalline phases.
Davies and his co-workers (1995)	MCM-41	~1000	Ion-exchange between the sodium cations in the kanemite and surfactant molecules (cetyltrimethylammonium cation) that subsequently pillar the kanemite. Thus, a two-dimensional silicate structure is transformed into a three-dimensional network by subsequent calcination of the kanemite-surfactant complex

Although there has been an ever increasing interest in the use of pillared clays as catalysts and molecular sieves in processes where larger pores are required the disadvantages are that the pore sizes and pore size distributions of pillared clays cannot be easily controlled.

Very recently, however, two independent research groups shook the zeolite community by reporting the syntheses and characterisation of a new and exciting class of mesoporous inorganic solids. Table 8 briefly summarizes the former information regarding new attempts to synthesize mesoporous materials in the early nineties. Figure 8 illustrates the X-ray diffractograms of these new mesoporous solids and their corresponding structures.

5.7 SYNTHESIS OF MESOPOROUS MATERIALS FROM KANEMITE

In 1990 Kurado[211] and his co-workers at Waseda University, Tokyo reported the synthesis of microporous materials by using the single layered polysilicate, kanemite and alkyltrimethylammonium cations as the starting materials. They were further able to alter the pore size of the material in a controllable manner by varying the chain length of the alkyl trimethyl ammonium cation.

Following on from this in 1992[212-214] they synthesised highly ordered mesoporous silicates in a similar manner utilizing a single layered polysilicate and alkyltrimethylammonium cations as the precursor species. These mesoporous materials possessed surface areas in the 1100 m^2/g range and nitrogen adsorption isotherms were type IV indicating the presence of mesopores. The structure of these materials was assigned to a unidimensional hexagonal array of uniform channels of diameter 30Å and furthermore the framework composition could be manipulated to introduce acidity into the material. The authors claimed that "The silicate layers of kanemite wind around the exchanged alkyltrimethylammonium cations. This causes the condensation of silanol groups on the adjacent silicate layers of kanemite, since the silicate layers of have the flexibility to wind due to its single layered structure." Since then, however, due to its poor phase and structure characterisation not much research has been conducted on the kanemite derived mesoporous material made by Kurado and his co-workers.

5.8 M41S MESOPOROUS MOLECULAR SIEVES PREPARED WITH LIQUID CRYSTAL TEMPLATES

In 1992 Beck and his co-workers at Mobil Oil Research and Development, New Jersey announced their latest discovery of the synthesis and characterisation of a new family of mesoporous materials[215-222] known as M41S. These materials invented by scientists at Mobil research labs have had outstanding repercussions on the scientific community and can be distinguished from other porous inorganic solids by the regularity of their large open pores. These types of pore sizes are normally possessed by amorphous or semi-crystalline materials.

Moreover, in this family of materials (designated as M41S) a regular arrangement and uniformity of sizes of pores is observed which is normally only found in zeolites. One of the materials invented by Mobil is called MCM-41 and is a mesoporous silica with an hexagonal arrangement of cylindrical pores, whose pore size, outstandingly, could be adjusted from 18-200Å, thus, most channels in this material are surrounded by six nearest neighbour channels at roughly the same distance. Obviously any defects would cause some violation of this hexagonal arrangement. MCM-41(hexagonal), MCM-48 (cubic)[222] and the lamellar phase comprise the M41S family (see Figure 8), although there are many members belonging to each hexagonal, cubic or lamellar phase due to the possibility of changing the size of the pore openings. Each member of this family exhibits a characteristic X-ray diffraction pattern as indicated in Figure 8. As in the case of zeolites, the framework bears a net negative charge and consists of silica tetrahedra. The materials exhibit a long range order as exemplified by the X-ray diffractograms but do not possess a short range order as in zeolites. The organic material (the structure directing surfactant molecule) occluded during the synthesis can be removed thermally or chemically to empty the pores of the hexagonal and cubic phase.

The literature indicates that since their discovery a lot of attention has been focused on further tailoring the structure, crystallinity, porosity and modifying the framework composition of the cubic, lamellar and hexagonal phases that comprise the M41S family of mesoporous molecular sieves. The phases are similar to lyotropic liquid crystal phases which are produced in surfactant/water mixtures[223-226].

The synthesis of these materials involves the use of alkyl trimethylammonium cationic surfactants which operate as templates together with anionic silicate species. As will be discussed later, much research has been devoted to innovate new synthetic variations based on the original synthesis proposed by Mobil chemists. This is mainly because of the presence of a wide range of flexibility of bond angles which a traditional zeolite framework does not possess. The synthetic diversification and a wide array of framework compositions will be discussed in the following sections.

Quite soon after their synthesis, various mechanisms were developed to explain and understand the mechanism of formation of these products. These mechanisms so far proposed will also be reviewed. Although X-ray diffraction patterns have been sufficient to identify the structure of these materials other techniques such as TEM and NMR have been extensively used to characterise the M41S family of materials and will also be discussed later.

The dimensions of the molecular sieves MCM-41 and MCM-48 with their large pores open up great opportunities for their application as catalysts for molecules too large to enter into zeolite pores. The adsorption isotherms of these materials[222] indicated capillary condensation within uniform pores (characteristic of mesopores) and BET surface areas in the region of 1040 m^2/g rendering them as desirable materials for separation and catalytic processes. The research conducted to isomorphously substitute (Ti, Al, Fe, Mn) various elements into the silicate framework in order to induce acidity and hence catalytic activity as well as to substitute the silicate with other metal oxides to prepare mesoporous inorganic oxide materials will also be briefly reviewed.

In order to get a clear overview the following areas will be dealt with:

(a) Optimization of synthesis and new synthetic adaptations;

(b) Gain insight into the synthesis mechanism;

(c) Adsorption behavior and extensive characterization;

(d) Isomorphous substitution, novel framework compositions, host-guest interactions and subsequent catalysis.

Before proceeding to the general review concerning the M41S family of materials it is important to mention that the very recent formation of MCM-41 from the single polysilicate kanemite by Davies[227] and his co-workers is a natural follow up to the work of Beck and Kuroda. As indicated in Table 6 this synthesis involves the transformation of a kanemite-cationic surfactant into the MCM-41 product which had a narrow, tailorable mesopore size distribution, BET surface area in the range of 1000 m^2/g and an average pore diameter of 40Å.

5.8.1 Optimization of synthesis and new synthetic adaptations

A summary of the method for synthesising the M41S family of materials as proposed by the team at Mobil would be to prepare a mixture of:

- At least one oxide (divalent, trivalent, tetravalent or pentavalent element oxides).

- An organic agent ($R_1R_2R_3R_4Q^+$, where Q = N or P and at least one of the R groups is aryl or alkyl group of from 6-36 carbon atoms or a combination).

- A solvent or a solvent mixture.

- The mixture has to be maintained under sufficient conditions of pH, temperature and time for formation.

The most basic preparation of these materials can be represented by the following formula:

$$SiO_2 - Al_2O_3 - (Na_2O) - TAA - RTMA_2O - H_2O$$

where TAA is a symmetrical tetraalkylammonium species such as an aqueous tetramethylammonium hydroxide solution, RTMA2O is a long chain alkyltrimethylammoium cation (surfactant molecule), SiO_2 is a source of silicate anions such as sodium silicate, a fumed silica or tetraethyl orthosilicate, Al_2O_3 is a source of aluminum anions such as aluminum sulphate, sodium aluminate or Catapal B.

Na_2O was present due to the addition of reactants such as sodium aluminate or sodium silicate.

A typical synthesis published by the group at Mobil of aluminosilicate MCM-41 is[222]:

Sodium aluminate (4.2g) was added slowly to a solution containing 16g of $C_{14}H_{29}(CH_3)_3NBr$ in 100 g of water. One hundred grammes of tetramethylammonium silicate solution, 25g of HiSil silica, and 14.2g of tetramethylammonium hydroxide solution (25 wt %) were then added with stirring. This combined mixture was then loaded into an autoclave and heated with stirring at 100°C for 24 hours. The resulting reaction mixture was cooled to room temperature; the resulting solid product was recovered by filtration on a Buchner funnel, washed with water and dried in air at ambient temperature.

Basically the surfactant molecules electrostatically interact with the silicate counterions in the reaction mixture to yield silicate-surfactant composite arrays which exhibit hexagonal, cubic or lamellar structures[223-226]. These are the thermally stable analogues of organic, lyotropic liquid crystalline phases. The addition of auxilliary organics[222] of varying polarity in the reaction mixture were also used to alter the pore diameter of the mesoporous inorganic phase. Organics such as alkylated aromatics and straight or branched chain hydrocarbons were able to increase the pore size of the mesoporous materials. Also a change in the hydrocarbon chain length of the surfactant molecule causes a variation in the pore diameter of the final products. For example the hexagonal phase of the mesoporous material synthesised using alkyltrimethylammonium surfactant cations of carbon chain length C_9-C_{16} yielded products that exhibited increasing pore sizes with increasing chain length.

By varying the starting molar composition, reaction time and temperature, source of chemicals and alkyl chain length of the surfactant molecule a certain type of phase can be formed (hexagonal, cubic, lamellar, microporous zeolite or amorphous phase) and can be further modified. For instance a new cubic phase, $Pm3n$, was synthesised by using cetylethylpiperdinium cationic surfactant which possessed a larger head group than that normally used[228].

By constructing a phase diagram[229] the synthesis can be optimised to yield a mesoporous product with a specific pore size, framework composition, extra framework composition, hydrothermal and thermal stability, catalytic activity, pore size distribution and phase purity. The degree of control in tailoring these mesoporous molecular sieves has surpassed that of microporous molecular sieves.

Developing on from this work various groups have optimised and very creatively developed the synthesis of M41S materials. It was Stucky et al. who recently reported a generalized approach to the synthesis of mesophases of metal oxides and cationic or anionic surfactants under a range of pH conditions[228]. They realised that in order to construct mesostructures it was important to consider the electrostatic relationship between the head group of the charged surfactant and the inorganic ions in solution. Silicate mesoporous materials were originally synthesised in basic solutions in which the silicate oligoanions (I⁻) are structured by the cationic surfactant (S⁺). The resulting organic/inorganic mesostructure can be represented by S⁺I⁻. So to summarise, the reaction pathways reported by Stucky *et al.* for the synthesis of mesostructured surfactant inorganic biphase arrays are:

- Inorganic species with a cationic surfactant which can be represented by S⁺I.

MCM-41, MCM-48, Antimony oxide, Tungsten oxide (pH<7).

- Cationic oxide species with anionic surfactants which can be represented by S⁻I⁺.

Alkylsulphonate surfactant and lead or iron oxides giving hexagonal and lamellar phases.

- Ionic inorganic species in the presence of similarly charged surfactant molecules. This strategy employs counterions of opposite charge to that of the head group and can be represented by:

    S+X-I+, where X- = Br-, Cl-, silicate (pH<2) and zinc phosphate (pH<3)

    S⁻M⁺I⁻, where M = Na⁺, K⁺, zinc oxide (pH >12.5)

Further examples of innovating diverse synthetic routes include the use of **gemini surfactants**[230]: these entailed the use of surfactants with two quaternary ammonium head groups separated by a methylene chain of variable length and with each head group attached to a hydrophobic tail. Such a synthesis resulted in the formation of a mesophase designated as SBA-2 that exhibits three-dimensional hexagonal ($P6_3/mmc$) symmetry, a large surface area and regular supercages that can be bimensionally tailored.

Neutral surfactants[231-233]: this neutral templating route was based on the hydrogen-bonding interactions and self-assembly between neutral primary amines (for example dodecylamine) micelles (S°) and neutral inorganic precursors (I°), as opposed to the electrostatic templating pathway originally used with cationic and anionic surfactants. Tanev[231] thus claimed that the S°I° ordered mesoporous materials produced, possessed smaller X-ray scattering domain sizes and desirably thicker framework walls that improve the thermal and hydrothermal stability and facilitate the recovery of the template by simple solvent extraction.

Mokaya and Jones[234], however, reported a synthesis route in contrast to the neutral inorganic framework/neutral primary amine where primary neutral amines act as templates for charged aluminosilicate precursors. Syntheses using hexadecylamine[233] as a neutral template resulted in the formation of trivalent metal (Al, Ga, B) containing mesoporous silicas from which the template could be easily removed by solvent extraction leaving behing thermally stable, calcinable mesoporous silicas containing tetrahedrally coordinated trivalent compounds.

Furthermore, following the use of neutral primary amines, mesoporous molecular sieves were also synthesised **using nonionic polyethylene oxide surfactants**[232]. The subsequent products consisted of disordered channel structures with uniform diameters ranging from 20Å to 58Å, depending on the size and structure of the surfactant molecule. **Room temperature**[235] formation of MCM-41 and MCM-48, synthesis of MCM-41 in **fluoride medium**[236] and **microwave synthesis**[237] of molecular sieve MCM-41 have all been reported in the recent literature.

5.8.2 Gain insight into the synthesis mechanism

In order to manipulate and fine tune the syntheses of these mesoporous materials it is important to gain a fundamental understanding of the reaction mechanism. It is clear from the literature that MCM type materials have certainly whet the appetite of many groups to undertake the challenge to understand how the materials are made on a molecular level. By using a variety of techniques independent groups have developed models to explain actually what takes place during the synthesis of these materials.

Although the synthesis of these materials is actually an "extension" of the hydrothermal synthesis of zeolites by using a different type of templating molecule and perhaps also the addition of an auxiliary organic such as trimethylbenzene to increase the pore size, understanding this novel synthesis mechanism has proved to be a great challenge to scientists. Till now three main schools of thought have been presented in the literature:

Beck and his co-workers have proposed a **liquid crystal templating mechanism**[221,222,238,239] (see Figure 9) in which the "inorganic material occupies the continous solvent (water) region to create inorganic walls between the surfactant cylinders (surfactant liquid crystal structure). It may be that encapsulation occurs because anionic inorganic species enter the solvent region to balance the cationic hydrophilic surfaces of the micelles. Alternatively, perhaps it is the introduction of the inorganic species themselves that mediates the liquid crystal ordering. M41S formation differs from that of normal zeolite synthesis principally on its timing..."

The group at Mobil suggested that silicate condensation is not the dominant factor for structure formation. Rather that the organisation of surfactant molecules into micellar liquid crystals serve as templates for the formation of the MCM-41 structure; that two possible pathways are proposed: firstly that the liquid crystal is intact before the silicate species are added

440

and secondly that the addition of the silicate results in the ordering of the subsequent silicate encased micelles.

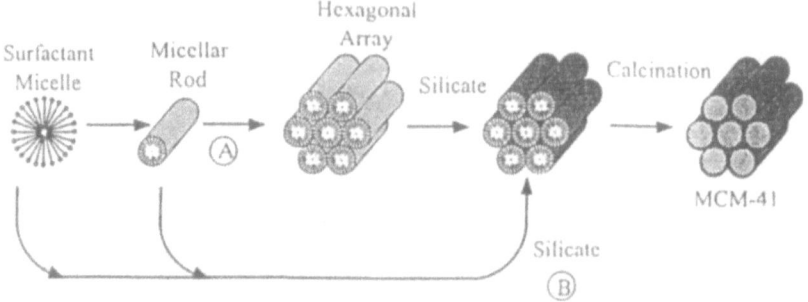

Figure 9: Liquid crystal templating mechanism proposed for the formation of MCM-41: (A) liquid crystal phase initiated and (B) silicate anion initiated[222].

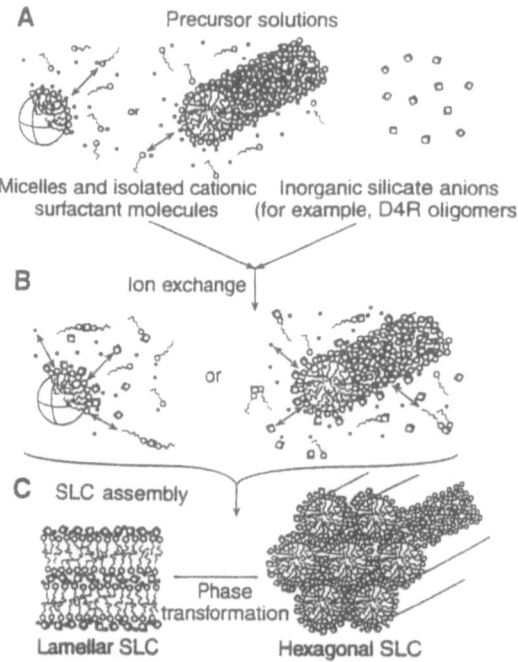

Figure 10: Schematic diagram of the cooperative organization of silicate-surfactant mesophases. (A) represents the organic and inorganic precursor solutions, (B) represents the situation after the two precursor solutions are mixed and (C) represents the multidentate interaction of the oligomeric silicate units with the surfactant molecules. The process of ion exchange and self assembly appear to occur on comparable time scales[241].

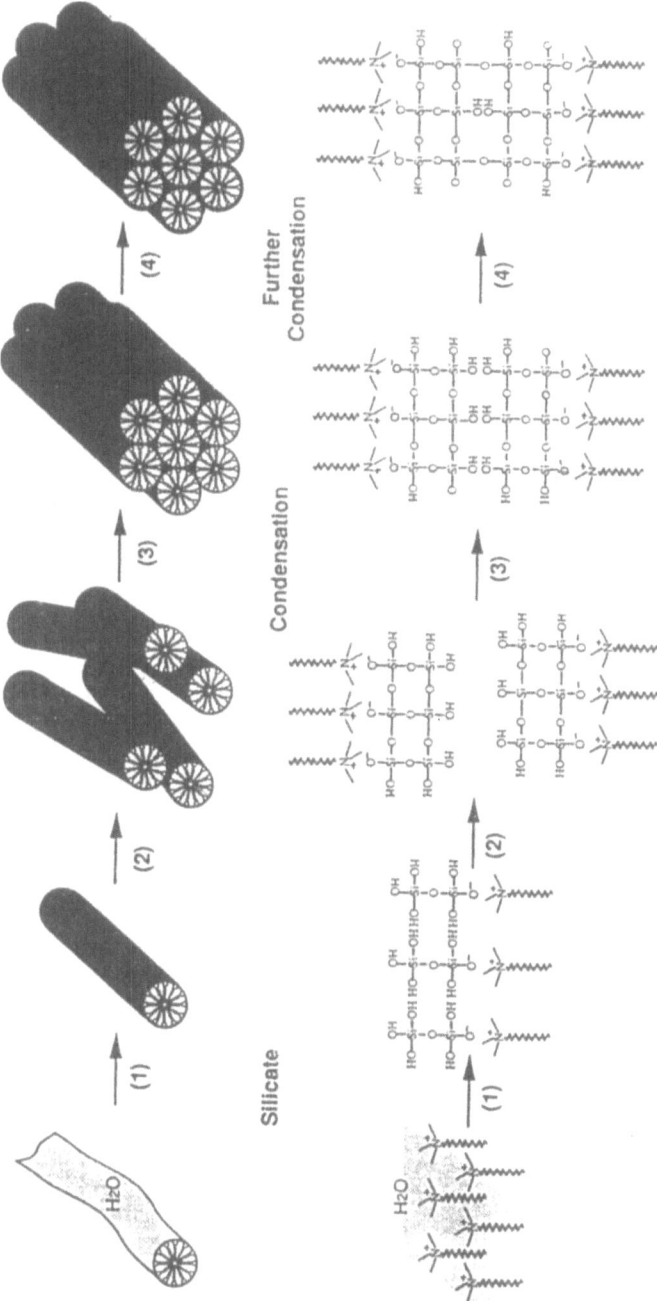

Figure 11: Mechanism proposed by Davies and his co-workers for the formation of MCM-41: (1) and (2) involve the random ordering of rod-like micelles and interaction with silicate species; (3) represents the spontaneous packing of the rods and (4) is the remaining condensation of silicate species upon final heating of the organic/inorganic composites[243].

Stucky and his co-workers presented a model that makes use of the cooperative organisation[228-230,240,241] of inorganic and organic molecular species into three dimensionally structured arrays (see Figure 10). They actually broke down the mechanism into more detail and identified three processes that take place: **multidentate binding** of the silicate oligomers to the cationic surfactant, preferential silicate **polymerization** in the interface region and **charge density matching** between the surfactant and the silicate. Furthermore, they stated that "in this model, the properties and structure of a system are determined by the dynamic interplay among ion-pair inorganic and organic species, so that different phases can be readily obtained through small variations of controllable synthesis parameters, including mixture composition and temperature. Nucleation, growth and phase transitions may be directed by the charge density, coordination and steric requirements of the inorganic and organic species at the interface and not necessarily by a preformed structure. A specific example is presented in which organic molecules in the presence of multiply charged silicate oligomers self assemble into silicatropic liquid crystals. "

Finally Davies and his co-workers[242,243] suggest that "randomly ordered rod-like organic micelles interact with silicate species to yield approximately two or three monolayers of silica encapsulation around the external surfaces of the micelles. Subsequently these composite species spontaneously assemble into the long-range ordered structure (hexagonal packing) characteristic of MCM-41 (see Figure 11). With further heating, the silicate species in the interstitial spaces of the ordered organic-inorganic composite phase continue to condense. Complete condensation of the silicate species is not possible as the SiO^- species are necessary for charge compensation of the occluded alkylammonium ions."

Although much information is available, a lot more research is being carried out to further elucidate and manipulate the syntheses and application of M41S family of mesoporous molecular sieves[242,245].

5.8.3 *Adsorption behaviour and extensive characterization*

Several techniques such as adsorption gravimetry, controlled rate-evolved gas analysis, argon porosimetry, water sorption and physisorption of nitrogen and oxygen on aluminosilicate MCM-41[246-251] indicate that it has well defined mesopores and promises to be a highly promising model adsorbent. The nitrogen isotherm is of unusual character[246-248]: it is a reversible Type IV isotherm. The water isotherm shows that the water-MCM-41 interactions are characterised by a Type V isotherm indicating an initial repulsive character followed by a capillary condensation step of the adsorbate. This highlights both hydrophobic and hydrophilic properties.

Purely siliceous MCM-41 materials with different pore sizes[248] whose nitrogen isotherms were measured showed that the type of isotherm obtained depended on the pore size. A Type I isotherm was obtained for "small pore sizes" and a reversible Type IV isotherm was obtained for "intermediate pore sizes". A typical Type IV isotherm was obtained for "large pore sizes". Therefore, the pore size was seen to effect the adsorbate condensation and hysteresis within MCM-41[252].

Further techniques to thoroughly characterise the physical and chemical nature of the M41S family including NMR[253-255], EPR[256] using transitional metal ion probes, IR studies[257] to distinguish the surface silanol groups, high resolution TEM[258-260] and XRD[252] of MCM-41 and MCM-48[261] and high temperature calorimetry[262] have been reported in the literature.

5.8.4 Isomorphous substitution, novel framework compositions, host-guest interactions and subsequent catalysis

As seen in the previous sections many contributions in the field of synthesis of these materials have been reported. However, the possible use of these materials in catalysis, sorption and host-guest chemistry has initiated research into identifying different routes to make the inert, pure silica MCM-41 and MCM-48 catalytically active. Early syntheses concentrated on the incorporation of titanium, aluminum and vanadium into the silicate framework as well as the formation of a number of other mesostructured oxides[228,253,263-266]. Based on the synthesis mechanism proposed by Stucky et al.[240,241] research groups attempted to synthesise analogous transition metal oxides that contain a regular array of uniform mesopores that would be desirable as high surface area catalysts especially for partial oxidation reactions and as hosts for quantum sized materials. Basically the uses of MCM-41 as an **oxidation catalyst, acid catalyst** and **base catalyst** will be discussed.

As reported by Corma et al., the incorporation of titanium into the silicate framework of MCM-41 could yield catalysts for the oxidation of organics under mild conditions[135,140,267].

By replacing the Si atom with a Ti atom no electric charge and hence no Brønsted activity is introduced but **MCM-41 based oxidation catalysts** for the oxidation of organics with H_2O_2 in solution, such as dehydrogenation of cyclohexanol to cyclohexanone, oxidation of phenol to catechol and hydroquinone, epoxidation of olefins and the oxygenation of alkanes are all processes which could be catalysed[263,264,268]. Such reactions are traditionally carried out by MEL and MFI type zeolites that contain isomorphously substituted Ti for Si in the framework. Therefore, in order to oxidise large molecules which cannot penetrate into the pores of the zeolites many groups have studied the synthesis, incorporation and location of titanium in the mesoporous silicate frameworks[263,264,269-273].

Corma *et al.* first reported[263,264,268] the preparation and properties of Ti containing MCM-41 and applied this as a preferable alternative catalyst to oxidise large and small molecules such as ∝- terpineol and norbornene with respect to the classical chromium and manganese catalysts which are used in solution and involve the hazardous use of perchloric acid. It was concluded from an in depth characterisation, such as UV-visible, EXAFS and XANES and subsequent catalytic testing that Ti was incorporated into the MCM-41 framework and that it was present in isolated positions[271a] by direct synthesis to yield the first ultralarge pore titanosilicate catalyst to oxidise bulky olefins under mild conditions using H_2O_2.

Thorough characterisation[271a] indicated that the Ti-MCM-41 showed a lower intrinsic activity with respect to the Ti-Beta or Ti-silicalite samples also used. Very recently it was shown that Ti-MCM-41 possesses high actvity and selectivity for carrying out the oxidation of thiols to sulfoxides and sulfones with both H_2O_2 and tert-butylhydroperoxide[271b].

Ti-MCM-41 has also been reported[269,273] to show exceptionally high selectivity and catalytic activity for the oxidation of 2,6-di-*tert*-butyl phenol to the corresponding quinone, however, the results were erroneos[269] due to an analytical problem. From this it is clear that care must also be taken when interpreting the catalytic results as acetone which can be used as a solvent in the hydroxylation of benzene with Ti-MCM-41 is oxidised at rates comparable with the benzene oxidation such that acetone oxidation products can be falsely interpreted as the benzene oxidation product, namely phenol.

In a water free environment the hydrophilic character of Ti-MCM-41 made it desirable for use in water-free reactions such as when using organic peroxides as oxidants[271a]. Also Ti-MCM-41 prepared in the absence of aluminium showed a high selectivity for the formation of epoxides. Therefore, Ti-MCM-41 was shown to be especially effective in diffusion limited oxidation reactions and opens up a new horizon of possibilities for the production of oxygenated fine chemicals. However, it has been reported that the intrinsic activity for the oxidation of molecules that can diffuse easily is higher for zeolites than for Ti-MCM-41. With respect to this it has been found that while it is a good catalyst for the oxidation of cyclododecanol and naphthol with 30% wt H_2O_2 it is a poor catalyst for the oxidation of *n*-hexane and primary amines as well as for the ammoximation of cyclohexane[271c]. It must also be noted that, due to the hydrophilic nature of the material the type of oxidising agent employed is limited such that the behaviour of materials rendered hydrophobic has to be explored.

Quite recently the synthesis of micro-mesoporous amorphous titanosilicates (MMATS) synthesised in the absence of nitrogenated organic bases has been reported270. The authors claim that these materials, designated as MMATS, possess a bimodal pore size distribution and activity and selectivity comparable to Ti-MCM-41 materials.

Latest developments in the substitution of titanium into the M41S silicate family indicate the very recent success of introducing Ti atoms into the MCM-48 silicate, which with respect to the mono-dimensional porous MCM-41 possesses a three dimensional array of mesopores. However, catalytic results and in depth characterization to identify whether the titanium is incorporated into the framework is pending[274]. Koyano et al.[275] also reported the synthesis of Ti-MCM-48 but on analysis of the turnover number of these materials for the epoxidation of cyclododecene it is clear that the conversion is lower than 1%. This research, however, indicates that more scientific investigations have to be conducted in order to develop Ti-MCM-48 as an efficient, active and selective catalyst.

M41S materials containing aluminium in the framework walls of the pores have been used as catalysts. Pure aluminosilicate mesoporous materials253,265,276-289 have been employed as **acidic catalysts** in aromatic alkylation, hydrocarbon cracking and upgrading of heavy oil reactions.

Isomorphous substitution with trivalent cations such as Al^{3+}, B^{3+} and Fe^{3+} causes the formation of Brønsted acid sites. By varying the nature and number of the trivalent framework cations the density and strength of the acid sites may be varied. Owing to their large pore size the first applications envisaged for mesoporous molecular sieves as **acid catalysts** were multiton processes such as cracking and hydrocracking. These applications were envisaged based on the discovery that MCM-41 aluminosilicates showed Brønsted acidity of medium strength[281] similar to that of amorphous silica/alumina.

A remarkable result was recently obtained using a large pore bifunctional titanium-aluminosilicate MCM-41 material[290] containing Ti^{4+} as oxidation sites and Al^{3+} which has an associated H^+ as the complementary cation. This catalyst was shown to carry out the one pot conversion of linalool to a cyclic furan and pyran hydroxy ethers with 100% selectivity. The ratio of furane to pyranes was very close to that obtained by the epoxidase enzyme. The high activity observed was due to the fast diffusion of products and reactants through the large pores of the Ti-MCM-41.

Characterisation techniques such as IR, pyridine adsorption, ^{27}Al NMR, ^{29}SI NMR, ^{13}C NMR and XRD have been extensively carried out on the MCM-41 samples synthesised with different Si/Al ratios in order to distinguish the degree of incorporation of tetrahedrally coordinated aluminium in aluminosilicate MCM-41.

To date the research presented by the various groups has culminated in the preparation of Al-MCM-41 samples containing tetrahedrally (framework) coordinated aluminium which after calcination is catalytically active. The syntheses have been optimised, by using various sources of aluminium or by preferably calcining in N_2 rather than in O_2, to maximise the incorporation of aluminium and also to increase the stability and catalytic activity of the Al-MCM-41.

However, independent groups such as Cheng et al.[285], Kolodziejski et al.[253], Janicke et al.[265], Corma et al.[281], Schmidt et al.[260,261] all observed that the tetrahedral aluminium is converted into a highly disordered aluminium species upon calcination. This suggests further research into modifying the calcination technique or possibly using solvent extraction to remove the template and hence avoid the perturbation of the Al environment.

In the case of catalytic cracking of vacuum Gasoil (FCC) it has been shown[281b] that aluminosilicate MCM-41 is active and selective and that its activity depends directly on the amount of Al^{IV} in the walls. However, after steaming at 750°C its activity strongly diminishes due to the relatively lower hydrothermal stability of Al^{IV} which is converted to Al^{VI}. Fortunately the hydrothermal stability handicap is not a such a serious problem for processes requiring lower catalyst generation temperatures such as the hydrocracking process.

Al-MCM-41, prepared by Corma et al.[281], was seen to possess medium acid strength. They also reported that calcination induces dealumination of the sample, thus causing a decrease in the Brønsted acidity and an increase in the Lewis acidity. In the chemical field, especially for the preparation of fine chemicals and commodities, there are many cases where mild acidity and large pores are required. In such cases the Al form of MCM-41 is a good candidate. An example is for the selective alkylations involving bulk reactants, such as 2,4-di-tert-butylphenol[283a]. Although the Al-MCM-41 sample exhibits an acidity that is lower than that of HY zeolites alkylation of 2,4-di-tert-butylphenol with cinnamyl alcohol to yield 6,8-di-tert-butyl-2-phenyl-2,3-dihydro [4H] benzopyran was observed to take place in the Al-MCM-41 pores and not in the smaller pores of HY zeolites. In an analogous manner, the mildly acidic Al-MCM-41 was also active and selective for converting large sized alcohols such as cholesterol, demantan-1-ol and 2-napthol into the corresponding tetrahydropyranyl ethers[283b].

The great combination of mild acidity and large pores have rendered these materials as catalysts for acetalisation reactions291 in which zeolites or even amorphous silica-alumina were traditionally used as catalysts and which yielded faster catalytic decay and lower selectivity with respect to Al-MCM-41. Furthermore, large pore zeolites were practically inactive when molecules such as diphenyl acetaldehide were reacted (Table 9).

TABLE 9: Influence of the catalyst pore size on the rate and conversion in the acetalization reaction 1, 2, and 3.

Catalysts	r_a (mol.h^{-1}.g^{-1}). 10^9			Conversion (%)^a		
	1	2	3	1	2	3
1-MCM-41	2500	2480	2340	98	90	80
1-βH	3300	2180	380	95	83	13
2-βH	3038	1340	180	94	55	8

^{a} 2h reaction time

1: n-heptanal; 2: 2-methyl-2-phenylacetaldehyde; 3: 3-diphenylacetaldehyde

Auguado *et al.*[288] synthesized aluminium containing MCM-41 at room temperature and found that it exhibited suitable catalytic properties for the conversion of low-density polyethylene into hydrocarbon feedstocks.

Another original attempt to control the properties of Al-MCM-41 from the synthetic point of view was reported by Mokaya and Jones[289] who employed the synthesis of Al-MCM-41 using primary amines to yield materials that they claimed to possess an acidity and catalytic activity higher than those of equivalent aluminosilicate MCM-41 synthesised with surfactant molecules. Mokaya also synthesised Al-MCM-41 with a Si/Al = 10 that was able to convert cumene *via* catalytic cracking into propene and benzene.

A number of patents have been written concerning the catalytic application of the aluminosilicate MCM-41 samples[268,276-280] including the the catalytic oligomerisation of olefin feedstock in paraffin which showed very high selectivity especially in the 40-250°C temperature range; as well as the Al-MCM-41 which in the presence of a reducing agent such as ammonia was used to reduce emissions of nitrogen oxides (NO_x) to the atmosphere. The hydrocracking of heavy wax over MCM-41 within an Al matrix and containing Ni and W showed higher yields than fluorinated NiW/Al$_2$O$_3$ catalysts when working at high conversions, and the lubes obtained on the MCM-41 based catalyst gave higher viscosities[278a]. In acid catalysed reactions such as propene oligomerisation the aluminium containing MCM-41 showed excellent activity, with very high yields of C$_9$ and C$_{12}$ olefins[278b].

Zhao and Goldfarb have very recently published the synthesis of Al-MCM-48 without actually giving any details about its synthesis or characterisation. Soon after, however, Schmidt and his coworkers[292] published the synthesis of mesoporous MCM-48 containing only tetrahedral aluminium (Si/Al = 22) that is stable to calcination such that during calcination the aluminium remains in its framework position.

To summarise and have a general overview of the research concerning heteroatom substitution into the silicate MCM-41 framework the following materials have been published in the literature and are discussed briefly:

- Boron modified MCM-41 mesoporous materials[293,294] ^{11}B NMR confirms the incorporation of boron into the lattice with the coexistence of several B sites in calcined samples. The stability of such samples was low such that B is partially removed by water vapour at room temperature.

- Mesoporous vanadium silicate molecular sieve[259,266,295,296], V-MCM-41, where Raman spectroscopy and ^{51}V NMR indicate that the vanadium is included into the silicate framework. Similar UV-visible bands 373-385 and 252-272 nm to those found in V-Silicalite (384 and 265 nm) were observed. V-mesoporous catalysts are apt for the hydroxylation of aromatic compounds and the sample reported was applied for the oxidation catalysis of 1-napthol and cyclododecane. However, it must be noted that a potential problem of such catalysts is the occurrance of leaching.

- Further MCM-41 derivatives in which the Si is claimed to be substituted or contain Cr, Sn, V, Mo, Mn (MCM-48 and layer phase as well), Fe and Ga have been recently reported[228,273,297-302]. It is clear, however, that extensive optimisation of the synthesis, characterisation and testing of the catalytic activity (redox catalysis) is required.

Many attempts to make mesostructured transition metal oxide compounds have been undertaken. The literature indicates attempts at synthesising mesostructured vanadium oxide[303], zirconium oxide[304,305], aluminium oxide[306], tin sulphide (lamellar)[307] and zirconium oxo phosphate[308] materials. Based on the synthesis mechanism published by Monnier *et al.* a cooperative, interface controlled condensation process allowed SchÅth and his coworkers to synthesize tungstate and molybdate mesostructured oxides from an anionic metal oxide/cationic surfactant system as well as lead and iron mesostructured oxides from the reversed transitional metal polycation/anionic surfactant system. However, one of the greatest problems encountered was that the metal oxide phases were predominantly layered and collapsed upon calcination unlike the analogous silicate and aluminosilicate systems.

Antonelli and Ying[309], however, have reported the novel sol-gel preparation of hexagonally packed mesoporous TiO_2 using alkylphosphate surfactants which could prove to be attractive adsorbents and catalysts and compare to the tungsten and antimony systems by being thermally stable and calcinable with only a partial structure collapse and a final surface area of 200 m^2/g. Reported by the same authors[310] is the synthesis of a stable hexagonally packed mesoporous niobium oxide molecular sieve through a novel ligand-assisted templating mechanism using a long chain amine surfactant molecule. The resulting high surface area, mesoporous niobium oxide material is claimed to be completeley stable to calcination and initiates the potential synthesis of Ti, Zr, Ta, Ce and Y mesoporous oxides in a similar manner which could be suitable for a wide range of catalytic processes.

In the case of calcined aluminosilicate MCM-41 the well defined hexagonal pore walls contain terminal hydroxyl groups which could react with organometallic compounds to provide the encapsulation of large catalytic species. Also the ion-exchange property of the aluminium allows the introduction of catalytically active transitional metal ions and clusters inside the MCM-41 support. Based on these facts MCM-41 has been used as a host for many catalytically active species such as:

- Thermally stable **trimethylstannyl molybdenum complex**[311] in MCM-41 with a possible high activity for olefin hydrogenation.

- **NiMo supported** on the mesoporous crystalline MCM-41[312] aluminosilicate possessing catalytic activity for the mild hydrotreating of vacuum gasoil. As stated earlier the low hydrothermal stability of the aluminosilicate MCM-41 is not a major problem for such hydrocracking processes. Thus, when the MCM-41 sample was loaded with Mo and Ni a highly active and selective hydrocracking catalyst was obtained. The hydrocracking activity was higher than that of amorphous silica-alumina with the HDS activity being much more desirable with a better HDN[312] (Figure 12).

- **Nanosize Pt clusters** using ion exchange of $Pt(NH_3)_4^{2+}$ inside the mesoporous channel of MCM-41 aluminosilicate[313,314] resulting in a highly active oxidation catalyst and a high activity for the hydrogenolysis of ethane.

- **Sodium and caesium cation-exchanged** mesoporous aluminosilicate MCM-41 - a mild selective, water stable and recyclable catalyst for the base-catalysed Knoevenagel condensation and acid-catalysed acetalisation and aldol condensation. This **base catalyst** was used for the condensation of benzaldehyde with ethylcyanoacetate and can also

450

be used for the condensation reaction of benzaldehyde with acetophenone to chalcone[126,315].

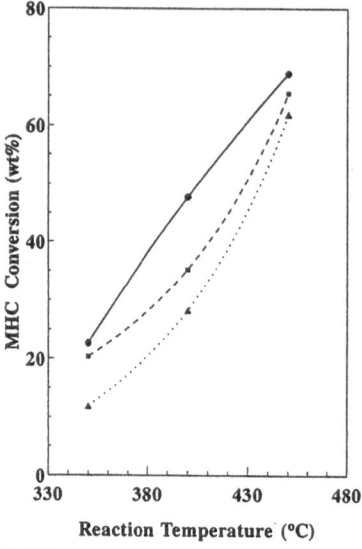

Very recently Lasperas *et al.*[316] produced **base catalysts** by anchoring alkylamines on zeolites so that catalysts for the Knoevenagel condensation reaction were yielded.

Figure 12: Hydrocracking (MHC) conversion of feed A as a function of reaction temperature for (l) NiMo/MCM-41, (s) NiMo/SiO$_2$-Al$_2$O$_y$, (n) NiMo/USY catalysts.

- **Manganese-oxo species grafted**[317] onto the internal walls of MCM-41 channels showed a high redox catalytic activity for hydrocarbon oxidation

- Heteropolyacid supported on MCM-41 which showed catalytic activity for the isomerisation of paraffins and the alkylation of aromatics such as the liquid-phase alkylation of 4-t-butylphenol (TBP) by isobutene and styrene[318]. **HPA/MCM-41** is more efficient than H$_2$SO$_4$ for liquid phase alkylations. Supported heteropolyacids possess adequate activities for reactions requiring acid sites with strong and medium acidic strength such as n-butane and n-hexane isomerisation alkylation of isobutane with 2-butene and alkylation of benzene with 1-tetradecene[319]. The good acid properties of these catalysts are related to the fact that the HPA is highly dispersed. However, in order to produce good catalysts the interaction of the HPA with the surface Al must be avoided such that for this reason pure silica MCM-41 structures are preferred.

- Introduction of **cupric ions**[320] into siliceous MCM-41 by liquid and solid state ion exchange.

- Encapsulation of **Fe_2O_3 nanoparticles**[321] in MCM-41

- **Chromium oxide** supported on MCM-41 behaves as a catalyst for the oligomerisation of alpha olefins to produce hydrocarbon oligomers useful as lubricants and lubricant additives[322].

- Attachment of **pseudotetrahedral $O_{3/2}V=O$ centres** on the walls of a MCM-48 support and their subsequent reactivity to water[323].

- The catalytic performance of **$BF_3/MCM-41$** system has been patented[324] and is analogous to the $BF_3/ZSM-5$ catalyst for the alkylation of isobutane/butene. However, this process is probably not the most adequate for this system.

- **Pd** supported MCM-41 was shown to be a better catalyst than Pd/USY and PD/SiO_2 for the hydrogenation of benzene, due to the degree of high dispersion of Pd[325].

- **V_2O_5-TiO_2** supported on MCM-41 was used as a catalyst for the selective reduction of NO_x with NH_3 and showed superior activity at high temperatures than V_2O_5-TiO_2/SiO_2[279].

Furthermore, within this field of host-guest chemistry mesoporous materials are also promising for the immobilisation of metalloporphyrin oxidation catalysts, Ti complexes and photosynthesisers311,[326-328].

Some novel applications of these materials include the preparation of conducting polyaniline filaments[329] in the 30Å wide hexagonal channel system of the aluminosilicate MCM-41 and the controlled in-situ polymerisation[330] of styrene, methyl methacrylate and vinyl acetate within the mesopore system of the transitional metal oxide MCM-41.

6. Conclusions

We have tried to show that zeolites, Zeotypes, and their extension to well defined mesoporous materials, offers to catalysis the always dreamed possibility of identifying and tailoring the active sites. Moreover, the posibility of controlling the pore geometry and hydrophilic-hydrophobic properties is approaching to zeolites and related materials to the specificity of enzymes. We are not far from the point in where multifunctional catalytic sites, including framework and extraframework species (transition metal complexes for instance) will really mimic the catalytic behaviour of enzymes. This will reach the point that even enantioselectivity will directly be achieved or, at least, induced.

452

REFERENCES

1. E.M. Flanigen, Stud. Surf. Sci. Catal., 58, 1 (1991)

2. S.T. Wilson, B.M. Lok, C.A. Mesina, T.R. Cannan, E.M. Flanigen, J. Am. Chem. Soc., 104, 1146 (1982).

3. B.M. Lok, C.A. Messina, R.L. Patton, R.T. Gajek, T.R. Cannan, E.M. Flanigen, J. Am. Chem. Soc., 106, 6092 (1984).

4. E.M. Flanigen, B.M. Lok, R.L. Patton, S.T. Wilson, Proc. 7th Int. Zeol. Conf. (Y. Murakami, et al., Eds.), Kodanska Elsevier, p. 103 (1986).

5. M.E. Davis, C. Montes, P.E. Hathaway, J.M. Garces, Stud. Surf. Sci. Catal., 49, 199 (1989).

6. Q. Huo, R.Xu, S. Li, Z. Ma, J.M. Thomas, R.H. Jones, A.M. Chippendale, J. Chem. Soc. Chem. Commun., 875 (1992).

7. M. Esterman, L.B. McCusker, C. Baerlocher, A. Merrouche, H. Kessler, Nature, 352, 320 (1991).

8. W.J. Mortier, Proc,. 6th Int. Zeolites Conf. p. 734 (1984).

9. V. Gutmann, "The donor Acceptor Approach to Molecular Interaction", Plenum Press New York 1978).

10. G.J. Gajda and J.A. Rabo in Acidity and Basicity of solids: Theory, Assessment and utility, Nato Asi Series, (J. Fraissard and L. Petrakis, eds), vol 444, p. 127, (1994).

11. M. Czjzek, H. Jobic, A.N. Fitch, T. Vogt, J. Phys. Chem., 96, 1535 (1992).

12. D. Freude, and J. Klinowsky, J. Chem. Soc. Chem. Commun. 1411 (1988).

13. J. Sauer, C.M. Kölmel, J. R. Hill and R.Ahlirchs Chem. Phys. Lett. 164, 193 (1989).

14. R.A. van Santen, Stud. Surf. Sci. Catal. 85, 273, (1994).

15. W.J. Mortier, J. Catal 55, 138 (1978).

16. G.J. Kramer, and R.A. van Santen, J. Am. Chem. Soc. 115, 2887 (1993).

17. D. Barthomeuf, Mat. Chem. Phys, 17, 49 (1987).

18. R. Carson, E.M. Cooke, J. Dwyer, A. Hinchliffe and P.J.O. Malley, Stud. Surf. Sci. Catal 46, 39 (1989).

19. J. Sauer, K.P. Schroeder, and J.R. Hill,.Am. Chem. Soc. Preprints 37 (2), 666 (1992).

20. W.J. Mortier, Stud. Surf. Sci. Catal., 28, 423 (1986), ibid 37, 253 (1988).

21. E.M. Flanigen, Stud. Surf. Sci. Catal., 58, 13 (1991).

22. J. Szabo, J. Parrotey, G. Szabo, J.C. Duchet, D.J. Cornet, J. Mol. Catal., 67, 79 (1991).

23. V.R. Choudary, Chem. Ind. Dev., 32 (1974).

24. H.H. Mooieweer, K.P. de Jong, B. Kraushaar - Czarnetzki, W.H.J. Storkan, B.C.H. Krutzen, Stud. Surf. Sci. Catal., 84, 2327 (1994).

25. M.W. Simon, S.L. Suib, C.L. O'Young, J. Catal, 147, 484 (1994).

26. W.Q. Xu, Y.G. Yin, S.L. Suib, C. L. O'Young, J. Catal, 150, 34, (1994).

27. C.L. O'Young, R.J. Pellet, D.G. Casey, J. R. Ugolini, R.A. Swicki, J. Catal, 151, 467 (1995).

28. R.J. Lawson, D.M. Richmond, G.J. Gadja, P.T. Barger US 5367101 (1994).

29. P.B. Venuto, P.S. Landis, Adv. Catal., 18, 259 (1968).

30. P.B. Venuto, Microporous Materials, 2, 297 (1994).

31. B. Coughlan, M.A. Keane, J. Catal., 138, 164 (1992).

32. D.B. Lukyanov, V.I. Shtral, Prepr. Am. Chem. Soc. Div. Petr. Chem., 36, 693 (1991).

33. F.G. Dwyer, in "Catalysis of Organic Reactions" (W.R. Moser, Ed.), Marcel Dekker, N.Y., p. 39 (1981).

34. W.W. Kaeding, L.B. Young, A.G. Prapas, Chemtech., 12, 556 (1982).

35. W.W. Kaeding, C. Chu, L.B. Young, B. Weinstein, S.A. Butter, J. Catal., 67, 159 (1981).

36. J.A. Brenan, W.E. Garwood, S. Yurchak, W. Lee, Proc. Int. Sem. on Alternative Fuels, 191 (1981).

37. L.B. Young, US Pat. 4301317 (1981).

38. C.B. Dart, M.E. Davis, Catal. Today, 19, 187 /1994).

39. Y. Sugui, T. Matsuzaki, T. Hanoka, K. Takenchi, T. Tokoro, G. Takenchi, Stud. Surf. Sci. Catal., 60, 303 (1991).

40. G.S. Lee, J.J. Maj, S.C. Rocke, J.M. Garcés, Catal. Lett., 2, 243 (1989).

41. A. Katayama, M. Toba, G. Takenchi, F. Mizukami, S. Niwa, S. Mitamura, J. Chem. Soc. Chem. Commun., 39 (1991).

42. B.W. Wojciechowski, and A. Corma, "Catalytic Cracking: Catalysts, Chemistry and Kinetics", Marcel Deker, New York (1984).

43. V.J. Frilette, W.D. Haag, R.M. Lago, J. Catal., 67, 218 (1981).

44. P.H. Schipper, F.G. Dwyer, P.F. Sparell, S. Mizrahi, J.A. Herbst, ACS Symp. Ser., 375, 64 (1988).

45. R. Madon, J. Catal., 129, 275 (1991).

46. T.G. Roberie, G.M. Young, D.S. Chin, R.F. Wormsbecher, E.T. Habib, NPRA Ann Meeting 1992, Washington DC, AM-92-43.

47. L. Bonetto, M.A. Camblor, A. Corma, and J. Pérez-Pariente, J. Appl. Catal., 82, 37 (1992).

48. L. Boneto, A. Corma, and E. Herrero (Proc. 10th Int. Zeolite Conf. 10, (1992).

49. M. Guisnet, F. Alvarez, G. Gianetto, G. Perot, Catal. Today, 1, 415 (1987).

50. A. Corma, J. Frontela, J. Lázaro, M. Pérez, Prepr. ACS Petrol. Div., 36, 833 (1991).

51. S.T. Sie, Stud. Surf. Sci. Catal., 85, 587, (1984).

52. J.W. Ward, Fuel Proc. Technol., 35, 55 (1993).

53. W. Souverijns, R. Parton, J.A. Martens, G.F. Froment, and P.A. Jacobs Catal. Lett., 37, 207, (1996).

54. F.G. Derouane, S.B. Abdul Hamid, I.I. Ivanova, N. Blom, and P.E. Hojlund Nielsen, J. Mol. Catal., 86, 371, (1994).

55. A. Corma, C.M. Zicovich-Wilson, P. Viruela, J. Phys. Chem., 98, 10863 (1994).

56. A. Corma, H. García, G. Sastre, P. Viruela, to be published.

57. D. Barthomeuf, J. Phys. Chem. 88, 42, (1984).

58. D. Barthomeuf; "Acidity and Basicity of Solids", Nato Asi Series, (J. Fraissard and L. Petrakis, Edit.) Kluwer Academic Publishers, 182, (1994).

59. A. Corma, R.M. Martín-Aranda, F. Sanchez, J. Catal., 126, 192 (1990).

60. M. A. Camblor, A. Corma, R.M. Martín-Aranda, J. Pérez-Pariente "Proceed 9th Intern. Zeol. Conf. Montreal" (R. Von Ballmoos, J.B. Higgins, M.M. J. Treacy, eds.) Butterworth-Minemann, Boston II, 647, (1993).

61. G.V. Gibbs, E. P. Meagher, J.V. Smith and J.J. Pluth, ACS Symposium Series, Washington DC 40, 19 (1977).

62. L. Uytterhoeven, D. Dompas and W. Mortier, J. Chem. Soc. Faraday Trans., 88, 2753, (1992).

63. L.R. M. Martens, W.J.M. Vermeiren, P.J. Grobet, and P.A. Jacobs, Stud. Surf. Sci. Catal. 31, 531, (1987).

64. Y.S. Park, Y.S. Lee, K.B. Yoon , Stud. Surf. Sci. Catal., 84 B, 901 (1994).

65. S. Bordiga, A. Ferrero, E. Giamello, G. Spoto, A. Zecchina, Catal. Lett., 8, 375, (1991).

66. P.H. Kasai, and R.J. Bishoper, J. Phys Chem. 77, 2308, (1973).

67. P.P. Edwards, M.R. Harrison, J. Klinowski, S. Ramdas, J.M. Thomas, D.C. Johnson, C.J. Page, J. Chem. Soc. Chem. Comm., 982 (1984);

68. P.P. Edwards, M.R. Harrison, J. Klinowski, S. Ramdas, J.M. Thomas, D.C. Johnson, C.J. Page, J. Solid State Chem., 54, 330 (1984).

69. P.J. Grobet, L.R.M. Martens, W.J. M. Vermeiren, D.R.C. Huybrechts, P.A. Jacobs, Z. Phys. D. Atoms, Molecules and Clusters 12, 37, (1989).

70. L.R. M. Martens, W.J. H. Vermeiren, P.J. Grobet, P.A. Jacobs; Stud. Surf. Sci. Catal., 28, 531, (1987).

71. L.R.M. Martens, W.J. M. Vermeiren, D.R. Huybrechts, P.J. Grobet, P.A. Jacobs, Proceed 9th, Int. Cong. Catal. Calgary, (M.J. Phillips, M. Ternan eds.); Chem. Institute of Canada, Otawa 1, 420, (1988).

454

72. P.E. Hathaway and M.E. Davis, J. Catal., 119, 497, 1989).

73. C.B. Dart and M.E. Davis, Catal. Today, 19, 151, (1994).

74. F. Yagi, H. Tsuji, H. Hattori, H. Kita, "Acid-Base Catalysis II", Kodansha (Tokyo) - Elsevier, 349, (1994).

75. H. Tauji, F. Yagi, H. Hattori, H. Kita, Proceeding of the 10th, International Congress on Catalysis, Budapest, Hungary, 1171 (1992).

76. G.V. Gibbs, E.P. Meagher, J. V. Smith, J.J. Pluth, ACS Symp. Series, Washington DC 40, 19, (1977).

77. T.L. Barr, Zeolites, 10, 760 (1990).

78. Y. Okamoto, Zeoraito, 10, 195 (1993).

79. Y. Okamoto, M. Ogawa, A. Maezawa, T. Imanaka, J. Catal., 112, 427 (1988).

80. M. Huang, A. Adnot, S. Kaliaguine, J. Am. Chem. Soc., 114, 1005 (1992).

81. M.M. Chehimi, M. Delamar, Analysis, 23, 291 (1995).

82. H. Hattori, Chem. Review, 95, 543 (1995).

83. S.J. Kulkarni, Indian J. Chem. Sect.: Inorg. Phys. Theor. Anal., 29A(11), 1125 (1990).

84. E.A. Paukshtis, N.S. Kotsarenko, L.G. Karakchiev, React. Kinet. Catal. Lett., 12, 315 (1979).

85. C. Mirodatos, P. Pichat, D. Barthomeuf, J. Phys. Chem., 80, 1335 (1976).

86. J.C. Lavalley, "Trends in Physical Chemistry", 2, 305 (1991).

87. A. Gervasini, A. Auroux, J. Therm. Anal., 37, 1737 (1991).

88. W. Przystajko, R. Fiedorow, I.G. Dalla Lana, Zeolites, 7, 477 (1987)

89. M. Berkani, J.L. Lemberton, M. Marczewski, G. Perot, Catal. Lett., 31, 405 (1995).

90. S. Huber, H. Knozinger, Chem. Phys. Lett., 244, 111 (1995).

91. J. Kotrla, J. Florian, L. Kubelkova, J. Fraissard, Collect. Czech. Chem. Comm., 60, 393 (1995).

92. J. Xie, S. Kaliaguine, Catal. Lett., 29, 281 (1994).

93. S. Bordiga, E. Garrone, C. Lambert, A. Zecchina, C.O. Arean, V.B. Kazanski, L.M. Kustov, J. Chem. Soc. Faraday Trans., 90, 3367 (1994).

94. .A. Kheir, J.F. Haw, J. Am. Chem. Soc., 116, 817 (1994).

95. A. Zecchina, R. Buzzoni, S. Bordiga, F. Geobaldo, D. Scarano, G. Ricchiardi, G. Spoto, Stud. Surf. Sci. Catal., 97, 213 (1995).

96. E.B. Uvarova, L.M. Kustov, V.B. Kazansky, Stud. Surf. Sci. Catal., 94, 254 (1995).

97. S. Dzwigaj, A. de Mallmann, D. Barthomeuf, J. Chem. Soc. Faraday Trans., 86, 431 (1990).

98. A. de Mallmann, D. Barthomeuf, Zeolites, 8, 292 (1988).

99. J.P. Shen, J. Ma, T. Sang, D.Z. Jiang, E.Z. Min, J. Chem. Soc. Faraday Trans., 90, 1351 (1994).

100. H. Hattori, Chem. Rev., 95, 537 (1995).

101. Y.N. Sidorenko, P.N. Galich, V.S. Gutyrya, V.G. Ill'in, I.K. Neimark, Dokl. Akad. Nauk. SSSR, 173, 132 (1967).

102. T. Yashima, H. Suzuki, N. Hara, J. Catal., 33, 486 (1974).

103. R.M. Dessau, Zeolites, 10, 205 (1990).

104. M. Berkani, J.L. Lemberton, M. Marczewski, G. Perot, Catal. Lett., 31, 405 (1995).

105. A. Corma, V. Fornés, R.M. Martín-Aranda, H. García, J. Primo, Appl. Catal., 59, 237 (1990).

106. A. Corma, R.M. Martín-Aranda, F. Sanchez, Stud. Surf. Sci. Catal., 59, 503 (1991).

107. A. Corma, R.M. Martín-Aranda, J. Catal., 130, 130 (1990).

108. A. Corma, R.M. Martín-Aranda, F. Sánchez, J. Catal., 126, 192 (1990).

109. R. Davis, Heterogeneous Chemistry Reviews, 1, 41 (1994).

110. C.B. Dartt, M.E. Davis, Catal. Today, 19, 151 (1994).

111. N. Giordano, L. Pino, S. Cavallaro, P. Vitarelli, B.S. Rao, Zeolites, 7, 131 (1987).

112. M. Itoh, T. Hattori, K. Suzuki, Y. Murakami, J. Catal., 79, 21 (1983).

113. Z.M. Fu, Y. Ono, Catal. Lett., 22, 277 (1993).

114. P.R. Hari Prasad Rao, P. Massiani, D. Barthomeuf, Catal. Lett., 31, 369 (1993).

115. Z.M. Fu, Y. Ono, J. Catal., 145, 166 (1994).

116. B.L. Su, D. Barthomeuf, Appl. Catal. A General, 124, 73 (1995).

117. J.B. Nagy, J.P. Lange, A. Gourgue, P. Bodart, Z. Gabelica, Stud. Surf. Sci. Catal., 20, 127 (1985).

118. A. Corma, "Symp. Proc. of the Material Research Society", 233, 17 (1991).

119. A.M. Efstathiou, S.L. Suib, C.O. Bennett, J. Catal., 135, 223 (1992).

120. H.B. Schwarz, H. Ernst, S. Ernst, J. Karger, T. Roser, R.Q. Snurr, J. Weitkamp, Appl. Catal. A, General, 130, 227 (1995).

121. B.L. Su, D. Barthomeuf, Appl. Catal. A, General, 124, 73 (1995).

122. I. Hannus, H. Forster, G. Tasi, I. Kiricsi, A. Molnar, J. Mol. Struct., 348, 345 (1995).

123. M. Huang, S. Kaliaguine, React. Kinet. Catal. Lett., 56, 21 (1995).

124. C.O. Veloso, J.L. Monteiro, E.F. Sousaaguiar, Stud. Surf. Sci. Catal., 84, 1913 (1994).

125. H. Tsuji, F. Yagi, H. Hattori, Chem. Lett., 1881 (1991).

126. M. Lasperas, H. Cambon, D. Brunel, I. Rodriguez, P. Geneste, Microporous Materials, 1, 343 (1993).

127. I. Rodriguez, H. Cambon, D. Brunel, M. Lasperas, P. Geneste, Stud. Surf. Sci. Catal., 78, 623 (1993).

128. F.P. Gortsema, B. Beshty, J.J. Friedman, D. Matsumoto, J.J. Sharkey, G. Wildman, T.J. Blacklock, S.H. Pan, in "Catalysis of Organic Reactions", J.R. Kosak and T.A. Johnson, Eds.), p. 445, Marcel Dekker Inc., N.Y. (1994).

129. R. Sheldon, Stud. Surf. Sci. Catal., 83, 407 (1994).

130. D.A. Young, US Pat. 3329481, (1967).

131. D.W. Breck, Zeolites Molecular Sieves, 322, (1974).

132. R.M. Barrer, Hydrothermal Chemistry of Zeolites, Academic Press, London, (1982).

133. M. Tamarasso, G. Perego, B. Notari, US Pat. 4410501 (1981).

134. J.S. Reddy, R. Kumar, P. Ratnasamy, Appl. Catal., 58, L1 (1990).

135. M. A. Camblor, A. Corma, A. Martínez, J. Pérez-Pariente, J. Chem. Soc., Chem. Comm., 589 (1992).

136. D.P. Serrano, H.X. Li, M.E. Davis, J. Chem. Soc., Chem. Comm., 745 (1992).

137. A. López, M.H. Tuiller, J.L. Guth, L. Delmotte, J.M. Popa, J. Solid State Chem., 102, 480 (1993).

138. A. Zechina, G. Spoto, S. Bordiga, F. Geobaldo, G. Petrini, G. Leofanti, M. Padovan, M. Mantegazza and P. Roffia, 10th. Int. Congr. Catal. Budapest, L. Guczi et al Edit, 719 (1993).

139. G. Bellusi, A. Carati, M.G. Clerici, G. Maddinelli, and R. Milini, J. Catal., 133 (1992).

140. T. Blasco, M.A. Camblor, A. Corma, J. Pérez-Pariente, J. Am. Chem. Soc., 115, 11806 (1994).

141. F. Geobaldo, S. Bordiga, A. Zecchina, E. Giamello, G. Leofanti, and G. Petrini, Catal. Lett., 16, 109 (1992).

142. G. Bellusi, A. Carati, M.G. Clerici, G. Maddinelli, R. Millini, J. Catal., 133, 220 (1992).

143. A. Tuel, Y. Ben Taarit, J. Chem. Soc., Chem. Commun., 1667 (1994).

144. B. Kraushaar-Czarnetzki, W.G.M. Hoogervorst, R.R. Andrea, C.A. Emeis, W.H.J. Stork, J. Chem. Soc., Faraday Trans., 87, 891 (1991).

145. J.A. Rossin, C. Saldarriaga, M.E. Davis Zeolites, 7, 295 (1987).

146. D.L. Vanoppen, D.E. De Vos, M.J. Genet, P.G. Ruxhet, and P.A. Jacobs, Ang. Chemie (completar) (1995).

147. R.A. Sheldon, J.D. Chen, J. Dakka, and E. Neeleman, New Developments in Selective Oxidation II, Cortés Corberan and Vic Bellon Edit. Elsevier Science B.V., 515 (1994).

148. K.M. Reddy, I.L. Mondrakovski, A. Sayari, J. Chem. Soc., Chem. Commun., 1491 (1994).

149. G. Perego, G. Bellussi, C. Corno, M. Tamarasso, F. Buonomo, A. Esposito in "Proceedings 7th. Int. Conf. Zeol." (Y. Murakami et al., Eds.) Kondansha-Elsevier, Amsterdam 129 (1987).

150. A. Esposito, M. Taramasso, C. Neri, F. Buonomo, U.K. Pat. 2116974 (1985).

151. M.A. Camblor, A. Corma, P. Esteve, A. Martínez, to be published.

456

152. P. Roffia, G. Leofanti, A. Cesana, M. Mantegazza, M. Padovan, G. Petrini, S. Tonti, P. Gervasutti, Stud. Surf. Sci. Catal., 55, 43 (1990).

153. F. Maspero, U. Romano, J. Catal., 146, 476 (1994).

154. A. Corma, P. Esteve, A. Martínez, Appl. Catal., July (1996).

155. G. Bellussi, A. Caratí, M.G. Clerici, G. Maddinelli, R. Milline, J. Catal., 33, 220 (1992).

156. A. Corma, P. Esteve, A. Martínez, S. Valencia, J. Catal., 152, 18 (1995).

157. T. Sato, J. Dakka, R.A. Sheldon, J. Chem. Soc., Chem. Commun., 1887 (1994).

158. T. Sato, J. Dakka, R.A. Sheldon, Stud. Surf. Sci. Catal., 84, 1853 (1994).

159. A. Corma, P. Esteve, A. Martínez, J. Catal., June (1996).

160. M. A. Camblor, A. Corma, S. Valencia, J. Chem. Soc., Chem. Comm., June (1996).

161. M.A. Camblor, M. Constantini, A. Corma, P. Esteve, L. Gilbert, A. Martínez, S. Valencia, Appl. Catal. A, General, 133, L185 (1995).

162. M.G. Clerici, G. Bellusi, and U. Románo, J. Catal., 129, 159 (1994).

163. P. Roffia, G. Leofanti, A. Cesana, M. Mantegazza, M. Padovan, G. Petrini, S. Tontí, V. Gervazutti, and R. Varagnolo, Chem. Ind. 72, 598 (1990).

164. D. R. C. Huybrechts, L. Bruycker and P. A. Jacobs, Nature, 945, 240 (1990).

165. J.S. Reddy, S. Sivasanker, P. Ratnsany, J. Mol. Catal., 70, 335 (1991).

166. M.G. Clerici, Appl. Catal., 68, 249 (1991).

167. B. Notari, Catal. Today., 18, 163 (1993)

168. T. Sen, M. Chatterjee, S. Sivasanker, J. Chem. Soc., Chem. Comm. 207 (1995).

169. P.R. H. Prasad Rao, A. V. Ramanswany, J. Chem. Soc. Chem. Commun, 1245 (1992).

170. G. Bellusi, M. Riguto, Stud. Surf. Sci. Catal., 85, 177 (1994).

171. A.V. Ramasnamy, S. Sivasanker, Catal. Lett. 22, 239 (1993).

172. P. Ratnasamy, and R. Kumar, Catal. Lett, 22, 227 (1993).

173. C. Marchal, A. Tuel, and Y. Ben Taarit, Stud. Surf. Sci. Catal., 78, 447 (1993).

174. S.S. Lin, and H.S. Weng, Appl. Catal., A. General, 105, 289 (1993); S.S. Lin, H.S. Weng, Appl. Catal., A. General, 118, 21 (1994).

175. M.P.J. Peeters, M. Busio, P. Leitjen, Appl. Catal., A. General, 118, 51 (1994).

176. B. Kraushaar-Czarnetzki, W.G.M. Hoogervorstand, W.H.J. Stork, Stud. Surf. Sci. Catal., 84, 1869 (1994).

177. C.C. Freyhardt, M. Tsapatsis, R.F. Lobo, K.J. Balkus Jr., M.E. Davis, Nature, 385, 295 (1996).

178. W.M. Meier, D.H. Olson, "Atlas of Zeolite Structure Types", 3rd Revised Edition, 1992, Butterworth-Heinemann.

179. S.T. Wilson, B.M. Lok, E.M. Flanigen, US Pat. 4310440 (1982).

180. R.M. Dessau, J.L. Schlenker, J.B. Aiggins, Zeolites, 10, 522 (1990).

181. J.W. Richardson Jr., E.T.C. Vogt, Zeolites, 12, 13 (1992).

182. M.E. Davies, C. Saldarriaga, C. Montes, J. Garces, C. Crowder, Nature, 331, 698 (1988)

183. M.E. Davies, C. Montes, J.M. Garces, ACS Symp. Ser., 398, 291 (1989).

184. M. Estermann, L.B. McCuster, CH. Baerlocher, A. Merrouche, H. Kessler, Nature, 352, 320 (1991).

185. A. Merrouche, J. Patarin, H. Kessler, M. Soulard, L. Delmotte, J.L. Guth, J.F. Jolly, Zeolites, 12, 226 (1992).

186. J.L. Guth, H. Kessler, P. Caullet, J. Hazm, A. Mewouche, J. Patarin, Proc. IXth Int. Zeolite Conference (R. von Ballmoos et al., Eds.), 215 (1993).

187. R.H. Jones, J.M. Thomas, J. Chen, R. Xu, Q. Huo, S. Li, Z. Ma, A.M. Chippindale, J. Solid. State Chem., 102, 5605 (1993).

188. L.B. McCuster, C. Baerlocher, E. Jahn, M. Bülow, Zeolites, 11, 308 (1991).

189. A. Corma, Proc. VIIIth Int. Zeolite Conf. (P.A. Jacobs and R.A. van Santen, Eds.), Elsevier, Amsterdam, 49 (1989).

190. P.B. Moore, J. Shen, Nature, 306, 356 (1983).

191. T.J. Pinnavaia, H. Kim, NATO ASI Series, Vol. 352, Kluwer Academic Publishers, p. 79 (1992).

192. T.J. Pinnavaia, T. Kwanrand, S.K. Yun, NATO ASI Series, Vol. 352, Kluwer Academic Publishers, p. 91 (1992).

193. "Pillared Layer Structures" (I.V. Mitchell, Ed.), Elsevier (1990).

194. R. Burch, Catalysis Today, 2, 1-185 (1988).

195. "Multifunctional Mesoporous Inorganic Solids", NATO ASI Series, Vol. 400, Kluwer Academic Publishers, p. 19 (1993).

196. P. Grange, J. Chem. Phys., 87, 1547 (1990).

197. A. Clearfield, A. Kuchenmeister, ACS Symp. Series, 499, 128 (1992).

198. A. Clearfield, R.M. Tindwa, Inorg. Nucl. Chem. Lett., 15, 251 (1979).

199. A. Clearfield, Comments Inorg. Chem., 10, 89 (1990).

200. C. Alberti, U. Costantino, in "Inclusion Compounds", Vol. 5 (J.L. Atwood, J.E.D. Davies, D.D. MacNicol, Eds.), Oxford University Press, Oxford, p. 132 (1991).

201. A. Clearfield, B.D. Roberts, Inorg. Chem., 27, 3237 (1988).

202. D.A. Burwell, M.E. Thompson, Chem. Mater., 3, 730 (1991).

203. G.L. Rosenthal, J. Coruso, Inorg. Chem., 31, 3104 (1992).

204. W.T. Reichle, Chemtech, 58 (1986).

205. A. Clearfield, Chem. Rev., 88, 125 (1988).

206. T. Kwon, T.J. Pinnavaia, Chem. Mater., 1, 381 (1989).

207. D.E. Vaughan, R.J. Lussier, J.S. Maaee, US Pat. 4176090 (1979).

208. A. Moini, T.J. Pinnavaia, Solid State Ionics, 26, 119 (1988).

209. J. Sterte, Clay & Clays Miner., 34, 658 (1986)

210. J. Sterte, Clay & Clays Miner., 39, 167 (1991).

211. K. Kuroda, T. Yanagisawa, T. Shimizu, C. Kato, Bull. Chem. Soc. Jpn., 63, 988 (1990).

212. K. Kuroda, T. Yanagisawa, T. Shimizu, C. Kato, Bull. Chem. Soc. Jpn., 63, 1535 (1990).

213. K. Beneke, G. Lagaly, Am. Mineralogist, 62, 763 (1977)

214. S. Inagaki, Y. Fukushima, K. Kuroda, J. Chem. Soc., Chem. Commun., 680 (1983).

215. J.S. Beck, C.T-W. Chu, I.D. Johnson, C.T. Kresge, M.E. Leonowicz, W.J. Roth, J.C. Vartuli, WO91/11390 (1991).

216. J.S. Beck, US Pat. 5057296 (1991).

217. J.S. Beck, C.T-W. Chu, I.D. Johnson, C.T. Kresge, M.A. Leonowicz, W.J. Roth, J.C. Vartuli, US Pat. 5108725 (1992).

218. C.T. Kresge, M.E. Leonowicz, W.J. Roth, J.C. Vartuli, US Pat. 5098684 (1992).

219. D.C. Calabro, K.D. Schmitt, J.C. Vartuli, US Pat. 5110572 (1992).

220. I.D. Johnson, J.P. McWilliams, US Pat. 5112589 (1992).

221. C.T. Kresge, M.E. Leonowicz, W.J. Roth, J.C. Vartuli, J.S. Beck, Nature, 359, 710 (1992).

222. J.S. Beck, J.C. Vartuli, W.J. Roth, M.E. Leonowicz, C.T. Kresge, K.D. Schmitt, C.T-W. Chu, D.H. Olson, E.W. Sheppard, S.B. McCullen, J.B. Higgins, J.C. Schlenker, J.Am. Chem. Soc., 114, 10834 (1992).

223. P. Ekwall, "Advances in Liquid xls.", Vol. I (G.H. Brown, Ed.), Academic New York (1971).

224. V. Cuzzati, "Biological Membranes" (D. Chapman, Ed.), Academic, New York, 71-123 (1968).

225. G.J.T. Tiddy, Phys. Rep., 57, 1-46 (1980).

226. P.A. Winsor, Chem. Rev., 68, 1-40 (1968).

227. C-Y. Chen, S-Q. Xiao, M.E. Davies, Microporous Materials, 4, 1 (1995).

228. Q. Huo, D. Margolese, U. Ciesla, P. Feng, T. Gier, P. Sieger, R. Leon, P. Petroff, F. Schüth, G. Stucky, Nature, 368, 317 (1994).

229. G.D. Stucky, A. Monnier, F. Schüth, Q. Huo, D. Margolese, D. Kumar, M. Crishnamurty, P. Petroff, A. Firouzi, M. Janiche, B.F. Chmella, Mol. Cryst. Liq. Cryst., 240, 187 (1994).

458

230. Q. Huo, R. Leon, P. Petroff, G.D. Stucky, Science, 268, 1324 (1995).

231. P.T. Tanev, T.J. Pinnavaia, Science, 267, 865 (1995).

232. S.A. Bagshaw, E. Prouzet, T.J. Pinnavaia, Science, 269, 1242 (1995).

233. A. Tuel, S. Gontier, Chem. Mat., 8, 114 (1996).

234. R. Mokaya, W. Jones, J. Chem. Soc., Chem. Commun., 981 (1996).

235. K.J. Edler, J.W. White, J. Chem. Soc., Chem. Commun., 155 (1995).

236. F.H.P. Silva, H.O. Pastore, J. Chem. Soc., Chem. Commun., 833 (1996).

237. C.G. Wu, T. Bein, J. Chem. Soc., Chem. Commun., 925 (1996).

238. J.C. Vartuli, C.T. Kresge, M.E. Leonowicz, A.S. Chu, S.B. McCullen, I.D. Johnson, E.W. Sheppard, Chem. Mater., 6, 2070 (1994).

239. J.S. Beck, J.C. Vartuli, G.J. Kennedy, C.T. Kresge, W.J. Roth, S.E. Schramm, Chem. Mater., 6, 1816 (1994).

240. A. Monnier, F. Schüth, Q. Huo, D. Kumar, D. Margolese, R.S. Maxwell, G.D. Stucky, M. Krishnamurty, P. Petroff, A. Firouzi, M. Janicke, B.F. Chmeika, Science, 261, 1299 (1993).

241. A. Firouzi, D. Kumar, L.M. Bull, T. Besier, P. Sieger, Q. Huo, S.A. Walker, J.A. Zasadzinski, C. Glinka, J. Nicd, D. Margolese, G.D. Stucky, B.F. Chmeika, Science, 267, 1138 (1995).

242. C-Y. Chen, H-X. Li, M.E. Davis, Microporous Materials, 2, 17 (1993).

243. C-Y. Chen, S.L. Burkett, H-X. Li, M.A. Davis, Microporous Materials, 2, 27 (1993).

244. C-F. Cheng, H. He, W. Zhou, J. Klinowski, Chem. Phys. Lett., 244, 117 (1995).

245. A. Steel, S.W. Carr, M.W. Anderson, J. Chem. Soc., Chem. Commun., 1571 (1994).

246. P.J. Branton, P.G. Hall, K.S.W. Sing, J. Chem. Soc., Chem. Commun., 1257 (1993).

247. O. Franke, G. Schulz-Ekioff, J. Rathously, J. Stárek, A. Zukal, J. Chem. Soc., Chem. Commun., 724 (1993).

248. P.L. Llewellyn, Y. Grillet, F. Schüth, H. Reichert, Microporous Materials, 3, 345 (1994).

249. R. Schmidt, M. Stöcker, E. Hansen, D. Akporiaye, O.H. Ellestad, Microporous Materials, 3, 443 (1995).

250. A. Saito, H. Foley, Microporous Materials, 3, 531 (1995).

251. P.L. Llewellyn, F. Schüth, Y. Grillet, F. Rouguerol, J. Rouguerol, K.K. Unger, Langmuir, 11, 574 (1995).

252. V.Y. Gusev, X. Feng, Z. Bu, G. Haller, J.A. O'Brien, J. Phys. Chem., 100, 1985 (1996).

253. W. Kolodziejski, A. Corma, M.T. Navarro, J. Pérez-Pariente, Solid State Nucl. Magn. Reson., 2, 253 (1993).

254. D. Akporiaye, E.W. Hansen, R. Schmidt, M. Stöcker, J. Phys. Chem., 98, 1927 (1994).

255. E.W. Hansen, R. Schmidt, M. Stöcker, D. Akporiaye, J. Phys. Chem., 99, 4148 (1995).

256. V. Luca, D.J. Maclachlan, R. Brauley, K. Morgan, J. Phys. Chem., 100, 1793 (1996).

257. J. Chen, Q. Li, R. Xu, F. Xiao, Angew. Chem. Int. Ed. Engl., 34, 2694 (1995).

258. V. Alfredsson, M. Keung, A. Monnier, G.D. Stucky, K.K. Unger, F. Schüth, J. Chem. Soc., Chem. Commun., 921 (1994).

259. A. Chenite, Y. Le Page, Chem. Mater., 7, 1015 (1995).

260. R. Schmidt, E.W. Hansen, M. Stöcker, D. Akporiaye, O.H. Ellestad, J. Am. Chem. Soc., 117, 4049 (1995).

261. R. Schmidt, M. Stöcker, D. Akporiaye, E. Heggelund,Törstad, A. Olsen, Microporous Materials, 5, 1 (1995).

262. I. Petrovic, A. Navrotsky, C-Y. Chen, M.E. Davies, J. Weitkamp, H.G. Karge, H. Pfeifer, W. Hölderich, Stud. Surf. Sci. Catal., 84, 677 (1994).

263. A. Corma, M.T. Navarro, J. Pérez-Pariente, F. Sánchez, Stud. Surf. Sci. Catal., 84, 969 (1994).

264. A. Corma, M.T. Navarro, J. Pérez-Pariente, J. Chem. Soc., Chem. Commun., 147 (1994).

265. M. Janicke, D. Kumar, G.D. Stucky, B.F. Chmeika, Stud. Surf. Sci. Catal., 84, 243 (1994).

266. K.M. Reddy, I. Moudrakovski, A. Sayari, J. Chem. Soc., Chem. Commun., 1059 (1994).

267. M.A. Camblor, A. Corma, J. Pérez-Pariente, Zeolites, 13, 82 (1993).

268. A. Corma, M.T. Navarro, J. Pérez-Pariente, Spa. Pat. 9301327 (1993).

269. P. Tanev, M. Chibwe, T.J. Pinnavaia, Nature, 368, 321 (1994).

270. A. Keshavaraja, V. Ramaswamy, H.S. Soni, A.V. Ramaswamy, P. Ratnaswamy, J. Catal., 157, 501 (1995).

271. a) T. Blasco, A. Corma, M.T. Navarro, J. Pérez-Pariente, J. Catal., 156, 65 (1995); (b) A. Corma, M. Iglesias, F. Sánchez, Catal. Lett., June (1996); (c) J.S. Reddy, A. Sayari, Appl. Catal., 128, 231 (1995).

272. M. Alba, Z. Luan, K. Klinowski, J. Phys. Chem., 100, 2178 (1996).

273. W. Zhang, J. Wang, P. Tanev, T. Pinnavaia, J. Chem. Soc., Chem. Commun., 979 (1996).

274. M. Morey, A. Davidson, G.D. Stucky, Microporous Materials, 6, 99 (1996).

275. K.A. Koyano, T. Tatsumi, J. Chem. Soc., Chem. Commun., 145 (1996).

276. N.A. Bhore, Q.N. Le, US Pat. 5134243 (1992).

277. Q.N. Le, R.T. Thomson, US Pat. 5134242 (1992).

278. (a) M.R. Apelian, T.F. Degran Jr., D.O. Marler, D.N. Mazzone, US Pat. 5264116 (1993); (b) Q.N. Le, R.T. Thomson, US Pat. 5134241 (1992).

279. J.S. Beck, R.F. Socha, D.S. Shihabi, J.C. Vartuli, US Pat. 5143707 (1992).

280. Q.N. Le, R.T. Thomson, US Pat. 5191144 (1993).

281. (a) A. Corma, V. Fornés, M.T. Navarro, J. Pérez-Pariente, J. Catal., 148, 569 (1994); (b) A. Corma, M. Grande, V. Martínez, A.V. Orchilles, J. Catal., 159, 375 (1996).

282. R. Schmidt, D. Akporiaye, M. Stöcker, O. Ellestad, J. Chem. Soc., Chem. Commun., 1493 (1994).

283. (a) E. Armengol, M. Cano, A. Corma, H. García, M.T. Navarro, J. Chem. Soc., Chem. Commun., 519 (1995); (b) K.R. Kloetstra, H. van Bekkum, J. Chem. Res., 26 (1995).

284. R. Borade, A. Clearfield, Catal. Lett., 31, 267 (1995).

285. Z. Luan, C-F. Cheng, H. He, J. Klinowski, J. Phys. Chem., 99, 10590 (1995).

286. Z. Luan, C-H. Cheng, W. Zhou, J. Klinowski, J. Phys. Chem., 99, 1018 (1995).

287. R. Mokaya, W. Jones, Z. Luan, M. Alba, J. Klinowski, Catal. Lett., 37, 113 (1996).

288. J. Aguado, D.P. Serrano, M.D. Romero, J.M. Escola, J. Chem. Soc., Chem. Commun., 725 (1996).

289. R. Mokaya, W. Jones, J. Chem. Soc., Chem. Commun., 983 (1996).

290. A. Corma, M. Iglesias, F. Sánchez, J. Chem. Soc., Chem. Commun., 1635 (1995).

291. M.J. Climent, A. Corma, S. Iborra, M.C. Navarro, J. Primo, J. Catal., in press (1996).

292. R. Schmidt, H. Junggreen, M. Stöcker, J. Chem. Soc., Chem. Commun., 875 (1996).

293. A. Sayari, C. Danumah, I.L. Moudrakovski, Chem. Mater., 7, n°5 (1995).

294. A. Sayari, I. Moudrakovski, C. Danumah, C. Ratcliffe, J.A. Ripmeester, K.F. Preston, J. Phys. Chem., 99, 16373 (1995).

295. J.S. Reddy, A. Sayari, J. Chem. Soc., Chem. Commun., 2231 (1995).

296. K.M. Reddy, I.C. Moudrakovski, A. Sayari, J. Chem. Soc., Chem. Commun., 1059 (1994).

297. N. Ulagappan, C.N.R. Rao, J. Chem. Soc., Chem. Commun., 1047 (1996).

298. T.M. Abdel-Fattah, T.J. Pinnavaia, J. Chem. Soc., Chem. Commun., 665 (1996).

299. T.K. Das, K. Chaudhari, A.J. Chandwadkar, S. Sivasanker, J. Chem. Soc., Chem. Commun., 2495 (1995).

300. D. Zhao, D. Goldfarb, J. Chem. Soc., Chem. Commun., 875 (1995).

301. Z.Y. Yuan, S.Q. Liu, T.H. Chen, J.Z. Wang, X-H. Li, J. Chem. Soc., Chem. Commun., 973 (1995).

302. C-F. Cheng, H. He, W. Zhou, J. Klinowski, J.A. Sousa Goncalves, L.F. Gladden, J. Phys. Chem., 100, 390 (1996).

303. V. Luca, D.J. MacLachlan, J.M. Hook, R. Withers, Chem. Mater., 7, 2220 (1995).

304. U. Ciesla, D. Demuth, R. Leon, P. Petroff, G. Stucky, K. Unger, F. Schüth, J. Chem. Soc., Chem. Commun., 1387 (1994).

460

305. J.A. Knowles, M.J. Hudson, J. Chem. Soc., Chem. Commun., 2083 (1995).

306. M. Yada, M. Machida, T. Kijima, J. Chem. Soc., Chem. Commun., 769 (1996).

307. J. Li, L. Delmotte, H. Kessler, J. Chem. Soc., Chem. Commun., 1023 (1996).

308. U. Ciesla, S. Schacht, G.D. Stucky, K.K. Unger, F. Schüth, Angew. Chem. Int. Ed. Engl., 35, 541 (1996).

309. D.M. Antonelli, J.Y. Ying, Angew. Chem. Int. Ed. Engl., 34, 2014 (1995).

310. D.M. Antonelli, J.Y. Ying, Angew. Chem. Int. Ed. Engl., 35, 426 (1996).

311. C. Huber, K. Moller, T. Bein, J. Chem. Soc., Chem. Commun., 2619 (1994).

312. A. Corma, A. Martínez, V. Martínez-Soria, J.B. Monton, J. Catal., 153, 25 (1995).

313. R. Ryoo, C.H. Ko, J.M. Kim, R. Howe, Catal. Lett., 37, 29 (1996).

314. U. Junges, W. Jacobs, I. Voigt-Martin, B. Krutzch, F. Schüth, J. Chem. Soc., Chem. Commun., 2283 (1995).

315. K.R. Kloetstra, H. van Bekkum, J. Chem. Soc., Chem. Commun., 1005 (1995).

316. M. Lasperas et al., private communication.

317. R. Burch, N. Cruise, D. Gleeson, S.C. Tsang, J. Chem. Soc., Chem. Commun., 951 (1996).

318. I.V. Kozhevnikov, A. Sinnema, R.J.J. Jansen, K. Pamin, H. van Bekkum, Catal. Lett., 30, 241 (1995).

319. C.T. Kresge, D.O. Marler, G.S. Rav, B.H. Rose, US Pat. 5366945 (1994).

320. A. Pöppl, M. Mewhouse, L. Kevan, J. Phys. Chem., 99, 10019 (1995).

321. T. Abe, Y. Tachibana, T. Uematsu, M. Iwamoto, J. Chem. Soc., Chem. Commun., 1617 (1995).

322. B.P. Pelrine, K.D. Schmidt, J.C. Vartuei, US Pat. 5105051 (1992).

323. M. Morey, A. Davidson, H. Eckert, G.D. Stucky, Chem. Mater., 8, 486 (1996).

324. Degnan Jr., K.J. Del Rossi, A. Hussain, A. Huss Jr. US Pat. 519114 (1993).

325. Boghard, C.T. Chu, T.H. Degnan, and S.S. Shih, US Pat. 5264641 (1993).

326. Chibwe, A. Barodawalla, T.J. Pinnavaia, 14 th, North Amer. Cat. Meeting of Cat. Soc., paper PB107 (1995).

327. T. Maschmeyer, F. Rey, G. Sankar, J.M. Thomas, Nature, 378, 159 (1995).

328. A. Corma, V. Fornés, H. García, M.A. Miranda, M.J. Sabater, J. Am. Chem. Soc., 116, 9767 (1994).

329. C-G. Wu, T. Bein, Science, 264, 1757 (1994).

330. P.L. Llewellyn, U. Ciesla, H. Decher, R. Stadler, F. Schüth, K.K. Unger, Stud. Surf. Sci. Catal., 84, 2013 (1994).

DEVELOPMENTS IN SILICA-SUPPORTED ORGANOMETALLIC CATALYSIS

Silsesquioxanes and Mesoporous MCM-41 Silicates

THOMAS MASCHMEYER,[a,*] JOHN M. THOMAS[a]
and ANTHONY F. MASTERS[b]

[a] *Davy-Faraday Research Laboratories, The Royal Institution of Great Britain,
21 Albemarle Street, London W1X 4BS, U.K.*
[b] *School of Chemistry, The University of Sydney,
N.S.W., 2006, Australia.*

Silsesquioxanes as models for silica surface reactions with organometallic complexes and heteroatom-substituted mesoporous MCM-41 silicas are discussed. Particular attention is given to catalytically active materials. Interactions of $Os_3(CO)_{10}$, titanocene dichlorides and chromocene with silica surfaces are investigated with silsesquioxanes, yielding new insights into silica surface silanol and siloxane reactivity. XAS (X-ray Absorption Spectroscopy) is used to study the UNIPOL (chromocene on silica) ethylene polymerisation catalyst, revealing the average structure of the surface species to be $[Cr(\eta^5-C_5H_5)(\mu-OSi\equiv)(OSi\equiv)]_2$. The synthesis and characterisation of 7 framework-substituted MCM-41 silicas as well as a titanium containing MCM-41 derivative in which the titanium was grafted onto the surface are presented. These materials are structurally characterised using XAS; the grafted Ti derivative is also described in terms of its catalytic activity for epoxidation reactions involving cyclohexene and pinene.

1. Silsesquioxanes – Siliceous Analogues of Silica Surfaces

In the following section the utility of polyhedral oligosilsesquioxanes as models for silica surfaces is examined, first from a structural (1.1) and then from a chemical reactivity (1.2) point of view.

1.1 SILSESQUIOXANES AS STRUCTURAL MODELS OF SILICA

In the last few years it has been shown that it is possible to develop soluble, molecular, well-characterised model systems (polyhedral oligosilsesquioxanes) designed to simplify the study of silica surface complexes [1-4]. These soluble molecular systems allow for both spectroscopic/structural comparisons as well as mechanistic/chemical studies. Silsesquioxanes are composed of polyhedral silicon-oxygen skeletons with organic or inorganic substituents on the silicon atoms. Two excellent reviews on silsesquioxanes by Baney, *et al.* and Vronkov, *et al.* are available [5,6]. Fully condensed frameworks have the general formula $(RSiO_{3/2})_n$ (n > 4, R = H, alkyl, aryl, halogen), whereas partially condensed compounds retain a OH group on one or several silicon atoms. Molecular model compounds for different silica surface group arrangements can be obtained by the synthesis of such incompletely condensed polyhedral oligosilsesquioxanes (POSS). These are readily accessible in experimentally useful quantities.

---

* Presenting Author.

C.R.A. Catlow and A. Cheetham (eds.), New Trends in Materials Chemistry, 461–494.
© *1997 Kluwer Academic Publishers.*

They can model a variety of surface silanol arrangements, ranging from strained siloxanes and isolated silanols to assemblies with up to four hydrogen-bonded silanols. Their three-dimensional framework is similar to some crystallographic phases of silica, but can also be altered by the substitution of some silanols to model a more disordered structure.

Intuitively, the three-dimensional frameworks of POSS may be expected to present much more reasonable model structures for bulk silica and silica surface sites than simple siloxanes and silanols (such as bistrimethylsiloxane or triphenylsilanol for example). On closer examination this is confirmed since the silsesquioxane, [(c-hexyl)$_7$Si$_7$O$_9$(OH)$_3$] (1), when viewed down its pseudo C_3-axis, resembles potential coordination sites available on the [111] octahedral face of β-cristobalite (Figure 1) [1].

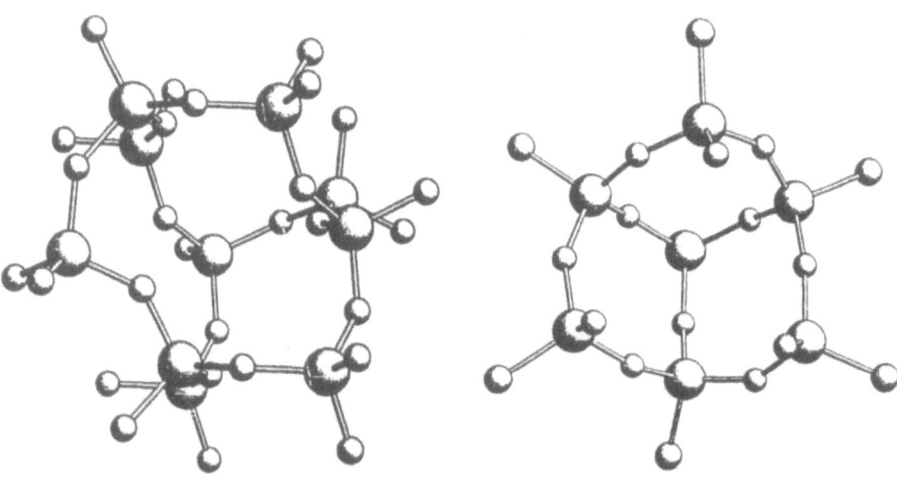

Figure 1. (a) Structure of [(c-hexyl)$_7$Si$_7$O$_9$(OH)$_3$], (1), Viewed Down Pseudo-C$_3$ Axis. For Clarity, Only Carbon Atoms Attached to Si Are Shown, (b) Plot of the Idealised [111] Octahedral Face of β-cristobalite.

The structures in figure 1 (a) and (b) are remarkably similar; both contain a siloxane rim of 6 silicon and 6 oxygen atoms each and have a SiO_3-unit in the centre. Additionally, the Si-Si distances in β-cristobalite and [(c-hexyl)$_7$Si$_7$O$_9$(OH)$_3$], (1), differ on average by only 0.14 Å, demonstrating the close structural match. The structural sketches in figure 2 display the principal silsesquioxanes discussed in this work.

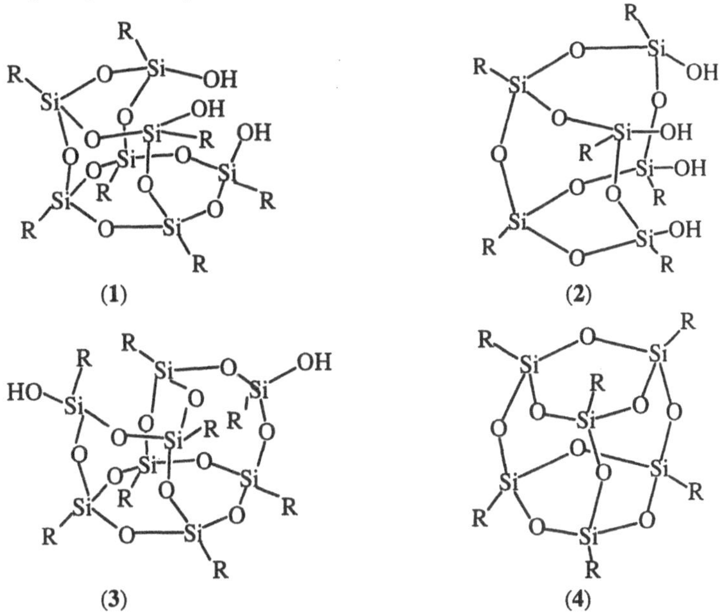

(1)   (2)

(3)   (4)

Figure 2. Structural Sketch of Some Basic Silsesquioxane Frameworks (R = alkyl).

1.2 USING SILSESQUIOXANES TO MODEL HETEROGENEOUS ORGANOMETALLIC CATALYSTS

1.2.1 *Reactions of Group VIII Trinuclear Carbonyl Clusters with Silsesquioxanes*

Silica supported organometallic complexes are implicated in a large number of catalytic reactions [7-9]. However, surface reactions are difficult to study and hence are less well understood than their homogeneous counterparts. Heterogeneous catalysts can either be studied directly by the examination of the catalyst surface or indirectly by using model compounds. As POSS can be employed as molecular models of a variety of silica surface sites, their interactions with organometallic precursors of silica-supported catalysts can be explored [10-12].

The anchoring of carbonyl cluster complexes to inorganic oxides *via* suitable ligands potentially allows controlled syntheses of particular species on the surface of the support. This technique has been applied mainly to silica gel, [13,14] although Evans *et al.* have extended this method by the use of HS(CH$_2$)$_3$Si(OMe)$_3$ yielding substituted clusters of the type [(μ-H)M$_3$(CO)$_{10}$(μ-S)(CH$_2$)$_3$Si(OMe)$_3$], M = Ru, Os, making a large variety of oxides accessible [15]. However, in contrast to directly anchored clusters, these materials have as yet not exhibited significant catalytic activity. Therefore, directly anchored clusters are of importance both because of their apparent simplicity and well-defined infra-red spectral response and because of their utility as catalysts used in reactions such as olefin hydrogenation, olefin isomerisation and

464

the water-gas-shift reaction [16-18]. Consequently, considerable effort has been devoted to define the nature of the surface-confined species produced for example by reaction of the cluster [Os$_3$(CO)$_{12}$] with silica [10,19-28]. The resulting material has been extensively studied and the infra-red, XPS, EXAFS, and electronic spectroscopic responses and catalytic performance widely reported. The near-quantitative extraction of [Os$_3$(CO)$_{10}$(μ-H)(μ-OH)] from silica-supported [Os$_3$(CO)$_{12}$] has been demonstrated, [29] but the influence of the extraction chemistry on all potential surface species has yet to be established.

The trinuclear surface clusters generated by depositing [Os$_3$(CO)$_{12}$] on silica are generally thought to be stable at room temperature; however, it is interesting to note that when supported on a lanthanum oxide the cluster breaks up at room temperature, possibly forming mononuclear tri- and dicarbonyl species [28]. Hence, conclusions drawn from the study of one oxide support do not necessarily apply to other oxides.

The structure of the material generated by depositing [M$_3$(CO)$_{12}$] (M = Ru, Os) on silica has not, despite many claims, been assigned unambiguously [42]. Nevertheless, there is general agreement that structures based on an [(X)(Y)M$_3$(CO)$_{10}$] unit are generated as the first observable product [21, 23, 24, 30-32]. Four main structural types have been proposed (cf. figure 3).

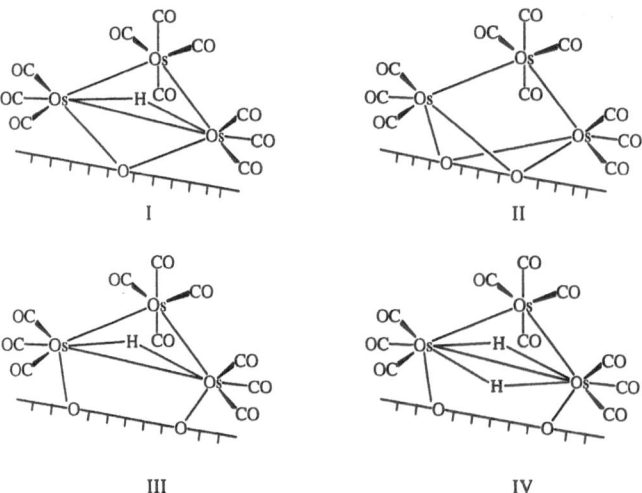

Figure 3. Four Principal Structures Commonly Proposed for [Os$_3$(CO)$_{12}$] Anchored to Silica.

The XPS technique as applied by Zanoni *et al.*[25] is not sufficiently sensitive to differentiate clearly between the various subtle structural changes shown in figure 3. However, calculational modelling [33], IR and Raman[16,23,34-39], EXAFS [16,24,40] and ^{13}C n.m.r. [41] studies have shown the presence of an hydride (hence discounting structure type II) and indicate the main surface species to be type I, although a detailed examination of those results reveals that a co-existence of types I, III and IV cannot altogether be ruled out [42].

The possibility of a bidentate anchoring to the silica surface as postulated in structure IV (cf. figure 3) was investigated by reacting [Os$_3$(CO)$_{12}$], (**5**), with an incompletely condensed silsesquioxane having the appropriate silanol geometry [43].

Up to four products were observed , the major one of which could be isolated and characterised by XPS, FT-IR, ^{1}H and ^{13}C n.m.r. as well as FABS and has been assigned as [(nor)$_6$Si$_6$O$_{10}${(μ-H)$_2$Os$_3$(CO)$_{10}$}], (**6**), in which one pair of SiOH groups has condensed and the other pair of SiOH groups has oxidatively added to [Os$_3$(CO)$_{12}$] with the elimination of carbon monoxide and the formation of a single hydride environment (cf. figure 4) [43].

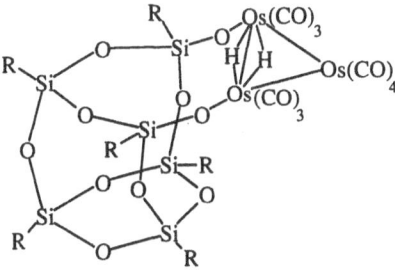

Figure 4. Proposed Structural Assignment of [(nor)$_6$Si$_6$O$_{10}${(μ-H)$_2$Os$_3$(CO)$_{10}$}], (**6**).

Shapley and Feher have reported a cyclodehydration of two adjacent silsesquioxane silanol groups during the formation of [(μ-H)Os$_3$(CO)$_{10}$(μ-O)Si$_7$O$_{10}$(C$_6$H$_{11}$)$_7$] [10]. They have therefore suggested that surface dehydration of non-calcined silica may be induced by the reaction of metal carbonyls with the surface. The formation of (**6**) is also interpreted in terms of silsesquioxane cyclodehydration accompanying the reaction between the cluster and the silsesquioxane. No dehydration of [(nor)$_6$Si$_6$O$_7$(OH)$_4$], (**2**), could be detected under these reaction conditions in the absence of [Os$_3$(CO)$_{12}$], (**5**), and no reaction between [Os$_3$(CO)$_{12}$], (**5**), and the fully condensed silsesquioxane [(C$_6$H$_{11}$)$_6$Si$_6$O$_9$], (**4**), could be detected either in the presence or absence of Me$_3$NO or H$_2$ [44]. It does not appear likely therefore that silsesquioxane dehydration precedes reaction with the cluster. Additionally, the reaction of [Ru$_3$(CO)$_{12}$], (**7**), with [((c-hexyl)$_8$Si$_8$O$_{11}$)(OH)$_2$], (**3**), did not proceed and [Ru$_3$(CO)$_{12}$], (**7**), could be recovered quantitatively. Direct co-ordination of silicon atoms to the cluster as reported by Braunstein *et al.* [45-47] and Adams [48] was also not evident in the reaction between [Os$_3$(CO)$_{12}$], (**5**), and the fully condensed silsesquioxane [(C$_6$H$_{11}$)$_6$Si$_6$O$_9$], (**4**). Thus, although there is significant evidence that some of the surface material generated by the chemisorption of [Os$_3$(CO)$_{12}$], (**5**), onto silica exists as [(μ-H)Os$_3$(CO)$_{10}$(μ-OSi$\equiv$)], the co-existence of [(μ-H)$_2$Os$_3$(CO)$_{10}$(OSi$\equiv$)$_2$] type species cannot be ruled out on the current evidence. Such species are likely to be less abundant since potentially 4 adjacent SiOH groups are needed versus the 3 adjacent ones for [(μ-H)Os$_3$(CO)$_{10}$(μ-OSi$\equiv$)] [19]. The proposed cyclodehydration raises the interesting possibility that surface dehydration at a site adjacent to a surface-confined metal carbonyl may provide a two-site mechanism for water-gas-shift catalysis (cf. scheme 1).

M--CO	<=======>	M + CO
SIL--OH + SIL--OH	<=======>	Si---O---Si + H$_2$O
Si---O---Si + CO$_2$ + H$_2$ + M	<=======>	SIL--OH + SIL--OH + M--CO
CO$_2$ + H$_2$	<=======>	CO + H$_2$O

Scheme 1. Potential Two-site Mechanism for Water-gas-shift Catalysis Involving a Site Adjacent to a Surface-confined Metal Carbonyl [42].

It may be suggested that a cyclodehydration step is necessary for the oxidative addition of a silanol to an Os-Os bond. This would be consistent with this proposed catalytic cycle for the water-gas-shift Reaction (cf. figure 5).

466

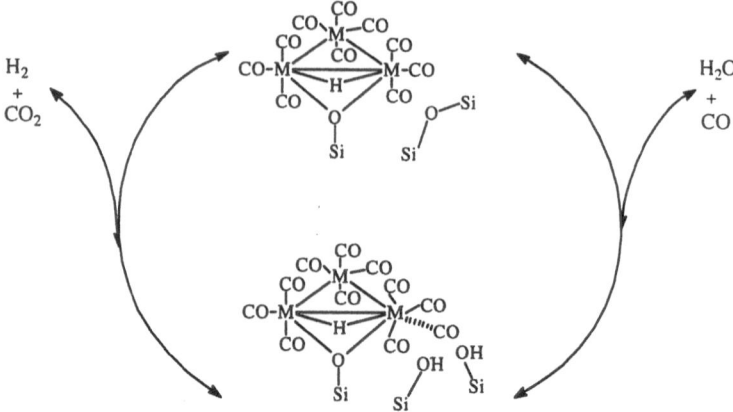

Figure 5. Potential Catalytic Cycle for the Water-Gas-Shift Reaction of the $[Os_3(CO)_{12}]$/Silica Catalyst

1.3   *Investigating the UNIPOL Catalyst*

Spectroscopic Studies of the Catalyst Surface. Highly active olefin polymerisation catalysts can be prepared *via* the treatment of oxide supports with organometallic transition metal complexes. One of the most potent catalysts for ethylene polymerisation (UNIPOL Catalyst by Union Carbide, Karol *et al.*, 1972 [53]) is obtained by reacting partially dehydrated silica with chromocene [49-65]. Chromocene in solution or as solid material does not catalyse ethylene polymerisation (except for an isolated report of catalysis in the presence of O_2 [67]), indicating the importance of the chromocene/silica interaction. The catalyst is thought to be formed by a chemical reaction between chromocene and the hydroxyl groups of the silica surface, during which cyclopentadienyl ligands are released [66]. The stoichiometry of this interaction and the composition of the resulting surface complexes depend on the type of support and the temperature of the support dehydration [53]. By increasing the temperature of silica dehydration from 25 to 800 °C the amount of cyclopentadiene evolved decreases from 1.65 to 0.90 moles of cyclopentadiene per mole of the reacted $[Cr(\eta^5\text{-}C_5H_5)_2]$.

However, there is still only a limited understanding of the active site(s) and/or the average chromium surface environment. Thus, the chromium oxidation state has variously been assigned as being from Cr(0) to Cr(VI) [57]. Different infra-red spectra have been reported following exposure of the catalyst to carbon monoxide and a variety of structures has been proposed to account for the active site. A common, often unstated, assumption underpinning most of these spectroscopic investigations is that the spectroscopic response is that of the active catalyst, i.e., that the active catalyst is the only, or at least predominant, surface species and that the active site is sensitive to the spectroscopic probe. Several investigations have probed the catalyst by carbon monoxide poisoning experiments [49-52,60,62,66] and it is significant that the careful work of Lunsford and Fu [50] has been interpreted as indicating that the catalytically active species accounts for less than 0.5 % of the surface confined chromium (the interpretations of these experiments were restricted to establishing the relative amount, but not the structure, of the active species).

Two of the more recent postulates of the active site, or at least of the species which interact(s) with carbon monoxide are shown below (figures 6 and 7) [62,66].

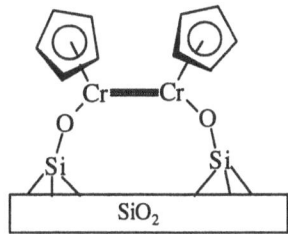

Figure 6. Dimeric Site Postulated by Rebenstorf *et al.*, [62].

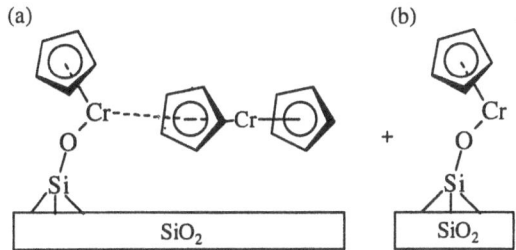

Figure 7. Mixture of Sites Proposed by Zecchina *et al.*, [66].

The surface dimer in figure 6 is an attractive suggestion, because a homogeneous molecular (although somewhat poor in terms of its CO substituents) analogue, $[Cr(\eta^5-C_5H_5)(CO)_3]_2$, is known to react with dihydrogen with cleavage of its metal-metal bonds and formation of two separate hydride complexes. Thus, such dimeric compounds could possibly form precursors of the polymerisation sites, these supported precursors being cleaved by the substrate.

The mixture of surface complexes postulated in figure 7 consists formally of one fourteen electron aggregate (cf. figure 7(a)) and one twelve electron species. This presents a very high degree of electronic and co-ordinative unsaturation which is rare in organometallic chemistry and is therefore surprising.

Clearly, there is some disagreement as to the existence and identity of the non-catalytically active chromium species which account(s) for the bulk of the surface confined chromium [49-52]. The main surface species have been variously assigned as monomeric $[Cr(\eta^n-C_5H_5)(OSi\equiv)_m]$ complexes, stabilised by (physisorbed or chemisorbed) chromocene [60] or as surface confined $[Cr(\eta^n-C_5H_5)(OSi\equiv)_m]_l$ clusters which vary in their degree of aggregation, l, according to the silica pretreatment [45-52]. These clusters are believed to be made up of dimeric sub-units. When the system is exposed to carbon monoxide, infra-red absorptions consistent with the presence of $Cr(\mu-CO)Cr$ moieties are observed. These chromium carbonyl fragments have been postulated to be derived from the direct interaction of carbon monoxide with preformed $[Cr(\eta^n-C_5H_5)(OSi\equiv)_m]_l$ clusters or by carbon monoxide-induced oligomerisation of the monomers.

A Cr K-edge XAS investigation utilising X-ray absorption near-edge structure (XANES) and extended X-ray absorption fine structure (EXAFS) into the general nature of the average chromium surface environment of the chromocene/silica system has enabled us to establish the average structure of the surface species [70a] and using silsesquioxanes modelling techniques allowed us to speculate about the active site and silica surface reactivity [70b]. The shape of the K-edges XANES can yield qualitative (and in some rare cases quantitative) information about the oxidation state of the metal studied. In figure 8 the edges of chromium compounds in oxidation states zero, I, II, III, V and VI are compared with the edge observed for the chromocene/silica system.

Cr K−edges

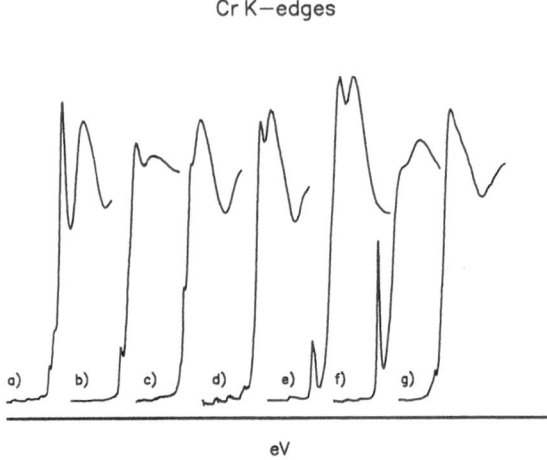

eV

Figure 8. K-edges of a) [Cr(CO)$_6$] = Cr(0), b) [Cr(η^5·C$_5$H$_5$)(CO)$_3$]$_2$ = Cr(I), c) Cr(II)acetate = Cr(II), d) [Cr(acac)$_3$] = Cr(III), e) Na[Cr(O)(ehba)$_2$] = Cr(V), f) [CrO$_3$] = Cr(VI) and g) Chromocene/Silica System [42, 70a].

This comparison conclusively confirms that the bulk of the surface chromium is not present as Cr(VI) nor as Cr(V), since the EXAFS of model compounds with these oxidation states exhibit distinctive pre-edge features which arise from inner core transitions, primarily from the 1s —> 3d transition. The edge of the chromium/silica system is fairly featureless and the major part of the surface chromium may be presumed to be Cr(III), the small pre-edge peak can be accounted for in a variety of ways from the other edges, thereby allowing no precise assignment.

The EXAFS can be fitted with either 2 or 3 oxygen atoms around the chromium absorber (cf. figure 9 and table 1).

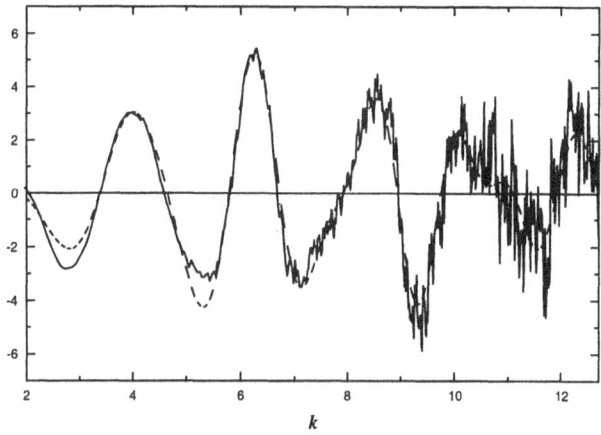

Figure 9. Cr K-edge k^3-weighted EXAFS of Silica-supported Chromocene (solid : exp.; broken, calc.) [70a].

However, the Debye-Waller factors exhibit an increase with distance in the solution containing 3 oxygen atoms which is not the case for the model solution containing 2 oxygen atoms. Therefore, the Debye-Waller Factors suggest the presence of 3 oxygen atoms around the chromium absorber to be more reasonable.

TABLE 1. EXAFS Analysis of Silica-supported Chromocene, cp = (η^5-C_5H_5)

Scatterer	Parameter	$[Cr(cp)(\mu\text{-}OR)]_2$	$[Cr(cp)(\mu\text{-}OR)(OR)]_2$
	Occupancy	5	5
Carbon	Distance	2.04	2.06
	DWF	0.006	0.001
	Occupancy	2	2 (μ-OR) and 1 (OR)
Oxygen	Distance	1.93	1.96 (μ-OR) and 1.86 (OR)
	DWF	0.003	0.006 0.006
	Occupancy	1	1
Chromium	Distance	2.98	2.98
	DWF	0.002	0.007
	E_0	-11 eV	-11 eV
	S_0^2	0.8	0.8
	R-Factor	15 %	15 %

The presence of a Cr-Cr vector is also established, the Cr-Cr distance being consistent with related molecular dimeric chromium compounds of oxidation state III [67,68]. The results of this analysis are consistent with known structures for the oxygen and chromium shells. Thus, the EXAFS data imply that most of the chromium exists as dimeric or oligomeric species. A survey of related crystallographically derived structures confirms the expected correlation between the chromium oxidation state and the Cr-Cr bond distance [42]. By this criterion the present data are more nearly consistent with the surface species being Cr(III)-Cr(III) dimers, an oxidation state in agreement with the shape of the absorption edge. Although the presence of the adjacent chromium centres on the surface also accords with the suggestion of chemi- or physisorbed chromocene in close proximity to monomeric surface species, [60,66] the Cr-Cr distance is too short to be accommodated by reasonable models of such an interaction. An oligomeric surface

470

species similar to $((\eta^5\text{-}C_5H_5)CrO)_4$ is also inconsistent with the fit to the data. However, a significant shortening of the chromium-carbon bond is apparent when compared to the average distances derived from related crystal structures. This suggests that the deviation is either real or that there is an inherent deficiency in the EXAFS technique, distorting distances of ring systems like cyclopentadienyl ligands. Therefore, a multiple scattering analysis of the EXAFS of the dimer, $[Cr(\eta^5\text{-}C_5H_5)(CO)_3]_2$, (8), was performed for comparison, yielding a Cr-C distance 0.02 Å shorter than the one derived from the crystal structure which is within the experimental error of EXAFS. Hence, the short bond distance observed for the Cr-C bond in the silica-supported chromocene system is unlikely to be an experimental artifact. It may be explained by the suggestion that silica surfaces are as electronegative as a CF_3-group [69], postulating that the decrease in electron density available from the oxygen atoms is compensated for by an increase in the strength of the chromium-cyclopentadienyl interaction, leading ultimately to a shorter bond length. The EXAFS data and chemical considerations taken together indicate that the average chromium environment is represented by the formulation $[Cr(\eta^5\text{-}C_5H_5)(O)_2]_2$, (9), as shown in figure 10 below. Given the existence of well characterised molecular species of the composition $[Cr(\eta^5\text{-}C_5H_5)(OR)_2]_2$, it is likely that surface compounds with a very similar stoichiometry, and possibly structure, exist on the surface.

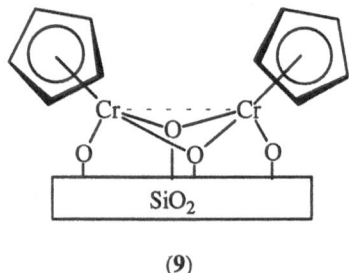

(9)

Figure 10. Representation of the Average Chromium Silica Surface Environment (9) of the UNIPOL Catalyst.

These data are consistent with the surface compound, (9), representing the average environment of chromium on the surface; (9) might be also expected to react with carbon monoxide, consistent with the results of carbon monoxide poisoning experiments. For example, the related dark red $[Cr(\eta^5\text{-}C_5H_5)(OBu^t)]_2$ dimer reacts with carbon monoxide to produce a non-volatile, greenish-blue hydrocarbon-insoluble compound of empirical formula $[Cr(\eta^5\text{-}C_5H_5)(OBu^t)(CO)_2]$ with infra-red absorptions consistent with the presence of bridging carbonyls [67].

However, in the context of Lunsford and Fu's interpretation of their data - that less than 0.5 % of the surface chromium sites are active for the polymerisation - it is important to emphasise that a surface species such as (9) most likely represents the bulk surface species, rather than the catalytically active site. Such a conclusion may explain the observations that molecular species of a similar formulation have been shown to be poor models of the supported catalyst [67] even though they exhibited stoichiometric activity in the reaction with small molecules such as carbon monoxide. Hence, monomeric surface species (as suggested initially by Karol) seem to be potential candidates for forming the basis of active sites.

An attempt was made to suppress the surface dimer formation using a bulky alkene to co-ordinatively saturate the surface monomers during the deposition process. The catalyst was prepared in presence of excess 2,3-dimethyl-1-butene. In this case, the colour changes were virtually the same as in the preparation of the chromocene/silica catalyst as were the XANES.

The EXAFS of the resultant material yielded distances and co-ordination numbers for the carbon, oxygen and chromium shells which were essentially the same as in the chromocene/silica case with the exception of an additional two carbons at 2.346 Å. This analysis would fit with the surface reaction proposed in figure 11, i.e. after the dimer formation the alkene reacts stoichiometrically with the chromium centres on the surface to form (10).

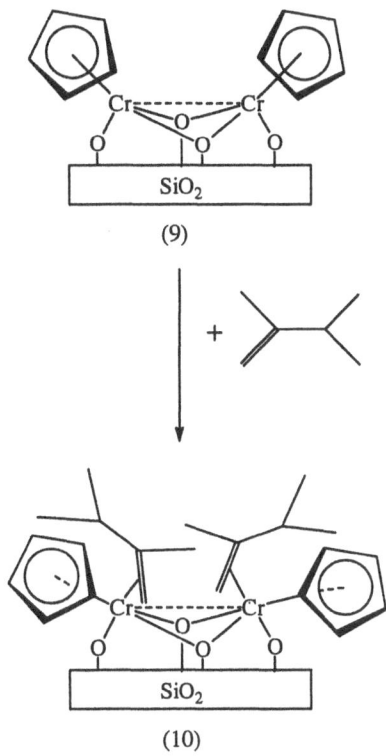

Figure 11. Proposed Surface Reaction of (9) with Alkene, Yielding (10).

It should also be possible to observe the carbons attached to the second carbon of the alkene in (10), but although the fit is improved, the improvement is not statistically significant. These results seem to suggest that the reaction between chromocene and the surface silanols is faster than with the alkene and thus predominant. Monomeric surface species are either generated by a small minority of the chromocene reacting with silanols but not dimerising or by reacting with the silica surface in a different way, e.g. via strained siloxanes.

In order to examine these possibilities modelling studies with partially and fully condensed silsesquioxanes were performed.

Modelling Reactions of Chromocene with Incompletely Condensed Silsesquioxanes. The interaction of chromocene with isolated surface silanols was studied by reacting chromocene with [(c-hexyl)$_8$Si$_8$O$_{11}$(OH)$_2$], (3), yielding a dark purple/black

472

highly air-sensitive, NMR and ESR silent solid, (11). The structure of this solid has been determined by EXAFS (cf. figure 12 and table 2).

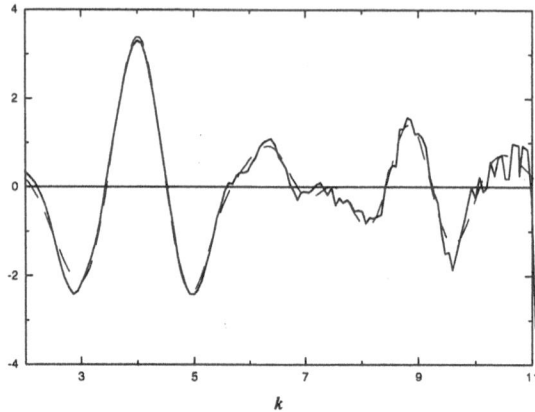

Figure 12. Cr K-edge k^3 weighted EXAFS Data of (11), (solid line: exp.; broken line: calc.).

The EXAFS data are consistent with a model solution that employs five different shells made up from five carbon atoms at 1.986 Å, one oxygen atom at 2.111 Å, three carbon atoms at 2.205 Å , two carbon atoms at 2.784 Å and one chromium atom at 2.917 Å.

TABLE 2. EXAFS Analysis of (11) when kept under Argon

Scatterer	Occupancy	Distance [Å]	Debye-Waller Factor [Å²]
C	4.8	1.986	0.010
O	1.1	2.111	0.010
C	3.0	2.205	0.009
C	2.4	2.684	0.009
Cr	0.9	2.917	0.018
E_0 : -12 eV	S_0^2 : 0.7	R : 9 %	

This solution can be rationalised by postulating the oxidative addition of a silanol group to the chromium with subsequent $\eta^5 \rightarrow \eta^3$ ring-slippage of one cyclopentadienyl ligand and dimerisation of the chromium centres (cf. figure12).

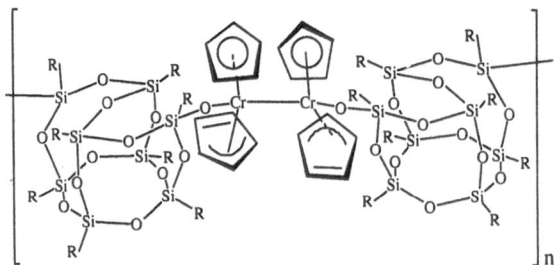

Figure 12. Proposed Structure of (11).

The proposed dimer is a 16 electron species containing formally Cr(III) centres. The Cr-Cr and Cr-O bond distances correspond to distances reported for chromium(III) dimers elsewhere [42]. The distance between the chromium atom and the (η^5-C_5H_5) ligand is shorter by about 0.20 Å than the average crystallographic distance for (η^5-C_5H_5) chromium systems. However, the same shortening in the bond length was observed in the analyses of EXAFS data of chromocene and chromocene/alkene systems on dehydrated silica (i.e. compounds (9) and (10)) [70b]. Thus, the silsesquioxane appears to provide a model at least for that aspect of the reaction of chromocene with silica. At this stage it is premature to speculate about the exact nature of such an interaction other than to say that it may have its origins in the electron withdrawing properties of the silica surface/siloxane cages.

The (η^5-C_5H_5) "allyl carbon"-chromium bond is about 0.12 Å shorter than the only crystallographically characterised allyl (η^3-C_5H_5) to chromium bond, i.e. that observed in the 18 electron species [$Cr(\eta^3$-$C_5Me_5)(\eta^5$-$C_5Me_5)(CO)_2$] (2.329 Å average) [73a]. However, as the proposed complex is a 16 electron species rather than an 18 electron species some difference in the bond lengths might be expected. The other two carbons of the proposed species (11) are 0.46 Å closer to the chromium centre than are the "distant" (C_5H_5) carbons of [$Cr(\eta^3$-$C_5Me_5)(\eta^5$-$C_5Me_5)(CO)_2$]. However, as a change of about 20° in the bonding angle of the (C_5H_5) is likely to have that effect, large variations in that distance would be expected for different geometries.

Following a short exposure to air the compound turns green and the EXAFS spectrum changes noticeably. The interatomic distances and coordination environment derived from the EXAFS analysis of the air-exposed material are essentially the same as those observed for the surface supported chromocene on silica, suggesting that the following reaction took place (figure 13).

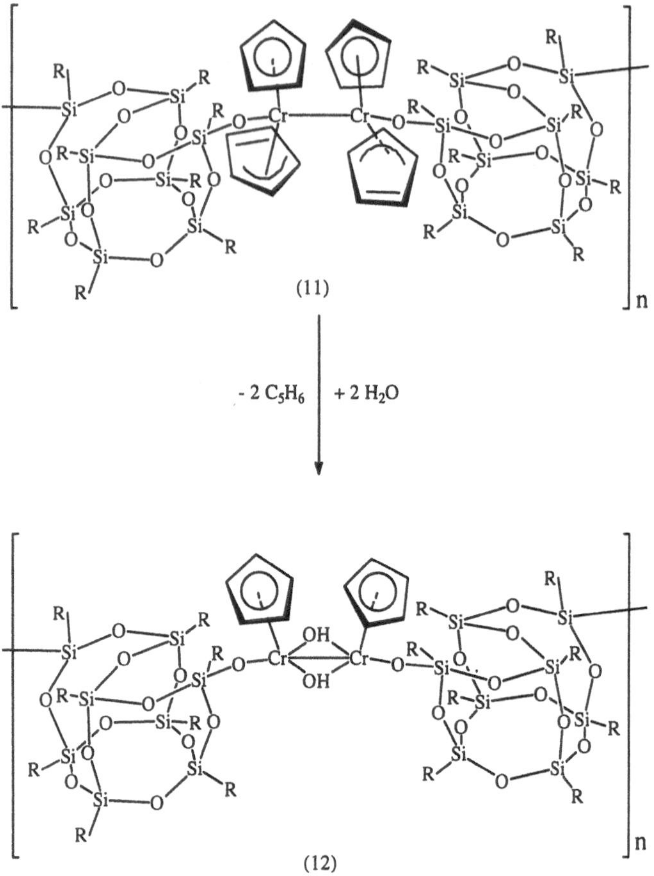

$- 2 \; C_5H_6$ $+ 2 \; H_2O$

Figure 13. Proposed Transformation of (11) to (12) After Brief Air-Exposure.

Reaction of Chromocene with Fully Condensed Silsesquioxanes. It is generally assumed that only the surface silanol groups are reactive towards chromocene [49-65]. However, due to the dehydration process, strained surface siloxanes are also present and these strained siloxanes may, contrary to current belief, play a non-innocent role. Certainly, examples of reactions of organometallic species with strained siloxane surface sites have been claimed in the literature [71,72]. This possibility has been tested in the present study by reacting [(c-hexyl)$_6$Si$_6$O$_9$], (4), a homogeneous analogue of the strained surface siloxanes, with chromocene. The highly air-sensitive, NMR and ESR silent reaction product, (13), was then characterised by EXAFS (figure 14 and table 3).

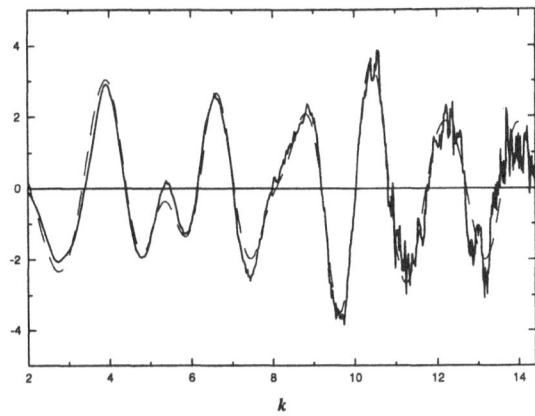

Figure 14. EXAFS Data of Compound (**13**), (solid line: exp., broken line: calc.).

The EXAFS data are consistent with a model solution that employs four different shells consisting of 5 carbon atoms at 1.990 Å, one oxygen atom at 2.073 Å, five carbon atoms at 2.259 Å and one chromium atom at 2.891 Å. Similarly to the solution of the EXAFS data of [(*cyclo*-C$_6$H$_{11}$)$_7$Si$_7$O$_{11}$(OSiMe$_3$)CrO$_2$], (**8**), no Cr-Si vectors can be assigned. However, the Cr-Cr vector is clearly identifiable.

TABLE 3. *EXAFS Analysis of (13)*

Scatterer	Occupancy	Distance [Å]	Debye-Waller Factor [Å^2]
O	0.9	1.990	0.004
C	4.9	2.073	0.004
C	4.9	2.259	0.010
Cr	1.2	2.891	0.011
E$_0$: -11 eV	S$_0^2$: 0.7	R: 15 %	

This solution can be rationalised by postulating the insertion of chromocene into one of the Si-O bonds of the silsesquioxane and a subsequent dimerisation of two chromium centres. Figure 15, below, is only one of a range of possibilities that would match the EXAFS derived parameters. However, at this stage of the investigations the emphasis should be placed on the observation that the siloxane bond has reacted with chromocene rather than on the exact structure of the resulting compound.

476

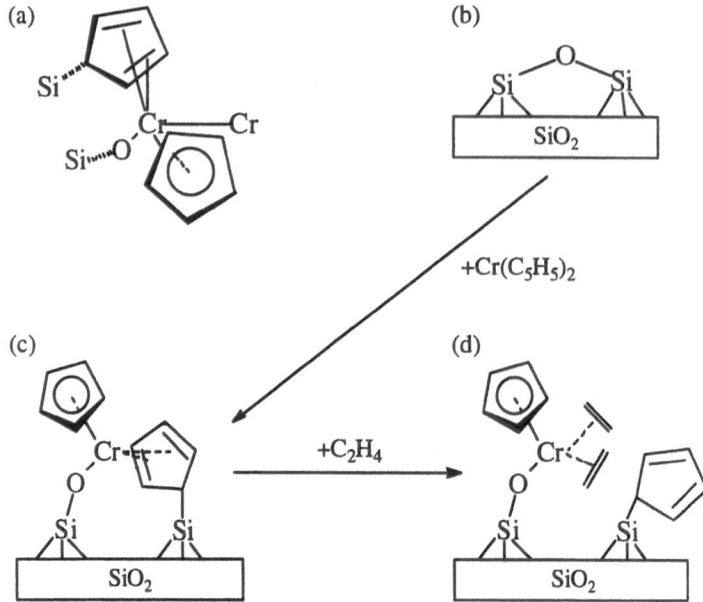

Figure 15. (a) Proposed Co-ordination Environment of (**13**) Consistent with EXAFS Data Analysis, (the (η^4-C_5H_5) ring structure, although containg two different Cr -C distances has been averaged, the silicon atoms could not be unambiguosly assigend); (b) Strained Surface Siloxane; (c) Insertion of Chromocene into Strained Surface Siloxane. Yielding the Proposed Active Site of theUNIPOL Catalyst; (d) Initial Interaction with Ethylene Showing Possible First Step in Polymerisation Mechanism.

The proposed co-ordination environment is consistent with a 17 or 18 electron species (depending on the Cr-Cr bond order) containing chromium centres of formal oxidation state II. The Cr-Cr distance (2.891 Å) corresponds to dimeric distances reported elsewhere, however, those distances are observed in formally chromium(III) dimers, the longest chromium(II) dimer Cr-Cr distance being 2.65 Å [67].

In general, there seems to be a trend which suggests an elongation in the Cr-Cr bond length corresponding to an increased formal oxidation state. However, this may be too simplistic a view, since in the case of $[Cr(\eta^5$-$C_5Me_5)(\mu^2$-$O)(O)]_2$ and $[Cr(\eta^5$-$C_5H_5)(CO)_3]_2$ the expected trend is reversed. In both cases the deviation could be explained by the respective lack or presence of a crowded co-ordination environment of the chromium centres. The larger than expected Cr-Cr bond distance of the proposed dimer might be rationalised analogously.

The Cr-C distance of the (η^5-C_5H_5) ligand is shorter by about 0.26 Å than the average crystallographic distance for such systems. However, this is consistent with the Cr-C distances derived by EXAFS for similar systems in the preceding sections (cf. compounds (**9**), (**10**), (**11**) and (**12**)).

The η^4-(C_5H_5) carbon-chromium bond is about 0.1Å longer than the corresponding bond in related systems, e.g. $[Cr(\eta^4$-$\{C_5H_4(CH_2)\})(CO)_3]$ (2.154 Å) [73b]. However, since it is proposed that the η^4-(C_5H_5) might be bound not only to the chromium, but also to the siloxane cage, the bonding environment could be much more restrained than that in

$[Cr(\eta^4-\{C_5H_4(CH_2)\})(CO)_3]$. Therefore, it is possible to postulate that the active site of the UNIPOL catalyst may be generated *via* an interaction with strained surface siloxanes (cf. figure 15(b)), giving rise to a sixteen electron species (cf. figure 15(c)) which can undergo structural rearrangements to allow for ethylene insertion (cf. figure 15(d)).

Conclusions It has been established that the major surface material produced by the reaction of chromocene with silica can be described as a dimeric species. This assignment is supported by the stoichiometric reaction of this material with alkenes which yields a product which also appears to have a dimeric structure. It has to be pointed out, however, that the EXAFS provides information only about the *average* surface environment and that there may well be a mixture of different surface species present, the dominant one being consistent with the structures (9) and (10) suggested by the EXAFS analyses.

This is an important consideration. The studies involving the model compounds have shown that chromocene reacts with models of different silica surface environments in different ways. Consequently, the implication is that different surface environments may also produce different chromocene-derived surface species. Interestingly, the trend favouring dimeric surface species seems to be paralleled by the reactions with the silsesquioxane model compounds as well, leading in all cases to dimeric chromium centres. In the case of the reactions with silsesquioxane model compounds, however, the mobility of the homogeneous molecular species in solution may have some influence on this trend. Chromocene, as expected, does react with silanols, but also with strained siloxane bonds. This has been an unanticipated result and has significant implications for silica surface chemistry in general as siloxane bonds largely have been presumed to play an innocent role in silica surface chemistry and it may be that this very reaction on the surface gives rise to the active catalytic centres (cf. figure 15(c)).

2. Mesoporous MCM-41 Derivatives, their Synthesis, Structure and Application to Catalysis

2.1 FRAMEWORK-SUBSTITUTED DERIVATIVES

Regular arrays of a siliceous unidimensional mesoporous material (commonly referred to as MCM-41) with pore sizes in the range of 20 to 100Å have been synthesised recently, [74,75] bridging the gap between uniformly networked microporous materials and macroporous (e.g. porous glasses and amorphous gels) materials. The largest microporous materials, reported thus far, are JDF-20 [76] and cloverite [77] which contain 20-membered rings with pore openings of ca 13Å. The mesoporous MCM-41 has attracted considerable attention due to its large uniform pore distribution, high surface area (up to some 800 m^2/g), distinct adsorptive properties, [78] and the possibilities for replacing the Si^{4+} ions (in small quantity) by other hetero-valent cations, thereby yielding potentially new materials for catalytic applications. Recently, isomorphous substitutions of Ti [79,81], V [82], Fe [93], Mn [84], Al [74,85] and B [86] into the MCM-41 structure have been reported. The large pores permit the catalytic conversion of bulky molecules which would otherwise be impossible using existing zeolitic materials owing to steric hinderances [87]. Hence, the oxidation of large olefins (e.g. norbornene, limonene) or substituted phenols by peroxides using titanium-containing MCM-41 as a catalyst is feasible[79].

We have prepared MCM-41 type mesoporous silicas in which one or more of the following elements are accommodated in framework sites: titanium, iron, chromium, vanadium, manganese, boron and aluminium [80]. It has been possible to characterise these materials using X-Ray absorption spectroscopy (XAS), yielding valence states, bond lengths and coordination numbers of the metal ion. XRD and FTIR were used as further aids to characterisation. Ti-containing MCM-41, as well as the Fe-, V- and Cr-containing variants, yield self-consistent, XAS-based, structural data of the respective metal-ion sites. Some of these (especially those containing Ti) are exceptionally good catalysts for the selective oxidation of large organic molecules such as limonene and norbornene.

Figure 16 displays an illustration which is derived from an idealised model for a siliceous mesopore, generated by computational methods and optimised by energy minimisation.

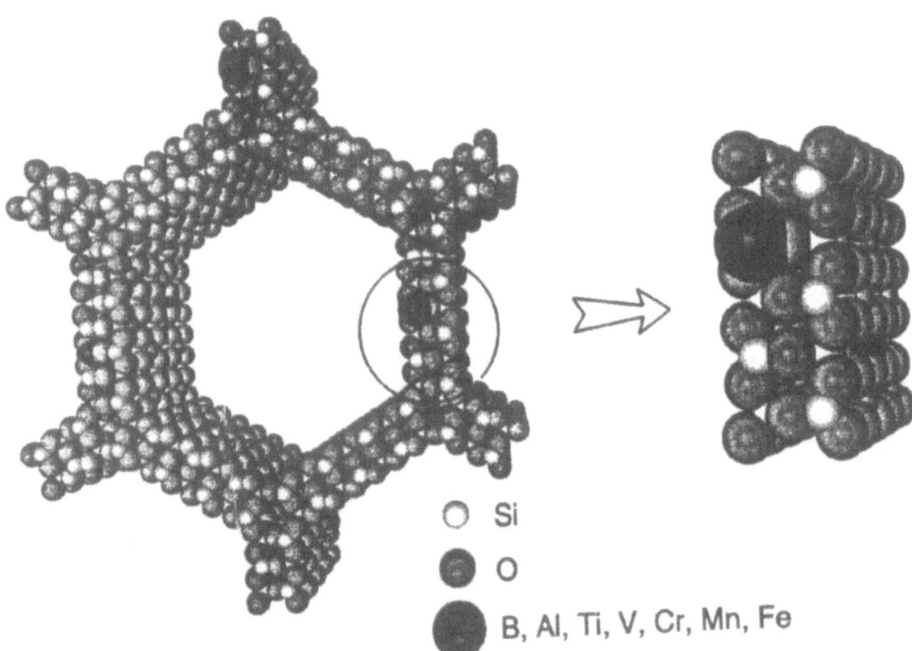

Figure 16. Idealised metal-containing MCM-41 structure generated by computer modelling methods.

In figure 17 we present the XRD patterns of siliceous MCM-41 and various heteroatom-substituted analogues.

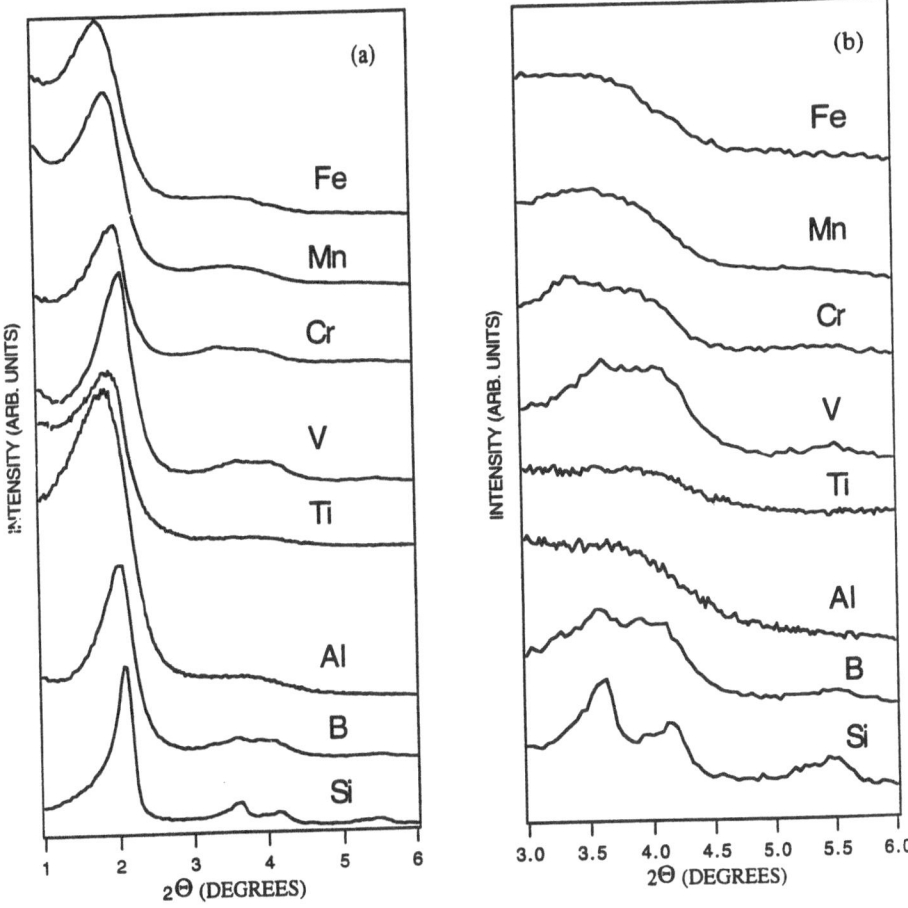

Figure 17. (a) XRD patterns of siliceous MCM-41 and heteroatom substituted analogues. (b) A narrow region of the XRD pattern is given in this figure to clearly show the (110) and (220) reflections.

The patterns are typical of the mesoporous material having a pore diameter in the range of 30Å. It is noticeable that the inclusion of heteroatoms into the MCM-41 structure results in a loss of definition in the diffraction peaks assigned to the (110) and (200) reflections and both appear as a broad peak centered around 24.5Å in d-space (see figure 17b). Analogous results have been shown for aluminium substituted MCM-41 [85]. In general, the similarity in the XRD patterns of all the metal-ion-substituted materials indicates that the differnt types of heteroatoms have no detectable impact on mesoporosity and only very little influence on the corresponding long-range order.

Of the various materials reported here, Ti-MCM-41 has been studied the most throughly by us and others [79,88]. The XANES spectra of the as-prepared (template-incarcerated) Ti-incorporated MCM-41 and of the calcined form show distinct differences in their pre-edge features (cf. figure 18).

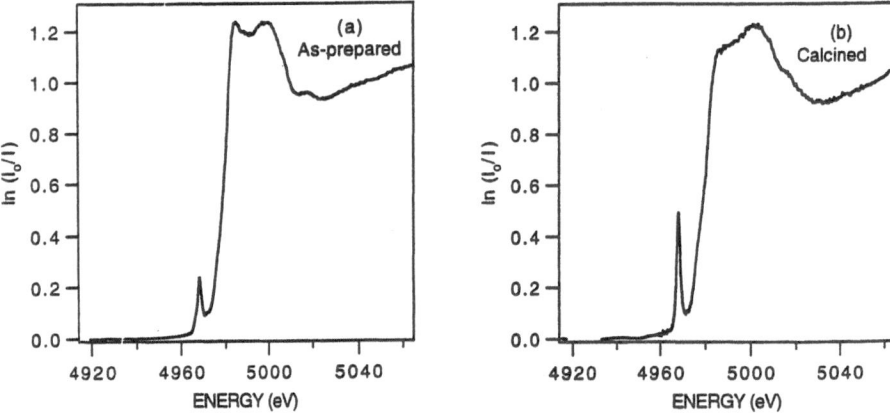

Figure 18. Characterisation by Ti K-edge X-ray absorption spectroscopy of Ti-MCM-41 are summarised in this figure. The XANES part of the XAS data of the as-prepared and calcined material are shown in (a) and (b), respectively.

The pre-edge peak intensity is known to be sensitive to the type of coordination [89], thus serving as a fingerprint in assessing the local structure around the titanium atoms. It is evident from figure 18 that this intensity in the calcined material is higher than that of the as-prepared material, signifying a change in the local structure. The EXAFS investigation of the calcined material clearly reveals the presence of a four-coordinated Ti-O environment with a Ti-O bond distance of 1.80Å; whereas the local structure around the titanium centres in the as-prepared material can be described as consisting of a distorted octahedral environment with Ti-O bond distances of 1.88 and 2.40 Å. The presence of a short Ti-O distance of 1.80Å upon calcination is consistent with the generation of a tetrahedral environment around the titanium centres and also with the increase in pre-edge intensity. In our earlier work we have shown that this activated (calcined) catalyst undergoes an increase in the coordination around titanium when subjected to catalysis (epoxidation of cyclohexene in presence of H_2O_2 with methanol as solvent [88]).

As microporous Fe-substituted ZSM-5 is active for several reactions (including the partial oxidation of methane to methanol [90]), we have compared the local structure around Fe^{3+}

in Fe-MCM-41 with Fe-ZSM-5 using XAS data [80]. The pre-edge feature of the as-prepared Fe-MCM-41 is similar to those of FePO4 and Fe-ZSM-5, both of which contain Fe^{3+} in tetrahedral coordination [91].

However, the pre-edge intensity of Fe-MCM-41 is slightly lower, possibly due to a distorted environment which is consistent with further analysis of the EXAFS data as it shows that the static disorder in Fe-MCM-41 is higher than that in Fe-ZSM-5 and FePO4. Nonetheless, the incorporation of the majority of Fe^{3+} into the Fe-MCM-41 framework can be inferred, since the Fe-O distance (1.84Å), coordination number (4.2) and colour (white) are closely similar to Fe-ZSM-5 and FePO4 [91].

Upon calcination (to remove the occluded template), the sample turned to a pale brown, indicating that some of the iron has come out of the framework, possibly forming highly dispersed oxide particles. Analysis of the corresponding EXAFS data, however, did not show any significant increase in the Fe-O distance, as would be expected for the formation of Fe_2O_3 in which Fe^{3+} (octahedral coordination) has a Fe-O distance of between 1.94 to 2.11Å. This indicates that a large amount of Fe^{3+} is still in the framework, although the static disorder is found to be higher than in the as-prepared material.

Several studies have been reported with the aim to substitute Cr^{3+} in the framework of the silicalite structure [92]. Although it is believed that the Cr^{3+} susbtitution takes place during synthesis, subsequent calcination to remove the organic template results in the formation of higher valent Cr^{6+}. We have synthesised Cr^{3+}-containing MCM-41 with a similar aim, i.e. to substitute Cr^{3+} for Si^{4+} in the framework [80]. The XANES spectrum of the as-prepared material is found to be similar to that of Cr_2O_3, indicating the chromium is indeed present in the material as Cr^{3+}. The average Cr-O distance (1.98 Å) obtained from EXAFS data is also similar to that of Cr_2O_3. To establish whether or not Cr^{3+} is actually in the framework, we performed the analysis of the second shell neighbours, similar to that of Ti-MCM-41, where only silicon atoms could be detected [17]. The present results indicate that the Cr^{3+} has chromium as its second shell neighbour; but a significantly lower coordination number compared to Cr_2O_3 suggests the presence of highly dispersed oxidic species.

When this material was subjected to calcination the sample turned yellow orange, similar to many of the Cr^{6+}-containing compounds and supported catalysts. Our XANES studies establish the presence of tetrahedrally coordinated Cr^{6+} species. However, the intensity of the pre-edge is significantly lower than that of the model compound which possesses perfect tetrahedral coordination[93]. Analysis of the EXAFS data revealed that it is very difficult to describe the Cr^{6+} environment in Cr-MCM-41 with a single average Cr-O distance. The results from the best fit yielded a distorted four-coordinated environment with Cr-O distances of 1.60 and 1.93 Å. This result is self-consistent with the lower pre-edge intensity, since the distorted environment gives an average Cr-O distance of 1.77Å which is higher than the 1.65Å found in the model compound, $(NH_4)_2CrO_4$. Similar such highly distorted environments are present in many other Cr^{6+} compounds having either two short and two long Cr-O bonds (as in CrO_3 which contains two short distances at 1.59Å and two long distances at 1.73Å) or three short and one long Cr-O bond (as in most of the dichromate salts) [94].

The IR spectrum of the calcined material clearly supports the presence of Cr=O in the Cr^{6+} species with the formation of a band at ca 910 cm^{-1}. Upon reduction in CO this band vanishes and subsequent CO adsorption at room temperature revealed the presence of the characteristic triplet band as observed in Cr/SiO_2, the ethylene polymerisation catalyst [95].

Although the synthesis of V-MCM-41 has been reported in the literature [82], to our knowledge, proper characterisation has not been performed thus far. XAS is ideally suited to characterise vanadium, since it is well established that the fine structure associated with XANES can be conlusively used as a fingerprint in ascertaining the nature of its coordination [96]. It can be surmised from the edge shift (ca 4 eV lower than the model compounds V_2O_5 and NH_4VO_3) and the fine structures in our XANES spectra (cf. figure 19) that vanadium is present in the 4+

oxidation state in the as-prepared material [96]. The EXAFS data could be fitted to a distorted octahedral model having short and long V-O distances of 1.62Å and 2.01Å, respectively.

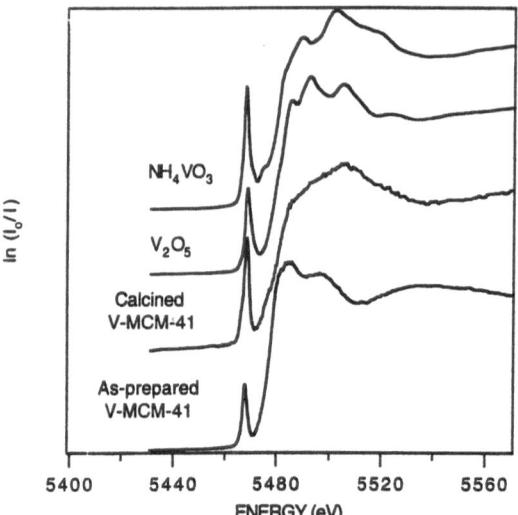

Figure 19. The XANES part of the V K-edge XAS data of the as-prepared and calcined materials is shown along with those of the model compounds V_2O_5 and NH_4VO_3.

Upon calcination the XANES features undergo dramatic changes, as can be seen in figure 19a. The shift in the edge position to higher energy compared to the as-prepared material and its similarity with the V^{5+} model sytstems indicates that V^{5+} is present in the calcined material. The EXAFS data of this calcined sample did not yield analysable spectra due to reduced fine structure oscillations in the EXAFS region. This lack of extended fine structure has also been noted [97] in other vanadium-containing catalysts and is likely to be due to large variations in V-O distances.

Aluminium and boron-containing MCM-41 materials have also been synthesised (their XRD patterns are shown in figure 17). IR spectra of these materials did not show any characteristic bands due to the formation of Brønsted acid sites as a result of the direct substitution of trivalent ions in place of Si^{4+}. It has been reported, based on NMR investigations, that the boron substituted material contains three and four-coordinated species when calcined [86] and that the heteroatom in the alumininum-containing material moves out of the framework when subjected to calcination [98]; although some studies suggest that aluminium ions (depending on the aluminium source used in the synthesis) are retained in the framework sites [85].

When subjected to calcination, titanium has been found to be the most stable in the framework. There are also indications that the majority of the iron remains in the framework, whereas vanadium and chromium are found to undergo an increase in their oxidation states to 5+ and 6+, respectively, and are subsequently present as highly dispersed oxides.

2.2   MCM-41 DERIVATISATION BY SURFACE ORGANOMETALLIC CHEMISTRY

It is not only possible to substitute hetero-elements in the framework of these mesoporous silicas, producing catalysts for, e.g. selective oxidations as we and others [99-102] have done, but their large channels permit, in principle, the direct grafting of complete metal complexes and organometallic moieties onto the inner walls of these solids. This opens routes to the preparation of novel catalysts consisting of large concentrations of accessible, well-spaced and structurally well-defined active sites.

We have produced several metallocene-derived catalyst precursors anchored to the inner walls of MCM-41, and converted them to powerful catalysts for epoxidations, oxidations and oxidative dehydrogenations.

There is currently great interest in titanium-containing zeolitic catalysts for selective oxidations. Ti ions incorporated into the framework sites of silicalite I and II (ie. TS1 and TS2 respectively introduced by the Enichem Company, [103]) as well as into the framework sites of ZSM-12, [104], ZSM-48, [105,106], zeolite β [107] and analogous microporous alumino (and silico) phosphates such as ALPO-5, ALPO-11 and SAPO-5 all show remarkable catalytic properties [108,109]. But one disadvantage of these titano-microporous catalysts is that their pore dimensions are too small to allow access to bulky reactants of the kind that dominate most of the chemical transformations which are of central importance in the fine-chemical and pharmaceutical industries. A significant step forward came recently [99, 100] (as described above) with the preparation and use of framework-substituted Ti-MCM-41 in which the Ti is incorporated during the MCM-41 synthesis.

Notwithstanding this improvement, the catalytic performance of the framework-containing Ti-MCM-41 is below what is desireable. Thus, the turnover frequencies of Ti-MCM-41 for the conversion of small molecules are distinctly inferior to those of TS-1, [103] and Ti-zeolite β, [100, 107, 110] ostensibly because access by the reactants to the Ti active centres is still somewhat limited since the latter are buried within the inner walls of the mesoporous solid. Other factors are also thought to play an important role in these reactions such as Coulombic and co-ordination effects, weak electron interaction and electronic confinement of molecules within the micropores (never greater than 7.5 Å dia.) [111].

By extending the concept of what has come to be known as interfacial co-ordination chemistry [112] and of surface organometallic chemistry, [113] we have circumvented the above-mentioned difficulties associated with Ti-MCM-41 by grafting titanocene onto the totally accessible inner surfaces of siliceous MCM-41 as schematised in figures 20 and 21 [114].

The resulting material, with well-separated, well-defined, high surface concentrations of Ti-containing active sites exhibits very high catalytic performance. Titanocene dichloride is superior to $TiCl_4$ or $Ti(OR)_4$ as a grafting reagent since, with either of the latter there is a marked tendency for oligomeric titano-oxo species and/or some anatase by-products to be formed during the grafting. This is circumvented by our method as the relatively stable cyclopentadienyl ligands protect the titanium centre and, hence, prevent either dimerisation and/or oligomerisation. Moreover, with $TiCl_4$, the copious quantities of evolved HCl are potentially damaging to the siliceous MCM-41, whereas using our methodology the evolved HCl is scavenged by an amine rendering it harmless.

484

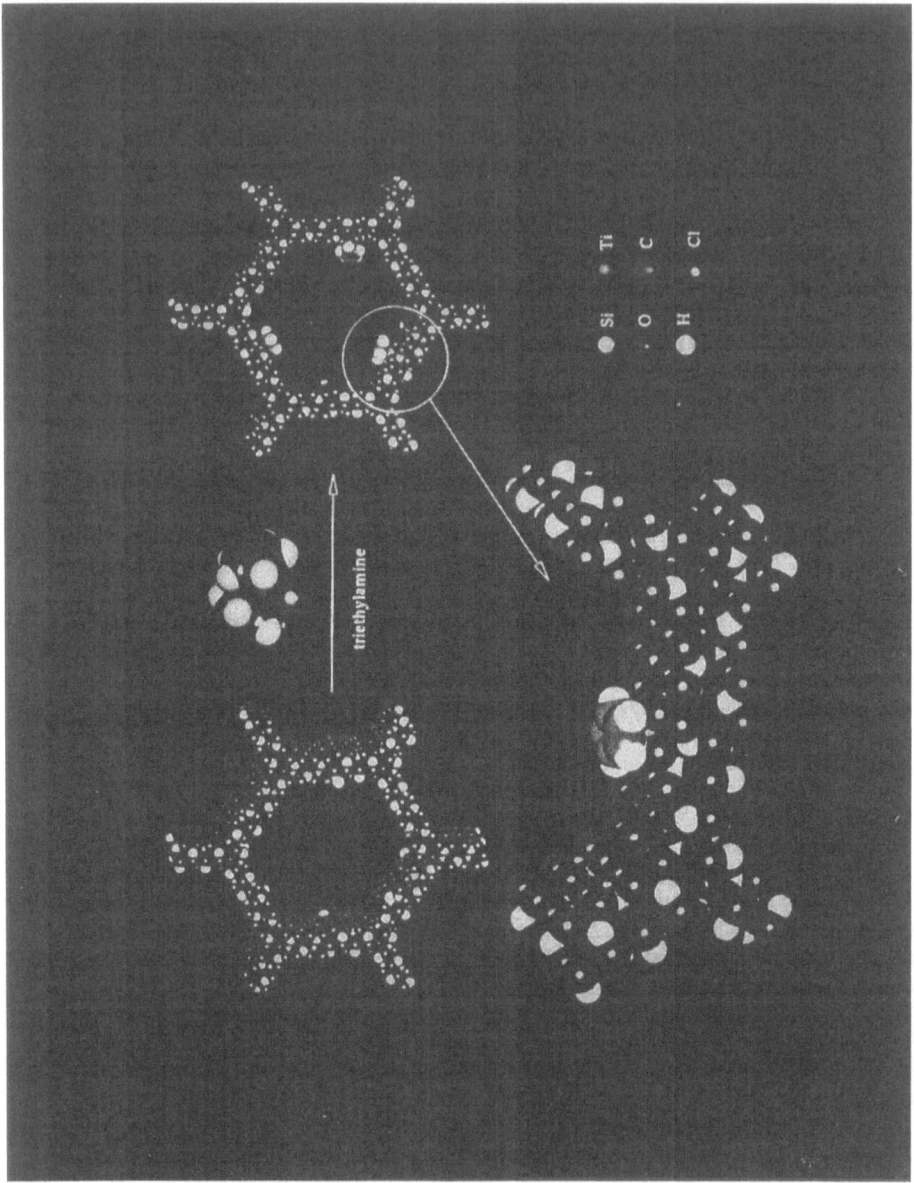

Figure 20. Computer generated illustration of the accommodation (diffusion/adsorption/chemisorption) of molecules of titanocene dichloride inside a pore (30Å dia.) of siliceous MCM–41. For simplicity, only a few of the pendant ≡Si–OH (silanol) groups, that make it possible to graft organometallic–moieties inside the mesoporous host, are shown.

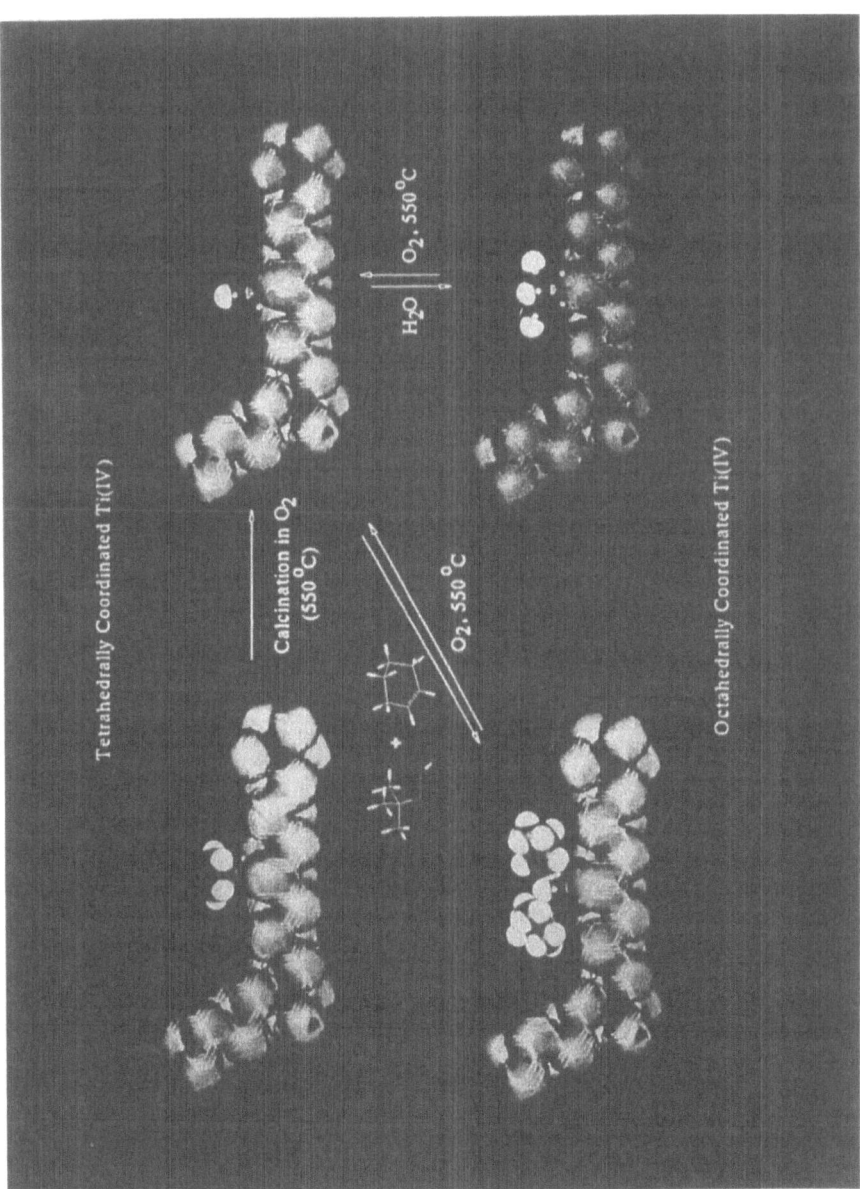

Figure 21. Schematic representation of the grafted organometallic–derived Ti–MCM–41 epoxidation catalyst: calcination, catalysis/regeneration and reversible reaction with water upon exposure to the atmosphere.

Titanocene dichloride was dissolved in chloroform and allowed to diffuse into MCM-41 for 30 minutes. The now red MCM-41 was exposed *in situ* to triethylamine to activate the surface silanols of the MCM-41. The colour of the suspension changed from red *via* orange to yellow over a period of two hours, signifying that the well-established substitution of the chloride with alkoxide/siloxide ligands had occurred [115]. After extensive washing with chloroform, the organic components (typically 12.7 w/w percentage) of this material were removed by calcination under dry oxygen leaving the white, powdery mesoporous catalyst (Si/Ti = 10, i.e. 6 w/w percent Ti), the structural integrity of which had remained intact as established by XRD.

All the intermediate materials produced in the catalyst preparation, as well as the catalyst itself during the course of epoxidation and subsequent regeneration were examined by Ti K-edge X-ray absorption spectroscopy using an *in situ* cell, [116] the experiments being performed on station 8.1 at the Daresbury Synchrotron Radiation Source (the SRS at Daresbury is operated at 2 GeV with a typical current of 200 mA. Station 8.1 was equipped with a Si(111) monochromator, ion chambers and a thirteen element solid state, Canberra, detector for measurement using transmission and fluorescence mode, respectively).

The pre-edge peak (due to a $1s \longrightarrow 3d$ transition) was used as a fingerprint in assessing the local structure around the Ti — analogous to procedures used in identifying metal ion environments of materials containing V and Cr [117,118]. Here the local structures around the anchored titanocene complex and, after calcination, around the corresponding Ti^{4+} active centre were determined with the aid of both XANES and EXAFS.

The pre-edge peak and the EXAFS of the titanocene dichloride adsorbed into MCM-41 were identical to those of the unanchored starting material, showing that the material is only loosely bound as symbolised in figure 20. In particular, we found identical chlorine and carbon neighbours in both free and MCM-41 adsorbed titanocene dichloride, using rigorous multiple-scattering treatment of the EXAFS data analysis [119,120]. Upon reaction with triethylamine, the X-ray absorption spectrum altered significantly. The change in type of neighbours (oxygen for chlorine) and their average bond distance (1.82Å compared with 2.33Å) clearly signify that anchoring had taken place. Moreover, detailed analysis of the EXAFS data disclosed that the two cyclopentadienyl (cp) rings (at ca 2.39Å) underwent considerable modification giving way to a single ring as opposed to the two in the starting material. The FT-IR spectrum of this anchored material shows that the C=C stretching mode in the region of 1400 cm^{-1} is strongly altered compared with that of the starting material, thereby corroborating the supposed modification, seen by EXAFS. The bands assigned to the cp ring (appearing at 1485 cm^{-1} and at 1365 cm^{-1} in the unanchored material) are shifted by 40 cm^{-1} to higher frequency upon anchoring, indicating stronger interaction between the titanium centre and its cp ligand: this is in line with expectation for a shorter Ti-cp distance [121]. Thus, on the basis of EXAFS and FT-IR, as well as from chemical considerations, we see that the Ti^{4+} ion is anchored via three oxygens (each linked to silicon) and one cp ring (cf. figure 21(b)).

Calcination in oxygen at 550°C removes the cp ligand (Fig. 21) , and the resulting material becomes catalytically active. *In situ* XAS measurements reveal distinct modification in the local structure around the anchored Ti^{4+} ion: there is an increase in pre-edge intensity (Fig. 22(b)) and the Ti-O co-ordination is close to four in the catalytically active state. (Comparable changes have been reported for Ti incorporated MCM-41 [101, 102], for titanium silicalite, TS-1, [122,123], and for titanium zeolite β, [124]; and there is little doubt that four co-ordinated Ti is the active centre for the selective oxidation of alkenes in the presence of a peroxide). Further detailed analysis of our data (applying multiple scattering techniques) established the local structure of the active Ti centre. The key features are shown in table 4 and figure 22 (along with the best fit for the EXAFS data and their associated Fourier transforms).

TABLE 4 Ti K-edge EXAFS derived local structural parameters

	Scatterer	Occupancy	Distance [Å]	σ^2 [Å^2]
Ticp$_2$Cl$_2$	Cl	2	2.33	0.003
	C	10	2.39	0.005
diffused	Cl	2	2.33	0.003
	C	10	2.39	0.005
anchored	O	3	1.82	0.008
	C	5	1.99	0.010
	Si	3	3.14	0.008
calcined	O	4	1.81	0.004
	Si	3	3.30	0.003
reactive	O	4	1.84	0.008
state of	O	1	2.25	0.008
catalyst	O	1	2.40	0.008
	Si	3	3.28	0.007

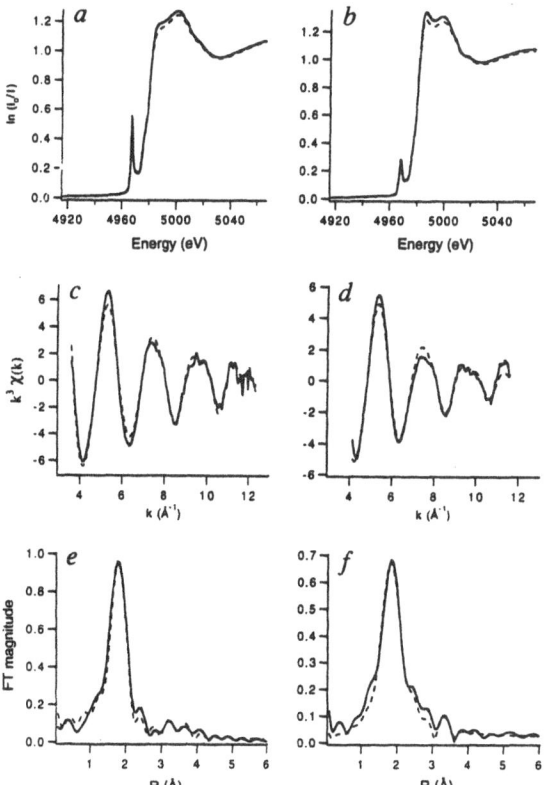

Figure 22. Summary of the characterization of the catalyst by X-ray absorption spectroscopy: (a) schematic representation of the anchored. active Ti^{4+} species and during catalysis. The XANES part of the Ti K-edge XAS data detailing the pre- and near-edge structures of the calcined catalyst (b) and during the course of catalytic reaction (c). The dotted curve in (b) and (c) represents the recalcined catalyst after catalytic reaction and the calcined catalyst exposed to atmorspheric conditions, respectively. Best fit to the Ti K-edge EXAFS data of the calcinated (activated) catalysts (d). during catalytic reaction (e) and their associated Fourier transforms (f) and (g). respectively (see table 4 for bond distances).

During the course of catalysis, the solid turned to pale yellow (similar to the framework-incorporated Ti-MCM-41 catalyst for epoxidation reported by us previously [101,102]. From the significant diminution in intensity of the pre-edge peak (figure 22(c)), and other chemical considerations we processed our raw EXAFS data in terms of distorted six-fold co-ordination taking into account second nearest neighbours. The results are summarised in figure 22 (d-g).

The catalytic oxidation reactions of cyclohexene and pinene with tetrabutylperoxyhydroxide (TBPH) were performed under argon following standard procedures [125]. At 40 ^{0}C and at a ratio of 1 : 1.2 cyclohexene to TBPH and a reactant to catalyst ratio of 1 : 0.05 by weight using an 8 percent solution of cyclohexene in chloroform, a maximum turnover frequency of 3 mmol C_6H_{10}/g cat./min. could be achieved. To our knowledge, this is the highest frequency of turnover reported on epoxidation of alkenes catalysed by Ti-containing mesoporous materials and represents an improvement by about one order of magnitude on previously described preparations [100]. Over a period of 60 minutes a 50 percent conversion of cyclohexene with 95 percent selectivity of TBPH towards epoxidation was recorded. Although the catalyst was essentially deactivated after 90 min., we believe that fine-tuning of its preparation and of the reaction conditions should further enhance its remarkable activity.

In general, the increased performance can be rationalised by noting that: all titanium centres are surface species; that there are no detectable amounts of oligomeric, and hence inactive, species present; and that the titanium precursor can diffuse thoroughly into the channels before grafting, thereby yielding highly dispersed catalytic sites, which, in turn, present maximum access to reactant molecules. Additionally, this new material is active in catalytic oxidations of bulky reactants (i.e. pinene and TBPH gave 20 percent conversion after 12 hours reacting), whereas microporous zeolitic materials are inactive owing to the narrowness of their pores.

Our catalyst is readily regenerable to its active state by re-calcining the expired form in oxygen at 550°C. The XRD and XAS data as well as the catalytic characterisation of the recycled catalyst show no detectable structural degradation nor loss in activity of any significance. The XAS data clearly show the re-formation of four co-ordinated titanium identical to that of the fresh catalyst (the XANES spectrum is shown as dotted curve in figure 22(b)). Upon exposure to atmospheric conditions a distorted six-fold co-ordinated titanium, similar to that of the reacted catalyst is seen by XAS. (XANES data of the calcined catalyst exposed to atmospheric conditions are shown as a dotted curve in figure 22(c)). However, this exposure does not deactivate the catalyst.

In summary, we have developed a methodology which allows the preparation of a new and highly promising family of metal-containing catalysts. In particular, we describe the synthesis of a Ti MCM-41 catalyst in which the active centres are highly dispersed and located on the surface of MCM-41 by using a new grafting process involving titanocene dichloride. We are currently extending this methodology to other metallocenes containing Mo, Cr, Zr, Co, Fe and V. In particular, in the case of the molybdenum derivative a very high dispersion of isolated, tetrahedral Mo(VI) centres could be achieved, underpinning the general utility of the methodology presented in this paper. Furthermore, comprehensive characterisation of the catalyst by EXAFS allows us to track the entire system from the embryonic stage of synthesis (i.e. diffusion/adsorption, figure 20) through to the active state of the catalyst during operation and after its regeneration (cf. figure 22). All this establishes a direct structure-functionality relationship by combining catalytic studies with *in situ* XAS studies.

The structural assignment made of the siliceous MCM-41 anchored titanocene is born out by modelling studies completed previously [42,127]. There, interactions of the complexes $[Ti(\eta^5-C_5H_5)Cl_3]$, $[Ti(\eta^5-C_5(CH_3)_5)Cl_3]$, $[Ti(\eta^5-C_5H_5)_2Cl_2]$ and $[Ti(\eta^5-C_5(C_6H_5)_5)Cl_2]$ with silsesquioxanes were examined as a means of modelling the reactions of these species with silica surfaces. The reaction of stoichiometric amounts of $[Ti(\eta^5-C_5H_5)Cl_3]$,

[Ti(η^5-C$_5$(CH$_3$)$_5$)Cl$_3$], [Ti(η^5-C$_5$H$_5$)$_2$Cl$_2$] and [Ti(η^5-C$_5$(C$_6$H$_5$)$_5$)Cl$_2$] with [(c-hexyl)$_7$Si$_7$O$_9$(OH)$_3$], (1), yielded in each case a single new species (under reaction conditions similar to those previously resported for the Ti-MCM-41 preparation). It was possible to unequivocally establish by ^{1}H, ^{13}C and ^{29}Si n.m.r., [42,44,126,127], that all these model surface compounds were isostructural. In each case the product was a monomeric species of the type [Ti(η^5-C$_5$R$_5$)((c-hexyl)$_7$Si$_7$O$_{12}$)] where the Ti(η^5-C$_5$R$_5$) moiety is co-ordination to the titanium centre along the C_3 axis of the silsesquioxane framework as shown in figure 23.

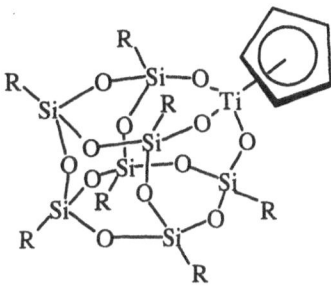

Figure 23. Structure of [Ti(η^5-C$_5$R$_5$)((c-hexyl)$_7$Si$_7$O$_{12}$)], R = H, methyl, phenyl.

Conclusions

What we have attempted to convey in this work is that there is a rational progression and a continuos intellectual and experimental path from model silsesquioxane frameworks on the one hand and real, viable silica supported (mesoporous) catalysts, derivatised with adroitly placed metal-ion active sites, on the other. The kinship between molecular silsesquioxane-derivatised species (as models) at the one extreme and the grafted, or tethered, heterogeneous catalysts on the other is striking. There is clearly abundant scope here to pursue further the marriage of organometallic chemistry and heterogeneous catalysis. (A pedagogic elaboration of this theme is contained in a recent review article by Emsley [128].)

490

References

1. Feher, F. J.; Newman, D. A. and Walzer, J. F. (1989) Silsesquioxanes as Models for Silica Surfaces, *J. Amer. Chem. Soc.*, **111**, 1741-1748.
2. Feher, F.j., Newman, D.A. (1990) Enhanced Silylation Reactivity for a Model Compound for Silica Surfaces, *J. Am. Chem. Soc.*, **112**, 1931-1936.
3. Maschmeyer, T., Masters, A.F., Hambley, T.W. (1992) The Synthesis and Characterisation of Norbornylsilsesquioxanes, *Applied Organometallic Chemistry*, **6**, 253-260.
4. Liu, F.Q., Schmidt, H.G., Noltemeyer, M., Freireerdbrugger, C., Scheldrick, G.M. and Roesky, H.W.. Synthesis and Structure of 8-membered Titanium Containing Siloxane Rings (1992) *Zeitschrift fuer Naturforschung, Section B - A Jounal of Chemical Sciences*, **47**, 1085-1090.
5. Baney, R.H.; Ioth, M; Sakakibara, A, Suzuki, T. (1995) Silsesquioxanes, *Chemical Reviews*, **95**, 1409-1430.
6. Vronkov, M.G. and Lavrent'yev, V.I. (1982) *Top. Curr. Chem.*, **102**, 199.
7. Ichikawa, M., Ed. (1984) *Tailored Metal Catalysts*, D. Reidel: Dordrecht.
8. Bond, G. C. (1990), *Heterogeneous Catalysis, Principles and Applications*; Oxford Science Publ.: Oxford.
9. Yermakov, Y. I.; Kuznetsov, B. N. and Zakharov, V. A. (1981) *Catalysis by Supported Complexes*; Elsevier: Amsterdam.
10. Feher, F. J.; Walzer, J. F. and Blanski, R. L. (1991) Olefin Polymerisation by Vanadium-containing Polyhedral Oligosilsesquioxanes, *J. Amer. Chem. Soc.*, **113**, 3618-3619.
11. Feher, F. J. and Walzer, J. F. (1991) Synthesis and Characterisation of Vanadium-containing Silsequioxanes, *Inorg. Chem.*, **30**, 1689-1694.
12. Feher, F. J. and Blanski, R. L. (1990) Polyhedral Oligometallasilsesquioxanes as Models for Silica-Supported Catalysts - Chromium Attached to 2 Vicinal Siloxy Groups, *J. Chem. Soc., Chem. Commun.*, 1614.
13. Brown, S. C. and Evans, J. (1978) Anchoring of Osmium Clusters to Silica, *J. Chem. Soc., Chem. Commun.*, 1063-1064.
14. Pierantozzi, R.; McQuade, K. J.; Gates, B. C.; Wolf, M.; Knoezinger, H. and Ruhmann, W. (1979), *J. Amer. Chem. Soc.*, **101**, 5436.
15. Evans, J. and Gracey, B. P. (1980), Generalised Cluster Anchoring to Oxide Supports, *J. Chem. Soc., Chem. Commun.*, 852-853.
16. Evans, J.; Alexiev, V. D.; Binsted, N.; Greaves, G. N. and Price (1987) On the Chemisorption of [Ru$_3$(CO)$_{12}$] and [Os$_3$(CO)$_{12}$] on Silica and Alumina R. J., *J. Chem. Soc., Chem. Commun.*, 395-397.
17. Collier, G.; Hunt, D. J.; Jackson, S. D.; Moyes, R. B.; Pickering, I. A.; Wells, P. B.; Simpson, A. F. and Whyman, R. (1983) Metal-Clusters .1. Characterisation of Materials Obtained by Impregnation of Os$_3$(CO)$_{12}$ and Os$_6$(CO)$_{18}$ onto Silica, Alumina and Titania, *J. Catal.*, **80**, 154-171.
18. Wells, P. B.; Hunt, D. J.; Jackson, S. D.; Moyes, R. B. and Whyman, R. (1982) Air Stability of Catalysts Derived from Osmium and Ruthenium Cluster Carbonyls, *J. Chem. Soc., Chem. Commun.*, 85-86.
19. Feher, F. J.; Liu, J. C.; Wilson, S. R. and Shapley, J. R. (1990) A Triosmium Cluster Siloxane Cage Complex - Synthesis and Structure of HOs(CO)$_{10}$[(μ-O)Si$_7$O$_{10}$(C$_6$H$_{11}$)$_7$], *Inorg. Chem.*, **29**, 5138-5139.
20. Hambley, T. W.; Maschmeyer, T. and Masters A. F. (1992) The Synthesis and Characterisation of Norbornylsilsesquioxanes, *Appl. Organomet. Chem.*, **6**, 253-260, 1992.
21. Basset, J. M.; Besson, B.; Moraweck, B. and Smith, A. K. (1980) I.R. and EXAFS Characterisation of a Supported Osmium Cluster Carbonyl, *J. Chem. Soc., Chem. Commun.*, 569-571, 1980.
22. Smith, A. K.; Besson, B.; Basset, J. M.; Psaro, R.; Fusi, A. and Ugo, R (1980), *J. Organomet. Chem.*. **192**, C31.
23. Deeba, M. and Gates, B. C. (1981) Mononuclear and Trinuclear Osmium Carbonyl Catalysts Supported on SiO$_2$, Al$_2$O$_3$, TiO$_2$ and ZnO, *J. Catal.*, **67**, 303-307.
24. Cook, S. L.; Evans, J.; Mc Nulty, G. S. and Greaves, G. N. (1986) Spectrocopic Studies on Adsorbed Metal-Carbonyls .3. Interaction of [Os3(CO)12] with Silica, Alumina and Titania, *J. Chem. Soc., Dalton Trans.*, **1**, 7-14.
25. Zanoni, R.; Carcini, V. and Abu-Samn, R. H., (1985) Osmium Carbonyl Clusters: XPS Characterisation of Some Catalyst Precursors, *J. Mol. Struct.*, **131**, 363.
26. Shapley, J. R.; Walter, T. H.; Frauenhoff, G. R. and Oldfield, E. (1988) Characterisation of a Silica-Supported Trinuclear Osmium Carbonyl Cluster by Magic Angle Spinning C^{13} NMR Spectroscopy, *Inorg. Chem.*, **27**, 2561-2563.

27. Evans, J.; Cook, S. L. and Greaves, G. N. (1983) Characterisation of Supported Trinuclear Osmium Clusters by Extended X-ray Fine Structure (EXAFS) Spectroscopy, *J. Chem. Soc., Chem. Commun..* 1287-1289.

28. Chakrabarty, D. K. and Desai, A. A., *Inorg. Chim. Acta*, **118**, 169, 1986.

29. Gross, D. C. and Ford, P. C. (1985) Nucleophilic Activation of Coordinated Carbon-monoxide .3. Hydroxide and Methoxide reactions with the Trinuclear Clusters $Fe_3(CO)_{12}$, $Ru_3(CO)_{12}$, $Os_3(CO_{12})$ - Implications with Regard to Catalysis of the Water-gas-shift Reaction, *J. Amer. Chem. Soc.*, **107**, 585-593, 1985.

30. Basset, J. M.; Theolier, A.; Choplin, A. and D'Ornelas, L. (1983) The Characterisation and Thermal Stability of a Cluster $HRu_3(CO)_{10}(O-Si\equiv)$ Grafted on a Silica Surface, *Polyhedron*, **2**, 119-121.

31. Evans, J. and Gracey, B. P. (1983) Standardised Tethering of Ru_3-Ru_6 Clusters to High-Surface Area Oxides *J. Chem. Soc., Chem. Commun.*, 247-249; Evans, J. and Gracey, B. P. (1982) Anchoring of Cobalt, Ruthenium and Osmium Carbonyls to Oxides by Pendant Thiol and Phosphine-Ligands, *J. Chem. Soc., Chem. Commun.*, 1123-1129.

32. Psaro, R.; Ugo, R.; Zanderighi, G. M.; Besson, B.; Smith, A. K. and Basset, J. M. (1981) Surface-Upported Metal Cluster Carbonyls - Chemisorption, Decomposition and reactivity of $Os3(CO)12$, $H_2Os_3(CO)_{10}$ and $Os_6(CO)_{18}$ Supported on Silica and Alumina and the Investigation of the Fischer-Tropsch Catalysis with These Systems, *J. Organomet. Chem.*, **213**, 215-247.

33. Hsu, L. Y.; Shore, S.; D'Ornelas, L.; Choplin, A. and Basset, J. M. (1988) Metal Support Interaction - $(\mu-H)(\mu-OSiEt_3)Os_3(CO)_{10}$ - A Molecular Analog of a Surface Complex and the Computer Modelled Surface Structres of $(\mu-H)(\mu-OSi\equiv)Os_3(CO)_{10}$ and $(\mu-OSi)_2Os_3(CO)_{10}$, *Polyhedron*, **7**, 2399-2403.

34. Howard, M. W.; Jayasooriya, U. A.; Kettle, S. F. A.; Powell, D. B. and Sheppard, N. (1979) The M-H Band Stretching Frequencies of μ_2-Bridged Metal Hydrides and Their Realtionship to the M-H-M Interbond Angle, *J. Chem. Soc., Chem. Commun.*, 18-20.

35. Andrews, J. R.; Kettle, S. F. A.; Powell, D. B. and Sheppard, N. (1982) Infra-red and Raman-spectra of $HOs_3(CH=CH_2)(CO)_{10}$, $H2Os_3(C=CH2)(CO)_9$ and $H_2Os_3(CO)_{10}$ - Wavenumbers Associated with Olefinic and Hydride Ligands and the Metal Skeleton, *Inorg. Chem.*, **21**, 2874-2877.

36. Evans, J. and McNulty, G. S. (1983) Spectroscopic Studies on C-2 Hydrocarbon Fragments .1. Vibrational Studies of Cluster-Bound Vinyl and Vinylidene Ligands, *J. Chem. Soc., Dalton Trans.*, 639-644.

37. Oxton, I. A. (1982) On the Infrared Spectrum of $(\mu-H)(\mu-D)Os_3(CO)_{10}$, *Inorg. Chem.*, **21**, 2877-2878.

38. Hartley, D.; Kilty, P. A. and Ware, M. J. (1968) The Vibrational Spectra and Structures of Some Derivatives of Triosmium Dodecacarbonyl, *J. Chem. Soc., Chem. Commun.*, 493-494.

39. Kettle, S. F. A. and Stanghellini, P. L. (1982) Metal Cluster Structural Data from Vibrational Frequencies on Os_3 Systems of C-2-epsilon Symmetry, *Inorg. Chem.*, **21**, 1447-49.

40. Evans, J. (1987), *Spectrochim. Acta*, **43A**, 1511.

41. Shapley, J. R.; Walter, T. H.; Frauenhoff, G. R. and Oldfield, E. (1991) Characterisation of Silica-Supported Osmium Carbonyl Clusters by Magic-Angle-Spinning C^{13} NMR-Spectroscopy, *Inorg. Chem.*, **30**, 4732-4739.

42. T. Maschmeyer (1994) Catalytic Centres - A Study of Structure and Functionality, *PhD Thesis*, The University of Sydney.

43. Thomas Maschmeyer, Anthony F. Masters and Anthony K. Smith (1996) Models of Surface-confined Osmiumcarbonyl Species, *Australian Journal of Chemistry*, (in press).

44. Smith, A. K. (1993) *Personal Communication.*

45. Braunstein, P.; Knorr, M.; Piana, H. and Schubert, U. (1991) Synthesis and Structure of Bimetallic Allyl, Alkoxysilyl Complexes, *Organometallics*, **10**, 828.

46. Braunstein, P.; Knorr, M.; Tiripicchio, A. and Tiripicchio, C. (1989) Complexes with an $\eta^2-\mu-^2-SiO$ Bridge - Structure of the Bimetallic Complex $[Fe(CO)_3(\mu-Si(OMe)_2OMe)(\mu-DPPM)PdCl]-(Fe-Pd)$ *Angew. Chem. Int. Ed. Eng.*, **28**, 1361-1363, 1989.

47. Braunstein, P. and Knorr, M. (1990) Occurence of an $\eta^2-\mu-^2-SiO$ Bridge in Bimetallic Complexes. *New J. Chem.*, **14**, 583-587.

48. Adams, R. D.; Cortopassi, J. E. and Pompeo, M. P. (1991) Unusual Coordination Properties of Trialkoxysilyl Groups in Metal Cluster Complexes - Synthesis, Structure and Reactivity of $Os_3(CO)9[\mu-^3-\eta-^3-Si(OEt)_3](\mu-H)$, *Inorg. Chem.*, **30**, 2960-2961.

49. Lunsford, J. H. and Fu, S. (1990) Chemistry of Organochromium Complexes on Inorganic Oxides Supports .1. Characterisation of Chromocene on Silica Catalysts, *Langmuir*, **6**, 1774.

50. Lunsford, J. H. and Fu, S. (1990) Chemistry of Organochromium Complexes on Inorganic Oxides Supports .2. Interactions of Carbon Oxides with Chromocene on Silica Catalysts, *Langmuir*, **6**, 1784.

492

51.   Lunsford, J. H. and Fu, S. (1991) Chemistry of Organochromium Complexes on Inorganic Oxides
        Supports .3. Interactions of Nitrogen-Oxides with Chromocene on Silica Catalysts, *Langmuir*, **7**, 1179.

52.   Lunsford, J. H. and Fu, S., (1991) Chemistry of Organochromium Complexes on Inorganic Oxides
        Supports .4. Study of Ethylene Polymerisation Over Chromocene on Silica Catalysts, *Langmuir*, **7**,
        1172.

53.   Karol, F. J.; Karapinka, G. L.; Wu, C.; Dow, A. W.; Johnson, R. N. and Carrick, W. L. (1972)
        Chromocene-based Catalysts for Ethylene Polymerisation: Scope of the Polymerisation, *J. Polym. Sci.*,
        A-1, **10**, 2621-2637, .

54.   Karol, F. J.; Brown, G. L. and Davison, J. M. (1973) Chromocene-based Catalysts for Ethylene
        Polymerisation: Kinetic Parameters, *J. Polym. Sci., A-1*, **11**, 413-424.

55.   Karol, F. J. and Wu, C. (1974) Chromocene-based Catalysts for Ethylene Polymerisation: Thermal
        Removal of the Cyclopentadienyl Ligand, *J. Polym. Sci., A-1*, **12**, 1549-1558.

56.   Karol, F. J. and Johnson, R. N. (1975) Ethylene Polymerisation Studies with Supported
        Cyclopentadienyl, Arene and Allyl Chromium Catalysts, *J. Polym. Sc.*, **13**, 1607-1618.

57.   McDaniel, M. P.; Freeman, J. W.; Wilson, D. R.; Ernst, R. D.; Smith, P. D. and Klendworth, D. D.
        (1987), *J. Polym. Sci., A-1*, **25**, 2063.

58.   Zakharov, V. A.; Semikolenova, N. V.; Nesterov, G. A.; Krjukova, N. and Ivanov, V. P. (1988),
        Makromol. Chem., **189**, 1739.

59.   Karol, F. J. (1983), *CHEMTECH*, **4**, 222.

60.   Zecchina, A.; Spoto, G.; Arean, C. O. and Platero, E. E. (1989), *J. Mol. Catal.*, **56**, 211.

61.   Zakharov, V. A.; Yechevskaya, L. G. and Bukatov, G. D. (1989), *Makromol. Chem.*, **190**, 559.

62.   Rebenstorf, B.; Jonson, B. and Larsson, R. (1982), *Acta Chem. Scand.*, **A36**, 695.

63.   Karol, F. J.; Wu, C.; Reichle, W. T. and Maraschin, N. J. (1979), *J. Catal.*, **60**, 68.

64.   Karol, F. J.; Munn, W. L.; Goeke, G. L.; Wagner, B. E. and Maraschin, N. J. (1978) *J. Polym. Sci.:
        Polym. Chem. Ed.*, **16**, 771.

65.   Rasmussen, D. M. (1972), *Chem. Eng.*, **92.**, 104.

66.   Zecchina, A.; Spoto, G. and Brodiga, S. (1989) *Faraday Discuss. Chem. Soc.*, **87**, 149.

67.   Chisholm, M. H.; Cotton, F. A.; Extine, M. W. and Rideout, D. C. (1979)
        Dicyclopentadienyldi-<u>tert</u>-butoxychromium. Preparation, Properties, Structure and Reactions with
        Small Unsaturated Molecules, *Inorg. Chem.*, **18**, 120.

68.   Shestakov, A. F.; Yanovsky, A. I. and Struchkov, Y. T. (1990), *J. Organomet. Chem.*, **384**, 279.

69.   Feher, F. J. and Budzichowski, T. A., (1989) *J. Organomet. Chem.*, **379**, 33.

70.   (a) Maschmeyer, T; Ellis, P.; Masters, A. F.; Smith, A. K and Joyner, R. (1996) *J. Mol. Cat. A.* (in
        press); (b) Maschmeyer, T and Masters, A. F. (1996) (submitted).

71.   Marks, T. J. and Toscano, P. J. (1986) *Langmuir*, **2**, 820.

72.   Feher, F. J.; Liu, J. C.; Wilson, S. R. and Shapley, J. R. (1990) *Inorg. Chem.*, **29**, 5138.

73.   (a) Brintzinger, H. H.; Raaij, E. E.; Zsolnai, L. and Huttner, G. (1989) *Z. anorg. allg. Chem.*, **577**, 217;
        (b) Koch, O.; Edelmann, F. and Behrens, U. (1982) *Chem. Ber.*, **115**, 1313.

74.   J. S. Beck, W. J. Vartuli, W. J. Roth, M. E. Leonowicz, C. T. Kresge, K. D. Schimitt, C. T. W. Chu, D.
        H. Olson, E. W. Sheppard, S. B. McCullen, J. B. Higgins and J. L. Schlenker (1992), *J. Am. Chem. Soc.*
        114 10834.

75.   Q. S. Hue, D. I. Margolese, U. Ciesla, P. Y. Feng, T. E. Gier, P. Sieger, R. Leon, P. M. Petroff, F. Schuth
        and G. D. Stucky (1994), *Nature*, **368**, 317.

76.   Q. Huo, R. Xu, S. Li, Z. Ma, J. M. Thomas, R. H. Jones and A. M. Chippindale (1992), *J. Chem. Soc.,
        Chem Commun.* 875.

77.   M. Estermann, L. B. McCusker, C. Baerlocher, A. Merrouche and H. Kessler (1991), *Nature*, **352**, 20.

78.   P. J. Branton, P. G. Hall and K. S. W. Sing (1993), *J. Chem. Soc., Chem. Commun.*, 1257; P. J.
        Branton, P. G. Hall, K. S. W. Sing, H. Reichert, F. Schuth and K. K. Unger (1994), *J. Chem. Soc.,
        Faraday Trans.*, **90**, 2965.

79.   P. T. Tanev, M. Chibwe and P. J. Pinnavaia (1994), *Nature*, **368**, 321, A. Corma, M. T. Navarro and J.
        Perez-Pariente (1994), *J. Chem. Soc. Chem. Commun.*, 147.

80.   Fernando Rey, Gopinathan Sankar, Thomas Maschmeyer, John Meurig Thomas, Robert Bell and Neville
        Greaves (1995) Synthesis and Characterisation by X-Ray Absorption Spectroscopy of a Suite of Seven
        Mesoporous Catalysts Containing Metal Ions in Framework Sites, *Topics in Catalysis* (in press).

81.   J. M. Thomas (1994), *Nature*, **368**, 289.

82.   K. M. Reddy, I. L. Moudrakouski and A. Sayari, J. (1994) Chem Soc. Chem. Commun. 1059.

83.   Z. Y. Yuan, S. Q. Liu, T. H. Chen, J. Z. Wang and H. X. Li (1995) *J. Chem. Soc., Chem. Commun.* 973.

84.   D. Y. Zhao and D. Goldfarb (1995), *J. Chem. Soc., Chem. Commun.*, 875.

85.   Z. H. Luan, C. F. Chen, W. Z. Zhou and J. Klinowski (1995), *J. Phys. Chem.*, **99**, 1018; M. Janicke, D. Kumar, G. D. Stucky and B. F. Chmelka (1994) *Stud. Surf. Scien. Catal.*, **84**, 243; R. B. Borade and A. Clearfield (1995), *Catal. Lett.*, **31**, 267.

86.   A. Sayari, C. Danumah and I. L. Moudrakovski (1995), *Chem. Mater.*, **7**, 813.

87.   K. R. Kloetstra and H. van Bekkum (1995), *J. Chem Soc., Chem. Commun.*, 1005; E. Armengol, M. L. Cano, A. Corma, H. Garcia and M. T. Navarro (1995), *J. Chem. Soc., Chem. Commun.*, 519.

88.   G. Sankar, F. Rey, J. M. Thomas, G. N. Greaves, A. Corma, B. R. Dobson and A. J. Dent (1994), *J. Chem. Soc., Chem. Commun.*, 2279.

89.   R. B. Gregor, F. W. Lytle, D. R. Sandstrom, J. Wong and P. Schultz (1983), *J. Non-Crystalline Solids*, **55**, 27.

90.   V. A. Durant, D. A. Walker, S. N. Gussaw and J. E. Lyons (1972) U. S. Pat. 4918249.

91.   D. W. Lewis, G. Sankar, C. R. A. Catlow, S. W. Carr and J. M. Thomas (1995), *Nucl. Instr. Methods Phys. Res. B*, **97**, 44.

92.   T. Chapus, A. Tuel, Y. Ben Taarit and C. Naccache (1994), *Zeolites*, **4**, 349.

93.   J. S. Stephens and D. W. J. Cruickshank (1970) A Re-investigation of the Crystal Structure of $(NH_4)_2CrO_4$, *Acta Cryst.* (1970), **26**, 437.

94.   G. Sankar and J. M. Thomas, (unpublished results).

95.   M. Nishimura and J. M. Thomas (1993), *Catal. Lett.*, **19**, 33; M. Nishimura, (1993) *Ph. D. Thesis*, University of London, U.K.

96.   J. Wong, F. W. Lytle, R. P. Messmer and and D. H. Maylotte (1984), *Phys. Rev. B*, **30**, 5596; R. Kozlowski, J. M. Thomas and R. F. Pettifer (1983), *J. Phys. Chem.*, **86**, 5176; R. Kozlowski, R. F. Pettifer and J. M. Thomas (1982), *J. Chem. Soc., Chem. Commun.*, 438; M. Nabavi, F. Taulelle, C. Sanchez and M. Verdaguer (1990), *J. Phys. Chem. Solids*, **51**, 1375; S. Asbrink, G. N. Greaves, P. D. Hatton and K. Garg (1986), *J. Appl. Cryst.*, **19**, 331.

97.   K. Inumaru, T. Okuhara, M. Misono, N. Matsubayashi, H. Shimada and A. Nishijima (1992), *J. Chem. Soc, Faraday Trans.*, **88**, 625.

98.   A. Corma, V. Fornes, M. T. Navarro and J. Perez-Pariente (1994), *J. Catal.*, 148, 569; C. Y. Chen. H. X. Li and M. E. Davis (1993), *Microporous Mater.* 2, 17; C. Y. Chen, S. L. Burkett, H. X. Li and M. E. Davis (1993), *Microporous Mater.*, **2**, 27.

99.   Tanev, P. T., Chibwe, M. and Pinnavaia, T. J. (1994), *Nature*, **368**, 321.

100.   Corma, A., Navarro, M. T. and Perez-Pariente, J. (1994), *J. Chem. Soc., Chem. Commun.*, 147.

101.   Sankar, G., et al. (1994), *J. Chem. Soc., Chem. Commun.*, 2279.

102.   Thomas, J. M. and Greaves (1994), G. N. *Science*, 265, 1675.

103.   Tamarasso, M., Parego, G. and Notari, B. (1983), *U. S. Pat. 4410501*.

104.   Tuel, A. (1995), *Zeolites*, **15**, 236.

105.   Serrano, D. P., Li, H. X. and Davis, M. E. (1992), *J. Chem. Soc., Chem. Commun.*, 745.

106.   Reddy, K. M., et al., *Catal. Lett.* (1994), **23**, 175.

107.   Camblor, M. A., Corma, A. and Perez-Pariente (1993), J., *Zeolites*, **13**, 82.

108.   Ulagappan, N. and Krishnasamy, U. (1995), *J. Chem. Soc., Chem. Commun.*, 373.

109.   Tuel, A. and Ben-Taarit, Y. (1994), *J. Chem. Soc., Chem. Commun.*, 1667.

110.   Corma, A., Esteve, P., Martinez, A. and Valencia, S. (1995), *J. Catal.*, **152**, 18.

111.   Zicovich-Wilson, C. M., Corma, A. and Viruela, P. (1994), *J. Phys. Chem.*, **98**, 10863.

112.   Che, M. (1993), *"New Frontiers in Catalysis"* (Ed. L. Guezi et al., Elsevier) Proc. 10th Inter. Confer. on Catalysis (July, 1992), 31.

113.   Scott, S. L. and Basset, J. M. (1994), *J. Mol. Catal.*, **86**, 5.

114.   Maschmeyer, T, Rey,F., Sankar, G, Thomas, J.M. (1995) Heterogeneous Catalysts Obtained by Grafting Metallocene Complexes onto Mesoporous Silica, *Nature*, **378**, 159.

115.   Wilkinson, G., Stone, F.G.A. and Abel, E.W., eds. (1982) Comprehensive Organometallic Chemistry, Oxford, Pergamon Press.

116.   G. Sankar, et al., *J. Nucl. Inst. Methods B* (in press).

117.   Waychunas, G.A (1987), *Am. Miner.* 72, 89; Wong, J., Lytle, F.W.,Messmer, R.P. and Maylotte, D.H. (1984), *Phys. Rev B* **30**, 5596.

118.   Nishimura, M.(1993), *PhD Thesis*, University of London.

119.   *(SPLINE and XFIT-Program)* Ellis, P.J. and Freeman, H. C., *J. Synchrotron Rad.* (in press).

120.   *(EXCURV92-Program)* Binstead, N., Campell, J.W.,Gurman, S.J. and Stephenson, P., SERC Daresbury Laboratories, 1992.

121.   Nakamoto, K. (1986), Infra-red and Raman Spectra of Inorganic and Coordination Compounds, (New York, John Wiley & Sons).

494

122. Bordiga, S., et al. (1994), *J. Phys. Chem.*, **98**, 4125.
123. Bordiga, S., et al., (1994) *Catal. Lett.*, **26**.
124. Blasco, T., Camblor, M. A., Corma, A. and Perez-Pariente, J. (1993), *J. Am. Chem. Soc.*, **115**. 11806.
125. Shriver, D.F. and Drezdon, M.A. (1986), The Manipulation of Air-sensitive Compounds, 2nd ed.. (Wiley-Interscience, New York).
126. Feher, F. J. (1986), *J. Amer. Chem. Soc.*, **108**, 3850.
127. Field, L. D.; Lindall, C. M.; Maschmeyer, T. and Masters, A. F. (1994), The Synthesis and Characterisation of Decaphenyltitanocene Dichloride, $[Ti(\eta^5-C_5(C_6H_5)Cl_2]$ and of $[Ti(\eta^5-C_5(C_6H_5)((C_6H_{11})_7Si_7O_{12})]$, the First Pentaphenyloligosilsesquioxane, *Aust. J. Chem.*. **47**, 1127.
128. Emsley, J. (1996) Metallocene Catalysts Go Heterogeneous, Chem. Britain., 1, 16.

SELF-ASSEMBLY IN CHEMICAL SYSTEMS

FRANÇISCO M. RAYMO, J. FRASER STODDART

School of Chemistry

University of Birmingham

Edgbaston, Birmingham B15 2TT, UK

1. Introduction

1.1. SELF-ASSEMBLY

The conventional synthetic methodologies employed in the chemical laboratory involve the manipulation of functional groups using reagents and catalysts to make and break covalent bonds in a controlled manner generating through stepwise transformations the targeted molecular species. On the contrary, numerous biological systems self-assemble[1,2] efficiently and precisely from the spontaneous combination of simple *instructed* molecular components. Molecular recognition is central to these processes and both covalent and noncovalent bonding interactions are involved in a cooperative manner. The stereoelectronic information imprinted in the modular subunits is, therefore, crucial for the entire process. Thermodynamic control is usually associated with self-assembly processes and, as a result, error checking and recovery are achieved. Thus, occasional synthetic errors are overcome and defective subunits containing insufficient information are rejected: purification is built into the synthesis. Learning the principles of natural self-assembly processes and applying them to unnatural organic

495

C.R.A. Catlow and A. Cheetham (eds.), New Trends in Materials Chemistry, 495–511.
© 1997 *Kluwer Academic Publishers.*

496

and inorganic syntheses provides a powerful methodology to construct precisely and efficiently complex wholly-synthetic molecular assemblies and supramolecular arrays. Indeed, in the last few years, a number of artificial organic[3-6] and inorganic[7-10] systems have been synthesised by means of self-assembling approaches mimicking those occurring in nature. Although the level of sophistication of these abiotic systems is still far from that witnessed in the biological world, thermodynamically-stable and relatively-complex architectures with engineered shapes and functions have been already self-assembled by many investigators.[1-10]

1.2. CATENANES AND ROTAXANES

Catenanes and rotaxanes (*Figure 1*) are discrete molecular compounds incorporating mechanically-interlocked components.[11-16] An [n]catenane is comprised of n macrocyclic components held together as links in a chain. An [n]rotaxane incorporates one dumbbell-shaped component bearing bulky stoppers at both ends encircled by n-1 macrocyclic components. The synthesis of catenanes and rotaxanes can be achieved, as illustrated schematically in *Figure 2*. A [2]catenane can be synthesised as a result of a clipping procedure involving the macrocyclisation of one of the two macrocyclic components in the presence of the second preformed macrocycle. A [2]rotaxane can be synthesised by employing threading, clipping, and slipping procedures. In the case of threading, insertion of a linear species — the *axle* — into the cavity of a preformed

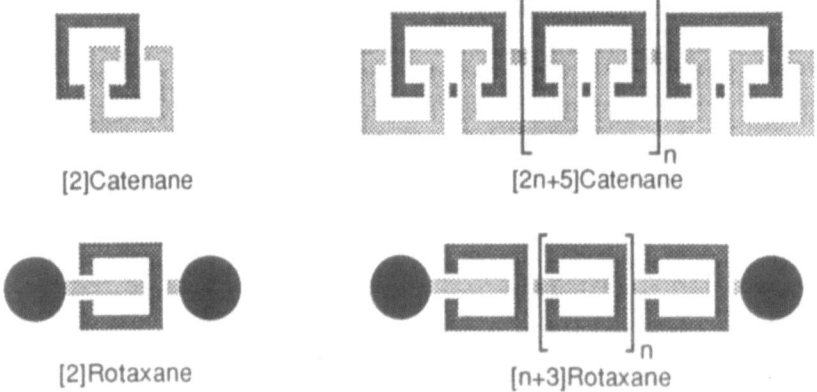

[2]Catenane   [2n+5]Catenane

[2]Rotaxane   [n+3]Rotaxane

Figure 1. Schematic Representation of Catenanes and Rotaxanes

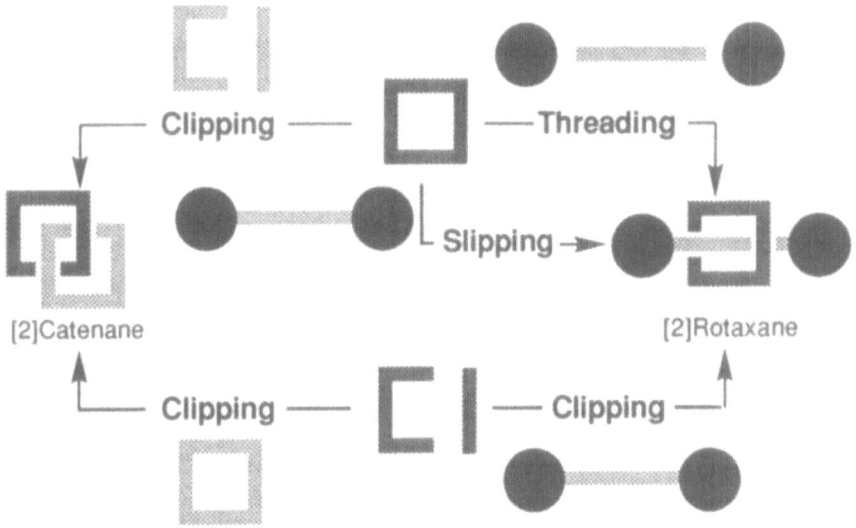

Figure 2. Synthetic Approaches to Catenanes and Rotaxanes

macrocycle — the *wheel* — followed by covalent attachment of two bulky groups at both ends of the *axle* affords the [2]rotaxane. In the case of clipping, the macrocyclisation of the cyclic component is performed in the presence of the preformed dumbbell-shaped component. In the case of slipping, dumbbell-shaped and macrocyclic components are preformed separately and then associated under the influence of an appropriate amount of thermal energy.

1.3. FROM PSEUDOROTAXANES TO CATENANES AND ROTAXANES

The macrocyclic polyether bis-*p*-phenylene-34-crown-10 **1** binds[17] (*Figure 3*) the bis(hexafluorophosphate) salt of paraquat **2** both in solution (K_a = 730 M^{-1} in Me$_2$CO at 25°C) and in the solid state. The resulting 1:1 complex possesses a pseudorotaxane-like geometry, *i.e.* the bipyridinium salt — the *axle* — is inserted through the cavity of the macrocyclic polyether — the *wheel* — with both ends of the *axle* protruding outside the cavity of the *wheel* on opposite directions. The noncovalent bonding interactions responsible for the self-assembly of such a supramolecular complex are mainly (i) π-π stacking interactions between the π-electron rich hydroquinone rings and the π-electron

498

deficient bipyridinium unit, as well as (ii) hydrogen bonding interactions between the polyether oxygen atoms and the protons in the α-positions (Me-N and α-bipy-CH-N) with respect to the nitrogen atoms on the bipyridinium unit. By reversing the role of the recognition sites, the bipyridinium-based macrocycle, cyclobis(paraquat-*p*-phenylene) **3**, as its tetrakis(hexafluorophosphate) salt, can be employed to bind,[18] in solution and in the solid state, acyclic hydroquinone-based polyethers, such as **4** (K_a = 2220 M^{-1} in MeCN at 25°C) with pseudorotaxane-like geometries. In addition to the π-π stacking interactions between the complementary aromatic units and the hydrogen bonding interactions between the polyether oxygen atoms and the α-protons of the bipyridinium units, T-type edge-to-face interactions between the hydroquinone ring protons and the π-clouds of the *p*-xylene rings separating the bipyridinium units are observed in these pseudorotaxane-like complexes. The geometries of these supramolecular complexes suggested the possibility of generating molecular compounds, such as rotaxanes by attaching covalently stoppers at the ends of the *axles*, as well as catenanes by interlocking two complementary *wheels* as illustrated in *Figure 3*. Indeed, a wide number of [n]catenanes and [n]rotaxanes have been self-assembled[19,20] recently by employing complementary π-electron rich and π-electron deficient components incorporating hydroquinone rings and bipyridinium units, respectively.

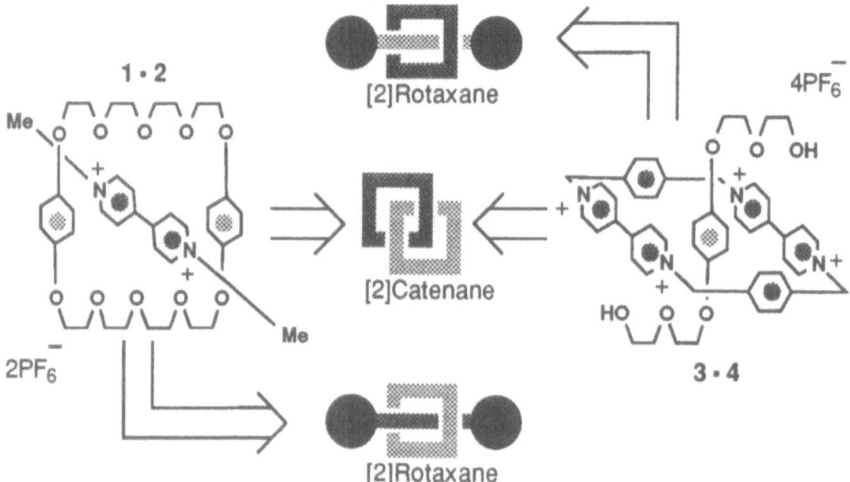

Figure 3. Genealogical Link between Pseudorotaxanes and Catenanes and Rotaxanes

2. Self-assembly of [n]Catenanes

Our first [2]catenane was self-assembled as illustrated in *Scheme 1*. Reaction of the bis(hexafluorophosphate) salt **5** with the dibromide **6** affords a tricationic intermediate, which is immediately bound by the hydroquinone-based macrocyclic polyether **1** and, in so doing, templates the macrocyclisation of the bipyridinium-based cyclophane, affording[18,21] the [2]catenane **8** in the amazing yield of 70%. Similarly, by employing a macrocyclic polyether incorporating 1,5-dioxynaphthalene units in place of the hydroquinone rings, the [2]catenane **9** can be self-assembled[22] (*Scheme 1*) in a yield of 51%. These [2]catenanes have been fully characterised by means of fast atom bombardment mass spectrometry, ^{1}H-NMR and ^{13}C-NMR spectroscopies, and elemental analysis. Furthermore, the interlocking of the two macrocyclic components has been unequivocally proved by X-ray crystallographic analysis. The solid state structures of both [2]catenanes show one of the two π-electron rich aromatic residues of the macrocyclic polyether located inside the cavity of the cyclophane with the other one positioned alongside one of the two bipyridinium units. As a result, a π-donor/π-acceptor stacking motif exists twice over within each catenane molecule. Furthermore, π-π stacking interactions are also achieved between complementary alongside recognition sites of adjacent catenanes. In this manner, the π-donor/π-acceptor stacking

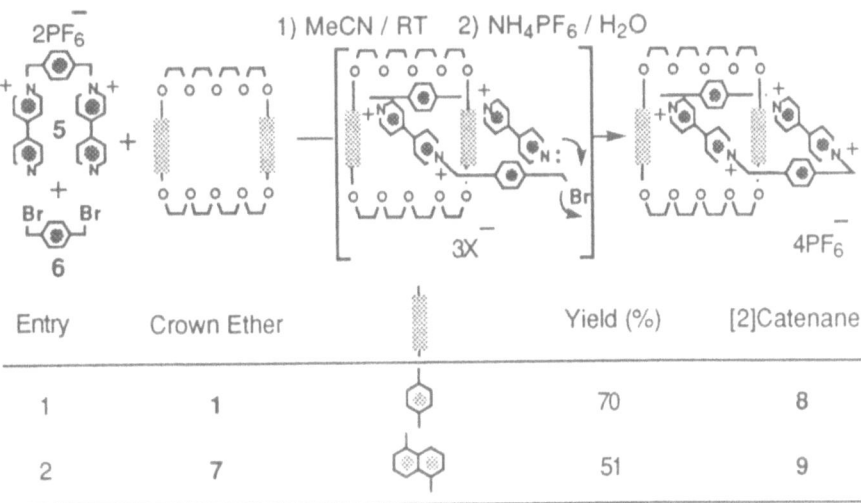

Entry	Crown Ether		Yield (%)	[2]Catenane
1	1		70	8
2	7		51	9

Scheme 1. Self-assembly of [2]Catenanes

500

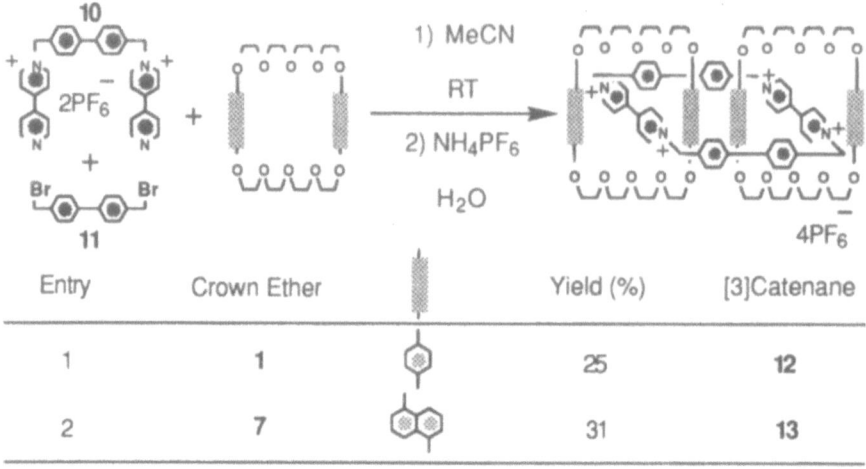

Entry	Crown Ether		Yield (%)	[3]Catenane
1	1		25	12
2	7		31	13

Scheme 2. Self-assembly of [3]Catenanes Incorporating an Enlarged Tetracationic Cyclophane

motif propagates infinitely along one of the crystallographic directions. By enlarging the size of either the macrocyclic polyether or the tetracationic cyclophane, catenanes incorporating a higher number of macrocyclic components can be self-assembled. The [3]catenanes **12** and **13** incorporate two π-electron rich macrocycles, comprising hydroquinone rings and 1,5-dioxynaphthalene units, respectively, and a tetracationic cyclophane in which the two bipyridinium units are separated by bitolyl spacers. The use of a longer spacer such as the bitolyl unit enlarges the cavity of the cyclophane allowing the inclusion of two π-electron rich aromatic residues. The self-assembly of these [3]catenanes was achieved[23,24] according to the route depicted in *Scheme 2*. Reaction of the bis(hexafluorophosphate) salt **10** with the dibromide **11** in the presence of either **1** or **7** affords the corresponding [3]catenanes **12** or **13**, respectively, in yields of 25 or 31%, respectively. Enlargement of the macrocyclic polyether component can be achieved at the same time as increasing the number of its aromatic recognition sites. Thus, the π-electron rich macrocyclic polyethers **14** and **15** (*Scheme 3*) incorporating three hydroquinone rings and three 1,5-dioxynaphthalene units, respectively, have been employed to self-assemble [3]catenanes incorporating two π-electron deficient cyclophanes and one π-electron rich macrocycle. Reaction of the bis(hexa-fluorophosphate) salt **5** with the dibromide **6** in the presence of either **14** or **15** affords the corresponding [3]catenanes 16[24,25] or 17,[26] respectively, in yields of 15% in each case. Although, ultrahigh pressure reactions conditions were required for the self-

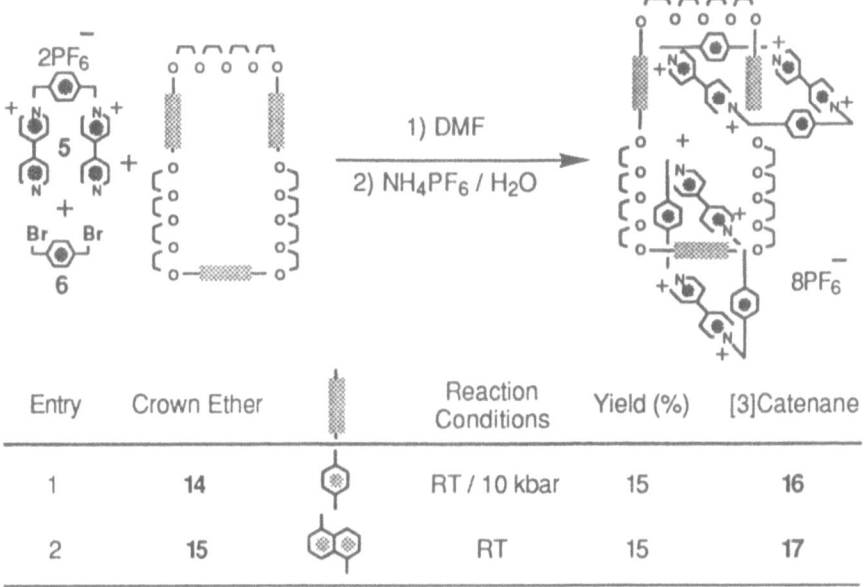

Entry	Crown Ether		Reaction Conditions	Yield (%)	[3]Catenane
1	14		RT / 10 kbar	15	16
2	15		RT	15	17

Scheme 3. Self-assembly of [3]Catenanes Incorporating Enlarged Macrocyclic Polyethers

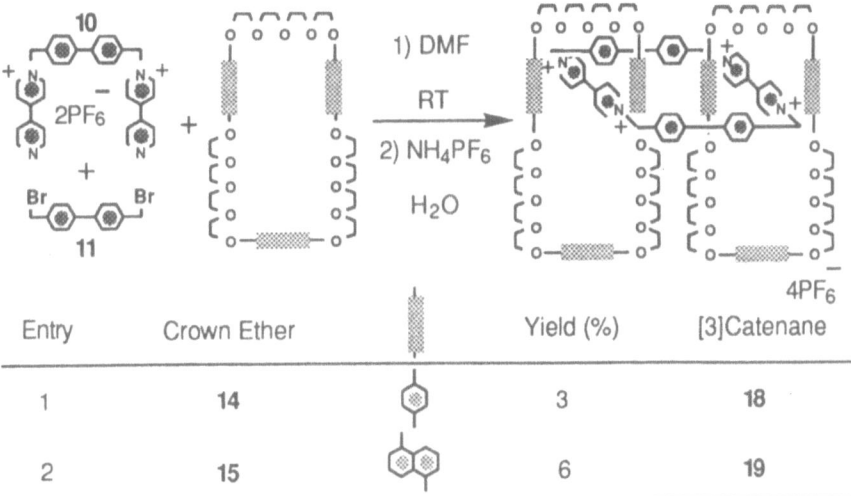

Entry	Crown Ether		Yield (%)	[3]Catenane
1	14		3	18
2	15		6	19

Scheme 4. Self-assembly of [3]Catenanes Incorporating Enlarged π-Electron Rich and π-Electron Deficient Macrocyclic Components

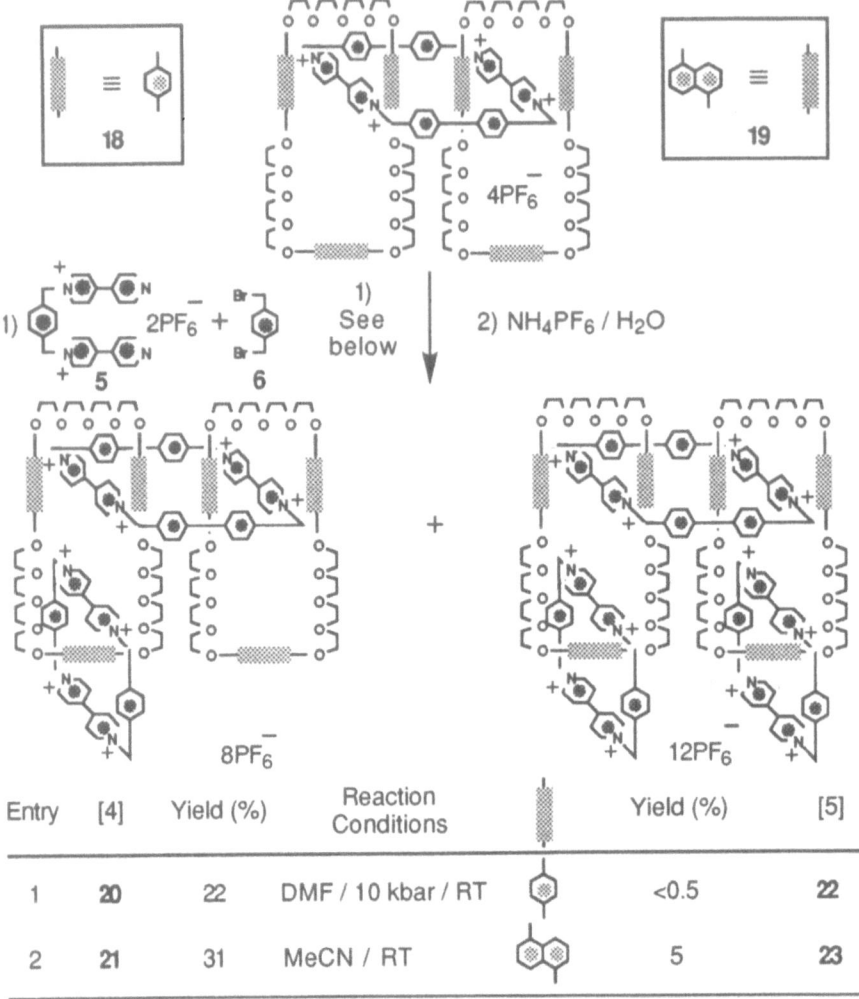

Entry	[4]	Yield (%)	Reaction Conditions		Yield (%)	[5]
1	20	22	DMF / 10 kbar / RT		<0.5	22
2	21	31	MeCN / RT		5	23

Scheme 5. Self-assembly of [4]Catenanes and [5]Catenanes

assembly of the [3]catenane **16**, incorporating the hydroquinone-based macrocycle, the [3]catenane **17**, incorporating the 1,5-dioxynaphthalene-based macrocycle, was obtained at ambient pressure. A similar approach was employed to self-assemble (*Scheme 4*) the [3]catenanes **18**[24,25] and **19**,[27] incorporating both enlarged π-electron rich and π-electron deficient macrocyclic components. The size as well as the presence of *free* recognition sites within the macrocyclic polyether components of both **18** and **19** suggest the possibility of employing these [3]catenanes as starting materials for the self-assembly of [4]- and [5]-catenanes. Reaction, under ultrahigh pressure conditions, of the bis(hexafluorophosphate) salt **5** with the dibromide **6** in the presence of the [3]catenane **18** affords[24,25] (*Scheme 5*) the [4]catenane **20** in a yield of 22% and the [5]catenane **22** in a very low yield (<0.5%) indeed. However, when the 1,5-dioxynaphthalene-containing [3]catenane **19** is employed, the [4]catenane **21** and the [5]catenane **23** self-assemble,[27] in yields of 31 and 5%, respectively, on performing the reaction at room pressure and temperature.

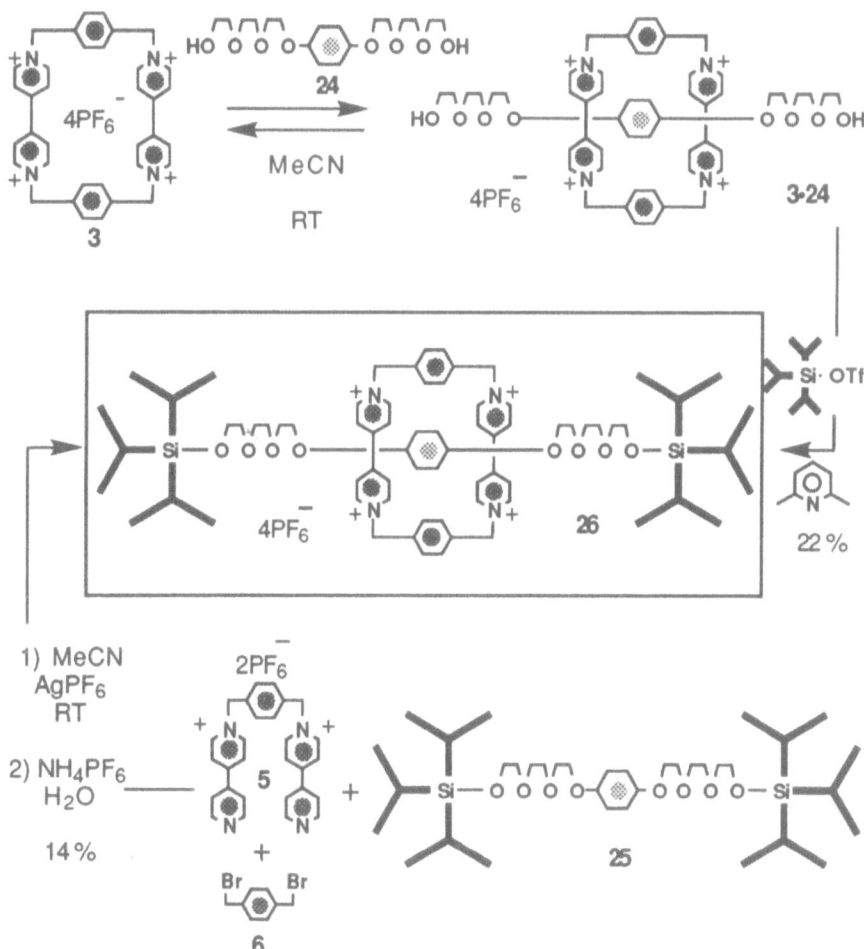

Scheme 6. Self-assembly of a [2]Rotaxane

3. Self-assembly of [n]Rotaxanes

Our first [2]rotaxane was self-assembled[18] by means of either a threading or a clipping procedure. Upon mixing (*Scheme 6*) in MeCN, solution the tetracationic cyclophane **3** and the acyclic polyether **24**, a supramolecular complex with pseudorotaxane-like geometry self-assembles spontaneously. Subsequent reaction with tri-*i*-propylsilyltriflate in the presence of lutidine affords the [2]rotaxane **26** in a yield of 22%, as a result of a threading procedure. Reaction of the bis(hexafluorophosphate) salt **5** with the dibromide **6** in the presence of the dumbbell-shaped compound **25** affords the [2]rotaxane **26** in a yield of 14%, as a result of a clipping procedure. By increasing the

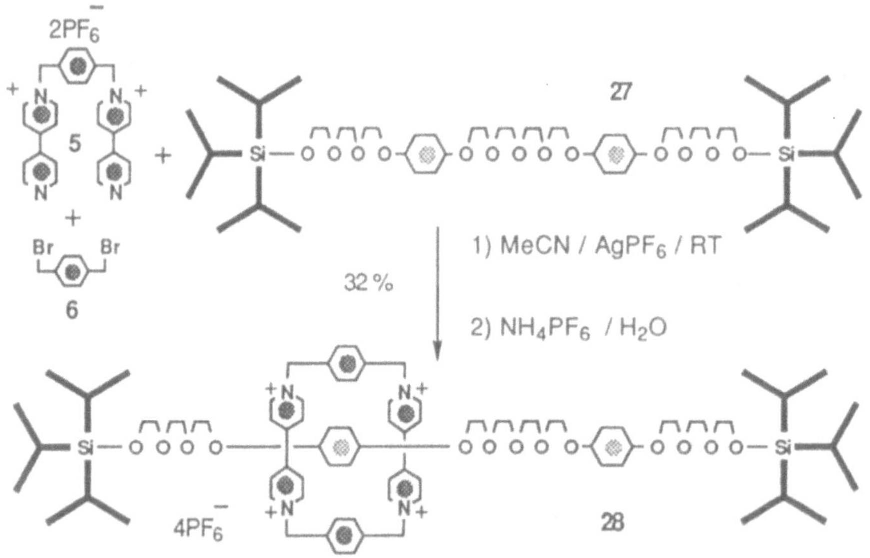

Scheme 7. Self-assembly of a *Molecular Shuttle* Incorporating a Bipyridinium-based Macrocycle

number of recognition sites incorporated within the dumbbell-shaped component, the [2]rotaxane **28** can be self-assembled[28] (*Scheme 7*) by employing the clipping methodology. The [2]rotaxane **28** incorporates two hydroquinone recognition sites within the dumbbell-shaped component but only one tetracationic macrocycle. As a result, the cyclophane is able to move from one hydroquinone recognition site to the other giving rise to a so-called *molecular shuttle*. The [2]rotaxanes **26** and **28** incorporate a π-electron rich dumbbell-shaped component and a π-electron deficient macrocyclic component. However, reversing the role of the recognition sites can be envisaged and, indeed, rotaxanes, incorporating bipyridinium units within their dumbbell-shaped components have been self-assembled with hydroquinone-based macrocycles by means of threading and slipping procedures. *Scheme 8* illustrates the self-assembly[29] of the [2]rotaxane **30** by employing a threading methodology. Reaction under ultrahigh pressure conditions of the bis(hexafluorophosphate) salt **5** with the tris(4-*t*-butylphenyl)methane-based chloride **29** affords a tricationic intermediate which is bound by the macrocyclic polyether **1**. Alkylation of the *free* nitrogen atom of the

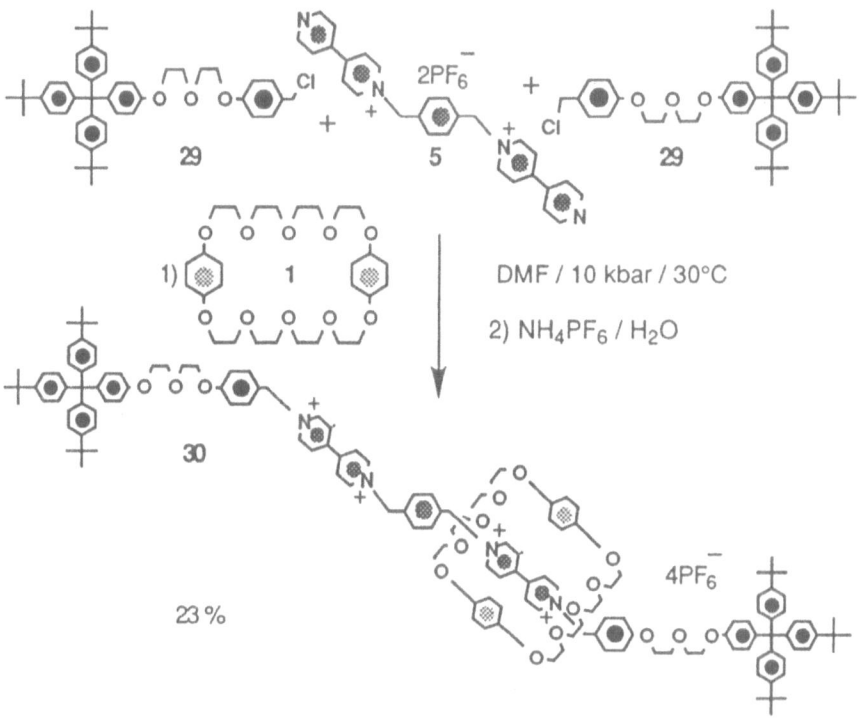

Scheme 8. Self-assembly of a *Molecular Shuttle* Incorporating a Hydroquinone-based Macrocycle

complexed tricationic intermediate by a second molecule of **29** gives the [2]rotaxane **30** in a yield of 23%. As a result of the presence of two bipyridinium recognition sites within the dumbbell-shaped component, the hydroquinone-based macrocyclic polyether moves back and forth from one bipyridinium *station* to the other. Thus, the [2]rotaxane **30** behaves as a *molecular shuttle*. The tris(4-*t*-butylphenyl)methane-based stoppers associated with the [2]rotaxane **30** are large enough to prevent the passage of the hydroquinone-based macrocyclic polyether over them. However, by replacing one of the three *t*-butyl substituents attached to the stoppers with less sterically hindering substituents, the *slippage* of the macrocycle polyether over them can be achieved by employing an appropriate amount of thermal energy. Upon heating MeCN solutions of the dumbbell-shaped compounds **31-33** and the macrocycle **1**, the corresponding

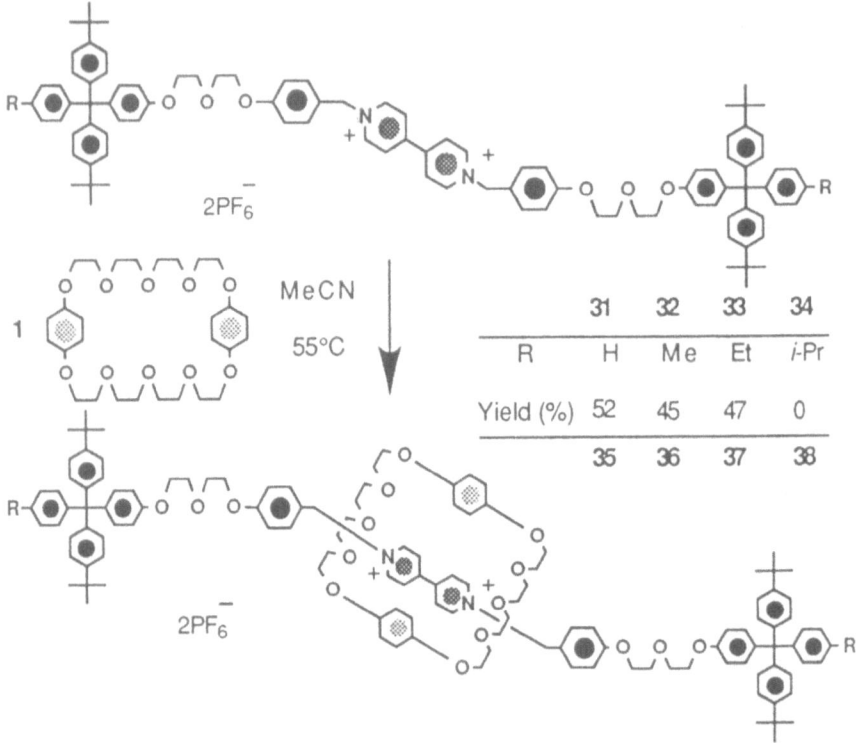

		31	32	33	34
	R	H	Me	Et	*i*-Pr
Yield (%)		52	45	47	0
		35	36	37	38

Scheme 9. Self-assembly of [2]Rotaxanes by Slipping

[2]rotaxanes **35-37** can be self-assembled[30] in yields of 52, 45, and 47%, respectively, as a result of a slipping procedure. When the dumbbell-shaped compound **34** bearing bis(4-*t*-butylphenyl)-4-*i*-propylphenylmethane-based stoppers is employed, under otherwise identical conditions, no rotaxane is formed. Thus, the steric barrier for the *slippage* of the macrocycle **1** over such tetraarylmethane-based stoppers is achieved on going from the Et- to the *i*-Pr-*substituted* stoppers. The slipping methodology has also

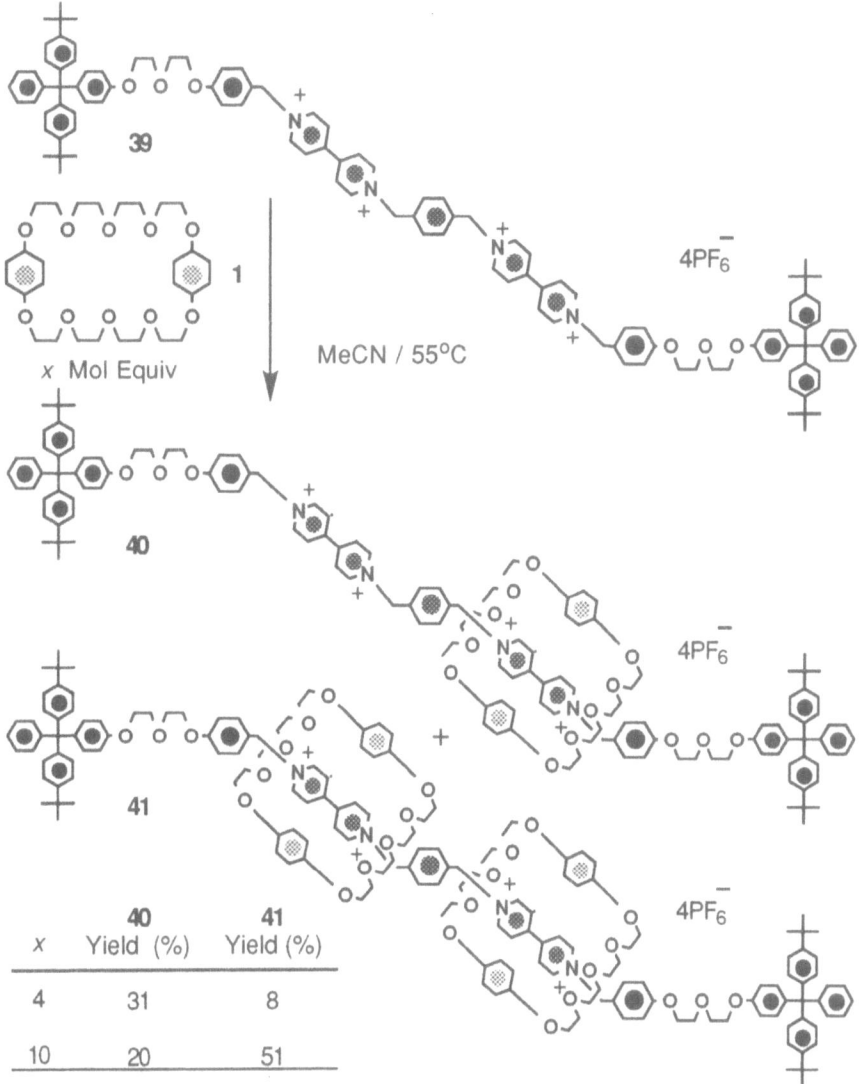

Scheme 10. Self-assembly of a [2]Rotaxane and a [3]Rotaxane by Slipping

been employed to self-assemble rotaxanes incorporating two bipyridinium recognition sites within their dumbbell-shaped components. By heating a MeCN solution containing the dumbbell-shaped compound **39** with four molar equivalents of the macrocyclic polyether **1**, the [2]rotaxane **40** — a *molecular shuttle* — and the

[3]rotaxane **41** self-assemble[31] (*Scheme 10*) in yields of 31 and 8%, respectively. On the contrary, by employing ten molar equivalents of the macrocycle **1**, under otherwise identical conditions, the yields of **40** and **41** are 20 and 55%, respectively. Thus, the ratio of the resulting rotaxanes can be controlled by changing the ratio between the starting macrocyclic and dumbbell-shaped components.

4. Conclusions

Self-assembly processes can be employed to generate complex wholly-synthetic chemical systems from simple complementary molecular components. The stereoelectronic information imprinted in the modular subunits determines precisely the form of the final architecture as a result of cooperative covalent and noncovalent bonding interactions. Thus, efficiency, a high degree of control, and error checking and recovery are ensured throughout the process. In the synthetic methodology developed by us to self-assemble molecular compounds such as catenanes, complementary π-electron rich hydroquinone- or 1,5-dioxynaphthalene-based components and π-electron deficient bipyridinium-based components, have been employed to generate a number of [2]-, [3]-, [4]-, and [5]-catenanes, as well as [2]- and [3]-rotaxanes. Noncovalent bonding interactions, such as π-π stacking and hydrogen bonding interactions between the complementary subunits, are the main driving forces for the self-assembly of such mechanically-interlocked molecular compounds. The methodology demonstrates how the concept of self-assembly — widely employed in nature — can be a powerful adjunct to conventional synthesis for the generation of abiotic chemical systems.

5. References

1.   Lindsey, J.S. (1991) Self-assembly in synthetic routes to molecular devices. Biological principles and chemical perspectives: a review, *New J. Chem.* **15**, 153-180.

2.   Philp, D. and Stoddart, J.F. (1996) Self-assembly in natural and unnatural systems, *Angew. Chem., Int. Ed. Engl.* **35**, In press.

3.   Whitesides, G.M., Mathias J.P., and Seto, C.T. (1991) Molecular self-assembly and nanochemistry: a chemical strategy for the synthesis of nanostructures, *Science* **254**, 1312-1319.

4.   Stoddart, J.F. (1991) From enzyme mimics to molecular self-assembly processes, *Chirality in Drug Design and Synthesis*, Ed. C. Brown, Academic Press, London, pp. 53-81.

5.   Whitesides, G.M., Simanek, E.E., Mathias, J.P., Seto, C.T., Chin, D.N., Mammen, M., and Gordon, D.M. (1995) Noncovalent synthesis: using physical-organic chemistry to make aggregates, *Acc. Chem. Res.* **28**, 37-44.

6.   Kumar, A., Abbot, N.L., Kim, E., Biebuyck, A., and Whitesides, G.M. (1995) Patterned self-assembled monolayers and meso-scale phenomena, *Acc. Chem. Res.* **28**, 219-226.

7.   Rong, D., Hong, H.G., Kim, Y.I., Kreuger, J.S., Mayer, J.E., and Mallouk, T.E. (1990) Electrochemistry and photochemistry of transition metal complexes in well-ordered surface layers, *Coord. Chem. Rev.* **97**, 237-248.

8.   Mallouk, T.E. and Lee, H. (1990) Designer solids and surfaces, *J. Chem. Ed.* **67**, 829-834.

9.   Ozin, G.A. (1992) Nanochemistry: synthesis in diminishing dimensions, *Adv. Mater.* **4**, 612-649.

10. Mann, S. (1993) Molecular tectonics in biomineralization and biomimetic materials chemistry, *Science* **365**, 499-505.

11. Schill, G. (1971) Catenanes, rotaxanes, and knots, Academic Press, New York,.

12. Walba, D.M. (1985) Topological stereochemistry, *Tetrahedron* **41**, 3161-3212.

13. Dietrich-Buchecker, C.O. and Sauvage, J.P. (1987) Interlocking of molecular threads: from the statistical approach to the templated synthesis of catenands, *Chem. Rev.* **87**, 795-810.

14. Dietrich-Buchecker, C.O. and Sauvage, J.P. (1991) Interlocked and knotted rings in biology and chemistry, *Bioorg. Chem. Front.* **2**, 195-248.

15. Gibson, H.W., Bheda, M.C., and Engen, P.T. (1994) Rotaxanes, catenanes, polyrotaxanes, polycatenanes and related materials, *Prog. Polym. Sci.* **19**, 843-945.

16. Amabilino, D.B. and Stoddart, J.F. (1995) Interlocked and intertwined structures and superstructures, *Chem. Rev.* **95**, In press.

17. Allwood, B.L., Spencer, N., Shahriari-Zavareh, H., Stoddart, J.F., and Williams, D.J. (1987) Complexation of paraquat by a bisparaphenylene-34-crown-10 derivative, *J. Chem. Soc., Chem. Commun.*, 1064-1066.

510

18. Anelli, P.L., Ashton, P.R., Ballardini, R., Balzani, V., Delgado, M., Gandolfi, M.T., Goodnow, T.T., Kaifer, A.E., Philp, D., Pietraszkiewicz, M., Prodi, L., Reddington, M.V., Slawin, A.M Z., Spencer, N., Stoddart, J.F., Vicent, C., and Williams, D.J. (1992) Molecular meccano. 1. [2]Rotaxanes and a [2]catenane made to order, *J. Am. Chem. Soc.* **114**, 193-218.

19. Philp, D. and Stoddart, J.F. (1991) Self-assembly in organic synthesis, *Synlett*, 445-458.

20. Amabilino, D.B. and Stoddart, J.F. (1993) Self-assembly and macromolecular design, *Pure Appl. Chem.* **65**, 2351-2359.

21. Ashton, P.R., Goodnow, T.T., Kaifer, A.E., Reddington, M.V., Slawin, A.M.Z., Spencer, N., Stoddart, J.F., Vicent, C., and Williams, D.J. (1989) A [2]catenane made to order, *Angew. Chem., Int. Ed. Engl.* **28**, 1396-1399.

22. Ashton, P.R., Brown, C.L., Chrystal, E.J.T., Goodnow, T.T., Kaifer, A.E., Parry, K.P., Philp, D., Slawin, A.M.Z., Spencer, N., Stoddart, J.F., and Williams, D.J. (1991) The self-assembly of a highly ordered [2]catenane, *J. Chem. Soc., Chem. Commun.*, 634-639.

23. Ashton, P.R., Brown, C.L., Chrystal, E.J.T., Goodnow, T.T., Kaifer, A.E., Parry, K.P., Slawin, A.M.Z., Spencer, N., Stoddart, J.F., and Williams, D.J. (1991) Self-assembling [3]catenanes, *Angew. Chem., Int. Ed. Engl.* **30**, 1039-1042.

24. Amabilino, D.B., Ashton, P.R., Brown, C.L., Córdova, E., Godínez, L.A., Goodnow, T.T., Kaifer, A.E., Newton, S.P., Pietraszkiewicz, M., Philp, D., Raymo, F.M., Reder, A.S., Rutland, M.T., Slawin, A.M.Z., Spencer, N., Stoddart, J.F., and Williams, D.J. (1995) Molecular meccano. 2. Self-assembly of [n]catenanes, *J. Am. Chem. Soc.* **117**, 1271-1293.

25. Amabilino, D.B., Ashton, P.R., Reder, A.S., Spencer, N., and Stoddart, J.F. (1994) The two-step self-assembly of [4]- and [5]-catenanes, *Angew. Chem., Int. Ed. Engl.* **33**, 433-437.

26. Amabilino, D.B., Ashton, P.R., Stoddart, J.F., Menzer, S., and Williams, D.J. (1994) The solid-state self-organisation of a self-assembled [2]catenane, *J. Chem. Soc., Chem. Commun.*, 2475-2478.

27. Amabilino, D.B., Ashton, P.R., Reder, A.S., Spencer, N., and Stoddart, J.F. (1994) Olympiadane, *Angew. Chem., Int. Ed. Engl.* **33**, 1286-1290.

28. Anelli, P.L., Spencer, N., and Stoddart, J.F. (1991) A molecular shuttle, *J. Am. Chem. Soc.* **113**, 5131-5133.

29. Ashton, P.R., Philp, D., Spencer, N., and Stoddart, J.F. (1992) A new design strategy for the self-assembly of molecular shuttles, *J. Chem. Soc., Chem. Commun.*, 1124-1128.

30. Ashton, P.R., Bělohradský, M., Philp, D., and Stoddart, J.F. (1993) Slippage — an alternative method for assembling [2]rotaxanes, *J. Chem. Soc., Chem. Commun.*, 1269-1274.

31. Ashton, P.R., Bělohradský, M., Philp, D., Spencer, N., and Stoddart, J.F. (1993) The self-assembly of [2]- and [3]-rotaxanes by slippage, *J. Chem. Soc., Chem. Commun.*, 1274-1277.

MOLECULAR MACHINES

FRANÇISCO M. RAYMO, J. FRASER STODDART

School of Chemistry

University of Birmingham

Edgbaston, Birmingham B15 2TT, UK

1. Introduction

1.1. NANOCHEMISTRY

Self-assembling nanoscale molecular assemblies and supramolecular arrays are widespread in the biological world.[1,2] Their construction is achieved precisely and efficiently by the combination of simple 'intelligent' modular components through a series of cooperative phenomena involving the making and breaking of covalent and noncovalent bonds. The stereoelectronic information possessed by the subunits is, therefore, crucial for the entire process dictating assembly and growth of the biotic nanostructure. Thus, size, topology, shape and properties of the final architecture are genealogically-directed.[3] These biological nanostructures exhibit specific functions and, indeed, recognition, catalysis, replication, transport, and light harvesting are some of the phenomena witnessed in biology. In other words, nature is able to engineer, construct, and use nanoscopic structures and superstructures exhibiting machine-like properties.[4] On the contrary, the production of artificial devices is approaching rapidly a physical limit to miniaturisation on account of practical problems associated with

C.R.A. Catlow and A. Cheetham (eds.), New Trends in Materials Chemistry, 513–528.

514

microfabrication. The top down approach (*Figure 1*) is not going to be enough if abiotic devices capable of mimicking their biological counterparts are to be realised. The need for a bottom-up approach, involving chemical synthesis atom-by-atom and molecule-by-molecule of artificial nanoscale molecular and supramolecular devices is required.[5,6] However, the conventional synthetic methodologies are insufficient alone for the construction of complex nanostructures having precise shapes and engineered functions. Thus, self-assembly processes have to be devised for the generation of wholly-synthetic functioning systems.[7] Indeed, a number of organic[8-11] and inorganic[12-15] artificial self-assembling systems have been generated recently, although, the level of sophistication is still far from that achieved by nature. The fast development of this research area will provide in the next future a generation of abiotic functioning materials, having defined shapes and properties and leading ultimately to molecular machinery.[5,6]

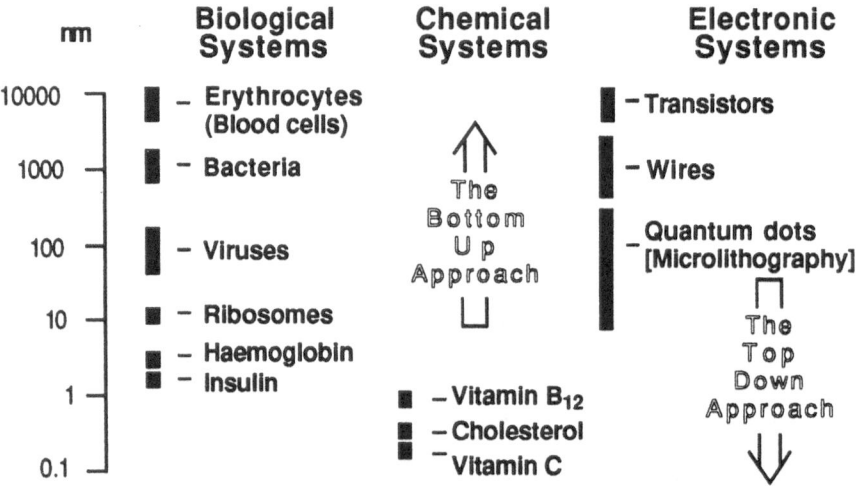

Figure 1. The Need for Nanochemistry

1.2. MECHANICALLY-INTERLOCKED MOLECULES

The features of mechanical-interlocking are widely employed at the macroscopic level. Chains, gears, and bearings are some examples of macroscopic devices functioning as a result of mechanical constriction. Recently, the idea of linking molecular components through a mechanical bond to construct thermodynamically-stable assemblies has aroused the interest of many investigators.[16-18] In particular, numerous synthetic methodologies to generate efficiently catenanes and rotaxanes have been devised. [19-21] These molecular compounds incorporate two or more components held together by means of a mechanical bond and, in some instances, also by noncovalent bonding interactions. Thus, motion of the interlocked components one with respect to the other occurs in a defined and discrete manner. These features suggest the possibility of designing controllable molecules in the shape of catenanes and rotaxanes. A schematic representation of a controllable [2]catenane is depicted in *Figure 2*. This [2]catenane incorporates an asymmetric macrocyclic component comprising two different units depicted as a shaded and an unshaded rectangle, respectively. Under ordinary conditions,

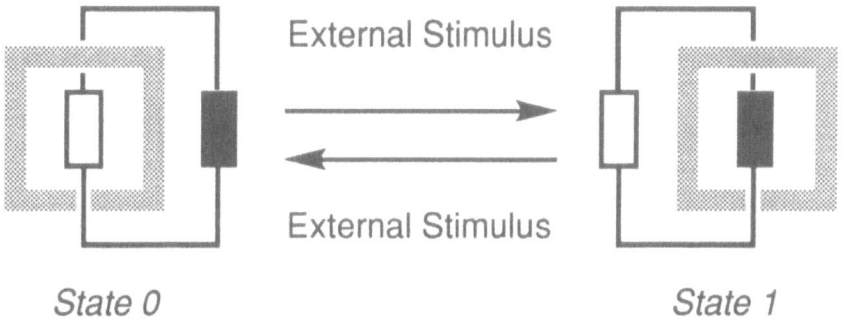

Figure 2. Schematic Representation of a Controllable [2]Catenane

the [2]catenane exists in the thermodynamically-stable *State 0* in which the unshaded unit of the asymmetric macrocyclic component is located inside the cavity of the other. A chemical, electrochemical, or photochemical external stimulus promotes the circumrotation of the asymmetric macrocycle through the other to locate inside the cavity of the latter the shaded unit affording *State 1*. Reverse interconversion from *State 1* to *State 0* can be, then, achieved through the intervention of yet another appropriate external stimulus. A similar molecular device can be envisaged in the shape of a [2]rotaxane (*Figure 3*). Two different units are incorporated within the dumbbell-shaped component of the [2]rotaxane. The thermodynamically-stable *State 0* is characterised by the location of the unshaded unit within the cavity of the macrocyclic component. An

516

appropriate external stimulus drives the macrocycles on to the shaded unit affording *State 1*. The reverse process can be achieved by a further external stimulus which forces the macrocycle to shuttle back to the unshaded unit. These mechanically-interlocked reversibly controllable molecular devices provide the opportunity of generating binary switches[22] at a molecular level. The interconversion between *State 0* and *State 1* can be controlled externally and provided that the two states can be easily differentiated, as a result of their spectroscopic output for example, the electronic molecular circuits can be envisaged leading ultimately to artificial chemical systems able to read, store, and process information — namely, molecular computers.

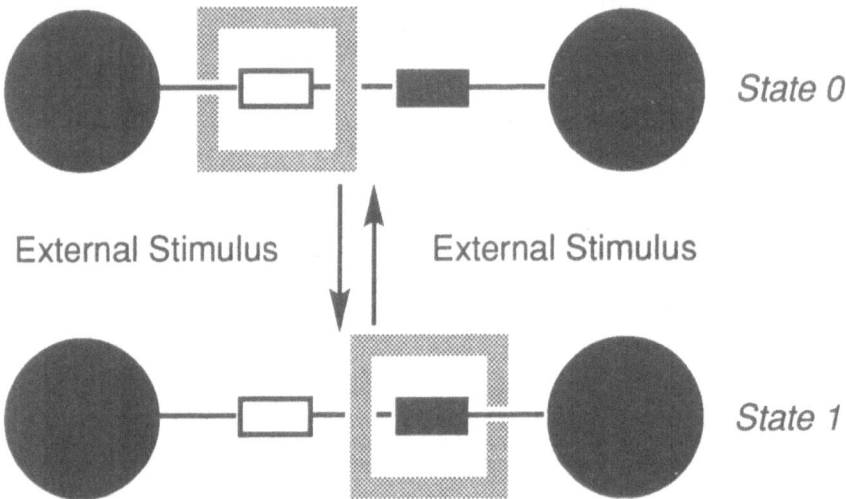

Figure 3. Schematic Representation of a Controllable [2]Rotaxane

2. Controllable Catenanes

A synthetic approach to self-assemble catenanes efficiently has been developed by us, recently.[23,24] The methodology relies upon the complementarity of π-electron rich macrocyclic polyethers and π-electron deficient tetracationic cyclophanes. The [2]catenane **4** can be self-assembled[25] as illustrated in *Scheme 1*. Reaction of the bis(hexafluorophosphate) salt **1** with the dibromide **2** affords a tricationic intermediate which is bound by the hydroquinone-based polyether **3**. The macrocyclic polyether, incorporated within the supramolecular complex, acts as a template, promoting the macrocyclisation of the bipyridinium-based cyclophane to yield, after counterion exchange, the [2]catenane **4**. The noncovalent bonding interactions responsible for the self-assembly of the [2]catenane **4** are (i) π-π stacking between the complementary aromatic units, (ii) hydrogen bonding between the polyether oxygen atoms and the protons located on the α-positions with respect to the nitrogen atoms on the bipyridinium units, as well as (iii) T-type edge-to-face interactions between the hydroquinone ring protons and the *p*-xylene spacers separating the bipyridinium units. These intercomponent noncovalent bonding interactions *live on* in the final [2]catenane, determining the rate of the dynamic processes (*Figure 4*) associated with **4** in solution. The circumrotation of the π-electron rich macrocyclic polyether component through the cavity of the π-electron deficient tetracationic cyclophane component (*Process I*), as well as the circumrotation of the π-electron deficient tetracationic cyclophane component

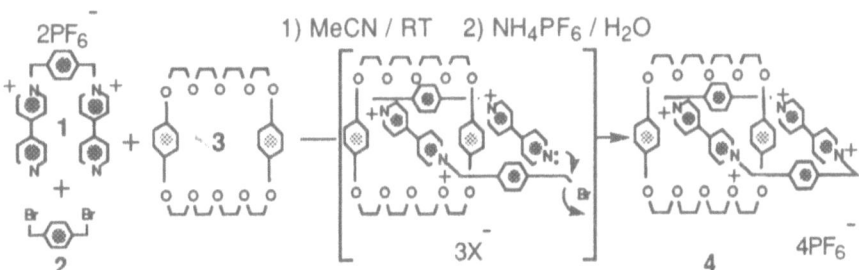

Scheme 1. Self-assembly of a [2]Catenane

through the cavity of the π-electron rich macrocyclic polyether component (*Process II*), are clearly revealed[25] by variable temperature ¹H-NMR spectroscopy. *Process I* is fast on the ¹H-NMR timescale at high temperatures. Thus, the ¹H-NMR spectrum of a CD_3SOCD_3 solution of **4** at 81°C reveals, for the hydroquinone ring protons, only one singlet centred on δ 4.57. Upon cooling the solution down to room temperature, *Process I* becomes slow on the ¹H-NMR timescale. As a result, the ¹H-NMR spectrum at room temperature shows two distinct singlets centred on δ 3.45 and 6.16, corresponding to the protons of the hydroquinone rings located *inside* and *alongside* the

518

cavity of the tetracationic cyclophane, respectively. The energy barrier associated with *Process I*, which was calculated by employing an approximation approach, corresponds to 15.6 kcal mol^{-1}. On the contrary, *Process II* is fast on the ^{1}H-NMR timescale at room temperature. Thus, the ^{1}H-NMR spectrum of a CD_3COCD_3 solution of the [2]catenane **4** at room temperature shows, for the protons attached to the α-positions with respect to the nitrogen atoms on the bipyridinium units, only one doublet centred on δ 9.34. On cooling the solution down to -40°C, however, two resonances centred on δ 9.23 and 9.41 are observed in the ^{1}H-NMR spectrum for the α-protons of the *inside* and *alongside* bipyridinium unit, respectively. The free energy of activation associated with *Process II* is 12.2 kcal mol^{-1}. Again, it was deduced by employing an approximation method. In order to generate a controllable [2]catenane, such as that

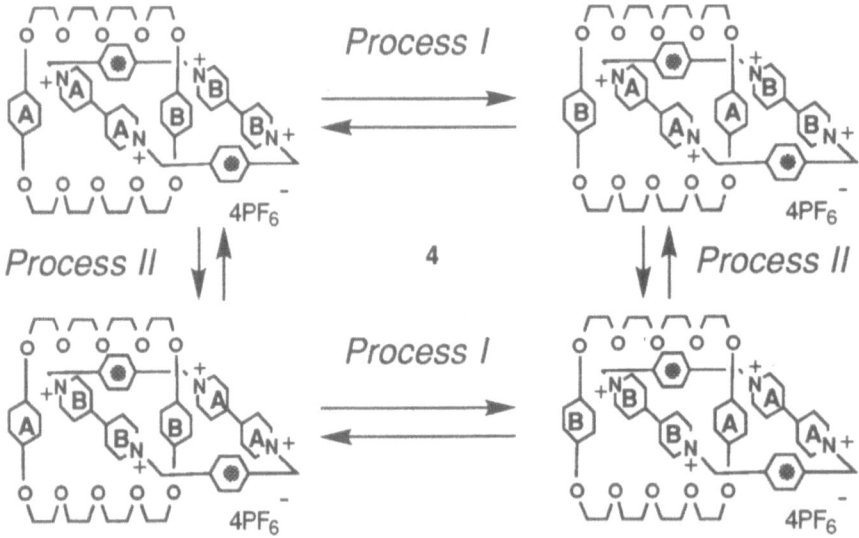

Figure 4. Dynamic Processes Associated with the [2]Catenane **4** in Solution

illustrated schematically in *Figure 2*, the possibility of desymmetrising the π-electron rich macrocyclic component was pursued. By replacing one of the two hydroquinone recognition sites incorporated within the macrocyclic polyether component with a 1,5-dioxynapthalene unit, the [2]catenane **5** was self-assembled.[26] The [2]catenane **5** exists in solution as the two translational isomers depicted in *Figure 5*. However, in

CD₃COCD₃ solution at 25°C, the ratio between the two isomers is 70:30 in favour of the translational isomers bearing the hydroquinone ring *inside* the cavity of the tetracationic cyclophane. By employing the more polar solvent CD_3SOCD_3, the ratio between the isomers can be reversed in favour of the one bearing the 1,5-dioxynaphthalene unit *inside* the cavity of the tetracationic cyclophane. Thus, the translational isomerism associated with the [2]catenane 5 in solution can be controlled by an external stimulus — namely, the polarity of the solvent. Asymmetric [2]catenanes can be generated by desymmetrising the π-electron deficient tetracationic cyclophane component. Replacing one of the two bipyridinium recognition sites with a (bispyridinium)ethylene unit affords[27] the [2]catenane 6. Two translational isomers (*Figure 6*) exist in solution as a result of the presence of two different recognition sites within the tetracationic cyclophane. However, at room temperature the circumrotation of the tetracationic cyclophane through the cavity of the macrocyclic

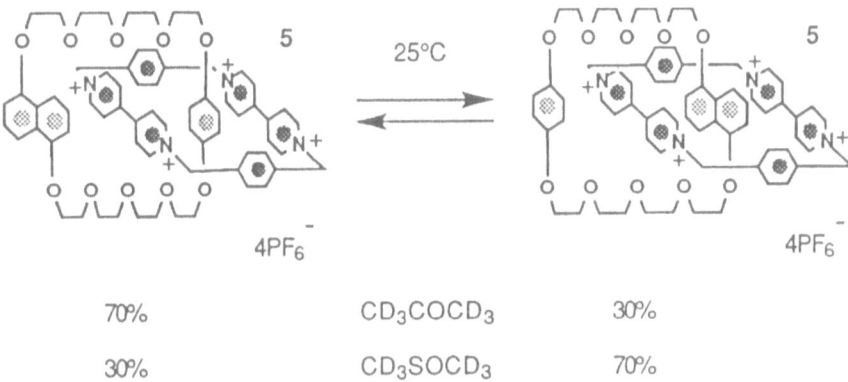

| 70% | CD₃COCD₃ | 30% |
| 30% | CD₃SOCD₃ | 70% |

Figure 5. Solvent Dependence of the Translational Isomerism Associated with the [2]Catenane 5

polyether is fast on the ¹H-NMR timescale. As a result, the two translational isomers cannot be distinguished at room temperature by means of ¹H-NMR spectroscopy. On cooling down to -60°C a CD₃COCD₃ solution of the [2]catenane 6, the two translational isomers can be clearly distinguished and an isomeric ratio of 92:8 in favour of the isomer bearing the bipyridinium recognition site *inside* the cavity of the macrocyclic polyether is observed. The bis(pyridinium)ethylene recognition site is located preferentially *alongside* the cavity of the π-electron rich macrocyclic polyether component. Furthermore, this unit possesses a reduction potential for its first reduction

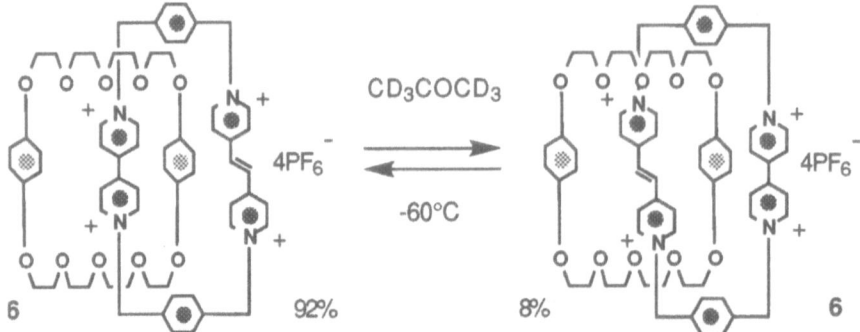

Figure 6. Translational Isomerism Associated with the [2]Catenane **6**

which is higher than that of the bipyridinium unit. As a result, the translational isomerism associated with the [2]catenane **6** can be controlled (*Figure 7*) electrochemically. Electrochemical reduction of the [2]catenane **6** involves a one-electron transfer on to the bipyridinium unit to generate a radical cation. In the case of

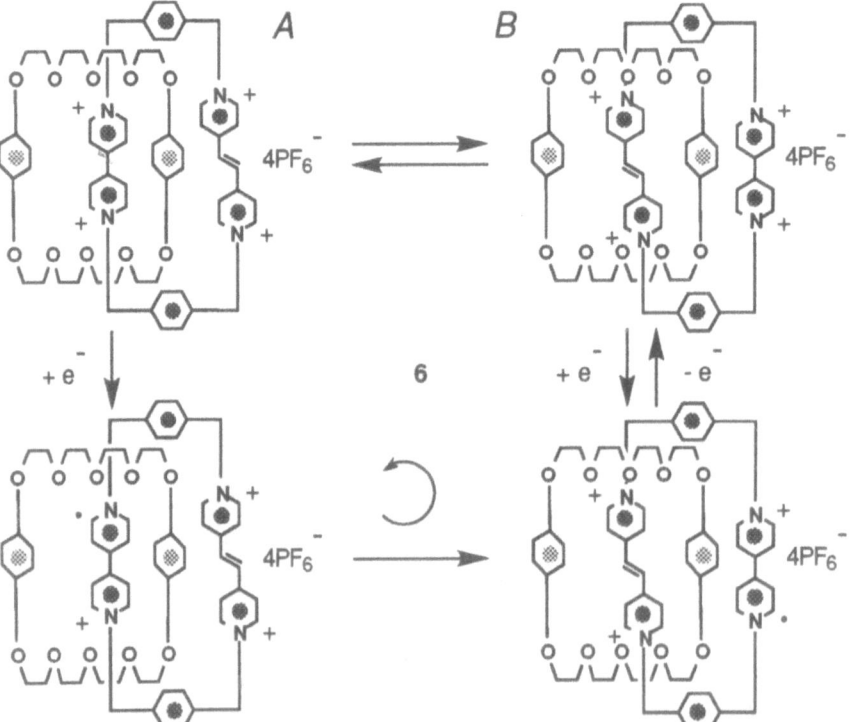

Figure 7. Electrochemical Control upon the Translational Isomerism Associated with the [2]Catenane **6**

the translational isomer *A*, the newly formed cationic species is located *inside* the cavity of the macrocyclic polyether. Thus, circumrotation of the tetracationic cyclophane through the cavity of the macrocyclic polyether occurs in order to accommodate *inside* it the dicationic bis(pyridinium)ethylene recognition site in place of the radical cation.

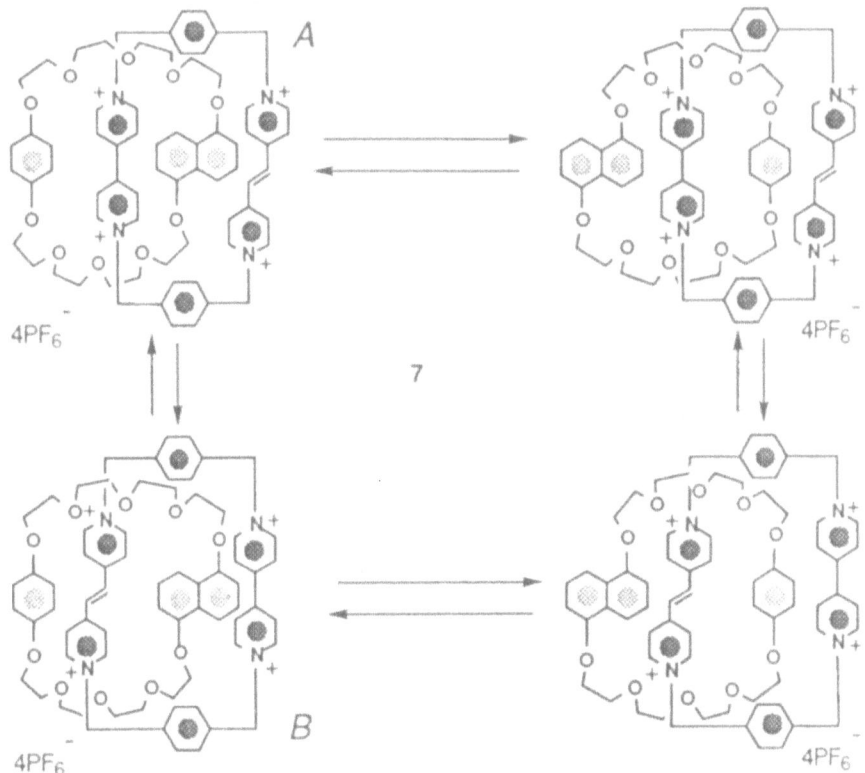

Figure 8. Translational Isomerism Associated with the [2]Catenane 7

Electrochemical oxidation of the resulting [2]catenane involves the regeneration of the dicationic bipyridinium recognition site and, as a result, the original equilibrium between the two translational isomers is restored. The [2]catenane 7[28] incorporates two different recognition sites — namely, one hydroquinone ring and one 1,5-dioxynaphthalene residue — within the π-electron rich macrocyclic polyether component, as well as two different recognition sites — namely, a bipyridinium and a bis(pyridinium)ethylene unit — within the π-electron deficient tetracationic cyclophane. As a result, the existence of the four translational isomers depicted in *Figure 8* can be predicted for the [2]catenane **7**. The circumrotation of the macrocyclic polyether through the cavity of the tetracationic cyclophane is slow on the ^{1}H-NMR timescale at room temperature and the ^{1}H-NMR spectrum of a CD_3COCD_3 solution of **7** shows the preferential inclusion of the 1,5-dioxynaphthalene unit *inside* the cavity of the π-electron deficient macrocycle. On cooling the solution down to -40°C, the circumrotation of the tetracationic cyclophane becomes slow on the ^{1}H-NMR timescale and only the signals associated with the isomers A and B are observed in the ^{1}H-NMR spectrum in a ratio of 95:5 in favour of A.

3. Controllable Rotaxanes

The [2]rotaxane **6** can be self-assembled[29] (*Scheme 2*) by reacting the bis(hexafluorophosphate) salt **1** with the dibromide **2** in the presence of the preformed dumbbell-shaped compound **5**. The dumbbell-shaped component of the [2]rotaxane **6** incorporates two hydroquinone recognition sites. As a result, the bipyridinium-based cyclophane moves (*Figure 9*) back and forth from one hydroquinone *station* to the other giving rise to a so-called *molecular shuttle*. This dynamic process can be followed

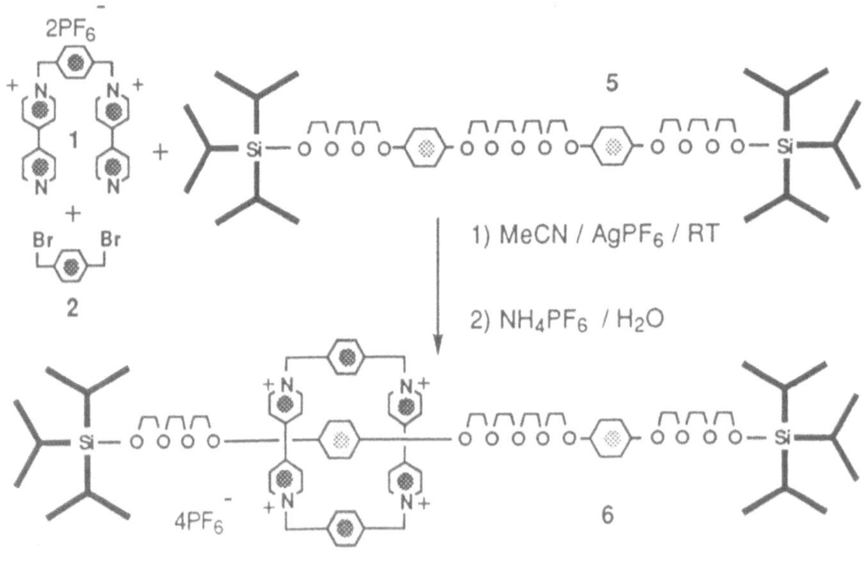

Scheme 2. Self-assembly of a [2]Rotaxane

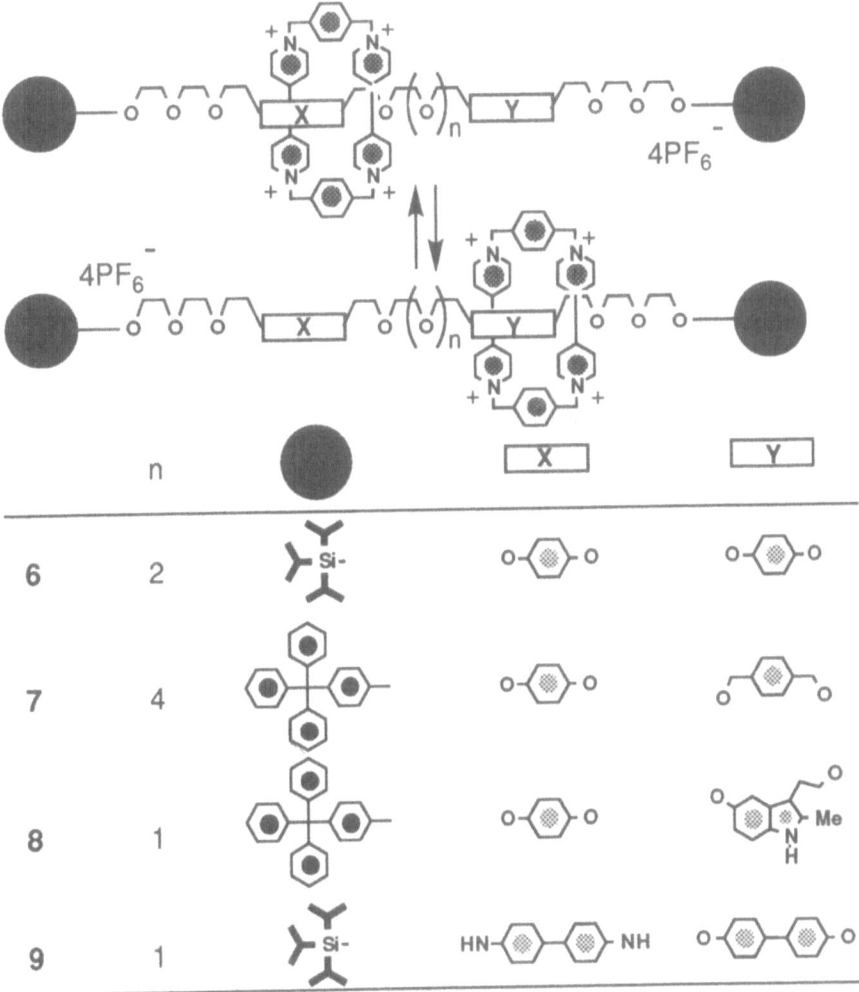

Figure 9. Molecular Shuttles in Action

easily by variable temperature ^{1}H-NMR spectroscopy. At elevated temperatures, the *shuttling* process is rapid on the ^{1}H-NMR timescale and the ^{1}H-NMR spectrum of a CD_3SOCD_3 solution of the [2]rotaxane **6** recorded at 140°C shows only one AA'BB' system centred on δ 5.16 for the protons of the two hydroquinone rings. On cooling a CD_3COCD_3 solution of the [2]rotaxane **6** down to -50°C, two distinct AA'BB' systems centred on δ 3.80 and 6.38 can be identified in the ^{1}H-NMR spectrum and these correspond to the *occupied* and *unoccupied* hydroquinone *station*, respectively. By employing an approximation method, the energy barrier associated with the process was found to be 13.0 kcal mol^{-1}. In order to generate controllable *molecular shuttles*, the possibility of self-assembling [2]rotaxanes incorporating two different *stations* within their dumbbell-shaped component was envisaged. The [2]rotaxane **7** incorporates[30] one hydroquinone and one *p*-xylene recognition site within its dumbbell-shaped component. It was hoped that the tetracationic cyclophane would reside preferentially on the more electron rich hydroquinone *station*. Oxidation of the hydroquinone ring occurs at a potential lower than that of the *p*-xylene unit, and thus, exclusive oxidation of the hydroquinone *station* should be achieved electrochemically. As a result, the tetracationic cyclophane should be forced to move away from the oxidised *station* and reside on the neutral *p*-xylene unit. Although, rapid equilibration between the two translational isomers occurs at +70°C in CD$_3$CN solution, at -40°C, both translational isomers can be detected by ^{1}H-NMR spectroscopy in a ratio of only 70:30 in favour of the isomer in which the hydroquinone *station* is located inside the cavity of the tetracationic cyclophane. In order to obtain a more favourable isomeric ratio, the *p*-xylene unit was replaced with an indole residue to generate the [2]rotaxane **8**.[31] The indole unit is more electron rich than the hydroquinone unit and possesses a lower oxidation potential. Hence, the possibility of generating preferentially in solution the translational isomer, having the indole unit located inside the macrocyclic component, was envisaged. Electrochemical oxidation of the indole *station* should, then, force the tetracationic cyclophane to move on to the hydroquinone *station*. However, presumably as a result of steric hindrance, the tetracationic cyclophane resides exclusively on to the less π-electron rich hydroquinone ring! Redesign of the *molecular shuttle* led[32] to the self-assembly of the [2]rotaxane **9** incorporating one benzidine and one biphenol recognition site within its dumbbell-shaped component. Variable temperature ^{1}H-NMR spectroscopy revealed the ratio between the two translational isomers to be 84:16 in favour of the isomer bearing the benzidine *station* inside the cavity of the cyclophane at -44°C in CD$_3$CN. Electrochemical oxidation of the benzidine *station* (*Figure 10*) forces the tetracationic cyclophane to move away from the now positively-charged benzidine unit. However, reduction of the benzidine unit restores the equilibrium situation. The shuttling process associated with the [2]rotaxane **9** can be controlled reversibly also by chemical means. Protonation of the nitrogen atoms of the benzidine unit with trifluoroacetic acid affords a dicationic *station*. As a result, the tetracationic cyclophane moves to the neutral biphenol *station*. Neutralisation with pyridine regenerates the neutral benzidine *station*, allowing the of the tetracationic cyclophane to *shuttle* back to the benzidine unit.

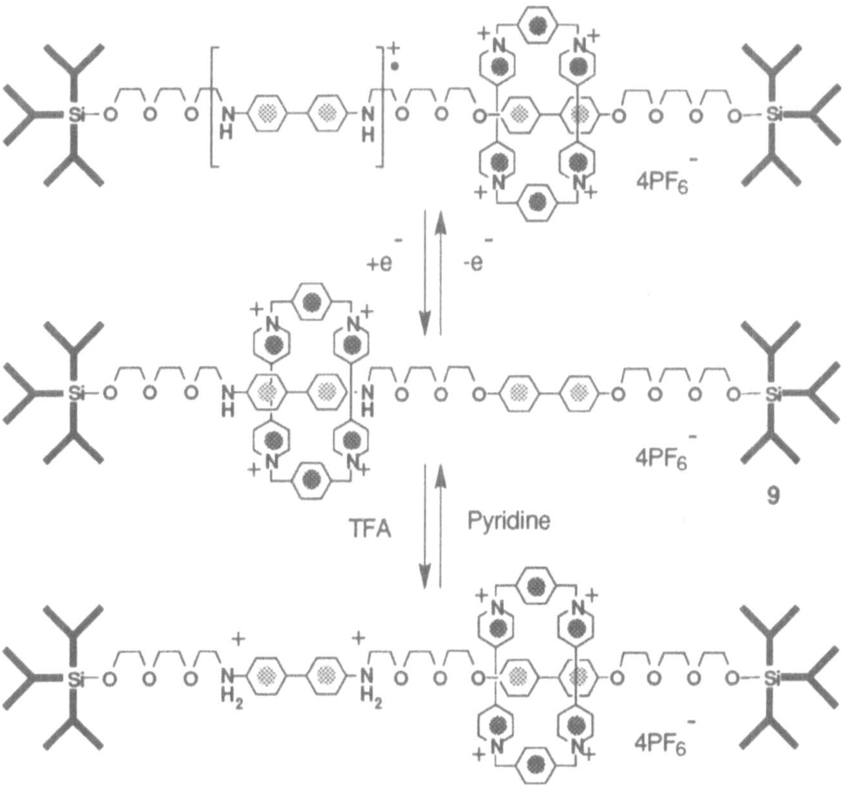

Figure 10. Chemical and Electrochemical Control upon the *Shuttling* Process Associated with the [2]Rotaxane **9**

4. Conclusions

We have developed a synthetic approach for self-assembling efficiently mechanically-interlocked molecular compounds such as catenanes and rotaxanes. The methodology relies upon the complementarity of π-electron deficient and π-electron rich components. Noncovalent bonding interactions such as π-π stacking and hydrogen bonding are the main driving forces of these self-assembling processes. The combined features of mechanical entanglement and discrete matching stereoelectronic geometries, associated with the catenanes and rotaxanes, offer the chance of generating controllable devices at a molecular level. The dynamic processes, involving the relative motions of the interlocked components constituting the catenanes and rotaxanes, can be subjected to a certain degree of control by means of external stimuli. Indeed, a series of chemically- and electrochemically-reversibly, controllable [2]catenanes and [2]rotaxanes have been self-assembled and their properties investigated. We believe that, in the near future, a number of even more sophisticated devices will be realised at the nanoscopic level. The growth of this research area is, in fact, bringing about a number of new molecular and supramolecular materials having sensory, switching, and device-like characteristics, and leading ultimately to the design and construction of a generation of molecular machines having defined shapes and sizes, as well as precisely engineered functions.

5. References

1. Lindsey, J.S. (1991) Self-assembly in synthetic routes to molecular devices. Biological principles and chemical perspectives: a review, *New J. Chem.* **15**, 153-180.

2. Philp, D. and Stoddart, J.F. (1996) Self-assembly in natural and unnatural systems, *Angew. Chem., Int. Ed. Engl.* **35**, In press.

3. Tomalia D. A. and Durst H. D. (1993) Genealogically directed synthesis: starbust/cascade dendrimers and hyperbranched structures, *Top. Curr. Chem.* **165**, 193-313

4. Urry, D.W. (1993) Molecular machines: how motion and other functions of living organisms can result from reversible chemical changes, *Angew. Chem., Int. Ed. Engl.* **32**, 819-841.

5. Drexler, K.E. (1994) Molecular nanomachines: physical principles and implementation strategies, *Annu. Rev. Biophys. Biomol. Struct.* **23**, 377-405.

6. Drexler, K.E. (1992) Nanosystems: molecular machinery, manufacturing, and computation, Wiley, New York.

7.   Whitesides, G.M., Mathias J.P., and Seto, C.T. (1991) Molecular self-assembly and nanochemistry: a chemical strategy for the synthesis of nanostructures, *Science* **254**, 1312-1319.

8.   Stoddart, J.F. (1991) From enzyme mimics to molecular self-assembly processes, *Chirality in Drug Design and Synthesis*, Ed. C. Brown, Academic Press, London, pp. 53-81.

9.   Whitesides, G.M., Simanek, E.E., Mathias, J.P., Seto, C.T., Chin, D.N., Mammen, M., and Gordon, D.M. (1995) Noncovalent synthesis: using physical-organic chemistry to make aggregates, *Acc. Chem. Res.* **28**, 37-44.

10. Kumar, A., Abbot, N.L., Kim, E., Biebuyck, A., and Whitesides, G.M. (1995) Patterned self-assembled monolayers and meso-scale phenomena, *Acc. Chem. Res.* **28**, 219-226.

11. Raymo, F.M. and Stoddart, J.F. (1996) Self-Assembly in Chemical Systems, *New Trends in Material Science*, Ed. C.R.A. Catlow, Kluwer, Dordrecht, Preceding article.

12. Rong, D., Hong, H.G., Kim, Y.I., Kreuger, J.S., Mayer, J.E., and Mallouk, T.E. (1990) Electrochemistry and photochemistry of transition metal complexes in well-ordered surface layers, *Coord. Chem. Rev.* **97**, 237-248.

13. Mallouk, T.E. and Lee, H. (1990) Designer solids and surfaces, *J. Chem. Ed.* **67**, 829-834.

14. Ozin, G.A. (1992) Nanochemistry: synthesis in diminishing dimensions, *Adv. Mater.* **4**, 612-649.

15. Mann, S. (1993) Molecular tectonics in biomineralization and biomimetic materials chemistry, *Science* **365**, 499-505.

16. Cram, D.J. (1992) Molecular container compounds, *Nature* **356**, 29-36.

17. Merkle, R.C. (1993) A proof about molecular bearings, *Nanotechnology* **4**, 86-90.

18. Shermann, J.C. (1995) Carceplexes and hemicarceplexes: molecular encapsulation-from hours to forever, *Tetrahedron* **51**, 3395-3422.

19. Dietrich-Buchecker, C.O. and Sauvage, J.P. (1991) Interlocked and knotted rings in biology and chemistry, *Bioorg. Chem. Front.* **2**, 195-248.

20. Gibson, H.W., Bheda, M.C., and Engen, P.T. (1994) Rotaxanes, catenanes, polyrotaxanes, polycatenanes and related materials, *Prog. Polym. Sci.* **19**, 843-945.

21. Amabilino, D.B. and Stoddart, J.F. (1995) Interlocked and intertwined structures and superstructures, *Chem. Rev.* **95**, In press.

22. Fabbrizzi L. and Poggi A. (1995) Sensors and switches from supramolecular chemistry, *Chem. Soc. Rev.* **24**, 197-202.

23. Philp, D. and Stoddart, J.F. (1991) Self-assembly in organic synthesis, *Synlett*, 445-458.

24. Amabilino, D.B. and Stoddart, J.F. (1993) Self-assembly and macromolecular design, *Pure Appl. Chem.* **65**, 2351-2359.

25. Anelli, P.L., Ashton, P.R., Ballardini, R., Balzani, V., Delgado, M., Gandolfi, M.T., Goodnow, T.T., Kaifer, A.E., Philp, D., Pietraszkiewicz, M., Prodi, L., Reddington, M.V., Slawin, A.M Z., Spencer, N., Stoddart, J.F., Vicent, C., and Williams, D.J. (1992) Molecular meccano. 1. [2]Rotaxanes and a [2]catenane made to order, *J. Am. Chem. Soc.* **114**, 193-218.

26. Ashton, P.R., Blower, M., Philp, D., Spencer, N., Ballardini, R., Ciano, M., Balzani, V., Gandolfi, M.T., Prodi, L., and McLean, C.H. (1993) The control of translational isomerism in catenated structures, *New J. Chem.* **17**, 689-695.

27. Ashton, P.R., Ballardini, R., Balzani, V., Gandolfi, M.T., Marquis, D.J.F., Pérez-Garcia, L., Prodi, L., Stoddart, J.F., and Venturi, M. (1994) The self-assembly of controllable [2]catenanes, *J. Chem. Soc., Chem. Commun.*, 177-180.

28. Ashton, P.R., Pérez-Garcia, L., Stoddart, J.F., White, A.J.P., and Williams, D.J. (1995) Controlling translational isomerism in [2]catenanes, *Angew. Chem., Int. Ed. Engl.* **34**, 571-574.

29. Anelli, P.L., Spencer, N., and Stoddart, J.F. (1991) A molecular shuttle, *J. Am. Chem. Soc.* **113**, 5131-5133.

30. Ashton, P.R., Bissell, R.A., Spencer, N., Stoddart, J.F., and Tolley, M.S. (1992) Towards controllable molecular shuttles - 1, *Synlett*, 914-918.

31. Ashton, P.R., Bissell, R.A., Górski, R., Philp, D., Spencer, N., Stoddart, J.F., and Tolley, M.S. (1992) Towards controllable molecular shuttle - 2, *Synlett*, 919-922.

32. Bissell, R.A., Córdova, E., Kaifer, A.E., and Stoddart, J.F. (1993) A chemically and elettrochemically switchable molecular shuttle, *Nature* **369**, 133-137.

INDEX

acid catalysis, 404
acid catalysts, 443
acidic solids, 88
activation energy, 304
adsorbate atom, 349
AgBr, 294, 296
AgCl, 316
AgI, 321
Al_2O_3, 4, 90, 383
$(AlO_2)_3$, 117
$AlCl_3$, 370
^{27}Al NMR, 116, 126
alkali halides, 297
alkyl diphosphonates, 174
alkylation, 409, 415
all-electron full potential, 215
all-electron muffin-tin, 215
ALLHKL, 44
alloys, 313
α-Al_2O_3, 165, 174, 232
α-$CrPO_4$, 47
α-SiO_2, 151, 165
AlPO-5, 167
aluminophosphates, 83
amorphous ionic conductors, 322
amorphous Silica-Alumina, 409
amorphous solids, 4
amorphous structures, 177
amperometric mode, 40
Amsterdam Density Functional code
     (ADF), 221
analytical microscopy, 60
Arrhenius equation, 304
atomic form factor, 22
atomic-sphere-approximation (ASA),
     217
augmented plane wave (APW), 224
auto-exhaust emissions, 393
Axilrod and Teller, 145

B-LYP correction, 184
B_2O_3, 117
$BaCuO_2$, 251
BaF_2, 316
Bardeen-Herring correlation factor, 305
base catalyst, 448
basic catalysts, 88

basic sites, 414
basic sites in zeolites, 412
$BaSO_4$, 174, 175
batteries, 331
BEDT-TTF, 7
$(BEDT-TTF)_2X$, 5
benzidine, 524
benzoquinone, 423
BET surface areas, 435
β-Ba_3AlF_9, 48
β-FeOOH, 62
β-Hf, 313
beta-PbF_2, 321
β-Ti, 313
β-Zr, 313
BFGS methods, 214
Bi_2O_3, 321
$Bi_2Sr_{2+x}Ca_{1-x}Cu_2O_8$, 14
$Bi_2Sr_4Cu_2CO_3O_7$, 254
$(Bi_2Sr_2CuO_6)_m(Sr_2CuO_2CO_3)_n$, 254
Bi_3ReO_8, 45
$Bi_4V_2O_{11}$, 324
BICOVOX, 325
bifunctional zeolite catalysts, 411
BIMEVOX, 324, 325
binary switches, 516
bipyridinium salt, 497
bismuth oxycarbonate, 260
Bloch function, 216, 219
Bloch's theorem, 216
Bond Order Conservation, 348
Boudouard reaction, 102
Bragg's law, 57
Bragg-equation, 20, 21
Brillouin zone, 213
Brønsted acid site, 186, 405, 412
Brønsted acidity, 444
Buckingham potential, 144, 145
bulk diffusion, 300
butene isomerization, 386

^{13}C-NMR, 499
C_{60}, 5, 6
C60 fullerenes, 36
Cacoxenite, 427
$CaCuO_2$, 265
CaF_2, 170, 316

529

534

536